Quantum Systems in Physics, Chemistry, and Biology

Progress in Theoretical Chemistry and Physics

VOLUME 30

More information about this series at http://www.springer.com/series/6464

Alia Tadjer · Rossen Pavlov (†)
Jean Maruani · Erkki J. Brändas
Gerardo Delgado-Barrio Editors

Quantum Systems in Physics, Chemistry, and Biology

Advances in Concepts and Applications

Springer

Editors
Alia Tadjer
Faculty of Chemistry and Pharmacy
University of Sofia
Sofia
Bulgaria

Rossen Pavlov (†)
Institute for Nuclear Research and Nuclear Energy
Sofia
Bulgaria

Jean Maruani
Laboratoire de Chimie Physique
CNRS and UPMC
Paris
France

Erkki J. Brändas
Department of Quantum Chemistry
Uppsala University
Uppsala
Sweden

Gerardo Delgado-Barrio
Consejo Superior de Investigaciones Científicas (CSIC)
Madrid
Spain

ISSN 1567-7354 ISSN 2215-0129 (electronic)
Progress in Theoretical Chemistry and Physics
ISBN 978-3-319-50254-0 ISBN 978-3-319-50255-7 (eBook)
DOI 10.1007/978-3-319-50255-7

Library of Congress Control Number: 2016959519

Printed on acid-free paper

This Springer imprint is published by Springer Nature
The registered company is Springer International Publishing AG
The registered company address is: Gewerbestrasse 11, 6330 Cham, Switzerland

PTCP Aim and Scope

Progress in Theoretical Chemistry and Physics

A series reporting advances in theoretical molecular and material sciences, including theoretical, mathematical and computational chemistry, physical chemistry and chemical physics and biophysics.

Aim and Scope

Science progresses by a symbiotic interaction between theory and experiment: theory is used to interpret experimental results and may suggest new experiments; experiment helps to test theoretical predictions and may lead to improved theories. Theoretical Chemistry (including Physical Chemistry and Chemical Physics) provides the conceptual and technical background and apparatus for the rationalization of phenomena in the chemical sciences. It is, therefore, a wide ranging subject, reflecting the diversity of molecular and related species and processes arising in chemical systems. The book series *Progress in Theoretical Chemistry and Physics* aims to report advances in methods and applications in this extended domain. It will comprise monographs as well as collections of papers on particular themes, which may arise from proceedings of symposia or invited papers on specific topics as well as from initiatives from authors or translations.

The basic theories of physics—classical mechanics and electromagnetism, relativity theory, quantum mechanics, statistical mechanics, quantum electrodynamics—support the theoretical apparatus which is used in molecular sciences. Quantum mechanics plays a particular role in theoretical chemistry, providing the basis for the valence theories, which allow to interpret the structure of molecules, and for the spectroscopic models, employed in the determination of structural information from spectral patterns. Indeed, Quantum Chemistry often appears synonymous with Theoretical Chemistry; it will, therefore, constitute a major part of this book series. However, the scope of the series will also include other areas of theoretical chemistry, such as mathematical chemistry (which involves the use of algebra and topology in the analysis of molecular structures and reactions); molecular mechanics, molecular dynamics, and chemical thermodynamics, which play an important role in rationalizing the geometric and electronic structures of molecular assemblies and polymers, clusters, and crystals; surface, interface, solvent, and solid state effects; excited-state dynamics, reactive collisions, and chemical reactions.

Recent decades have seen the emergence of a novel approach to scientific research, based on the exploitation of fast electronic digital computers. Computation provides a method of investigation which transcends the traditional division between theory and experiment. Computer-assisted simulation and design may afford a solution to complex problems which would otherwise be intractable to theoretical analysis, and may also provide a viable alternative to difficult or costly laboratory experiments. Though stemming from Theoretical Chemistry, Computational Chemistry is a field of research in its own right, which can help to test theoretical predictions and may also suggest improved theories.

The field of theoretical molecular sciences ranges from fundamental physical questions relevant to the molecular concept, through the statics and dynamics of isolated molecules, aggregates and materials, molecular properties and interactions, to the role of molecules in the biological sciences. Therefore, it involves the physical basis for geometric and electronic structure, states of aggregation, physical and chemical transformations, thermodynamic and kinetic properties, as well as unusual properties such as extreme flexibility or strong relativistic or quantum-field effects, extreme conditions such as intense radiation fields or interaction with the continuum, and the specificity of biochemical reactions.

Theoretical Chemistry has an applied branch (a part of molecular engineering), which involves the investigation of structure-property relationships aiming at the design, synthesis and application of molecules and materials endowed with specific functions, now in demand in such areas as molecular electronics, drug design or genetic engineering. Relevant properties include conductivity (normal, semi- and super-), magnetism (ferro- and ferri-), optoelectronic effects (involving nonlinear response), photochromism and photoreactivity, radiation and thermal resistance, molecular recognition and information processing, biological and pharmaceutical activities, as well as properties favoring self-assembling mechanisms and combination properties needed in multifunctional systems.

Progress in Theoretical Chemistry and Physics is made at different rates in these various research fields. The aim of this book series is to provide timely and in-depth coverage of selected topics and broad-ranging yet detailed analysis of contemporary theories and their applications. The series will be of primary interest to those whose research is directly concerned with the development and application of theoretical approaches in the chemical sciences. It will provide up-to-date reports on theoretical methods for the chemist, thermodynamician or spectroscopist, the atomic, molecular or cluster physicist, and the biochemist or molecular biologist who wish to employ techniques developed in theoretical, mathematical and computational chemistry in their research programs. It is also intended to provide the graduate student with a readily accessible documentation on various branches of theoretical chemistry, physical chemistry, and chemical physics.

Preface

This volume collects 24 selected papers from the scientific contributions presented at the Twentieth International Workshop on Quantum Systems in Chemistry, Physics, and Biology (QSCP-XX), organized by Alia Tadjer and Rossen Pavlov in the Golden Sands resort near Varna, Bulgaria, on September 14–20, 2015. Over 90 scientists from 26 countries attended this meeting. The participants discussed the state of the art, new trends, and future evolution of methods in molecular quantum mechanics, and their applications to a broad variety of problems in chemistry, physics, and biology.

The high-level attendance attained in this conference was particularly gratifying. It is the renowned interdisciplinary nature and friendly feeling of QSCP meetings that make them so successful discussion forums.

The Golden Sands resort is 18 km north of the city of Varna and near the Golden Sands National Park. The name comes from an old legend of pirates burying a huge golden treasure at the waterfront north of Varna. The land took revenge on the rascals and transformed that gold into wonderful sand. It is a magnificent resort, the largest on the northern Black Sea coast, with its richly wooded hills, a white golden beach, and clear blue sea. The old-growth forests between ancient Odessos and Dionysopolis were mentioned by Pliny as the home of mythical dwarfs visited by the Argonauts. There are famous mineral water springs, already known to the Thracians and Roman Caesars. Golden Sands suits both sun seekers and sightseers visiting Varna, the third largest city in Bulgaria and pearl of the Black Sea coast.

Details of the Varna meeting, including the scientific and social programs, can be found on the web site: http://ntl.inrne.bas.bg/qscp2015. Altogether, there were 17 morning and afternoon sessions, where 53 plenary talks were given, and two evening poster sessions, with 21 flash presentations for a total of 35 displayed posters. We are grateful to all the participants for making the QSCP-XX workshop a stimulating experience and a great success. QSCP-XX followed the traditions established at previous workshops:

QSCP-I, organized by Roy McWeeny in 1996 at San Miniato (Pisa, Italy);
QSCP-II, by Stephen Wilson in 1997 at Oxford (England);

QSCP-III, by Alfonso Hernandez-Laguna in 1998 at Granada (Spain);
QSCP-IV, by Jean Maruani in 1999 at Marly-le-Roi (Paris, France);
QSCP-V, by Erkki Brändas in 2000 at Uppsala (Sweden);
QSCP-VI, by Alia Tadjer in 2001 at Sofia (Bulgaria);
QSCP-VII, by Ivan Hubac in 2002 near Bratislava (Slovakia);
QSCP-VIII, by Aristides Mavridis in 2003 at Spetses (Athens, Greece);
QSCP-IX, by J.-P. Julien in 2004 at Les Houches (Grenoble, France);
QSCP-X, by Souad Lahmar in 2005 at Carthage (Tunisia);
QSCP-XI, by Oleg Vasyutinskii in 2006 at Pushkin (St Petersburg, Russia);
QSCP-XII, by Stephen Wilson in 2007 near Windsor (London, England);
QSCP-XIII, by Piotr Piecuch in 2008 at East Lansing (Michigan, USA);
QSCP-XIV, by G. Delgado-Barrio in 2009 at El Escorial (Madrid, Spain);
QSCP-XV, by Philip Hoggan in 2010 at Cambridge (England);
QSCP-XVI, by Kiyoshi Nishikawa in 2011 at Kanazawa (Japan);
QSCP-XVII, by Matti Hotokka in 2012 at Turku (Finland);
QSCP-XVIII, by M.A.C. Nascimento in 2013 at Paraty (Brazil);
QSCP-XIX, by Cherri Hsu in 2014 at Taipei (Taiwan).

The lectures presented at QSCP-XX were grouped into seven topics, in the field of *Quantum Systems in Chemistry, Physics, and Biology*, ranging from concepts and methods in Quantum Chemistry, through Relativistic Effects in Quantum Chemistry, Atoms and Molecules in Strong Electric and Magnetic Fields, Reactive Collisions and Chemical Reactions, Molecular Structure, Dynamics, and Spectroscopy, and Molecular and Nano-materials, to Computational Chemistry, Physics, and Biology.

The width and depth of the topics discussed at QSCP-XX are reflected in the contents of this volume of proceedings in *Progress in Theoretical Chemistry and Physics*, which includes five parts:

I. Quantum Methodology (five chapters);
II. Structure and Dynamics (six chapters);
III. Atomic and Molecular Properties (six chapters);
IV. Biochemistry and Biophysics (four chapters);
V. Fundamental Theory (three chapters).

In addition to the scientific program, the workshop had its usual share of cultural events. There was a tour of the resort, a visit to the Archaeological Museum and Thracian Treasures in Varna, the Palace and Botanical Garden in Balchik, the Stone forest and Aladja monastery, and the palace and winery of Euxinograd. Foreign participants successfully joined in a show of folk songs and dances. The award ceremony of the CMOA Prize and Medal took place during the banquet in the conference venue: *Berlin Green Park Hotel.*

The CMOA Prize for junior scientists was shared between two selected nominees: Henryk Witek (Taiwan) and Sachar Klaiman (Germany). Two other nominees: Anela Ivanova (Bulgaria) and Olga Khetselius (Ukraine), received a certificate of nomination and a gift. The prestigious CMOA Medal for senior scientists was awarded to Prof. Hardy Gross (Max-Planck Institute of Microstructure

Physics in Halle, Germany). Following a QSCP tradition, the venue of the next workshop was announced: Vancouver, BC, Canada, in July 2016.

We are most grateful to the members of the Local Organizing Committee: Yuliya Mutafchieva, Zhivko Stoyanov, Martin Ivanov, Chavdar Velchev, and Mitko Gaidarov, as well as to the Sofia University young team, particularly Galia Madjarova and Georgi Stoychev, for their work and dedication, which made the stay and work of the participants both pleasant and fruitful. Last but not least, we thank the members of the International Scientific and Honorary Committees for their invaluable expertise and advice.

We hope the readers will find as much interest in consulting these proceedings as the participants in attending the meeting.

The Editors

Contents

Obituary

Sir Harold W. Kroto (1939–2016)

Professor Harry Kroto, who shared the Nobel Prize for Chemistry in 1996 with Rick Smalley and Robert Curl, has died at the age of 76 after a long illness. His 1996 Nobel award was in recognition of his work on fullerenes—commonly referred to as 'buckyballs', a pure form of carbon that presented itself in the form of a polyhedron composed of 60 carbon atoms.

Harry was born Harold Walter Krotoschiner on 7th October 1939 in the small market town of Wisbech, Cambridgeshire, England. Both parents had come to Britain as refugees from Nazi Germany in the 1930s. His Jewish father was held on the Isle of Man as an 'enemy alien' for the duration of the war. Harry and his mother moved to Bolton, Lancashire, in 1940. This is where the family eventually settled at the end of World War II, when his father joined them again.

The winning of the Nobel award is a reflection of Harry as an exceptionally talented all-rounder interested and skilled in many things. As a high school student he grew increasingly interested in chemistry, physics, and mathematics, and applied to study Chemistry at Sheffield University, where he obtained a first class degree. As an undergraduate he excelled at various extracurricular activities such as music, athletics, tennis, and magazine design. His interests shifted to quantum chemistry and spectroscopy. He remained at Sheffield and obtained his doctorate (under Richard Dixon) for a work on spectroscopy of free radicals produced by flash photolysis. It was this interest in spectroscopy and molecular structure that formed a strong theme throughout much of his later work.

I first met Harry when I moved to the University of Sussex in 1982, where he directed a group doing microwave rotation spectroscopy. One of my first discussions with him was about the origin of the microwave cosmic background radiation (CBR), which had become of interest because of Fred Hoyle's speculations about organic molecules in interstellar dust and the origin of life on Earth. I was struck by his friendliness, informality, and intellectual vigor. In the late 1970s the long-chained polyvinylcyanides HC_5N, HC_7N and HC_9N were found by radio-astronomy in cold dark clouds of interstellar regions. The hunt for their origin indicated that they were being ejected out of red giant carbon stars. To discover the structure of molecules ejected from red giant carbon stars became of great interest to Harry, and this lead him naturally to the discovery of the fullerenes. In 1985 carbon-cluster experiments were undertaken with laser oblation of a graphitic target to form a carbon plasma.

Prior to the Nobel Prize winning work of Kroto, Smalley and Curl, carbon was known to exist in a pure form only as diamond and graphite. Through their work on fullerenes—vaporized carbon formed in inert gas then a plasma—it was found that clusters of 60 atoms were the most stable, and in such experiments the stable pure-carbon species, C_{60}, was discovered. Kroto's knowledge of graphic design no doubt helped him (and his colleagues) to propose that a specific symmetrical structure made up of carbon clusters, C_{60}, was composed of 20 hexagonal sides and 12 pentagonal ones, a shape now almost commonplace in science textbooks. Indeed, such a structure, which they referred to as "truncated icosahedron cage", had already found its place in modern architecture—a dome designed by American architect Buckminster Fuller for the 1967 Montreal World Exhibition, hence the name given to C_{60}—Buckminster fullerene!

Work on fullerenes initially met with a skeptical reception, particularly from organic chemists, following publication of the initial research in *Nature*. I have clear memories of distinguished organic chemists verbally and aggressively rejecting the idea of C_{60}, which to them was an unstable strained ball of carbon atoms. Yet, Harry's tenacity prevailed. With his student Jonathan Hare and others, he conducted further investigations to demonstrate the existence of both C_{60} and C_{70}. Later, related closed shell carbon cages were discovered in soot particles from diesel engines and flames, and are likely to be a component of polluting particulate matter, which is now an important health concern for the populations of big cities. NASA's telescopes also noted "buckyballs" in the remnants of a white-dwarf star. Harry's research also had implications for materials science.

Harry's impact at the University of Sussex was especially felt in the field of microwave spectroscopy, his study of carbon chains in dust clouds of interstellar space being especially significant. He was a vibrant, humorous, and very well liked colleague. His insistence on doing something worthwhile to the very best of his ability was a key to the way he worked.

Harry spent much of his academic career working in British universities. After working some time in North America in the early sixties, he joined in 1967 the University of Sussex at their newly established School of Molecular Sciences, being offered a postdoctoral position. Sussex was a new university—very interdisciplinary, liberal, and attracting outstanding academics on a beautiful rural campus; it became known as 'Oxbridge by the Sea', and this atmosphere suited Harry.

Becoming a lecturer soon after, he became a professor in 1985, and from 1991 to 2001 he was Royal Society Research Professor. Harry lectured widely at the University of Sussex. Elsewhere he often gave a very popular lecture: 'C_{60} Buckminster fullerene: the Celestial Sphere that fell to Earth', which was always very well received. From 2002 to 2004, he was President of the Royal Society of Chemistry. Discouraged at what he saw as a lack of backing for science in British universities, Harry took up a post at Florida State University in Tallahassee. In 2004, he returned an honorary degree to Exeter University in protest at the closure of its Chemistry department.

Harry's passion for science was also evident in the number of projects and initiatives that he helped to establish. In 1995, he formed the Vega Science Trust, a resource for scientists to discuss and broadcast their work, which he set up with the help of BBC producer Patrick Reams. He also created GEOSET, an online resource for teachers of science, where they could freely download outstanding pieces of teaching material. Harry's dedication and love for science continued to the end: in 2014, he and his wife Margaret set up the Prize for the Innovative Use of Technology in Science Learning, an award for children aged 11–18 based anywhere in the world, for the best STEM video.

Harry retired in Sussex in 2015. Knighted in 1996, he won numerous awards, including the Michael Faraday Prize in 2001 or the Longstaff Medal of the Royal Society in 1993. He is survived by his wife Margaret and his two sons Stephen and David.

He will be missed by us all.

Lawrence J. Dunne, May 2016

The photograph shows Harry Kroto at Sussex in 1996
(© The Royal Society/Anne Purkiss)

Obituary

Yves Chauvin (1930–2015)

Professor Yves Chauvin was a French organic chemist and Nobel Laureate. He became Honorary Director of Research at *Institut Français du Pétrole* after his retirement in 1995, and a member of the *Académie des Sciences* in 2005. In the early 1970s he became known for his work on deciphering the process of olefin metathesis, for which he was awarded the Nobel Chemistry Prize in 2005, along with Robert H. Grubbs and Richard R. Schrock.

I met Yves Chauvin at a dinner organized by Countess Marie de la Bouillerie in her mansion in Tours in November 2007, an initiative of our common friend Claude Gaudeau de Gerlitz, then treasurer of CMOA. Yves was a very friendly and modest person, who had initially been embarrassed to receive the prestigious Nobel award. However, he accepted to help us by joining the Editorial Board of this PTCP book series to replace Ilya Prigogine, who had passed away.

Yves Chauvin was born on 10 October 1930 at Menin in Flanders, on the border between Belgium and France. His parents were from Beaumont-la-Ronce in Touraine, France. He used to spend his holidays there, in his grandparents' large family house. Most of his ancestors were small farmers. His grandmother was fond of painting and of playing the piano, which she studied with Chabrier. His father was an electrical engineer, who was sent to Ypres and then to Menin after World War I, to help rebuild the electrical network in this war-ravaged province. Yves had five brothers and sisters and they had quite a strict upbringing. He went to preschool in Flanders then to primary school in France, which meant that he crossed the border every day.

He was not a brilliant student, even at chemistry. He chose chemistry rather by chance, because he firmly believed that one can become passionately involved in one's work whatever it is. Various circumstances, mainly due to military service, prevented him from doing a Ph.D., and he often regretted it.

He then took a job in industry. But the fact that process development consisted primarily of copying what already existed, with no possibility of exploring other fields, prompted him to resign. He discovered this was a very common attitude among managers: “Do what everyone else does and change as little as possible; at least we know it will work.” It was the opposite of his way of thinking: “If you want to find something new, look for something new!” There is a certain amount of risk in this attitude, but he believed you are so happy when you succeed that it is worth taking the risk.

Yves Chauvin joined the *Institut Français du Pétrole* in 1960 and managed to focus his work on what he thought would be most interesting.

The oil industry essentially uses heterogeneous catalysis: cracking, reforming, hydrodesulphurization, hydrogenation, etc., but that was not what interested him. At that time, nothing much was being done in France on coordination chemistry, organometallics or homogeneous catalysis by transition metals, and Yves Chauvin was fascinated by the achievements in Italy (G. Natta), Great Britain (J. Chatt), Germany (Max-Planck-Institute in Mülheim), and the United States. As a result, he unwittingly became the French specialist in these disciplines, which brought him into contact with the various activities of the *CNRS* committees. He spent the best part of his time on applied chemistry. This was how he came to develop two homogeneous catalysis processes.

The first one, which uses a nickel-based catalyst, was called “Dimersol” and exists in two basic process versions.

The “gasoline” version consists of dimerizing propene to the high-octane iso-hexenes. There is, quite often, an excess of propene, especially in oil refineries that do not have petrochemicals, as in the USA. In the early 2000s, there were 35 plants in operation, including 18 in the USA, with a combined annual output of 3.5 million tons. It was the first and only time that coordination catalysis had been used in refining.

The “chemical” version consists of dimerizing n-butenes to iso-octenes, basic inputs for plasticizers, using the *oxo* reaction. In the 2000s, production levels stood at 400,000 tons per year.

The second process that Chauvin developed, using a titanium-based catalyst, was called “Alphabutol.” It consists of dimerizing ethylene to 1-butene, the co-monomer of low-density linear polyethylene. The benefits of that process, not obvious at first, stem from a number of causes. In the early 2000s, there were 20 plants operating worldwide, with a combined output of 400,000 tons per year. Many more were built afterwards.

Yves Chauvin also worked on batteries and, in particular, on the non-aqueous electrolytes used to extend their electrochemical window. He thought it would be a good idea to use these electrolytes, which belong to the class of ionic liquids, as catalyst solvents. These liquids feature very low vapor pressure and virtual non-solubility in hydrocarbons, paving the way for a biphasic catalysis. The mixture of alkylimidazolium and aluminum chlorides forms a liquid with a very low melting point. It proved to be a first-rate solvent for nickel-based dimerization (“Dimersol”) catalysts. The reaction volume required for a biphasic system is ten times smaller than for a homogeneous system: this is important for safety, for refineries do not like to have large volumes in reaction because they are potential bombs; likewise for nickel consumption. This new process, dealt with in a PhD project in 1990, saw the light thanks to the inventiveness and determination of Hélène Olivier-Bourbigou, who took over Yves Chauvin's Laboratory.

What applied chemistry taught Pr Chauvin is the need for absolute solidarity between the research laboratory and the downstream side: testing, marketing, setting up an industrial plant; same enthusiasm and same determination, especially when everything seems to go wrong!

In Chauvin's view, there is no difference between fundamental research and applied research. The design of "processes" took up about three-quarters of his working time. However, he also took an interest in other aspects of coordination chemistry: palladium and rhodium catalyses, asymmetric aminoacid synthesis. After retiring in 1995, he became Emeritus Director of Research in J.-M. Basset's Laboratory at the Lyon School of Chemistry, Physics, and Electronics.

Yves Chauvin died at the age of 84 on 27 January 2015. Following his will, he was buried in his parents' village cemetery near Tours. He is survived by his wife and two sons and five grandsons.

Jean Maruani, May 2016
Adapted from *Les Prix Nobel, The Nobel Prizes 2005*
Editor Karl Grandin (Nobel Foundation), Stockholm, 2006

Yves Chauvin at Tours in 2006

Obituary

Christiane Bonnelle (1930–2016)

Professor Christiane Bonnelle was a French chemical physicist and, from 1979 through 1991, the Director of the prestigious *Laboratoire de Chimie Physique*, founded by Nobel Laureate Jean Perrin in 1926. She was born on 17 May 1930 in Paris, the second of three children in a 'Protestant High Society' family. She studied first at *Lycée Molière* and then at *Lycée de Tulle*, where her family took refuge during WW-II. Later she entered the *Sorbonne*, where she received a *Licence ès Sciences* in 1954 and a *Doctorat ès Sciences* in 1964.

From 1954 through 1960 she worked at the French *CNRS*; then she followed a university cursus at the *Sorbonne*, where she became Full Professor in 1974. In 1964 she received the Bronze Medal of *CNRS* for outstanding thesis and in 1991 she was named 'Man of the Year' by the Cambridge *International Biographical Centre*. During her career she published close to 150 research papers and several reviews and books and directed more than 60 doctorate theses.

Christiane Bonnelle was essentially an X-ray spectroscopist, a field in which she quickly became recognized as a world expert. She was especially a pioneer in the difficult topic of the signature of chemical surroundings of elements in X-ray spectra. Her main areas of research are outlined in the following.

- **Synchrotron radiation**. In 1963, under the direction of Yvette Cauchois, she obtained the first X-ray spectra on a synchrotron, that of Frascati in Italy. These results opened a brand-new field in spectroscopy, which underwent considerable development in the world. In France two machines are fully dedicated to the uses of this radiation: *ESRF* at Grenoble and *SOLEIL* at Saclay.
- **The Sun X-radiation**. En 1966, with Gilles Sénémaud and Yvette Cauchois, she obtained the first X-ray spectra of the Sun with a spectrograph boarded on a space rocket. The recent impressive development of X-ray spectrometry in astrophysics follows from this pioneering experiment.

- **Lanthanides and Uranides**. The *Laboratoire de Chimie Physique* has developed a tradition of X-ray investigation of elements that are little studied due to their very complex electronic structure. During her carrier, Christiane Bonnelle has contributed to the setup of a corpus of X-ray spectroscopic data for the rare-earths and the actinides, which are of major interest for being involved in the conception of electronic devices and the recycling of nuclear fuel.

  In 2015, after several years of thorough search and critical analysis of previous and original research data, Christiane Bonnelle and Nissan Spector published the landmark monograph: *Rare-Earths and Actinides in High Energy Spectroscopy*, in the present book series.
- **Condensed matter**. Christiane Bonnelle devoted a great part of her work to the X-ray spectroscopy of aggregates, interfaces, amorphous silicon, and multi-layer devices that are used in X-ray optics. She obtained significant results in all these fields, which led to further development. Members of her group were part of the Grenoble team that validated the X-ray diffraction results of Nobel Laureate Daniel Shechtman on quasicrystals.
- **Instruments and Applications**. An important aspect of Christiane Bonnelle's activity was her search for novel tools and applications. She especially initiated and designed the construction of prototype instruments such as *IRIS* (Instrument for Research on Interfaces and Surfaces), which has become the *Shallow Probe* in industrial applications.

Christiane Bonnelle is acknowledged by her colleagues and coworkers as a person of deep passion, rigor and honesty, whose dynamism, open-mindedness and readiness to share her expertise was a great incentive to those who had the luck to work with her. Even after she officially retired in 2000, she came to her office every day, working from morning to evening.

When the *CMOA* of *CNRS*, former laboratory of Prof. Raymond Daudel, was split into several units after he retired, Christiane Bonnelle, who was then the Director of the *Laboratoire de Chimie Physique*, accepted to host our group. She offered us facilities to settle and time to get acquainted with fields that were not familiar to us. In return, we stimulated the development of computer-oriented theoretical methods in a laboratory that was then mostly experimental.

Once in a while, I had the pleasure to enter her office and chat about scientific but also philosophical, artistic, and even political matters. She particularly enjoyed my recent research on the Dirac electron and my interest in the connections between science and music. In spite of her puritan education, she also had a great sense of humor. Her sudden illness and decease came as a loss for the scientific community, but also brought a shock to her colleagues and friends. The last time I phoned her home, she was waiting for her son to drive her to Southern France, where she would be surrounded by the love and care of her family.

Christiane Bonnelle died at the age of 86, on 21 August 2016. She is survived by a son, Marc, two grand-daughters, Valérie and Caroline, and two great-grand-children from Valérie, who works as a neuroscientist near Oxford.

Christiane Bonnelle was a great lady in the world of chemical physicists. She will be remembered and missed.

Jean Maruani, September 2016

With the cooperation of Nissan Spector, Alfred Maquet, Dominique Weigel, Jean-Michel André and Philippe Jonnard, and the approval of Marc Bonnelle

Obituary

Rossen Pavlov (1947–2016)

Rossen Lubenov Pavlov was a Bulgarian scientist who made his entire career at the Bulgarian Academy of Sciences, first at the Central Laboratory of Biophysics (from 1978 onwards) and then (from 1985 to this day) at the Institute of Nuclear Research and Nuclear Energy (*INRNE*). He was born on 8 January 1947 in Sofia, the second of two children in a middle-class family (his father was a landowner and a professor in agronomy) and, as most bourgeois people in those days, he had French as a second language. One-third of our 30 common papers were in French.

Rossen Pavlov graduated from high school in 1965 and from the Department of Physics of Sofia University in 1974. At the Central Laboratory of Biophysics he became involved in the experimental and theoretical investigation and design of hormones, steroids, and enzymes related to bioregulation and aging. He made the first quantum computations on melatonin, serotonin, and derivatives. He also investigated the possible effects of geophysical and astrophysical parameters on bioregulation processes.

Some time later he became the scientific secretary of Julia Vassileva-Popova, the then almighty spouse of the deputy-minister of Foreign Affairs. As such, he was acquainted with the highest-rank political figures in the country. He became a private consultant of the Minister of Culture Lyudmila Zhivkova, the daughter of President Todor Zhivkov. When Lyudmila deceased (allegedly due to a KGB plot), he left his former position and was introduced at *INRNE* by his old friend Yavor Delchev, a nephew of the famous nonconformist reporter Wilfred Burchett.

There he joined a group headed by Ivan Petkov, which had acquired a world-recognized expertise in the investigation of nuclear shell structure by using density functional theory and the Strutinsky partition technique. He became involved in the development of exact functionals by the use of the local-scaling transformation scheme. That is where our roads crossed.

I first made his acquaintance in 1986, at a meeting that I organized in Paris, together with Pr Imre Csizmadia (the founder of *WATOC*), for the retirement of Pr Raymond Daudel (the founder of the *CMOA* of *CNRS*). Among the participants there was a plethora of world-famous scientists, including eight Nobel Laureates. This event brought me in touch with a number of people who were to play a crucial role in the *QSCP* network: Bulgarian Academician Stefan Christov, Pr Roy McWeeny from Pisa and, last but not least, that young, strange Bulgarian fellow who stuttered in a stressing way but could make the ladies faint by practising the old-fashioned hand-kissing, his long blond hair covering his bearded face while his slender body was bowing down.

The following year I went for the first time to Sofia, where I was invited in a *IUPAC* Congress. There I met Rossen again, and we started a cooperation that lasted over 20 years.

Rossen Pavlov launched a number of Franco-Bulgarian projects, first between the *CNRS* and *BAS*, then between the Université Pierre et Marie Curie and Sofia University Saint-Kliment Ohridski. Later he moved to European projects involving laboratories from Bulgaria and France but also from Spain (Yves Smeyers, followed by Gerardo Delgado-Barrio), Italy (Roy McWeeny, one of the founders of the reduced density matrix formalism, a keynote in our project), England (Stephen Wilson, proposed by Sonia Rouve, a friend of Rossen Pavlov who was teaching scientific English to Bulgarians at the British Council), Sweden (Hans Agren, whom I met while working with Anders Lund at Linköping, followed by Erkki Brändas), as well as from Russia and later from Belgium and Germany. The *QSCP* annual workshops originate from this European network.

The picturesque chain of events that led to the *QSCP* network, and an illustrated overview of the *QSCP* meetings from 1996 through 2008, is described in an earlier volume of this book series edited by Piotr Piecuch: Jean Maruani, *Progr. Theor. Chem. & Phys.* B-19 (2009), pp. 3–32. In recognition of the role played by Bulgarians in this network, the *CMOA* Promising Scientist Prize, announced at the banquet dinner of the Uppsala meeting, was first awarded in Sofia in 2001.

Rossen Pavlov was a person who could bring you in touch with anyone you wished. After I arranged for some European grants to various Bulgarian institutes—and gave a few lectures at Sofia University (on exotic symmetries), he managed through Pr Peter Raychev, Jordan Stamenov (then the Director of *INRNE*) and Matey Mateev (a former Minister of Education) to have me nominated for a *Doctorate Honoris Causa* at Sofia University. Rossen received his *Doctorat ès Sciences* from the Université Pierre et Marie Curie in 1999.

Rossen Pavlov had a real passion for mathematical physics but, as many other Bulgarian scientists, he was plagued by material—often trivial—problems, which prevented him from developing his full potential. He liked to seek connections between biophysics, molecular physics, nuclear physics and even, in recent years, astrophysics (in cooperation with Nicolas Chamel from Belgium, with whom he wrote fifteen papers). He also was a man of culture, extending to other scientific fields as well as to literature and history.

Rossen Pavlov was deeply devoted to science, though occasionally fascinated by colleagues who had succeeded in business. He could be irritating but was capable of lasting friendship. He was an unconventional, crazy kind of fellow. One day, together with my wife Marja and Rossen's friend Sonia Rouve, we climbed up Vitosha Mountain near Sofia and, somewhere on the way downwards, we got lost in the darkening forest. We were exhausted and discouraged when suddenly Rossen appeared in the nocturnal wilderness, like some ethereal moonlit spirit!

Rossen liked to establish human links. He introduced his sister Diana to my wife and endeavored to make us feel home. He loved his country and tried to make others love it too. Once he insisted for me to join him at his father's former properties near Belovo. Years later, while preparing *QSCP-X* at Carthage (the last workshop he attended abroad), he accompanied me to my parents' former home at Ariana, a suburb of Tunis that has an exact homonym in Sofia district.

The family house where he lived with his mother was cluttered with books and papers, records by beloved Elvis Presley, cats and kittens up and down the steps, and had a garden full of quinces and raspberries. He liked solitary meditation in the midst of nature, but also knew how to organize group events and to be sharply efficient when needed. He was a guiding spirit, introducing other colleagues to the group, facilitating meetings, arranging agreeable locations—at the mountain or the seaside—for scientific workshops. He initiated a number of projects that he let others develop. The Varna 2015 *QSCP* workshop, organized with Alia Tadjer, was his last initiative. We could see how happy he was in the midst of his scientific family and how his generous nature had a contagious effect on all of us.

Those who have known him will find inspiration and strength in his memory.

Rossen Pavlov died from a brain stroke at the age of 69, on 9 September 2016. He is survived by a daughter, Yuliana (an English teacher) from his wife Radka Sungarska, and by two grandchildren, Alexander and Katerina.

Jean Maruani, September 2016
With the cooperation of Sonia Rouve and Alia Tadjer
and the approval of Yuliana Pavlova

Presenting the Hosts' *Radioisotopes Production Laboratory* at the *Institute for Nuclear Research and Nuclear Energy*

D. Tonev, N. Goutev, L.S. Georgiev, A. Nikolov

Abstract The Institute for Nuclear Research and Nuclear Energy at the Bulgarian Academy of Sciences has recently started the construction of an accelerator laboratory in Sofia. The cornerstone of the laboratory is a TR24 cyclotron, which provides a proton beam with a variable energy between 15 and 24 MeV and current of up to 0.4 mA. The TR24 accelerator allows for the production of a large variety of radioisotopes for medical applications and development of radiopharmaceuticals. The new cyclotron facility will be used for research in radiopharmacy, radiochemistry, radiobiology, nuclear physics, solid-state physics, applied research, new materials and for education in all these fields including nuclear energy.

Keywords Cyclotron, PET/CT, Radiopharmacy

Introduction

The project "Cyclotron laboratory of INRNE-BAS" has been started in 2012 by the Institute for Nuclear Research and Nuclear Energy (INRNE) at the Bulgarian

---

D. Tonev
Institute for Nuclear Research and Nuclear Energy,
Bulgarian Academy of Sciences, Sofia, Bulgaria

N. Goutev
Institute for Nuclear Research and Nuclear Energy,
Bulgarian Academy of Sciences, Sofia, Bulgaria

L.S. Georgiev
Institute for Nuclear Research and Nuclear Energy,
Bulgarian Academy of Sciences, Sofia, Bulgaria

A. Nikolov
Institute for Nuclear Research and Nuclear Energy,
Bulgarian Academy of Sciences, Sofia, Bulgaria

Academy of Sciences (BAS) under the auspices of the Council of Ministers of Republic of Bulgaria in order to solve a longstanding medical problem for the Bulgarian population—the shortage of short lived cyclotron radioisotopes for medical imaging and diagnostics. The limited funds released by the Bulgarian National Health Insurance Fund and the high price of the radioisotopes currently imported from abroad makes the *positron-emission tomography/computer tomography* (PET/CT) technology [1] available only for less than 1/3 of the Bulgarian patients who need urgent medical diagnostics. The aim of the Cyclotron laboratory of INRNE-BAS is to provide domestic production of such medical radioisotopes, which would meet the need for PET/CT isotopes for all PET/CT laboratories in the country, and eventually for those in the south-east region of the Balkan Peninsula.

Two generous financial grants have made the project possible: one from the United States Department of Energy (DOE) with a donation of $3,000,000 USD and a second from the Nuclear Power Plant "Kozloduy" with a donation of $2,000,000 USD.

The INRNE-BAS project for a new cyclotron laboratory has been included in the 2014 updated "National Roadmap for Scientific Infrastructure" of Bulgaria. At the beginning of 2015 the project was identified as a top priority of the Bulgarian–American relations in the field of science and education. For several years the management team of INRNE-BAS has been actively working to insure further government support for the construction of the building of the cyclotron laboratory.

The new cyclotron laboratory as a novel part of the existing infrastructure of INRNE-BAS would consist of:

- a specialized building that should meet the regulatory requirements in the field of radiation safety and good manufacturing practices (GMP) in the pharmaceutical industry;
- a bunker in which the cyclotron TR-24 will be installed;
- a sector for applied research and development in radiopharmacy with enhanced educational functions;
- a sector for the production of 18F-FDG, called Fluorodeoxyglucose, and in future, of other radiopharmaceuticals for clinical use;

Table 1 Nuclear reactions and half-lives $T_{1/2}$ of the most commonly produced PET radioisotopes for medical imaging

Radioisotope	Reaction	$T_{1/2}$ (min)
^{18}F	$^{18}O(p,n)^{18}F$	110
^{15}O	$^{14}N(p,n)^{15}O$	2
^{13}N	$^{16}O(p,\alpha)^{13}N$	20
^{11}C	$^{14}N(p,\alpha)^{11}C$	10

Fig. 1 A side view of the INRNE-BAS cyclotron during the tests in Vancouver in August 2015

The radioisotopes are produced by nuclear reactions [2], such as those shown in Table 1, when a target is irradiated with an intense proton beam. For example the isotope ^{18}F is produced by irradiation of a target of ^{18}O enriched water with protons and in the (p, n) nuclear reaction, shown in the first line in Table 1, one neutron is emitted, transforming the ^{18}O into ^{18}F.

The radioisotope ^{18}F is a positron emitter [1] with half-life of approximately 110 min which is used to produce Fluorodeoxyglucose (FDG), a form of glucose, which is injected into the blood system and is most noticeably accumulated in tumors. When a positron is emitted from the FDG it quickly annihilates with the first electron it encounters emitting two gamma quanta propagating in opposite directions, which are detected in PET scanners by multiple gamma detectors, measuring at the same time the difference in the arrival times in order to determine the point of the annihilation. After that computer tomography is used to reconstruct a three-dimensional image of the tumor. Currently, the resolution of the PET/CT is less than 1 mm, which is practically sufficient for the purposes of the computer tomography [1].

The manufacturing of the cyclotron TR24 purchased by INRNE-BAS has been completed in the middle of 2015 and the tests at site of the manufacturer have been successfully accomplished in August 2015, see Fig. 1 for a side view of the cyclotron TR24 before its transportation to Bulgaria.

There are many academic institutions in Europe where new accelerator centers are under development. TR24 is the cyclotron model of the accelerator which has recently been installed at CYclotron pour la ReCherche et lEnseignement (Cyrce) of the Institut Pluridisciplinaire Hubert Curien (IPHC) in Strasbourg, France. Soon the Institute of Radiopharmaceutical Cancer Research of Helmholtz-Zentrum Dresden Rossendorf (HZDR) in Germany will operate a new (modified) TR24. Another TR24 is installed in Rzez, Czech Republic. On the other hande, the Laboratori Nazionali di Legnaro at INFN, which uses a 70 MeV cyclotron, is planing to perform similar activities to ours [4].

For the time being, before installing the INRNE-BAS cyclotron in Sofia, Bulgaria which has a working nuclear power plant and intends to install more nuclear power plant units, is having no working research nuclear facility, neither a research reactor, nor an accelerator. With the new facility we will be able to restart our experimental research program not only in the field of nuclear physics, but also in many interdisciplinary fields related to nuclear physics. An excellent presentation of the accelerator physics is given in the work [3].

Objectives of the INRNE-BAS Cyclotron Project

The main purpose of the INRNE-BAS cyclotron project is to allow for regular supply of 18F-FDG to 8 regional PET/CT center, within five years, at a low price providing "full cost recovery", with a small profit margin which will be used to sustain the laboratory. The revenue from the sale of radioisotopes and radiopharmaceuticals will be fully reinvested in the research program and in the development of the laboratory. This will enable more hospitals in Bulgaria to purchase new PET/CT equipment as well as to increase the throughput of the existing scanners. The wider access to PET/CT will definitely improve the health and extend the life of the patients suffering from cancer.

Another objective of the INRNE-BAS cyclotron project is to ensure high quality training of nuclear pharmacists and to complete the chain from research to end products with clinical applications due to the availability of the GMP compliant production sector.

The TR24 cyclotron, shown in Fig. 1, whose parameters correspond to the technical specification in the public procurement organized by INRNE-BAS, has been chosen with the intention that in the next twenty years INRNE-BAS will be able to produce a wide range of radioisotopes which are used in medicine with a relatively low initial investment and moderate maintenance costs. In more detail, the proton energies and beam current of TR24, whose control and management system is shown in Fig. 2, makes it possible to produce commercial quantities of radioisotopes with traditional and expected future applications in medicine such as PET radioisotopes ^{18}F, ^{124}I, ^{64}Cu, ^{68}Ge/^{68}Ga, as well as single-photon-emission computer tomography (SPECT) radioisotopes ^{123}I, ^{111}In, ^{67}Ga, ^{99m}Tc and therapeutic alpha-emitters: ^{225}Ac/^{213}Bi, ^{230}U/^{226}Th.

Fig. 2 Control and management system of the cyclotron manufactured for INRNE-BAS

Next, from the viewpoint of the applied research, the radioisotopes produced by the cyclotron will be used for the development of new radiolabelled molecules for diagnostics [5] and therapy purposes [6]. A longer term goal of the laboratory is to become involved in preclinical studies and clinical trials through collaborations and research projects with academic partners and private companies.

From scientific point of view the cyclotron laboratory in INRNE-BAS will open new possibilities for experimental nuclear structure research in Bulgaria. The plan is to investigate level-schemes of excited nuclei as well as to measure lifetimes of excited nuclear states. The proton beam will be used to bombard different targets in order to cause nuclear reactions and the radiation generated from the reactions will be registered with a multidetector system. The data obtained could be analyzed to reveal new details of the nuclear structure.

On the other hand, it is possible to measure excited nuclear states lifetimes in the femtosecond region with the TR24 cyclotron using the Doppler-shift attenuation method (DSAM) following the inelastic proton scattering reaction. Such an approach is presented in details in the work of Hennig et al. [7]. In the experimental setup used in the Institute for nuclear physics in Cologne University, the proton in the exit channel of the $(p,p'\gamma)$ reaction is detected with charged-particle detectors placed close to the target. This approach uses the centroid-shift version of the DSAM method and profits from the coincident detection of scattered protons and de-exciting γ-rays.

Following the approach of Hennig et al. thirty lifetimes of excited states in ^{96}Ru have been determined in [8]. A similar experimental setup will be installed in the cyclotron laboratory of INRNE-BAS in order to measure lifetimes for low-lying states in nuclei which could be approached by the $(p,p'\gamma)$ reaction. The cyclotron laboratory will also allow us to continue our research in the fields of phase transitions in nuclear physics [9], chirality in nuclei [10, 11] and nuclear structure of light nuclei [12].

Last but not least, the cyclotron laboratory will be extensively used for education and training of physicists, chemists, pharmacists and biologists for radioisotopes production, radiochemistry, quality control of radiopharmaceuticals and their uses in medical imaging, as well as training of specialists in nuclear energy. More general practical topics like radiation protection and dosimetry will also be covered by the training and education program of the laboratory. Most of the training and education activities will be carried out by involving the students in the ongoing research projects of the laboratory. Special attention will be paid to the training of students which plan to work in the field of nuclear energy.

Characteristics of the Accelerator TR24

The cyclotron TR24, purchased by INRNE-BAS, is manufactured by Advanced Cyclotron Systems Inc. (ACSI), located in Vancouver, Canada. The TR-24 cyclotron of INRNE-BAS, shown in Figs. 1 and 2, has the following characteristics:

- it accelerates negatively charged hydrogen ions H^- and the proton beam is extracted with the help of carbon foil which can be moved in radial direction;
- the proton beam has variable energy; the minimum energy of the proton beam is 15 MeV while the maximum energy of the proton beam is 24 MeV;
- the proton beam current is 400 μA, however, it is upgradeable to 1000 μA;

- it provides simultaneous extraction of two proton beams;
- it has an external CUSP ion source.

Because of the very high current of up to 400 μA, upgradeable to 1 mA, and because of the variable proton energy ranging from 15 to 24 MeV, the cyclotron TR24 can produce commercial quantities of more than 15 isotopes, see Section "Objectives of the INRNE-BAS Cyclotron Project" for more details, by irradiating simultaneously two high current targets placed at the exit ports or along high current beam lines. At the beginning one Y-like beam line will be mounted at one of the exit ports of our cyclotron for greater flexibility. Currently, there are two target stations and three targets for the production of 18F-FDG available in the Cyclotron laboratory of INRNE-BAS.

Status of the Bulgarian National Cyclotron Center

In 2014 the Institute for Nuclear Research and Nuclear Energy has successfully completed the public procurement tender for the production of the cyclotron TR24 and signed a contract for the delivery of the cyclotron, manufactured by the Canadian company Advanced Cyclotron Systems Inc. The cyclotron produced for INRNE-BAS passed successfully the low energy tests in the factory in August 2015 and an illustration of the tests is shown in Fig. 3.

The cyclotron TR24 was delivered to INRNE-BAS on the 12 January 2016 as illustrated in Fig. 4.

The building for the cyclotron laboratory, with a bunker for the cyclotron and the hot cells for the production of 18F-FDG, will be constructed nearby the Institute for Nuclear Research and Nuclear Energy in Sofia. Currently, INRNE-BAS is involved in the feasibility study for the building of the Cyclotron laboratory.

Fig. 3 INRNE-BAS cyclotron during acceptance tests in Vancouver, Canada in 2015

Fig. 4 Delivery of the cyclotron to INRNE-BAS

Acknowledgements The project has been supported by the Department of Energy of USA, the Nuclear Power Plant "Kozloduy", the Bulgarian Ministry of Education and Science. This project has been supported by the Bulgarian Science Fund under Contract No. DFNI-E 01/2 and by a NUPNET-NEDENSAA project funded by the Bulgarian Ministry of Education and Science. The authors acknowledge the sincere support for the project from colleagues and friends.

Bulgaria was chosen for the home of the twentieth edition of the QSCP workshop series in recognition of its contribution for setting up the QSCP network, which originated from a Franco-Bulgarian collaborative project started on Bulgarian part by the INRNE researchers Dr. Rossen Pavlov (1947–2016) and Dr. Yavor Delchev (1944–2006).

Dr. Rossen Pavlov (1947–2016)

Dr. Yavor Delchev (1944–2006)

References

1. Bailey DL, Townsend DW, Valk PE et al. (2005) Positron emission tomography: basic sciences, Springer-Verlag, NJ, ISBN: 1-85233-798-2
2. Krane K (1988) Introductory Nuclear Physics, John Wiley & Sons ISBN: 0-471-85914-1
3. Dinev D (2006) Particle accelerators Marin Drinov Academic Publishing House ISBN: 13:978-954-322-067-0
4. de Angelis G, Prete G, Andrigetto A, for the SPES collaboration, et al. (2016) EPJ Web of Conferences A 107:01001
5. Yordanov AT, Llop J, Stimson D et al. (2011) J Nucl Med 52(1):1643
6. Yordanov AT, Hens M, Pegram C et al. (2007) Nucl Med Biol 34:173
7. Hennig A et al. (2015) Nucl Instrum Meth Phys Res A 794 171
8. Hennig A et al. (2015) Phys Rev C 92:064317
9. Tonev D et al. (2004) Phys Rev C 69:034334
10. Tonev D et al. (2006) Phys Rev Lett 96:052501
11. Tonev D et al. (2014) Phys Rev Lett 112:052501
12. Tonev D et al. (2002) Phys Rev C 65:034314

Part I
Quantum Methodology

Towards the Inclusion of Dissipative Effects in Quantum Time-Dependent Mean-Field Theories

L. Lacombe, M. Vincendon, P.M. Dinh, P.-G. Reinhard and E. Suraud

Abstract We discuss various theoretical approaches to describe dissipative dynamics in finite systems. We focus in particular on the case of irradiated clusters and molecules. We recall semi-classical approaches successful in the case of simple metallic clusters at high excitation. We then focus on dedicated quantum approaches. We discuss in particular two recently proposed approaches, a relaxation time ansatz (RTA) and a stochastic extension of mean field dynamics (STDHF). We present a brief outline of the theories and show some examples of applications.

1 Introduction

The analysis of dissipative effects in finite quantum systems is a long standing question [1], which has been and is addressed in various field of physics and chemistry. In recent years, it has been studied in growing detail in clusters and molecules excited by intense laser fields [2–5]. But it is also observed in the case of ballistic electron transport in nano systems [6], as well as for thermalization in trapped Fermi gases [7]. The interest in the molecular case has especially grown in the last years [4, 8], thanks to progress in the analysis of electronic emission (energy-, angular-resolved distributions...), through elaborate imaging techniques such as Velocity Map Imaging

Electronic supplementary material The online version of this chapter (doi:10.1007/10.1007/978-3-319-50255-7_1) contains supplementary material, which is available to authorized users.

L. Lacombe · M. Vincendon · P.M. Dinh · P.-G. Reinhard · E. Suraud (✉)
Laboratoire de Physique Théorique, Université de Toulouse, CNRS, UPS, Toulouse, France
e-mail: suraud@irsamc.ups-tlse.fr

P.-G. Reinhard
Institut für Theoretische Physik, Universität Erlangen, 91058 Erlangen, Germany

L. Lacombe · E. Suraud
Physics Department, University at Buffalo, The State University New York, Buffalo 14260, USA

A. Tadjer et al. (eds.), *Quantum Systems in Physics, Chemistry, and Biology*, Progress in Theoretical Chemistry and Physics 30,
DOI 10.1007/978-3-319-50255-7_1

[8, 9]. Thermalization effects also focused several investigations in nuclear physics, especially in induced nuclear fission or nuclear collisions [10, 11]. For example, in the case of nuclear collisions, the kinetic energy of the projectile is redistributed into thermal energy of the compound target + projectile system (possibly leading to the formation of truly "hot nuclei") which, in turn, de-excites via thermal emission and/or fragmentation. The scenario points out the role of "elementary collisional events" between system's constituents which we will from then on call "collisional correlations", the notion being taken from Fermi liquid theory [12] with incoherent reduction of two-body correlations to two-Fermion collisions.

From a theoretical point of view, nuclear physics devoted major efforts since three decades to address the question of hot nuclei. Semi-classical methods [11, 13, 14] were the stronghold, in the line of similar problems in particular in plasma physics [15], but also with investigations in terms of Molecular Dynamics approaches combining quantum features with a semi-classical treatment of dynamical correlations [16, 17]. Still, there exists to date no complete quantum approach in the field, in spite of numerous formal investigations [11, 18, 19]. The case of clusters and molecules has been more recently addressed with first attempts in terms of semiclassical approximations [20, 21]. These semi-classical approaches are based on the Vlasov equation complemented by a Boltzmann-like collision term, evaluated according to the Uehling-Uhlenbeck approximation [22], which accounts for Pauli-blocking factors in the course of elementary collisions between constituents of the system. These approaches are, nevertheless, bound to simple metal clusters, where the soft effective potentials characteristics of metallic bond can be well treated with a semi-classical approach. The applicability towards other binding situations, in particular highly studied covalent systems such as C_{60} [5] is more questionable. In addition, (semi) classical approaches are mostly tailored for high excitations such as those delivered by very intense laser pulses [3] where these methods can be justified and where the scenario is essentially a blowing up of the system. In the important case of less violent excitations quantum effects cannot be neglected. In such cases, the typical situation can be sketched as follows. In a typical irradiation process in clusters and molecules the confining potential delivered by the ionic background stabilizes the system for some time. Therefore, ionization becomes a highly efficient way to dissipate energy. As a consequence, the actual stored excitation energy is heavily regulated by electronic emission. This leads to a highly perturbed (far away from any equilibrium) system, but with moderate remaining excitation energy. This is a situation where semi-classics badly fails, and which obviously requires to account for quantum features. One should, nevertheless, cite some attempts to treat dynamical correlations semi-classically on top of quantum mean field [23] in the spirit of Time-Dependent Density Matrix approaches [24].

Fully quantum mechanical treatments have also been performed in schematic model systems [25] as well as in the time-dependent configuration-interaction (TD-CI) method [26]. One should also mention the standard extension of TDDFT which integrates current, namely Time Dependent Current Density Functional Theory (TDCDFT), which has proven very useful for analysing the width of collective modes [27]. This a priori suggest that TDCDFT might provide a ready to go approach to the inclusion of dissipative features on top of TDDFT. It nevertheless turns out that, even within keeping aside formal difficulties, especially concerning energy, and not mentioning the computational difficulties in handling the full TDCDFT in realistic 3D simulations, TDCDFT raises some fundamental difficulties when addressing an excitation regime beyond the linear domain. Indeed, systematics of relaxation time extracted from TDCDFT computations show that the theory leads to relaxation times independent of the excitation energy, which is a clear pathological behavior [28]. These elaborate treatments are limited to very small systems and low excitations. The need for a robust quantum theory addressing the regime of moderate to high excitation energies thus persists.

The present contribution addresses some directions of investigation of how to include dissipative effects in the case of the irradiation of clusters and molecules. After a brief reminder of the standard mean-field framework (Sect. 2.1), we shall successively consider a semi-classical approach (Sect. 2.2), a quantum relaxation time ansatz (Sect. 3), a stochastic extension of mean field (Sect. 4), and a simplification thereof in direction of a quantum kinetic equation. Although we shall focus the applications to cases of irradiated clusters and molecules, we should stress the fact that the formal developments we shall present can be applied to a wider variety of physical systems under non-linear excitations.

2 From Mean-Field to Semi-Classical Kinetic Equations

2.1 *The Mean Field as a Starter*

The dynamics of irradiated clusters and molecules has already been widely addressed within the robust and rather versatile Time-Dependent Density Functional Theory (TDDFT) [29]. The Time-Dependent Local-Density Approximation (TDLDA) [30–32] thereof certainly provides one of the best compromises in the domain of quantum dynamics and has been applied to several dynamical regimes from the linear one up to highly non-linear processes [4, 33, 34]. But TDLDA is an effective mean-field approach which misses by construction dissipative effects from electron-electron collisions. This practically limits the range of applicability of TDLDA to moderate excitations.

As a starting mean field theory, we therefore take TDDFT at the level of TDLDA treated in the real time domain [30, 31]. For further formal discussions it is worth re-expressing mean field dynamics in terms of the one-body density matrix $\hat{\rho}$. The (natural orbitals) representation of the one-body density operator

$$\hat{\rho} = \sum_{\alpha=1}^{\infty} |\phi_\alpha\rangle W_\alpha \langle\phi_\alpha| \quad . \tag{1}$$

It allows to recast the mean field propagation under the form

$$i\partial_t \hat{\rho} = \left[\hat{h}[\varrho], \hat{\rho}\right] \tag{2}$$

with the Hamiltonian $\hat{h}[\varrho]$ computed from the diagonal part ϱ of the density matrix $\hat{\rho}$. Details on the mean-field picture can be found in Supplemental material, Sect. A.

2.2 The VUU Approach

The semi-classical approximation to the TDLDA dynamics outlined in Sect. 2.1 is known as the Vlasov-LDA theory. It can be formally obtained through Wigner transform which provides the natural starting point for expansions in orders of $\hbar$. At lowest order in $\hbar$, this leads to the phase space distribution $f(\mathbf{r}, \mathbf{p}, t)$ as the semi-classical analog of $\hat{\rho}$, and the Vlasov equation as the semi-classical analog of Eq. (13). One can thus formally write

$$\left\{ \begin{array}{c} \hat{\rho} \\ i\partial_t \hat{\rho} = \left[\hat{h}[\varrho], \hat{\rho}\right] \end{array} \right\} \overset{\hbar \to 0}{\longrightarrow} \left\{ \begin{array}{c} f(\mathbf{r}, \mathbf{p}, t) \\ \partial_t f = \{h[f], f\} \end{array} \right\} \tag{3}$$

where the semi-classical Hamiltonian possesses the same density dependence as its quantum mechanical counterpart but now with the density of matter computed from the phase space distribution as $\varrho(\mathbf{r}, t) = \int d\mathbf{p} f(\mathbf{r}, \mathbf{p}, t)$.

The interest in the semi-classical approximation is that it provides a simple framework to complement mean field dynamics by a collisional contribution in the form of a collision term "à la Boltzmann", and at the same time by properly accounting for Pauli blocking effects in the form of the Uehling-Uhlenbeck form [22]. This finally leads to a kinetic equation which can be written as

$$\partial_t f = \{h(\mathbf{r}, \mathbf{p}, t), f(\mathbf{r}, \mathbf{p}, t)\} + I_{\text{coll}}[f] \tag{4}$$

The collision term reads, for the phase space distribution $f(\mathbf{r}, \mathbf{p}_1, t)$

$$I_{\text{coll}}[f_1] \sim \int d\mathbf{p}_2 d\mathbf{p}_3 d\mathbf{p}_4 \delta(\sum_i \mathbf{p}_i)\delta(\sum_i \varepsilon_i)\frac{d\sigma}{d\Omega}$$
$$[f_3 f_4(1-f_1)(1-f_2) - f_1 f_2(1-f_3)(1-f_4)] \tag{5}$$

where we have used the compact notation $f_i = f(\mathbf{r}, \mathbf{p}_i, t)$ and where $d\sigma/d\Omega$ labels the elementary cross section associated to the residual interaction not contained in the mean field. It is important to note here that the collision term is computed at a given real space point. Note also that, at this point, one enforces conservation of both, momentum and single particle energy (which reduces to kinetic energy for a local potential).

As a first example of application, we consider the irradiation of a moderate size metal cluster by a moderately intense laser. Effects of dissipation can particularly

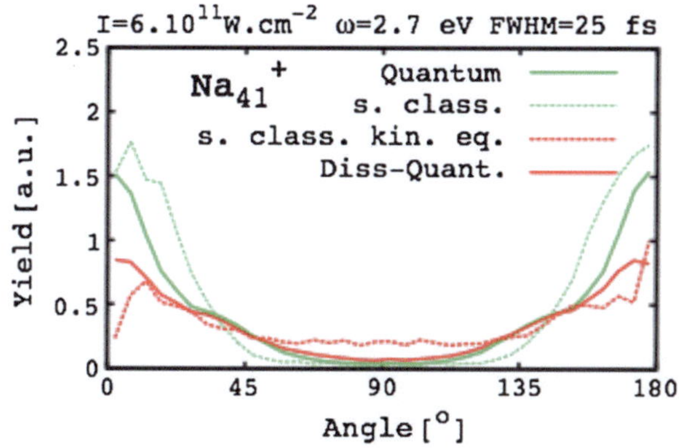

Fig. 1 Photo-angular distribution of Na_{41}^{+} following a laser irradiation (laser parameters on the figure). We compare mean fields approaches (quantum TDLDA and semi-classical Vlasov) to a semi-classical kinetic equation approach (VUU) and the dissipative quantum RTA

be identified on the Photoelectron-Angular Distribution (PAD) of emitted electrons. Figure 1 shows such an observable in Na_{41}^{+} and compares various levels of theory. The reference point is the quantum TDLDA approach (Sect. 2.1) which delivers the expected distribution with a sizable alignment of emission along laser polarization axis. The semi-classical version of TDLDA, the Vlasov LDA approximation is also plotted in Fig. 1 and delivers a very similar PAD with preferential forward/backward emission along laser polarization axis. The situation is clearly different in the case of the VUU approach which delivers a much more isotropic PAD. This reflects the fact that collisional correlations have led to some thermalization of the electron cloud. Therefore, there is a balance between alignment along laser polarization (which is accounted for at mean field level and which mostly occurs at short times) and isotropic emission (which is accessible at VUU level and which typically occurs at longer times). At least at a qualitative level, it is clear that VUU delivers the expected dissipative effects. It remains to be seen whether the case is accidental or the semi-classical approximation does blur some aspects. Therefore, the next step is to try to see how to include dissipative effects at a quantal level.

3 A Quantum Relaxation Time Ansatz

As mentioned in Sect. 2.2, collisions are local, changing for a given $\mathbf{r}$ only the momentum distribution at this specific point. This leads to establish local conservation laws [35]: collisional relaxation then preserves local density $\varrho(\mathbf{r}, t)$, local current $\mathbf{j}(\mathbf{r}, t)$, and local kinetic energy $E_{\mathrm{kin}}(\mathbf{r}, t)$ (for a local potential). As a consequence, the collision term drives the system towards a local and instantaneous equilibrium which can be characterized by local density, local current and local kinetic energy $f_{\mathrm{eq}}(\mathbf{r}, \mathbf{p}; \varrho, \mathbf{j}, E_{\mathrm{kin}})$. Equilibration at global scale is achieved at slower pace by interplay with the long range transport resulting from mean-field (Vlasov) propagation. Therefore one can simplify the full VUU equation in terms of a Relaxation Time

Ansatz (RTA) which reads [36–39]

$$\partial_t f = \{h,f\} - \frac{1}{\tau_{\rm relax}} \left(f(\mathbf{r},\mathbf{p},t) - f_{\rm eq}(\mathbf{r},\mathbf{p};\varrho,\mathbf{j},E_{\rm kin})\right) \tag{6}$$

In the latter equation, the constraints $(\varrho, \mathbf{j}, E_{\rm kin})$ are fully local in space and time. The RTA approximation provides a simple way to generalize the picture to the quantum case.

The equation of motion for $\hat{\rho}$ including dynamical correlations, namely the quantum kinetic equation generalizing the mean field propagation Eq. (13) and/or generalizing to the quantum world the semi-classical VUU equation, reads in general [40, 41]

$$\mathrm{i}\partial_t \hat{\rho} = \left[\hat{h}, \hat{\rho}\right] + \hat{I}[\hat{\rho}] \quad . \tag{7}$$

where the right-hand-side is the quantum-mechanical collision term. The RTA approximation Eq. (6) can in turn be imported to the quantum case as

$$\mathrm{i}\partial_t \hat{\rho} = \left[\hat{h}, \hat{\rho}\right] - \frac{1}{\tau_{\rm relax}} \left(\hat{\rho} - \hat{\rho}_{\rm eq}[\varrho, \mathbf{j}, E]\right) \ , \tag{8}$$

A key ingredient in Eq. (8) is the density operator $\hat{\rho}_{\rm eq}$ of the thermal equilibrium for local density $\varrho(\mathbf{r}, t)$, current distribution $\mathbf{j}(\mathbf{r}, t)$ and total energy $E(t)$ at a given instant time t, all ingredients being functionals of $\hat{\rho}$. The other key quantity in Eq. (8) is the relaxation time $\tau_{\rm relax}$, which is presently evaluated in semi-classical Fermi liquid theory and which depends on instantaneous excitation energy. Details on the evaluation of both $\hat{\rho}_{\rm eq}$ and $\tau_{\rm relax}$ can be found in Supplemental material, Sect. B.

A first example of application of this quantum RTA approximation is shown in Fig. 2 where we consider a simple initial boost excitation of the electron cloud of the metal cluster Na_{40}, in order to explore the optical response of the system [33, 42]. Note that the considered case is well beyond the linear response regime with a total ionization of order unity. This is a typical target case for a quantum description, as discussed above. The lower panel of Fig. 2 compares ionization in time for TDLDA and RTA. As expected, RTA delivers a significantly smaller ionization than TDLDA, which reflects the fact that a fraction of the deposited energy (via the initial time boost) is "dissipated" towards internal degrees of freedom of the system rather dominantly released via ionization. The upper panel of Fig. 2 delivers a complementary message in terms of the Fourier transform (time to energy) of the dipole response of the electron cloud [42]. TDLDA and RTA are once again compared. One can spot several interesting differences between both theories. First, it should be noted that dominant peaks are red shifted in RTA as compared to TDLDA. This simply results from the fact that TDLDA delivers more ionization than RTA. The second remark concerns the overall structure of the spectra which are globally similar but for the fact that major peaks are suppressed in RTA (as compared to TDLDA) or even suppressed in the case of satellite peaks (as, e.g., the structures around 4 eV). This is typical of dissipative features which tend to wipe out detailed structures and spread

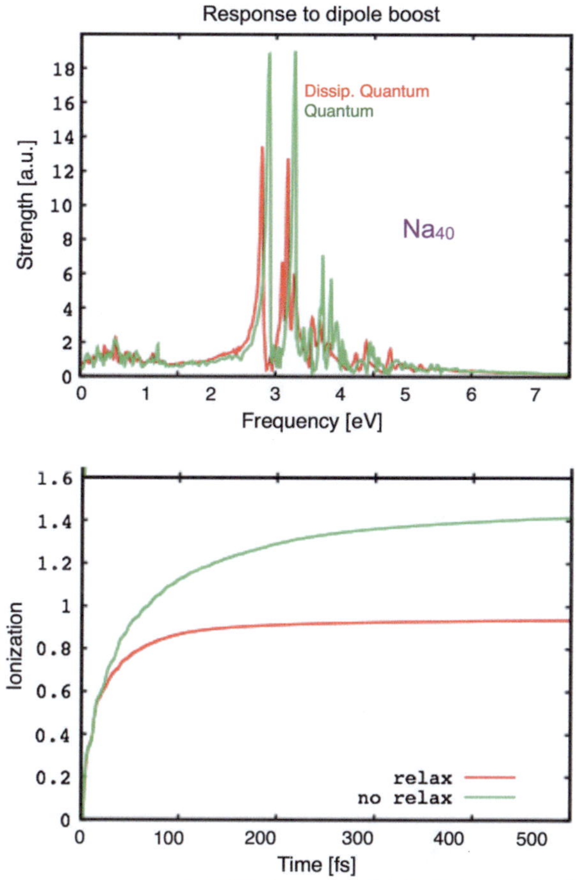

Fig. 2 Time-energy Fourier transform of the electronic dipole (*upper panel*) and ionization as a function of time (*lower panel*) in Na_{40} excited by an instantaneous boost, computed in TDLDA ("no relax" or "quantum") and RTA ("relax" or "dissip. quantum")

the spectrum. As in the previous VUU case, we observe that the RTA delivers the expected features of a dissipative theory.

It is also interesting to compare VUU to RTA. We thus come back to the previously discussed Fig. 1 which displays the PAD of a laser irradiated sodium cluster. We have indicated on this figure the PAD computed with the quantum RTA. We find a clearly qualitative, and to a large extent quantitative, agreement between VUU and quantum RTA. At closer look, we remark that VUU delivers a slightly more isotropic PAD than quantum RTA, which is especially visible in the transverse direction. The effect is not necessarily genuine and might results from the intrinsic error bars which blur standard VUU simulations [43]. Altogether, the basic message from the comparison is, nevertheless, the fact that quantum RTA looks rather close to VUU. One might thus question the interest of the quantum RTA as compared to VUU. A clearcut answer to this question is provided in Fig. 3 where we have made a systematics of laser irradiations varying the laser frequency. Several key quantities are plotted as a function of laser frequency. We explore the optical response in the upper panel. The middle panel shows the anisotropy parameter β_2. This parameter provides a compact

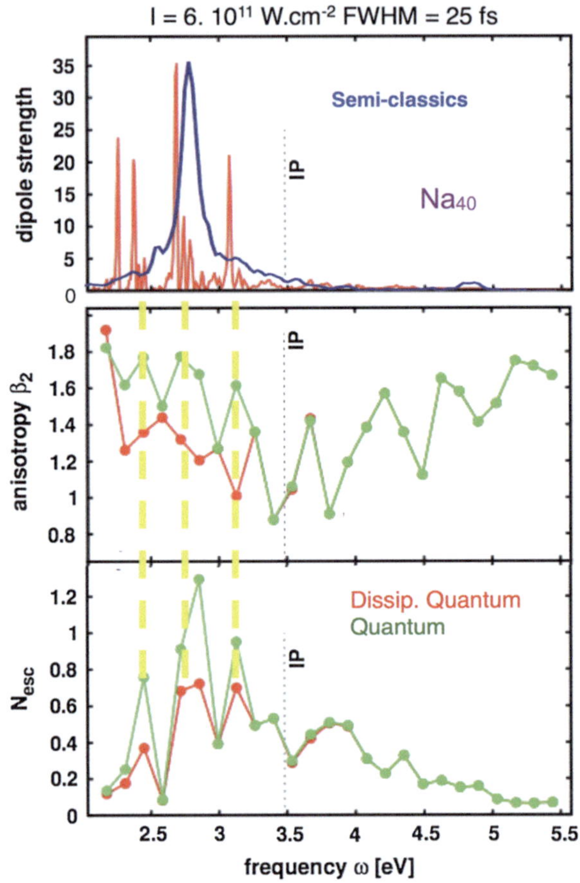

Fig. 3 Comparison of TDLDA ("quantum") and RTA ("dissip. quantum") observables in laser irradiated Na_{40} as a function of laser frequency ω. *Bottom* Total ionization N_{esc}. *Middle* Anisotropy parameter β_2 characterizing the PAD. *Top* Optical response calculated in RTA and semi-classically. For the sake of clarity the TDLDA optical response is not indicated. As already seen in Fig. 2, differences between RTA and TDLDA optical response are visible but not large. The vertical dashes indicate the value of the ionization potential (IP)

measure of the shape of the PAD, quantitatively in the one-photon regime, qualitatively in the multi-photon regime. Indeed, in the one-photon ionization regime, the PAD cross section can be written as $d\sigma/d\Omega \propto (1 + \beta_2 P_2(\theta))$ where P_2 is the second order Legendre polynomial and θ the emission angle with respect to the laser polarization axis. The bottom panel finally displays the total ionization. Very clearly, β_2 and ionization signals significantly depend on the spectral features with sizable peaks corresponding to the major peaks identified in the optical spectrum. One here faces a resonant enhancement of ionization and associated reduced alignment of emission. Very clearly as well, RTA does strongly differ from TDLDA at resonances, but the difference basically vanish when off-resonant. The mechanism would be about the same for the semi-classical approximation, typically with enhanced ionization at resonance. The semi-classical optical response is added for completeness in the upper panel of Fig. 3 and shows that the resonance landscape is much simplified in the semi-classical approximation which implies significantly different semi-classical ionization and anisotropy behaviors as compared to the quantum case. This demonstrates the interest of a quantum approach with respect to the semi-classical approximation, even in the forgiving case of simple metal clusters.

4 Stochastic Time-Dependent Hartree Fock (STDHF)

4.1 Standard Stochastic TDHF

The RTA is an interesting step beyond TDLDA but suffers from the standard limitations of RTA, namely the fact that it somewhat drives the system towards a dedicated (even if time-dependent) asymptotic state. While it is probably well adapted to the late time evolution of the system (rather close to actual asymptotic equilibrium), it may induce bias at shorter times. It is thus interesting to explore other options at the quantum mechanical level. A line of approach is provided by stochastic approaches. Indeed, according to the fluctuation-dissipation theorem, fluctuations scale with excitation energy (or dissipation). This implies that a mere kinetic equation is probably insufficient at very high energy. A possible solution is to consider more elaborate approaches in terms of a Langevin treatment, which leads to stochastic kinetic equations in which the Boltzmann equation is complemented by a stochastic collision term [44, 45]. The quantum analog of a stochastic quantum kinetic equation is also conceivable formally (very involved as well), but remained practically impossible to solve except for some semi-classical applications in nuclear dynamics [11, 46–48].

An alternative quantum mechanical approach, which can be reduced to such a stochastic kinetic equation, was proposed in [19] and provides a formally simple framework for true quantum approaches. The method is coined Stochastic Time-Dependent Hartree Fock (STDHF) and represents the dynamics in terms of an ensemble of pure mean field states propagated in parallel and allowing an (on the fly) account of collisional correlations. The formalism as well as its realization are simple. The difficultics lies in the computational cost because of the necessary large size of the ensemble to reach statistically reliable estimates. It is thus, as such, not yet applicable to realistic 3D computations. First realizations of STDHF were nevertheless recently obtained in 1D [49, 50] and provide a proof of principle of the method. The next step will be to simplify the approach to be able to apply it to realistic 3D cases. STDHF dynamics is outlined in Supplemental material, Sect. C.

4.2 From STDHF to Extended TDHF

Stochastic TDHF is an appealing theory which integrates both dissipative effects (via collisional correlations) and associated fluctuations via the stochastic ensemble. It is thus a highly elaborate approach covering expected behaviors of the dynamics in the dissipative regime. As such, it can provide a benchmark for other, simpler, theories. As already mentioned, it has been successfully applied in simple 1D model systems of covalent molecules in [49, 50]. A direct application to realistic full 3D TDLDA calculations is presently beyond computational possibilities. Indeed, the quality of the stochastic ensemble representation highly depends on the size of the ensemble. Below typically 100 events, statistical error bars may blur the quality of the estimates. This is clearly a severe hindrance for the applicability of STDHF as such. The major issue here concerns mean field propagation. By construction STDHF propagates an ensemble of mean field states. A single STDHF calculation thus requires at least 100 mean-field calculations, not mentioning the marginal cost of the evaluation of transition rates for each member of the ensemble. As such STDHF is thus at least costing two orders of magnitude the cost of a mere TDLDA calculations, whence a strong computational limitation. Even more so, in the case of moderate excitation energies, transition probabilities become even smaller and so increasingly difficult to sample, thus almost hampering any stochastic ensemble description of such situations. There is a practical need for simplified approaches, especially ones allowing to avoid the ensemble strategy. Any such approach will, of course, sacrifice the benefit of the ensemble description concerning fluctuations. But the simplification may allow to address realistic 3D cases as well as moderate excitation energies.

We have seen above that STDHF might be reduced to a stochastic kinetic equation. The simpler version thereof is the corresponding quantum kinetic equation which takes the form of Eq. (7). In its full version, the corresponding collision term $\hat{I}[\hat{\rho}]$ is a complicated operator which can hardly be solved in practice. It becomes simpler when assuming that the one-body density matrix entering $\hat{I}[\hat{\rho}]$ remains diagonal, in other words, of the form Eq. (14), with time-dependent weights W_α. The effect of the collision integral then reduces to a rate equation for the W_α, which reads

$$\frac{dW_\alpha}{dt} \sim \sum_\beta \sum_\gamma \sum_\delta \left|V_{\alpha\beta\gamma\delta}\right|^2 [(1 - W_\alpha)(1 - W_\beta)W_\gamma W_\delta - W_\alpha W_\beta(1 - W_\gamma)(1 - W_\delta)]\delta\left(\varepsilon_\alpha + \varepsilon_\beta - \varepsilon_\gamma - \varepsilon_\delta\right) \quad . \tag{9}$$

where the ε_i label the single particle energies. This approximation is usually coined Extended TDHF (ETDHF) and was explored in particular in nuclear physics [18]. As a first step, it is thus interesting to test this oversimplified picture and compare it to STDHF.

Due to the lack of benchmark calculations, ETDHF has never been fully tested in realistic dynamical scenarios, even less so in electronic systems. We test it here for a molecular 1D model system and compare it to STDHF. The setup we have used here is similar to that in [49, 50]. We mock up a 1D molecular/cluster system by a mean field Hamiltonian (in x representation, using $\hbar = 1$)

$$\hat{h} = -\frac{\Delta}{2m} + V_{\text{ext}}(x) + \lambda\varrho(x)^\sigma \tag{10}$$

The external potential $V_{ext}(x)$ has been chosen as a Woods-Saxon profile $V_{ext}(x) = V_0/(1 + \exp((x - x_0)/a))$ in which we take $V_0 = -5$ Ry, $x_0 = 15$ a$_0$, $a = 2$ a$_0$, $\lambda = 5$ Ry.a_0^2 and $\sigma = 2$. The potential is complemented by a confining harmonic oscillator which ensures soft reflecting boundary conditions. This allows to focus on thermal effects, without interference with ionization. Finally, the residual interaction V is taken as a zero range term with intensity 1.3 Ry.a_0. The set of parameters has been tuned to reproduce a typical molecular setup with energies in the Ry range, see [49, 50] for details.

The comparison between STDHF and ETDHF is done in Fig. 4. The one-particle entropy S is plotted as a function of time for two excitation energies E^*. The excitation energy is delivered at initial time in the form of a particle-hole excitation. The ensuing dynamics is followed by STDHF and ETDHF. The comparison shows that, for a small E^*, both approaches lead to very similar results. However, at a higher E^*, they deliver significantly different results, both in terms of the asymptotic values and the temporal profile (which indicates different relaxation times). The comparison is interesting in several respects. First, it shows that, at small excitations, the oversimplified ETDHF approach provides a reliable picture, which is a welcome feature. Comparisons on other observables would also be interesting. Second, the higher energy case shows that ETDHF misses some pieces of the dissipative dynamics included in STDHF (even in an integrated quantity such as the entropy). The fact that the ETDHF asymptotic value of S is significantly lower than the STDHF one indeed tells that some dissipation is missing (a pure TDHF evolution with fixed values of W_α at 1 or 0 exactly delivers a zero entropy in time). This probably stems from the (strong) assumption of diagonal density matrix underlying Eq. (9). Indeed, it should be noted here that the energy matching condition in Eq. (9) is ill defined outside the basis in which the one-body Hamiltonian is diagonal. Because of the dynamical evolution, this basis cannot by construction diagonalize the one-body density matrix. The effect seems harmless at low excitation but leads to problems at higher excitation, which is quite reasonable. There is thus a clear need for an improvement over the simple ETDHF approach as soon as the excitation energy becomes sizable.

5 Conclusions

We have presented in this paper a few directions of investigations of dissipative effects in irradiated clusters and molecules. The inclusion of such dissipative features in finite systems at a quantum level is a long pending question. It has been addressed in some detail in nuclear physics with a strong semi-classical bias. The semi-classical picture was also explored in simple metal clusters but is hardly applicable in general in the electronic case. In view of the experiments addressing these questions in clusters and molecules, there is an urgent need for the development of a robust quantum approach in these systems, integrating, at some level, dissipative features.

We have proposed here especially two strategies. The first one is based on a relaxation time ansatz of the quantum collision term of quantum kinetic equations. It

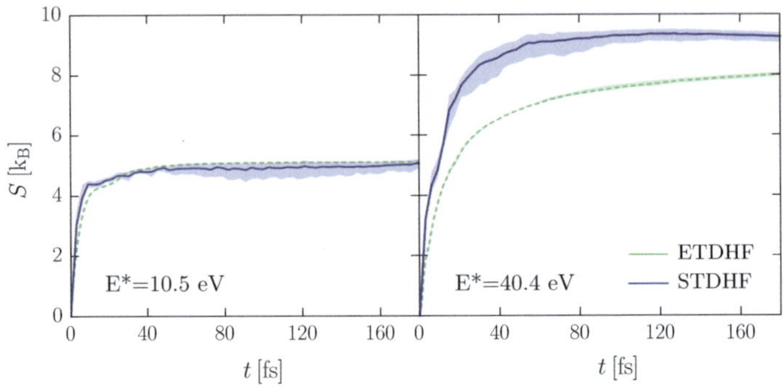

Fig. 4 Comparison of time evolution of entropy S (in Boltzmann constant unit k_B) computed with STDHF (*full curve*) and ETDHF (*dashes*) approaches, for two excitation energies E^* as indicated. The shaded areas about the STDHF curves are the *error bars* from the fluctuations within the STDHF ensemble

allows to explore several aspects of relaxation dynamics in molecular systems and clusters. We have, in particular, explored the impact of dissipation on angular distributions (PAD) of emitted electrons. We have seen that, in general, dissipation is especially active close to resonances where its impact is most visible. This was clearly seen in total ionization and PAD. The second approach (STDHF) we have presented consists in using a stochastic ensemble representation of the dynamics by considering a stochastic extension of mean field dynamics. This approach has conceptually a much wider spectrum of applications and is highly elaborate, even describing fluctuation dynamics. It is unfortunately still quite expensive numerically speaking which strongly limits applications in realistic cases. Still, STDHF can be tested in simplified 1D models and then provides an invaluable benchmark of dissipative dynamics beyond mean field. We have then tested the standard Extended TDHF (ETDHF) approach in comparison to STDHF and found that it performs acceptably well at moderate excitations while the simplifications raise difficulties at larger excitation energies. The development of an improved approach, in between ETDHF and full STDHF, but still practical in realistic computations thus remains a challenging task. Work along that line is in progress.

Acknowledgements We thank Institut Universitaire de France, European ITN network CORINF and French ANR contract MUSES for support during the realization of this work. It was also granted access to the HPC resources of CalMiP (Calcul en Midi-Pyrénées) under the allocation P1238, and of RRZE (Regionales Rechenzentrum Erlangen).

References

1. Bohr N, Wheeler JA (1939) Phys Rev 56:426
2. Hansen K (2013) Statistical physics of nanoparticles in the gas phase. Springer, Netherlands, Amsterdam
3. Saalmann U, Siedschlag C, Rost JM (2006) J Phys B 39
4. Fennel T, Meiwes-Broer K-H, Tiggesbäumker J, Reinhard P-G, Dinh PM, Suraud E (2010) Rev Mod Phys 82:1793
5. Kjellberg M, Johansson O, Jonsson F, Bulgakov AV, Bordas C, Campbell EEB, Hansen K (2010) Phys Rev A 81:023202
6. Chen G (2005) Nanoscale energy transport and conversion: a parallel treatment of electrons, molecules, phonons, and photons. Oxford University Press, New York
7. Dalibard J (1999) Collisional dynamics of ultra-cold atomic gases. In Inguscio M, Stringari S, Wieman C, (eds) Proceedings of the international school of physics-enrico fermi, course CXL, IOS Press, Amsterdam, p 321
8. Wopperer P, Dinh PM, Reinhard P-G, Suraud E (2014) Phys Rep. arXiv:1407.4965
9. Pinaré JC, Baguenard B, Bordas C, Broyer M (1999) Eur Phys J D 9:21
10. Wada T, Carjan N, Abe Y (1992) Nucl Phys A 538:283
11. Abe Y, Ayik S, Reinhard P-G, Suraud E (1996) Phys Rep 275:49
12. Kadanoff LP, Baym G (1962) Quantum statistical mechanics: green's function methods in equilibrium and nonequilibrium problems. Frontiers in physics, Benjamin, New York
13. Bertsch GF, Das S Gupta (1988) Phys Rep 160:190
14. Bonasera A, Gulminelli F, Molitoris J (1994) Phys Rep 243:1
15. Serene JW, Rainer D (1983) Phys Rep 101:221
16. Feldmeier H (1990) Nucl Phys A 515:147
17. Ono A, Horiuchi H (2004) Prog Part Nucl Phys 53:501
18. Goeke K, Reinhard P-G (1982) Springer, Heidelberg
19. Reinhard P-G, Suraud E (1992) Ann Phys (NY) 216:98
20. Domps A, Reinhard P-G, Suraud E (1998) Phys Rev Lett 81:5524
21. Fennel T, Bertsch GF, Meiwes-Broer K-H (2004) Eur Phys J D 29:367
22. Uehling E-A, Uhlenbeck G-E (1933) Phys Rev 43:552
23. Rajam AK, Raczkowska I, Maitra NT (2010) Phys Rev Lett 105:113002
24. Requist R, Pankratov O (2010) Phys Rev A 81:042519
25. Dutta A, Trefzger C, Sengupta K (2012) Phys Rev B 86:085140
26. Krause P, Klamroth T, Saalfrank P (2007) J Chem Phys 127
27. Ullrich CA (2012) Time-dependent density-functional theory: concepts and applications. Oxford University Press, New York
28. Escartin JM, Vincendon M, Romaniello P, Dinh PM Reinhard P-G, Suraud E (2015) J Chem Phys 142:084118
29. Reinhard P-G, Suraud E (2004) Introduction to cluster dynamics. Wiley, New York
30. Gross EKU, Kohn W (1990) Adv Quant Chem 21:255
31. Gross EKU, Dobson JF, Petersilka M (1996) Top Curr Chem 181:81
32. Marques MAL, Maitra NT, Nogueira FMS, Gross EKU, Rubio A (2012) Fundamentals of time-dependent density functional theory. Lecture notes in physics, vol 837. Springer, Berlin
33. Calvayrac F, Reinhard P-G, Suraud E, Ullrich CA (2000) Phys Rep 337:493
34. Marques MA, Ullrich CA, Nogueira F, Rubio A, Burke K, Gross EK (2006) Time dependent density functional theory. Springer, Berlin
35. Gütter K, Reinhard P-G, Toepffer C (1988) Phys Rev A 38:1641
36. Bhatnagar PL, Gross EP, Krock M, Phys Rev 94(1954):511
37. Kohler S (1980) Nucl Phys A 343:315
38. Kohler S (1982) Nucl Phys A 378:181
39. Wong CY, Davies KTR (1983) Phys Rev C 28:240
40. Reinhardt H, Reinhard P-G, Goeke K (1985) Phys Lett B 151:177

41. Goeke K, Reinhard P-G, Reinhardt H (1986) Ann Phys (N.Y.) 166:257
42. Calvayrac F, Reinhard P-G, Suraud E (1997) Ann Phys (N.Y.) 255:125
43. Reinhard P-G, Suraud E (1995) Ann Phys (NY) 239:193
44. Bixon M, Zwanzig R (1969) Phys Rev 187:267
45. Zwanzig R (1973) J Stat Phys 9:215
46. Ayik S, Gregoire C (1988) Phys Lett B 212:269
47. Randrup J, Burgio GF, Chomaz Ph (1992) Nucl Phys A 538:393
48. Napolitani P, Colonna M (2013) Phys Lett B 726:382
49. Suraud E, Reinhard P-G (2014) New J Phys 16:063066
50. Slama N, Reinhard P-G, Suraud E (2015) Ann Phys (NY) 355:182

On the Ordering of Orbital Energies in the ROHF Method: Koopmans' Theorem versus Aufbau Principle

B.N. Plakhutin, A.V. Novikov, N.E. Polygalova and T.E. Prokhorov

Abstract The restricted open-shell Hartree-Fock (ROHF) method in its different formulations is a standard tool used by quantum chemists for studying open-shell systems. This work presents a discussion of specific difficulties which arise in the ROHF method in those cases where the orbital energies violate the Aufbau principle (AP). The AP violations are often treated in the literature as a deficiency of both a ROHF method leading to these violations and of the respective computational results. We summarize examples of different AP violations and analyze them from the viewpoint of both Koopmans' theorem (KT) and experimental ionization potentials. We show that the main source of AP violations is a specific ordering of the ROHF orbital energies based on the orbital occupancies. In those cases, where the orbital energies obey KT, the violations of the AP reflect the physical reality. To overcome computational difficulties which arise in the iterative SCF procedure, we describe a simple and effective *orbital-energy scaling technique* which enables one to perform ROHF computations of systems violating the Aufbau principle.

Keywords Aufbau principle • Koopmans' theorem • ROHF method • Endofullerene $N@C_{60}$ • Atom Mn

1 Introduction

In the iterative procedure of solving the Hartree-Fock (HF) equation for the closed-shell systems [1], the orbital energies ε_i are usually ordered by their values,

---

Electronic supplementary material The online version of this chapter (doi:10.1007/978-3-319-50255-7_2) contains supplementary material, which is available to authorized users.

B.N. Plakhutin (✉) · A.V. Novikov · N.E. Polygalova · T.E. Prokhorov
Laboratory of Quantum Chemistry, Boreskov Institute of Catalysis,
Russian Academy of Sciences, Novosibirsk 630090, Russian Federation
e-mail: Plakhutin@catalysis.ru

A. Tadjer et al. (eds.), *Quantum Systems in Physics, Chemistry, and Biology*,
Progress in Theoretical Chemistry and Physics 30,
DOI 10.1007/978-3-319-50255-7_2

$\varepsilon_1 < \varepsilon_2 < \varepsilon_3 < \cdots$, and electrons fill the lowest energy orbitals. At convergence, the closed-shell and virtual orbital energies, ε_k and ε_v, respectively, obey the condition

$$\varepsilon_k < \varepsilon_v \tag{1}$$

which is usually called the Aufbau principle (AP). The HF orbital energies (1) also satisfy the physically meaningful relationships following from Koopmans' theorem [2]

$$\varepsilon_k = -I_k \tag{2}$$

$$\varepsilon_v = -A_v \tag{3}$$

where I_k and A_v are the vertical ionization potential (IP) and electron affinity (EA), respectively, defined in the "frozen" orbital approximation (see below). A fulfillment of Koopmans' relationships (2)–(3) shows that the orbital energies ε_k and ε_v are well-defined quantities, and this substantiates the use of the AP (1) for ordering of the orbitals in the closed-shell HF method.

In the restricted open-shell HF (ROHF) method [3] and in its different reformulations and generalizations [4–11], the full orbital space $\{\phi_i\} = \{\phi_k\} \oplus \{\phi_m\} \oplus \{\phi_v\}$ is separated into three orbital subspaces, closed-shell $\{\phi_k\}$, open-shell $\{\phi_m\}$, and virtual $\{\phi_v\}$, characterized by the respective orbital occupancies f_i,

$$f_k = 1;\ 0 < f_m < 1;\ f_v = 0 \tag{4}$$

The ROHF orbitals are usually ordered by their occupancies, i.e., the open-shell orbitals lie above the closed-shell (doubly occupied) orbitals and below the virtual ones. Within each of these subspaces, the orbitals are ordered by the values of the orbital energies. Based on this specific orbital ordering, it is often assumed that the ROHF orbital energies ε_i must obey the Aufbau principle, which in this case takes the form

$$\varepsilon_k < \varepsilon_m < \varepsilon_v \tag{5}$$

In practice, the ordering of the ROHF orbital energies derived for the ground state of different open-shell systems usually obeys the AP (5), however, a number of cases is known where the orbital energies violate the condition (5). The point is that the latter cases are sometimes treated as a deficiency of both a ROHF method leading to this violation and of the respective computational results [12, 13]. In the clear form this point of view was presented in Ref. [13]: "*The Aufbau principle states that electrons will fill the lowest energy orbitals first for the ground state. Computational results that predict otherwise must be view skeptically*".

This work presents a discussion of the Aufbau principle (5) and of its violations in the ROHF method. We summarize examples of different AP violations derived with different versions of the ROHF method and analyze them from the viewpoint of both Koopmans' theorem (KT) and experimental IPs. The main result of our treatment is that in those cases, where the orbital energies obey KT, the violations of the AP reflect the physical reality. To overcome computational difficulties which arise in the iterative SCF procedure, we describe a simple and effective *orbital-energy scaling technique* which enables one to perform ROHF computations of systems violating the Aufbau principle.

The paper is organized as follows. Section 2 presents a brief review of the present state of the art of the ROHF method and of its special (*canonical*) form [14–17], in which the orbital energies obey Koopmans' theorem. In Sect. 3 we discuss different kinds of AP violations in the ROHF method. A description of the *orbital-energy scaling technique* is presented in Supplementary materials.

2 Theory

2.1 Total One-Electron Hamiltonian in the ROHF Method

The present-day formulation of the ROHF method is based on the familiar representation for the total electronic energy [3]:

$$E_{ROHF}\,(X) = 2\sum_i f_i\, h_{ii} + \sum_i \sum_j f_i f_j\,(2a_{ij}J_{ij} - b_{ij}K_{ij}) \tag{6}$$

that involves only the Coulomb J_{ij} and exchange K_{ij} integrals and does not involve the 3- and 4-indexed electron repulsion integrals $\langle ij\,|\,kl\rangle$. In Eq. (6) and below, f_i is the occupation number (4), a_{ij} and b_{ij} are coupling coefficients characterizing the state and configuration of an open-shell system X under consideration [3]. Application of the variational principle to the energy functional (6) gives the generalized Hartree-Fock equation [3]:

$$\hat{R}|\phi_i\rangle = \varepsilon_i|\phi_i\rangle \tag{7}$$

where $\hat{R}$ is the total one-electron ROHF Hamiltonian. The most general definitions of $\hat{R}$ valid for an arbitrary unique state (non-repeating term symbol) arising from an arbitrary many-open-shell electronic configuration were derived by Dyadyusha and Kuprievich [4], and, later and independently, by Hirao and Nakatsuji [5]. The definitions [4, 5] are completely equivalent from the viewpoint of the variational principle [18], however, they are not equal to each other, and this point deserves discussion. The definitions [4, 5] can be presented in the following form:

$$\hat{R} = \hat{R}_{(1)} + \hat{R}_{(2)} + \hat{R}_{(3)} \tag{8}$$

where

$$\hat{R}_{(1)} = \sum_i \omega_i [(I-\rho)\hat{F}_i \rho^i + \rho^i \hat{F}_i (I-\rho)] \tag{9}$$

$$\hat{R}_{(2)} = \sum_i \sum_j \lambda_{ij} \rho^j (\hat{F}_i - \hat{F}_j) \rho^i \tag{10}$$

The term $\hat{R}_{(3)}$ which differs in the two definitions [4, 5] is discussed just below. In Eqs. (9)–(10): $\rho^i = |\phi_i\rangle\langle\phi_i|$, $\rho = \sum_{(k)} \rho^k + \sum_{(m)} \rho^m$, *and* $(I-\rho) = \sum_{(v)} \rho^v$ are the projectors; ω_i and λ_{ij} are arbitrary nonzero numbers (($\lambda_{ij} = -\lambda_{ji}$); and $\hat{F}_i$ is the Fock operator for the orbital ϕ_i [1],

$$\hat{F}_i = f_i[\hat{h} + \sum_j f_j (2a_{ij} \hat{J}_j - b_{ij} \hat{K}_j)] \tag{11}$$

The term $\hat{R}_{(1)}$ (9) expresses the variational conditions between occupied and virtual orbitals, while $\hat{R}_{(2)}$ (10) represents the variational conditions among occupied orbitals [4, 5]. All of these conditions are satisfied at convergence (see also below). For closed-shell systems, $\hat{R}_{(2)} \equiv 0$, since the Fock operators (11) for the closed-shell orbitals are all the same. The term $\hat{R}_{(3)}$ derived by Dyadyusha and Kuprievich [4] takes the form

$$\hat{R}_{(3)} = \sum_i \rho^i \hat{Q} \rho^i \tag{12}$$

where $\hat{Q}$ is an arbitrary nonvanishing Hermitian operator. The term $\hat{R}_{(3)}$ derived by Hirao and Nakatsuji [5] takes another form

$$\hat{R}_{(3)} = \sum_i \rho^i \hat{F}_i \rho^i + \sum_u \sum_w \sum_i \rho^u \hat{F}_i \rho^w \tag{13}$$

where the indices u and w refer to the virtual orbitals. To clarify the definitions (12) and (13), we present the matrix of the total ROHF Hamiltonian (8), $\hat{R}_{ij}$, defined in the molecular orbital basis $\{\phi_i\} = \{\phi_k\} \oplus \{\phi_m\} \oplus \{\phi_v\}$ For simplicity, we express this symmetric (Hermitian) matrix in terms of matrix elements of operators $\hat{R}_{(1)}$, $\hat{R}_{(2)}$, and $\hat{R}_{(3)}$,

$$\begin{array}{c|c:c:c|} & c & o & v \\ c & \hat{R}_{(3)} & \hat{R}_{(2)} & \hat{R}_{(1)} \\ \hdashline o & & \hat{R}_{(3)} & \hat{R}_{(1)} \\ \hdashline v & & & \hat{R}_{(3)} \end{array} \tag{14}$$

where the letters c, o, and v designate the closed-shell (c), open-shell (o), and virtual (v) orbital subspaces, respectively, i.e., $c = \{\phi_k\}$, $o = \{\phi_m\}$, and $v = \{\phi_v\}$

At convergence, matrix elements of operators $\hat{R}_{(1)}$ and $\hat{R}_{(2)}$ vanish,

$$\begin{aligned}\hat{R}_{kv} &= (\hat{R}_{(1)})_{kv} = 0\\ \hat{R}_{mv} &= (\hat{R}_{(1)})_{mv} = 0\\ \hat{R}_{km} &= (\hat{R}_{(2)})_{km} = 0\end{aligned} \tag{15}$$

and this provides a fulfillment of all the conditions imposed by the variational principle in the case of open-shell systems [3–5].

From the conditions (15) it follows that operator $\hat{R}_{(3)}$ standing in diagonal blocks of (14) cannot be defined from the variational principle [4, 5]. It is noteworthy that operator $\hat{R}_{(3)}$ should be nonvanishing, otherwise all eigenvalues of $\hat{R}$ (8) which are equal to eigenvalues of $\hat{R}_{(3)}$ will be equal to zero, and hence, all eigenvectors of $\hat{R}$ will be degenerate.

This arbitrariness in the choice of $\hat{R}_{(3)}$ reveals the way to take into account the additional conditions necessary to also satisfy Koopmans' theorem. As first shown by Plakhutin et al. [14], a proper form of the operator $\hat{R}_{(3)}$ whose eigenvalues obey Koopmans' relationships of the kind (2) and (3) can be derived from Kramers' variational condition underlying Koopmans' theorem [2]. A detailed discussion of Kramers' condition and of the approach [14] based on this condition is presented in Sect. 2.3.

Here we point out that the original definitions of $\hat{R}_{(3)}$ given by Eqs. (12) or (13) are not quite satisfactory [11, 19]. Offdiagonal matrix elements of $\hat{R}_{(3)}$ inside of the diagonal blocks $\hat{R}_{cc}$ and $\hat{R}_{oo}$ of the matrix (14) were postulated [4, 5] to be equal to zero, and this presents a rigid restriction which does not follow from the variational principle. A general form of $\hat{R}_{(3)}$ following from the treatment [4, 5] is as follows [19]:

$$\hat{R}_{(3)} = \sum_k \sum_l \rho^k \hat{R}_{(cc)} \rho^l + \sum_m \sum_n \rho^m \hat{R}_{(oo)} \rho^n + \sum_u \sum_w \rho^u \hat{R}_{(vv)} \rho^w = \sum_s \sum_{i_s} \sum_{j_s} \rho^{i_s} \hat{R}_{(ss)} \rho^{j_s} \tag{16}$$

where i_s and j_s are the numbers of orbitals from the orbital subspace s $(s = c, o, v)$, and $\hat{R}_{(ss)}$ are *arbitrary nonvanishing* Hermitian operators.

2.2 *High-Spin Half-Filled Open-Shell (HSHFOS) Systems*

Here we consider in more detail the particular case of open-shell systems which is represented by high-spin half-filled open-shell (HSHFOS) systems. Within a one configuration approximation underlying the ROHF method, the HSHFOS systems are described by a one determinant wave function (similar to that in the unrestricted

HF (UHF) method [20]) and are characterized by the following values of Roothaan's open-shell coefficients f, a, and b [3],

$$f_m = f = 1/2,\ a_{mn} = a = 1,\ b_{mn} = b = 2 \tag{17}$$

The set of the Fock operators (11) characterizing HSHFOS systems is reduced to the two operators, $\hat{F}_c$ and $\hat{F}_o$, which are specific for the closed-shell and open-shell orbitals, respectively,

$$\begin{aligned} \hat{F}_c &= \hat{h} + (2\hat{J}_c - \hat{K}_c) + f(2\hat{J}_o - \hat{K}_o) \\ \hat{F}_o &= f[\hat{h} + (2\hat{J}_c - \hat{K}_c) + f(2a\hat{J}_o - b\hat{K}_o)] \end{aligned} \tag{18}$$

If arbitrary coefficients ω_i and λ_{ij} in Eqs. (9) and (10) are chosen in the form $\omega_k = \omega_c = 1$, $\omega_m = \omega_o = 2$, and $\lambda_{km} = \lambda_{co} = 2$, the operators (9) and (10) take the form $\hat{R}_{(1)} = \hat{R}_{cv} = \hat{F}_c$, $\hat{R}_{(1)} = \hat{R}_{ov} = 2\hat{F}_o$, and $\hat{R}_{(2)} = 2(\hat{F}_c - \hat{F}_o)$, respectively, and the symmetric ROHF Hamiltonian matrix (14) for HSHFOS systems takes the familiar form

$$\begin{array}{c|c:c:c|} & c & o & v \\ \hline c & \hat{R}_{(cc)} & 2(\hat{F}_c - \hat{F}_o) & \hat{F}_c \\ \hdashline o & & \hat{R}_{(oo)} & 2\hat{F}_o \\ \hdashline v & & & \hat{R}_{(vv)} \end{array} \tag{19}$$

The diagonal operators $\hat{R}_{(ss)}$ ($s = c$, o, or v) in Eq. (19) are different in different versions of the ROHF method [3, 6–10]. As shown by Montgomery [21], the diagonal operators $\hat{R}_{(ss)}$ in the methods [3, 6–10] can be expressed in the same form

$$\hat{R}_{(ss)} = 2\{A_{(ss)}\hat{F}_o + B_{(ss)}(\hat{F}_c - \hat{F}_o)\} \tag{20}$$

where $A_{(ss)}$ and $B_{(ss)}$ are some coefficients (for details, see the manual [21] to the GAMESS suite of programs [22]). The values of these coefficients derived by Montgomery are presented in Table 1. For completeness, Table 1 also presents the

Table 1 Coefficients $A_{(ss)}$ and $B_{(ss)}$ of Eq. (20) for high-spin half-filled open-shell systems

$A_{(cc)}$	$B_{(cc)}$	$A_{(oo)}$	$B_{(oo)}$	$A_{(vv)}$	$B_{(vv)}$	ROHF Hamiltonian	
−1/2	3/2	1/2	1/2	3/2	−1/2	Roothaan	Ref. [3]
1/2	1/2	1	0	–	–	Roothaan and Bagus	Ref. [23]
1/3	2/3	1/3	1/3	2/3	1/3	McWeeny and Diercksen	Ref. [6]
1/2	1/2	1	0	1	0	Davidson	Ref. [7]
1/2	1/2	1	0	0	1	Binkley, Pople and Dobosh	Ref. [8]
1/2	1/2	1/2	1/2	1/2	1/2	Guest and Saunders	Ref. [9]
1/2	1/2	1	0	1/2	1/2	Faegri and Manne	Ref. [10]
1/2	1/2	1/2	0	1/2	1/2	Euler equations	Ref. [1]

$A_{(ss)}$ and $B_{(ss)}$ coefficients specific for the atomic ROHF method developed by Roothaan and Bagus [23] and similar coefficients characterizing the ROHF method based on direct solving the Euler equations [1] (for details, see Ref. [14]).

2.3 Koopmans' Theorem in the Closed-Shell HF and ROHF Methods

In all of the ROHF methods [3–11] treated above, the orbital energies do not obey Koopmans' theorem (KT) [2] (a fulfillment of KT was not presupposed in the derivations [3–11]). For the first time, the special (*canonical*) form of the ROHF method for HSHFOS systems, in which the orbital energies ε_k, ε_m, and ε_v satisfy the KT relationships of the kind (2) and (3) has been developed in Ref. [14]. The further elaboration and extension of the approach [14] has been given in Refs. [15–17]. Here we analyze the variational conditions underlying KT in both closed-shell HF and ROHF methods mainly following Refs. [14–17].

Before treating the validity of KT in the ROHF method, we recall that the formulation of KT in the closed-shell HF method in the form of the relationships (2) and (3) does not reveal the main (variational) essence of this theorem [14, 24]. This point deserves discussion. The validity of Koopmans' relationships (2)–(3) is due to the use in the closed-shell HF equation [1]

$$\hat{F}|\phi_i\rangle = \varepsilon_i|\phi_i\rangle \tag{21}$$

of the specific form of the one-electron effective Hamiltonian

$$\hat{F} = \hat{h} + (2\hat{J}_c - \hat{K}_c) \tag{22}$$

This form of the HF Hamiltonian derived by Fock [1] and called the *Fock operator* is the particular form of the total one-electron Hamiltonian $\hat{R}$ (8). Different particular forms of $\hat{R}$ for the closed-shell systems generate the same orbital subspaces $\{\phi_k\}$ and $\{\phi_v\}$ and the same energy $E_{HF}(X)$, however, only eigenvalues of the Fock operator (22) obey the KT relationships (2) and (3). It is why the form of the one-electron Hamiltonian in the closed-shell HF method given by Eq. (22) is called *canonical*.

To gain a better understanding of the difference between $\hat{F}$ (22) and other possible forms of the closed-shell HF Hamiltonian, we consider the expressions for I_k (2) and A_v (3) defined in Ref. [2] in the "frozen" orbital approximation (FOA):

$$I_k = E_{frozen}(X_k^+) - E_{HF}(X) \tag{23}$$

$$A_v = E_{HF}(X) - E_{frozen}(X_v^-) \tag{24}$$

$$E_{frozen}(X_j^{\mp}) = \langle \Psi(X_j^{\mp}) | \hat{H} | \Psi(X_j^{\mp}) \rangle, \tag{25}$$

where $E_{HF}(X)$ is the HF energy of a closed-shell system X under consideration; $X_j^{\pm}$ is an ion having a hole or an extra electron in the orbital ϕ_j; $\Psi(X_j^{\pm}) = \Psi_{frozen}(X_j^{\pm})$ is a one-determinant wave function of $X_j^{\mp}$ formed using the same ("frozen") HF orbitals $\{\phi_i\} = \{\phi_k\} \oplus \{\phi_v\}$ optimal for X; and $\hat{H}$ is the total many-electron Hamiltonian. The energy of ions $X_j^{\mp}$ (25) is defined in the FOA, i.e., in a non-variational manner, and hence, depends on the choice of "frozen" orbitals $\{\phi_k\}$ and $\{\phi_v\}$ defined up to a unitary transformation in the respective orbital subspaces.

The fundamental result underlying KT is that the orbitals $\{\phi_i\} = \{\phi_k\} \oplus \{\phi_v\}$ derived as eigenvectors of the *canonical* HF Hamiltonian $\hat{F}$ (22) for the initial (non-ionized) system X provide also the minimal (stationary) value of $E_{frozen}(X_j^{\mp})$ with respect to the choice of "frozen" orbitals $\{\phi_k\}$ and $\{\phi_v\}$. This crucial result providing the variational meaning of KT in the closed-shell HF method [1] has been proved by Kramers [25].

It follows from here that the *canonical* HF orbitals $\{\phi_i\} = \{\phi_k\} \oplus \{\phi_v\} = \{\phi^{(c)}\} \oplus \{\phi^{(v)}\}$ for a closed-shell system X obey three variational conditions. The first of them is the variational principle for the total energy

$$\delta E_{HF}(X)[\{\phi_i\}] = 0 \tag{26}$$

while two other ones represent the Kramers' condition [25] which is imposed upon the orbitals of the respective (ionized) electronic shell in both cation X_k^+ and anion X_u^-. These two particular forms of Kramers' condition [25] can be presented as follows [14, 17]:

$$\delta E_{frozen}(X_k^+)[\{\phi^{(c)}\}] = 0 \quad (\phi_k \in \{\phi^{(c)}\}) \tag{27}$$

$$\delta E_{frozen}(X_u^-)[\{\phi^{(v)}\}] = 0 \quad (\phi_u \in \{\phi^{(v)}\}) \tag{28}$$

where the orbitals of non-ionized electronic shells, i.e., the virtual orbitals $\{\phi^{(v)}\}$ in X_k^+ and closed-shell orbitals $\{\phi^{(c)}\}$ in X_u^-, remain "frozen" in the respective variational procedures (27) and (28).

In line with the above treatment (see Eq. (15) and the respective text), the variational principle (26) defines only the operator $\hat{R}_{(cv)}$ standing in the off-diagonal block of the closed-shell HF Hamiltonian matrix. The diagonal operators $\hat{R}_{(cc)}$ and $\hat{R}_{(vv)}$ cannot be defined from the variational principle, however, they can be defined from the conditions (27) and (28), respectively [14, 16]. It is easy to show that $\hat{R}_{(cv)}$ derived from the condition (26) takes the form $\hat{R}_{(cv)} = \omega\hat{F}$, where ω is an arbitrary nonzero coefficient, while $\hat{R}_{(cc)}$ and $\hat{R}_{(vv)}$ derived from the conditions (27) and (28) take the form $\hat{R}_{(cc)} = \hat{R}_{(vv)} = \hat{F}$. It follows from here that Kramers' conditions (27)–(28) are satisfied in the canonical closed-shell HF method [1], and this provides a physical meaning of the KT relationships (2) and (3).

The variational conditions (27)–(28) can be extended to open-shell systems in order to derive a special (*canonical*) form of the ROHF Hamiltonian [14], whose eigenvalues (orbital energies) obey KT. For the first time, this extension has been given in Refs. [14, 16] for the particular case of HSHFOS systems. A generalization of the approach [14, 16] to open-shell systems with arbitrary orbital occupancies and arbitrary total spin S including orbitally degenerate systems, such as atoms with configuration p^N, will be published elsewhere [26].

In HSHFOS systems under study, the full orbital space $\{\phi_i\}$ is divided into three subspaces, $\{\phi_i\} = \{\phi_k\} \oplus \{\phi_m\} \oplus \{\phi_v\} = c \oplus o \oplus v$ characterized by different orbital occupancies (4), and hence, there are six different one-electron processes $X \to X^{\pm}_{j,\sigma}$ which lead to formation of six ions: $X^{+}_{k,\alpha}$, $X^{+}_{k,\beta}$, $X^{+}_{m,\alpha}$, $X^{-}_{m,\beta}$, $X^{-}_{v,\alpha}$, *and* $X^{-}_{v,\beta}$, where $\sigma = \alpha$ or β is the spin of a removed (attached) electron. The full set of the variational conditions imposed upon the canonical ROHF orbitals in HSHFOS systems involves the variational principle for the total energy

$$\delta E_{ROHF}(X)[\{\phi_i\}] = 0 \tag{29}$$

and six conditions, each of which is imposed upon the energy of the respective ion defined in the FOA. These six conditions can be presented in the form [14, 17]:

$$\delta E_{frozen}(X^{\pm}_{j,\sigma})[\{\phi^{(s)}\}] = 0, \ (j \in s) \tag{30}$$

where $s = c$, o, or v, and, $\sigma = \alpha$ or β

In line with the above treatment (see Eq. (15) and the respective text), the variational principle (29) defines only the operators $\hat{R}_{(co)}$, $\hat{R}_{(cv)}$, and $\hat{R}_{(ov)}$ standing in the off-diagonal blocks of the matrix (19). The fundamental meaning of the conditions (30) is that each of them defines the respective diagonal operator $\hat{R}^{\sigma}_{(ss)}$ in Eq. (19), whose eigenvalues obey Koopmans' theorem. An explicit form of operators $\hat{R}^{\sigma}_{(ss)}$ is presented below.

As proved in Refs. [14, 16], if the ROHF orbitals $\{\phi_i\} = \{\phi_k\} \oplus \{\phi_k\} \oplus \{\phi_v\}$ used to evaluate the energies $\delta E_{frozen}(X^{\pm}_{j,\sigma})$ in Eq. (30) obey the variational principle (29), the eigenvalues of each of the operators $\hat{R}^{\sigma}_{(ss)}$ derived from the respective condition (30) obey the following Koopmans' relationship:

$$\begin{aligned} \hat{R}^{\alpha}_{(cc)}&: \quad \varepsilon^{\alpha}_{k} = -I^{\alpha}_{k} \\ \hat{R}^{\alpha}_{(oo)}&: \quad \varepsilon^{\alpha}_{m} = -I^{\alpha}_{m} \\ \hat{R}^{\alpha}_{(vv)}&: \quad \varepsilon^{\alpha}_{v} = -A^{\alpha}_{v} \end{aligned} \tag{31}$$

and

$$\begin{aligned} \hat{R}^{\beta}_{(cc)}&: \quad \varepsilon^{\beta}_{k} = -I^{\beta}_{k} \\ \hat{R}^{\beta}_{(oo)}&: \quad \varepsilon^{\beta}_{m} = -A^{\beta}_{m} \\ \hat{R}^{\beta}_{(vv)}&: \quad \varepsilon^{\beta}_{v} = -A^{\beta}_{v} \end{aligned} \tag{32}$$

It follows from (31)–(32) that the conditions (30) define two different operators, $\hat{R}^{\alpha}_{(ss)}$ and $\hat{R}^{\beta}_{(ss)}$, for each orbital subspace s. These two operators provide a fulfillment of KT for two respective one-electron processes which are possible in the electronic shell s. From here it follows that the conditions (29)–(30) define *two different ROHF Hamiltonians*, which we denote $\hat{R}^{\mathrm{I}}$ and $\hat{R}^{\mathrm{II}}$ [17].

The matrices of these two Hamiltonians take the same form (19) and differ from each other only in diagonal blocks. The triples of diagonal operators $\hat{R}^{\sigma}_{(ss)}$ $(s = c = o,\ v)$ entering each of the Hamiltonians $\hat{R}^{\mathrm{I}}$ and $\hat{R}^{\mathrm{II}}$ can be different [14, 16, 17], for example,

$$\hat{R}^{\mathrm{I}} = (\hat{R}^{\beta}_{(cc)},\ \hat{R}^{\alpha}_{(oo)},\ \hat{R}^{\alpha}_{(vv)}) \tag{33a}$$

$$\hat{R}^{\mathrm{II}} = (\hat{R}^{\alpha}_{(cc)}, \hat{R}^{\beta}_{(oo)}, \hat{R}^{\beta}_{(vv)}) \tag{33b}$$

or

$$\hat{R}^{\mathrm{I}} = (\hat{R}^{\alpha}_{(cc)}, \hat{R}^{\alpha}_{(oo)}, \hat{R}^{\alpha}_{(vv)}) \tag{34a}$$

$$\hat{R}^{\mathrm{II}} = (\hat{R}^{\beta}_{(cc)}, \hat{R}^{\beta}_{(oo)}, \hat{R}^{\beta}_{(vv)}) \tag{34b}$$

(See also Table 2). This specific ambiguity of $\hat{R}^{\mathrm{I}}$ and $\hat{R}^{\mathrm{II}}$ follows from that the three diagonal operators of a ROHF Hamiltonian are, by definition, independent of each other [see also Eq. (16)]. Below we will specify the diagonal blocks of $\hat{R}^{\mathrm{I}}$ and $\hat{R}^{\mathrm{II}}$ by referring to the respective definitions (33a), (33b) or (34a), (34b). In the particular

Table 2 Coefficients $A^{\sigma}_{(ss)}$ and $B^{\sigma}_{(ss)}$ of Eq. (37) characterizing a one-electron process $X \to X^{\pm}_{j,\sigma}$ $(j \in s)$ in HSHFOS systems with the total electronic spin S

$A^{\sigma}_{(cc)}$	$B^{\sigma}_{(cc)}$	$A^{\sigma}_{(oo)}$	$B^{\sigma}_{(oo)}$	$A^{\sigma}_{(vv)}$	$B^{\sigma}_{(vv)}$	Canonical ROHF Hamiltonian	
$X \to X^{+}_{k,\beta}$		$X \to X^{+}_{m,\alpha}$		$X \to X^{-}_{v,\alpha}$			
0	1	1	0	1	0	Plakhutin, Gorelik and Breslavskaya	Ref. [14]
$X \to X^{+}_{k,\alpha}$		$X \to X^{-}_{m,\beta}$		$X \to X^{-}_{v,\beta}$			
$\frac{2S+1}{2S}$	$\frac{-1}{2S}$	0	1	$\frac{-1}{2S}$	$\frac{2S+1}{2S}$	Davidson and Plakhutin	Ref. [16]

case, where the triples of diagonal operators $\hat{R}^{\sigma}_{(ss)}$ are chosen in the form (34a), (34b), the ROHF Hamiltonians $\hat{R}^{\mathrm{I}}$ and $\hat{R}^{\mathrm{II}}$ can be designated $\hat{R}^{\alpha}$ and $\hat{R}^{\beta}$, respectively [17].

The eigenvectors and eigenvalues of $\hat{R}^{\mathrm{I}}$ and $\hat{R}^{\mathrm{II}}$ represent *two sets of the canonical ROHF orbitals and orbital energies* [14, 16]. These characteristics can be derived by solving two different and independent (uncoupled) HF equations [16, 17],

$$\hat{R}^{\mathrm{I}}|\phi_i^{\mathrm{I}}\rangle = \varepsilon_i^{\mathrm{I}}|\phi_i^{\mathrm{I}}\rangle \tag{35}$$

$$\hat{R}^{\mathrm{II}}|\phi_i^{\mathrm{II}}\rangle = \varepsilon_i^{\mathrm{II}}|\phi_i^{\mathrm{II}}\rangle \tag{36}$$

A detailed discussion of the structure of solutions of Eqs. (35)–(36) is presented in Ref. [17]. Here we note that each of these equations can be solved using the usual iterative procedure. In this work, Eqs. (35) and (36) are solved using the general ROHF algorithm for HSHFOS systems implemented in the GAMESS program [21, 22] and the *orbital-energy scaling technique* described in Supplementary materials. When treating the AP violations in different open-shell systems (see Sect. 3) we present either both sets of the orbital energies derived by Eqs. (35)–(36) or only one of them derived by Eq. (35) with the use of the diagonal operator $\hat{R}^{\mathrm{I}}$ (33a) [14].

As shown in Refs. [14, 16], the diagonal operators $\hat{R}^{\sigma}_{(ss)}$ (31)–(32) derived from the variational conditions (30) can be expressed in the above form (20)

$$\hat{R}^{\sigma}_{(ss)} = 2\{A^{\sigma}_{(ss)}\hat{F}_o + B^{\sigma}_{(ss)}(\hat{F}_c - \hat{F}_o)\} \tag{37}$$

where $A^{\sigma}_{(ss)}$ and $B^{\sigma}_{(ss)}$ are characteristic coefficients of the *canonical* ROHF method. It is worth noting that despite of a formal analogy between the definitions (37) and (20), there is a principal difference between them. The operators $\hat{R}^{\sigma}_{(ss)}$ (37) are derived from the physically meaningful conditions (30), each of which provides a fulfillment of KT for the respective one-electron process $X \rightarrow X^{\pm}_{j,\sigma}$ (31)–(32), while the operators $\hat{R}_{(ss)}$ (20) and their eigenvalues have not a physical meaning. The numerical values of $A^{\sigma}_{(ss)}$ and $B^{\sigma}_{(ss)}$ (37) are presented in Table 2.

2.4 CI-Based Formulation of Koopmans' Theorem in the ROHF Method

Canonical ROHF orbitals and orbital energies derived from the extended set of the variational conditions [Eqs. (29) and (30)] can also be derived by an alternative method [15] based on a limited configuration interaction (CI) approach. This allows one to give an alternative (CI-based) formulation of KT in the ROHF method [17].

In this section we present only those formulas of the method [15, 17] which will be used in the treatment below, and refer the reader to the original work for details.

The essence of the approach [15] is that a system X under study is treated within a ROHF method, while ions $X^{\pm}_{j,\sigma}$ are treated within the ORMAS-CI (Occupation Restricted Multiple Active Space-CI) method [27]. The starting orbitals in the CI treatment are ROHF orbitals $\{\phi_i\}$ which are optimal for X and are derived by *any* ROHF method. The approach [15, 17] compares the CI energy of ion $X^{\pm}_{j,\sigma}$ derived with the ORMAS-CI method, $E_{CI}\left(X^{\pm}_{j,\sigma}\right)$, and the energy $E_{frozen}\left(X^{\pm}_{j,\sigma}\right)$ evaluated using the "frozen" ROHF orbitals $\{\phi_i\}$.

It has been shown [15] that the minimal (stationary) value of $E_{frozen}\left(X^{\pm}_{j,\sigma}\right)$, as the latter is defined by the condition (30), is equal to the energy $E_{CI}\left(X^{\pm}_{j,\sigma}\right)$,

$$E_{frozen}\left(X^{\pm}_{j,\sigma}\right)_{\min} = E_{CI}\left(X^{\pm}_{j,\sigma}\right) \tag{38}$$

Based on this result, it has been proved that the *canonical* ROHF orbital energies $\varepsilon^{\sigma}_{j}(X)$ for a system X under study obeying the KT relationships (31) and (32), and the energies $E_{CI}\left(X^{\pm}_{j,\sigma}\right)$ are connected by the fundamental relationship [15, 17]

$$\varepsilon^{\sigma}_{j}(X) = \pm\left\{-E_{CI}\left(X^{\pm}_{j,\sigma}\right) + E_{ROHF}(X)\right\} \tag{39}$$

where the two upper or the two lower signs must be taken together.

In the present work, the relationship (39) will be used for independent verification of the results derived with the canonical ROHF method in those cases where the orbital energies obeying KT (31)–(32) violate the Aufbau principle (5).

3 Violations of the Aufbau Principle

The well-known deficiency of the ROHF method [3–11] is that the orbital energies are defined ambiguously [3]. This ambiguity can lead and, in some cases, does lead to violations of the AP (5) [17]. Moreover, an ambiguity of the ROHF orbital energies is sometimes treated as a single source of AP violations (see, for example, Ref. [13]). In this section we discuss violations of the AP which arise in ROHF calculations of two open-shell systems, endofullerene N@C_{60} (I_h) and atom Mn. The AP violations in these two systems have been revealed many years ago (see below), however, as pointed out in Refs. [15, 28], these violations cannot be treated as well-substantiated, since the ROHF orbital energies derived for these systems generally do not obey KT. In this work we perform a detailed study of the discussed AP violations using two different ROHF approaches presented in Sects. 2.3 and 2.4.

3.1 Atom N and Endofullerene $N@C_{60}$ (I_h)

The geometry and electronic structure of endofullerene $N@C_{60}$ (I_h) has been studied by different experimental and theoretical methods (for review and bibliography, see Ref. [29]). By now, the main features of the electronic structure of $N@C_{60}$ are well established. The ground state of this molecule is ^4A_u arising from the electronic configuration $[\ldots\, 6h_u^{10}\, 7t_{1u}^3]$. The three-fold degenerate open-shell orbitals $7t_{1u}$ are practically pure atomic $2p_N$ orbitals, and so the encapsulated nitrogen keeps its atomic electronic configuration $[\ldots\, 2p^3]$. The highest in energy closed-shell orbitals $6h_u$ of $N@C_{60}$ are essentially the orbitals of the cage C_{60} (I_h).

Together with this, the quantum chemical data for $N@C_{60}$ derived in previous years with different quantum chemical methods were essentially different (for discussion and bibliography, see Refs. [28–30]). Of particular interest is the ROHF results [30] derived using the quantum chemical program Turbomole [31]. From computations [30] it follows that the orbital energy $\varepsilon\,(7t_{1u})$ and a number of closed-shell orbital energies ε_k including $\varepsilon\,(6h_u)$ for $N@C_{60}$ obey the relationship $\varepsilon\,(6h_u) > \varepsilon_k$, and thus, violate the AP (5). It was assumed in Ref. [30] that the ROHF orbital energies derived with program Turbomole [31] obey KT, and hence, the discussed AP violation is well-substantiated. However, these two related assumptions [30] deserve discussion.

To clarify the AP violation in $N@C_{60}$ following from computations [30], this work presents the orbital energies $\varepsilon\,(7t_{1u})$ and $\varepsilon\,(6h_u)$ for $N@C_{60}$ derived with different ROHF methods. (The other orbital energies for $N@C_{60}$ are not presented explicitly since those values are of little interest in a treatment below). To gain a better understanding of the source of the discussed AP violation [30], we also present the ROHF orbital energies for the free nitrogen atom N. All computations were performed with the GAMESS program using the coefficients $A_{(ss)}$ and $B_{(ss)}$ from Table 1 and similar coefficients $A^{\sigma}_{(ss)}$ and $B^{\sigma}_{(ss)}$ from Table 2. In the iterative SCF procedure for $N@C_{60}$ these coefficients were scaled as described in Supplementary materials. Computations of $N@C_{60}$ (and, for comparison, of C_{60}) were performed using the geometry of the cage C_{60} taken from Ref. [30]: $R_{CC} = 1.4507$ A (C–C bond in regular pentagons) and $R_{CC} = 1.3906$ A (a shorter C–C bond in hexagons). Icosahedral symmetry was retained in all calculations.

The results of calculations for atom N and endofullerene $N@C_{60}$ are presented in Table 3. To compare the full orbital energy spectra for $N@C_{60}$ derived with different ROHF methods, Fig. 1 presents a schematic representation of these spectra and compares them with the respective data for two constituents of $N@C_{60}$ (I_h), i.e., for fullerene C_{60} (I_h) and atom N.

An inspection of the data presented in Table 3 allows one to make a number of conclusions. First, we note that the open-shell orbital energies $\varepsilon\,(2p)$ for atom N and the orbital energies $\varepsilon\,(7t_{1u})$ for $N@C_{60}$ derived with different ROHF methods vary over wide limits. Both these variations and a similar one for the closed-shell orbital energy $\varepsilon\,(2s)$ in atom N are caused by ambiguity of the orbital energies in the

Table 3 ROHF orbital energies (eV) for atom N ($^4S, 1s^2 2s^2 2p^3$), and endohedral N@C_{60}(^4A_u, … $6h_u^{10}\,7t_{1u}^3$), derived with the use of different versions of the ROHF method, and experimental IPs for atom N. Basis set 6-31G (5*d*). Open-shell orbital energies are marked in bold

	McWeeny and Diercksen (Ref. [6])	Roothaan (Ref. [3])	Euler equations (Ref. [1])	Faegri and Manne (Ref. [10])	Roothaan and Bagus (Ref. [23])	Canonical ROHF[a] (Ref. [14])	IPs and cation states (experim.) (Ref. [32])
N							
3*s*	+26.352	+23.685	+26.886	+26.886	–	+25.285	
3*p*	+24.448	+20.860	+25.166	+25.166	–	+23.013	
2*p*	**−3.218**	**−4.827**	**−7.696**	**−15.392**	**−15.392**	**−15.392**	14.534 (^3P)
2*s*	−23.798	−14.439	−25.667	−25.667	−25.667	−20.055	20.33 (^5S)
1*s*	−425.084	−423.179	−425.467	−425.467	−425.467	−424.319	
E_{total} (a.u.)	−54.382051						
N@C_{60} $(I_h)^b$							
$7t_{1u}$ ($2p_N$)	**−3.436**			**−15.701**		**−15.701**	
$6h_u$	−7.596			−7.595		−7.595	
E_{total} (a.u.)	−2326.196557			−2326.196557		−2326.196557	

[a]For brevity, we here present only one set of the canonical ROHF orbital energies derived by the method [14], i.e., ε_k^β, ε_m^α, and ε_v^α (for details, see Eq. (33a) and Table 2). The orbital energies ε_{2s}^β and ε_{2p}^α are related via KT to the IPs I_{2s}^β and I_{2p}^α, respectively, presented in the last column

[b]For comparison, we also present the orbital energy $\varepsilon(6h_u)$ and experimental IP $I(6h_u)$ for molecule C_{60} (I_h): $\varepsilon(6h_u) = -7.593$, $I(6h_u) = 7.61$ [33]

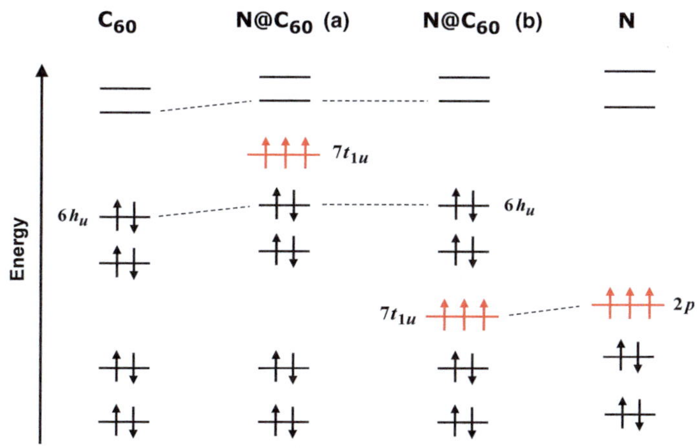

Fig. 1 Schematic representation of the spectra of ROHF orbital energies for fullerene $C_{60}(I_h)$, atom N, and endofullerene $N@C_{60}(I_h)$. The orbital energies for $N@C_{60}$ presented in the panel (*a*) were derived using the ROHF method [6]. The orbital energies for $N@C_{60}$ presented in the panel (*b*) and the orbital energies for atom N were derived using the canonical ROHF method [14]. (For details, see Table 3 and the respective text)

ROHF method [3]. Second, we point out that the open-shell orbital energies $\varepsilon\,(2p)$ for atom N and $\varepsilon\,(7t_{1u})$ for $N@C_{60}$ derived with the same ROHF method are close to each other. The latter agrees with the well-established result that the open-shell orbitals $7t_{1u}$ in $N@C_{60}$ are practically pure atomic $2p_{\mathrm{N}}$ orbitals (for discussion and bibliography, see Ref. [30]).

The closest agreement between computed orbital energies for atom N and the respective experimental IPs is achieved in the canonical ROHF method [14]. This result is expected since the orbital energies [14] obey Koopmans' theorem. Of particular interest is that the orbital energy $\varepsilon\,(2p) = -15.392\,\mathrm{eV}$ for atom N derived with the methods [10, 23] is equal to the respective energy derived with the canonical ROHF method [14]. This equality follows from that the diagonal operators $\hat{R}_{(oo)}$ in methods [10, 23] (and also in methods [7, 8]) are equal to the diagonal operator $\hat{R}^{\alpha}_{(oo)}$ in the canonical ROHF method [14] (for details, see Tables 1 and 2), and hence, the eigenvalues $\varepsilon\,(2p)$ derived with the methods [7, 8, 10, 23] are equal to each other and are equal to the orbital energy $\varepsilon\,(2p)$ in the method [14]. It follows from here that the open-shell orbital energies ε_m for HSHFOS systems, such as $N@C_{60}$ or atom Mn, derived by the methods [7, 8, 10, 23] satisfy the KT relationship $\varepsilon_m = -I^{\alpha}_m$ (31), although a fulfillment of KT was not presupposed in the derivations [7, 8, 10, 23].

As compared to this, the closed-shell orbital energy $\varepsilon\,(2s)$ for atom N derived by the methods [4–10, 23] is not equal to the respective canonical ROHF orbital energy [14], and hence, the closed-shell orbital energies in the methods [4–10, 23] do not obey KT. This conclusion clarifies the assumption [30] that the ROHF orbital energies for $N@C_{60}$ derived with program Turbomole [31] obey KT.

The ROHF method used in computations [30] is the method [10] developed by Faegri and Manne and implemented in program Turbomole. The latter means that the assumption [30] is correct only partially.

From Table 3 it follows, that the canonical ROHF orbital energies $\varepsilon\,(6h_u)$ and $\varepsilon\,(7t_{1u})$ in N@C$_{60}$ obey the relationship $\varepsilon\,(6h_u) > \varepsilon\,(7t_{1u})$ and hence, violate the AP (5). This result based on Koopmans' theorem confirms the prediction [30]. Unfortunately, this result cannot be verified by direct comparison of these orbital energies with the respective experimental IPs. To our best knowledge, the observed IPs $I\,(7t_{1u})$ and $I\,(6h_u)$ for N@C$_{60}$ are unavailable in the literature. However, taking into account that the canonical ROHF method [14] obeying KT gives good estimates for both open-shell and highest closed-shell orbital energies in atom N, we believe that the AP violation in N@C$_{60}$, following from the canonical ROHF calculations, can be considered as well-substantiated.

To complete this treatment, we should comment the form of the orbital energy spectrum for N@C$_{60}$ presented in the panel (*b*) of Fig. 1. This spectrum shows that the open-shell energy level $\varepsilon\,(7t_{1u})$ lies below a number of closed-shell energy levels. If all the canonical ROHF orbital energies in N@C$_{60}$ are ordered by their values $((\varepsilon_1 < \varepsilon_2 < \varepsilon_3)\ldots$, i.e., are presented in the form shown in the panel (*b*) of Fig. 1, the highest occupied five-fold degenerate orbitals $6h_u$ and open-shell orbitals $7t_{1u}$ take the numbers $k = 181, 182, \ldots, 185$, and $m = 143, 144, 145$, respectively. It follows from here that the canonical open-shell ROHF orbitals $7t_{1u}$ in N@C$_{60}$ lie on the energy scale below the *forty* closed-shell orbitals.

3.2 Atom Mn

Violations of the Aufbau principle (5) in ROHF computations of 3*d* transition-metal atoms have been revealed several decades ago [34, 35]. As follows from the numerical HF computation [35] of atom Mn in its ground state $(^6S, \ldots\, 4s^2\, 3d^5)$, performed using Roothaan and Bagus' atomic ROHF method [23], the highest closed-shell orbital energy ε_{4s} and open-shell orbital energy ε_{3d} are equal to −0.2479 and −0.6388 (a.u.), respectively. These orbital energies satisfy the relationship $\varepsilon_{4s} > \varepsilon_{3d}$, and hence, violate the AP (5). The same violation follows from the data [34]. It should be noted, however, that this AP violation cannot be treated as well-substantiated. As follows from Tables 1 and 2, only the open-shell orbital energy ε_{3d} for atom Mn derived with the method [23] obeys KT.

This work presents a detailed treatment of the discussed AP violation in atom Mn. Table 4 presents orbital energies for atom Mn derived with different ROHF methods including the canonical ROHF method. These data are compared with both experimental IPs and orbital energies derived by the fundamental relationship (39).

Before discussing the data of Table 4 we note that the AO basis set for atom Mn [36, 37] used in this work gives the ROHF energy close to the HF limit [35], and so, all the orbital energies in Table 4 can be immediately compared with each other.

Table 4 ROHF orbital energies and experimental IPs for atom Mn (^6S, … $4s^2 3d^5$), and CI energies of ions $Mn^{\pm 1}$. Basis set (23*s*, 15*p*, 11*d*) from Refs. [36, 37]. Energy values in a.u. Violations of the Aufbau principle (5) are marked in bold

	Roothaan (Ref. [3])	McWeeny and Diercksen (Ref. [6])	Numerical HF (Ref. [35])	Canonical ROHF (Refs. [14, 16])[a]		$\pm\{-E_{CI}(Mn^{\pm 1}_{j,\sigma}) + E_{ROHF}(Mn)\}$[b,c] (Refs. [15 and 17])				IPs and cation states (Ref. [32])
				I (α)	II (β)	$\sigma=\alpha$		$\sigma=\beta$		
Virtual						Mn^{-1}		Mn^{-1}		
4*p*	+0.2424	+0.2841	–	+0.2678	**+0.3249**	+0.2678	^7P	**+0.3249**	^5P	
5*s*	+0.0765	+0.0824	–	+0.0802	**+0.0876**	+0.0802	^7S	**+0.0876**	^5S	
Open						Mn^{+1}		Mn^{+1}		
3*d*					**+0.3578**			**+0.3578**	^5D	
	–0.1406	–0.0937	**–0.6388**	**−0.6388**		**−0.6388**	^5D			0.5231 (^5D)
Closed						Mn^{+1}		Mn^{-1}		
4*s*	–0.2153	–0.2424	**–0.2479**		−0.2316			−0.2316	^7S	0.2732 (^7S)
				−0.2707		**−0.2707**	^5D			0.3150 (^5D)
3*p*	−1.9903	−2.3981	−2.4795	−2.8210	−2.2351	−2.8210	^5P	−2.2351	^7P	
3*s*	−3.4162	−3.7500	−3.8166	−4.0964	−3.6165	−4.0964	^5S	−3.6165	^7S	
2*p*	−24.6824	−24.7908	−24.8126	−24.9047	−24.7473	−24.9047	^5P	−24.7473	^7P	
2*s*	−28.9777	−29.0874	−29.1095	−29.2022	−29.0435	−29.2022	^5S	−29.0435	^7S	
1*s*	−240.5325	−240.5337	−240.5340	−240.5350	−240.5332	−240.5350	^5S	−240.5332	^5S	
E_{total}	−1149.866 215		−1149.866 251	−1149.866 215						

[a]The two sets of the orbital energies, I (α) and II (β), were derived by Eqs. (35) and (36), in which the diagonal parts of $\hat{R}^{I}$ and $\hat{R}^{II}$ are formed as shown in Eqs. (34a) and (34b), respectively

[b]For details, see Eq. (39) and the respective text

[c]The CI energies of $Mn^{\pm 1}$ were derived with the method ORMAS-CI [27]. The ROHF orbitals used to construct the CI matrices for $Mn^{\pm 1}$ were taken from the ROHF computation of the neutral atom Mn performed with McWeeny and Diercksen's method [6]

The details of a computational procedure used in computations of atom Mn are analogous to those used in computations of N@C_{60}.

From Table 4 it follows that the canonical ROHF method [14, 16] generates two different sets of orbitals and orbital energies, and so, there is some similarity between one-electron characteristics derived with the canonical ROHF [14, 16] and UHF [20] methods (see also below). The orbital energies ε_{3d} and ε_{4s} of Ref. [35] violating the Aufbau principle (5) split in the canonical ROHF method into two respective pairs of orbital energies, $(\varepsilon^{\alpha}_{3d}\,\text{and}\,\varepsilon^{\beta}_{3d})$ and $(\varepsilon^{\alpha}_{4s}\,\text{and}\,\varepsilon^{\beta}_{4s})$ where $\varepsilon^{\beta}_{3d} > \varepsilon^{\beta}_{4s} > \varepsilon^{\alpha}_{4s} > \varepsilon^{\alpha}_{3d}$. At first we consider three latter orbital energies which can be compared with the respective IPs presented in Table 4. It is easy to see that although the agreement between IPs estimated via KT and observed IPs in these three cases is just qualitative, the order of orbital energies $\varepsilon^{\beta}_{4s} > \varepsilon^{\alpha}_{4s} > \varepsilon^{\alpha}_{3d}$ derived with the canonical ROHF method [14, 16] corresponds to that following from the experimental data.

Taking this into account we conclude that a fulfillment of the relationship $\varepsilon^{\alpha}_{4s} > \varepsilon^{\alpha}_{3d}$, which is the counterpart of the relationship $\varepsilon_{4s} > \varepsilon_{3d}$ [34, 35], is well-substantiated, and a violation of the AP (5) in atom Mn represented by these relationships *reflects the physical reality*.

Table 4 also reveals the second violation of the Aufbau principle in the spectrum of the canonical ROHF orbital energies [14, 16]. The open-shell orbital energy ε^{β}_{3d} and the respective virtual ones, ε^{β}_{5s} and ε^{β}_{4p}, obey the relationship $\varepsilon^{\beta}_{m} > \varepsilon^{\beta}_{v}$, and thus, violate the AP (5). This second violation deserves discussion. The orbital energies ε^{β}_{3d}, ε^{β}_{5s} and ε^{β}_{4p} are related via KT (31)–(32) to the vertical electron affinities (EAs). To our best knowledge, the respective experimental EAs for atom Mn are unavailable in the literature. On the other hand, the KT estimates of EAs derived within the Hartree-Fock approach usually do not agree with experimental EAs even qualitatively, and this presents a fundamental drawback, which is inherent to all Hartree-Fock methods based on a one configuration approximation (see, for example, Ref. [16]). In this situation, we should proe that the second AP violation in atom Mn taking the form $\varepsilon^{\beta}_{m} > \varepsilon^{\beta}_{v}$ is not the result of a deficiency of the canonical ROHF method [14, 16].

For this purpose, Table 4 presents the ROHF orbital energies derived using the fundamental relationship (39) [15, 17]. The orbital energies (39) are, by definition, ***independent*** of the ROHF method used, in the sense that the energy $E_{ROHF}(X)$ entering the definition (39) is the same in all ROHF methods, while the CI energies of ions $X^{\pm}_{j,\sigma}$ can be derived using the ROHF orbitals $\{\phi_i\} = \{\phi_k\} \oplus \{\phi_m\} \oplus \{\phi_v\}$, which are derived for a system X under study by *any* ROHF method.

A comparison between the orbital energies (39) and all other sets of orbital energies for atom Mn presented in Table 4 shows that the orbital energies (39) are equal in all cases to the canonical ROHF orbital energies. It allows one to conclude that the violation of the AP (5) for atom Mn, represented by the relationship $\varepsilon^{\beta}_{m} > \varepsilon^{\beta}_{v}$, is not caused by any deficiency of the canonical ROHF method [14–17] and thus is well-substantiated.

3.3 Discussion

The above treatment of different AP violations allows one to consider in more detail the sources of these violations. At first, we recall that in the previous (non-canonical) ROHF methods [3–11], the orbital energies are defined ambiguously [3]. Due to this ambiguity, the orbital energies can violate the Aufbau principle (5). Moreover, AP violations predicted by different ROHF methods *for the same system* can be essentially different, i.e., either $\varepsilon_k > \varepsilon_m < \varepsilon_v$ or $\varepsilon_k < \varepsilon_m > \varepsilon_v$ (for more details, see Ref. [17]). It is obvious that such violations do not reflect any reality and, in accordance with Glaesemann and Schmidt [13], must be view skeptically.

In the canonical ROHF method [14–17], the orbital energies obey the physically meaningful KT relationships (31)–(32). These orbital energies are obtained either as the eigenvalues of the canonical ROHF Hamiltonians (35)–(36) [14, 16] or by Eq. (39), which is the essence of the alternative (CI-based) formulation of Koopmans' theorem [15, 17]. The violations of the Aufbau principle (5) following from the canonical ROHF computations are well-substantiated.

To clarify the source of the latter violations, we compare the canonical ROHF orbital energies ε_i^σ $(\sigma = \alpha, \beta)$ for atom Mn presented in Table 4 with the respective UHF orbital energies derived in Ref. [15] (we denote the UHF orbital energies ω_i^σ). For brevity, we here consider only the orbital energies with $\sigma = \alpha$, since a treatment of the case $\sigma = \beta$ is quite similar. A comparison between ε_i^α of Table 4 and ω_i^α of Ref. [15] derived using the same AO basis set [36, 37] shows that the respective orbital energies are approximately equal to each other in all cases, i.e., $\varepsilon_i^\alpha \approx \omega_i^\alpha$.

The full set of the ROHF orbital energies, $\{\varepsilon_i^\alpha\}$, is divided into three different subsets, $\{\varepsilon_i^\alpha\} = \{\varepsilon_k^\alpha\} \oplus \{\varepsilon_m^\alpha\} \oplus \{\varepsilon_v^\alpha\}$, while the UHF orbital energies $\{\omega_i^\alpha\}$ are divided into two subsets, $\{\omega_i^\alpha\} = \{\omega_j^\alpha\} \oplus \{\omega_v^\alpha\}$, where, in this case, the index j refers to the occupied UHF orbitals. For the particular case of atom Mn, these two different divisions result in that the ROHF open-shell orbital energy ε_{3d}^α forms a separate subset $\{\varepsilon_m^\alpha\}$, while the UHF open-shell orbital energy ω_{3d}^α belongs to the subset $\{\omega_j^\alpha\}$.

Since in the UHF method all the orbital energies are ordered by their values, the orbital energies ω_{3d}^α and ω_{4s}^α belonging to the same subset obey the Aufbau principle, i.e., $\omega_{3d}^\alpha < \omega_{4s}^\alpha$. In the ROHF method, the respective orbital energies ε_{3d}^α and ε_{4s}^α belong to *different* subsets. Since the ordering of orbital energies in the ROHF method is performed separately in each subset [see Eq. (4)], the orbital energies ε_{3d}^α and ε_{4s}^α violate the AP (5), (see also Table 4). A similar explanation can be done for other AP violations displayed in Tables 3 and 4.

Based on these considerations, we conclude that the single source of violations of the AP (5) in the canonical ROHF method [14–17] is the specific separation of the occupied ROHF orbitals into the closed-shell and open-shell orbital subspaces, characterized by different orbital occupancies (4). This separation follows from the classic Roothaan's equations [3], in line with which the variational conditions imposed upon the open-shell and closed-shell orbitals are different.

It should be noted here that violations of the Aufbau principle appearing in the canonical ROHF method [14–17] lead to well-known difficulties in a computational procedure (see also Supplementary materials to this paper), and thus, can be considered as a deficiency of the method [14–17] itself. In this respect, the UHF method [20], which is free of AP violations, seems more preferable than the method [14–17]. However, this small preference of the UHF method becomes negligible if we compare the methods [14–17, 21] in more detail.

The canonical ROHF solutions (orbitals, orbital energies, and total wave function) obey the full set of the variational and symmetrical conditions, which are imposed upon a Hartree-Fock (HF, ROHF, UHF) solution by two fundamental theorems, KT [2] and Brillouin's theorem [38] (for discussion, see Ref. [17]). In contrast to this, a fulfillment of the theorems [2, 38] in the UHF method [20] is *purely formal* [15, 17], since UHF wave functions for both a system X under study and its ionized $X^{\pm}_{j,\sigma}$ and excited $X_{j\rightarrow v}$ states generally suffer from spin contamination. The latter means that UHF wave functions do not correspond to well-defined quantum states, as is required by both Koopmans' and Brillouin's theorems.

It is worth noting that two discussed methods become completely equivalent *from the physical viewpoint* if the UHF method is presented in its special (*canonical*) form developed in Ref. [17], in which all the deficiencies of the original UHF method [20] were eliminated.

Acknowledgements We thank Prof. Ernest R. Davidson for many valuable discussions. This work was conducted within the framework of budget project No. 0303-2016-0001 for Boreskov Institute of Catalysis and was supported by the Russian Foundation for Basic Research under Grant No. RFBR 15-03-00830.

References

1. Fock VA (1930) Zs f Phys 61:126
2. Koopmans TA (1934) Physica (Amsterdam) 1:104
3. Roothaan CCJ (1960) Rev Mod Phys 32:179
4. Dyadyusha GG, Kuprievich VA (1965) Theor Exp Chem 1:262
5. Hirao K, Nakatsuji H (1973) J Chem Phys 59:1457
6. McWeeny R, Diercksen G (1968) J Chem Phys 49:4852
7. Davidson ER (1973) Chem Phys Lett 21:565
8. Binkley JS, Pople JA, Dobosh PA (1974) Mol Phys 28:1423
9. Guest MF, Saunders VR (1974) Mol Phys 28:819
10. Faegri K, Manne R (1976) Mol Phys 31:1037
11. Bouscasse L, Jaffé HH (1979) J Chem Phys 71:4969
12. Tsuchimochi T, Scuseria GE (2010) J Chem Phys 133:141102
13. Glaesemann KR, Schmidt MW (2010) J Phys Chem A 114:8772
14. Plakhutin BN, Gorelik EV, Breslavskaya NN (2006) J Chem Phys 125:204110
15. Plakhutin BN, Davidson ER (2009) J Phys Chem A 113:12386
16. Davidson ER, Plakhutin BN (2010) J Chem Phys 132:184110
17. Plakhutin BN, Davidson ER (2014) J Chem Phys 140:014102

18. Carbó R, Riera JM (1978) A general SCF theory. Lecture notes in chemistry, vol 5. Springer, Berlin
19. Plakhutin BN (2014) J Struct Chem 55:1001
20. Pople JA, Nesbet RK (1954) J Chem Phys 22:571
21. http://www.msg.ameslab.gov/GAMESS/GAMESS.html
22. Schmidt MW, Baldridge KK, Boatz JA et al (1993) J Comput Chem 14:1347
23. Roothaan CCJ, Bagus PS (1963) Methods Comput Phys 2:47
24. Huzinaga S (1983) *Metod Molekulayrnyh Orbitalei (Method of Molecular Orbitals).* Mir, Moscow (translated from Japanese to Russian)
25. As pointed out by Koopmans in his Nobel prize lecture: http://www.nobelprize.org/economics/laureates/1975/koopmans-autobio.html, the proof of this result has been given by Kramers
26. Plakhutin BN (2012) Book of abstracts. In: XVII international workshop on quantum systems in chemistry and physics, Turku, Finland, 2012. Åbo Akademi University, p 23
27. Ivanic J (2003) J Chem Phys 119:9364; ibid (2003) 119:9377
28. Plakhutin BN, Breslavskaya NN, Gorelik EV, Arbuznikov AV (2005) J Molec Struct: THEOCHEM 727:149
29. Yang S, Wang CR (eds) (2014) Endohedral fullerenes. From fundamentals to applications. World Scientific, Singapore
30. Greer JC (2000) Chem Phys Lett 326:567
31. Ahlrichs R, Bär M, Häser M, Horn H, Kölmel C (1989) Chem Phys Lett 162:165
32. Grigor'ev IS, Meilikhov EZ (eds) (1991) Fizicheskie velichiny (Physical quantities), Handbook, (Energo-Atom-Izdat, Moscow) (in Russian)
33. Zimmerman JA, Eyler JR, Bach SBH, McElvany SW (1991) J Chem Phys 94:3556
34. Clementi E, Roetti C (1974) At Data Nucl Data Tables 14:177
35. Froese Fisher C (1977) The Hartree-Fock method for atoms: a numerical approach. Wiley, New York
36. Partridge H (1989) J Chem Phys 90:1043
37. http://www.emsl.pnl.gov/forms/basisform.html
38. Brillouin LN (1934) J Phys Radium 5:413

Spin Effects in sp^2 Nanocarbons in the Light of Unrestricted Hartree-Fock Approach and Spin-Orbit Coupling Theory

Elena F. Sheka

Abstract The paper presents the first reference to the *post factum* similarity of nonrelativistic unrestricted Hartree-Fock formalism and Dirac-Fock spin-orbit theory when applying for the consideration of spin effects characteristic for sp^2 nanocarbon open-shell molecules (fullerenes, carbon nanotubes, and graphene). The origin of the similarity as well as interchangeability of the approaches when determining either UHF or SOC peculiarities and parameters are discussed.

1 Introduction

The nineteenth century was marked by the discovery of a highly peculiar molecule named by M. Faraday in 1825 as benzene which has become one of the pillars of the modern organic chemistry. The twentieth century was enriched in 1926 by the Hückel explanation of the benzene molecule peculiarity which has formed the grounds of the π-electron theory of aromaticity that was the ground of the modern quantum chemistry in general. The twenty first century has faced a conflict between compositions of condensed benzenoid units and the aromaticity theory. This conflict makes to abandon the benzene-aromaticity view on the compounds formed by the condensed benzenoid rings and stimulates finding conceptually new approaches for their description. The current paper is an attempt to answer this demand. The author suggests peculiar spin effects to be laid in the foundation of the conflict resolution. The concept has arisen on the basis of a set of extended quantum chemical computational experiments performed by the author as well as by other scholars, results of which have found a convincing empirical support. In the way of the concept formation, one had to answer the question how nonrelativistic

E.F. Sheka (✉)
Theoretical Physics and Mechanics Department,
Peoples' Friendship University of Russia, 117198 Moscow, Russia
e-mail: sheka@icp.ac.ru

A. Tadjer et al. (eds.), *Quantum Systems in Physics, Chemistry, and Biology*,
Progress in Theoretical Chemistry and Physics 30,
DOI 10.1007/978-3-319-50255-7_3

formalism of tools used in the course of the above computational experiments is able to exhibit pure relativistic issues. The solution of this and other problems met in the way is described in the current paper.

2 About Open-Shell Molecules in General

The term 'open-shell molecule' covers a large set of species differing quite considerably. It was firstly attributed to radicals while generalizing the Roothaan iterative method of determining LCAO molecular orbitals when the number of α electrons with spin up is not equal to the number of β electrons with spin down ($N_\alpha \neq N_\beta$) [1] thus making the unrestricted Hartree-Fock (UHF) approach one of powerful tools of quantum chemistry. For a long time the term was associated with radicals or molecules with odd number of electrons. Fifty years after the first introduction, the term was applied to open-shell singlet diradicals based on olygocenes ($N_\alpha = N_\beta$) [2]. At the same time the application of the UHF formalism to singlet-ground-state fullerenes revealed their polyradical character [3–5] thus attributing the latter to the family of singlet open-shell molecules. These molecules form a particular class, inside of which sp^2 nanocarbons, such as polycyclic aromatic hydrocarbons [2, 6–17] (referred to as either linear acenes or polyacenes and olygoacenes), fullerenes [3–5, 18–25], carbon nanotubes [26–29], graphenes [30–42], graphdyines [43], and so forth occupy a special place. It was the first case in molecular physics when the species were distinguished not at experimental but computational level by using spin-unrestricted formalism, mostly UHF one.

Conceptually, UHF approach releases the constraint that electrons of opposing spins occupy the same spatial orbital, due to which different spatial orbitals are used for α and β spin. Besides, the UHF solution depends on whether the UHF wf is an eigenfunction of the total spin-squared operator $\hat{S}^2$ or not. Thus, applying UHF approach to closed-shell molecules one gets solutions that are identical to those of either restricted open Hartree-Fock (ROHF) ($N_\alpha \neq N_\beta$) or RHF ($N_\alpha = N_\beta$) (see a number of numerous examples in [44]). The solutions are spin pure and the UHF wf satisfies the total spin-squared operator $\hat{S}^2$ as well. In contrast, the UHF solutions for the open-shell molecules drastically differ from both ROHF and RHF ones. The difference consists in lowering energy and a remarkable spin contamination that is a result of the UHF wf not to be more the eigenfunction of the operator $\hat{S}^2$.

The two classes of UHF solutions do not follow from the inner logic of the UHF formalism but are dependent on molecular object under consideration [45]. Not paying attention to the latter, the spin contamination of the UHF solutions, which casts doubts on the purity of spin multiplicity of the molecule ground state, quite often is attributed to the method disadvantage that leads to erroneous results (see [46] and references therein) due to which singlet diradicals are classified as problematic species [47]. At the same time, the UHF spin contamination is usually considered as pointing to enhanced electron correlation that requires configuration interaction (CI) schemes for its description. Argued, that if UHF approach as the first stage of the CI ones is improved towards a complete CI theory, one would expect removing the spin contamination.

However, two extended computational experiments performed with interval of 15 years fully discard this expectation. The first was carried out back in 2000 on 80 molecules studying the HF solution instability [48]. The molecule set covered a large variety of species including valence saturated and unsaturated compounds, fully carbonaceous and containing heteroatoms. The instability feature was studied in relation with the electronic correlation, the vicinity of the triplet and singlet excited states, the electronic delocalization linked with resonance, the nature of eventual heteroatoms, and the size of the systems. It was shown, that for most conjugated systems, the RHF wf of the singlet fundamental state presents so-called triplet instability [49] that differs by value. The largest effect was observed for aromatic hydrocarbons.

The second experiment has been performed just recently for 14 polyaromatic hydrocarbons by using a number of different CI approaches such as UHF, UMP2, QCISD(T), and UDFT [16]. Results, concerning spin contamination (ΔS^2) of the molecules, are well consistent for the first three techniques while the latter in the case of UDFT was practically null. Actually, the data are dependent on the approach in use. However, the difference within either HF- or DFT-based CI approaches occurred to be not as big as that between the HF and DFT approaches of the same level. The first consequence is due to the fact that triplet states are mainly responsible for the HF instability, thus allowing a considerable wf truncation. Since the UHF formalism is fully adapted to the consideration of the triplet instability [50] the UHF results deviate from the higher approaches of the CI theory no more than 20%. Once wf-based, HF consideration is preferential since the DFT one is much less adapted to the consideration of delicate peculiarities connected with the correlation of electron of different spins (see fundamental Kaplan's comments [51] and the latest comprehensive review [52]). In support of the said above, Fig. 1 portrays the data related to the total number of effectively unpaired electrons N_D which can be considered as a qualitative measure of the spin contamination and which for the singlet state is [53]

$$N_D = 2\Delta\hat{S}^2. \tag{1}$$

Here, $\Delta\hat{S}^2$ is the deviation of squared spin from the exact value. Presented data are related to a number of olygoacenes and were obtained by using both UHF semi-empirical codes [54] and density matrix renormalization group (DMRG) algorithms [6].

According to the UHF AM1 algorithm implemented in the CLUSTER-Z1 codes, used in the current study,

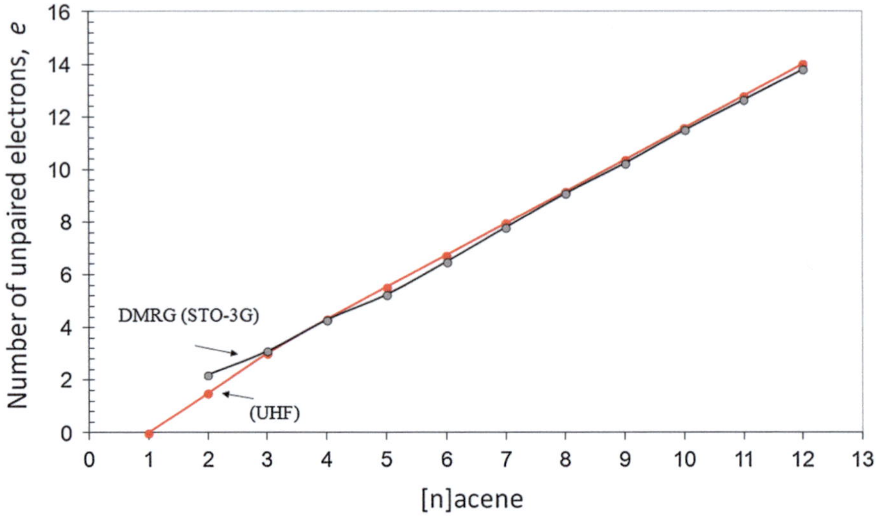

Fig. 1 Total number of effectively unpaired electrons in polyacenes calculated by using DMRG (STO-3G) [6] and UHF (current paper) formalism. No scaling of the data

$$N_D = 2\left(\frac{N^\alpha + N^\beta}{2} - \sum_{i,j=1}^{NORBS} P_{ij}^{\alpha} * P_{ij}^{\beta}\right). \tag{2}$$

Here P_{ij}^{α} and P_{ij}^{β} are the matrix elements of the relevant electron density matrices. The summation in Eq. 2 runs over all spinorbitals.

In conformity with the DMRG algorithm [6], N_D is determined as

$$N_D = \sum_i n_i(2 - n_i). \tag{3}$$

Here, n_i is the occupation number of the *i*th natural orbital that ranges from 0 to 2. The derivation of both Eqs. 2 and 3 corresponds to the same basic concept on effectively unpaired electrons suggested in [53, 55]. As seen in Fig. 1, thus obtained N_D values well coincide in both cases evidencing that the UHF capacity is quite high to be used as the computationally affordable approach. Similar picture is obtained in all the cases where UHF data can be compared with those obtained in the framework of higher-level CI approaches. Oppositely, in all known cases UDFT leads to underestimated data that are in conflict with empirical ones.

3 Reality of sp^2 OSMs UHF Peculiarities

There are a lot of experimental evidences that the UHF-originated peculiarities of open-shell molecules are real. Thus, addressing Fig. 1, it is necessary to remain the hampered availability of longer acenes, with pentacene being the latest well-characterized. In recent years substantial progress has resulted in the synthesis of n-acenes up to n = 9 by matrix isolation techniques [10, 39]. Nevertheless, these higher acenes are very reactive; for example, heptacene was found to be stable only for 4 *h* in a poly(methyl methacrylate) matrix. To overcome the stability problems, larger acenes were functionalized by adding protecting groups which inhibit the native high reactivity of the acenes.

In addition to the indirect N_D manifestation discussed above, the last decade has provided convincing direct evidences of the N_D existence. Figure 2b presents AFM atom-resolved images of two sp^2 molecules recorded in Zürich Research Laboratory of IBM Research [56, 57]. The first image portrays pentacene while the second is related to the smallest possible five-ring olygoacene named olympicene in commemorating London's Olympic Games 2012. The two images are accompanied with the calculated distributions of effectively unpaired electrons over the molecules atoms [37] (so called N_{DA} maps where $N_D = \sum_A N_{DA}$ [21]) presented in Fig. 2a. According to the experimental set up, the brightness of pentacene AFM image is the highest on the atoms of two edge pairs corresponding to the least force of the attraction of the oxygen tip atom. In contrast, brightness on the calculated molecule portraits is the highest at the central atoms that are the most active and are characterized by the largest attractive force due to which experimental images and calculated maps should be strongly brightness-inverse, which is really seen in Fig. 2.

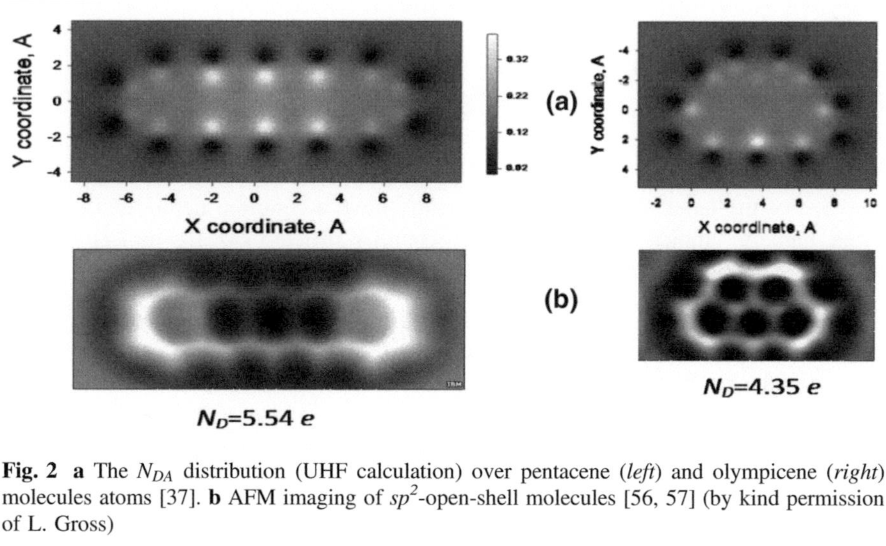

Fig. 2 **a** The N_{DA} distribution (UHF calculation) over pentacene (*left*) and olympicene (*right*) molecules atoms [37]. **b** AFM imaging of sp^2-open-shell molecules [56, 57] (by kind permission of L. Gross)

The other quite numerous observations of effectively unpaired electrons are related to graphene. First of all it is necessary to mention graphene bubbles. Raised above the substrate and mechanically deformed areas of graphene in the form of bubbles are found on different substrates [58, 59]. Typically, the bubbles are seen as bright spots on dark background formed by not wrinkled graphene film. Evidently, the spots exhibit places on graphene film with the largest electron density. The density excess is caused by the film local deformation (curving) which is usually accompanied with increasing the number of effectively unpaired electrons [60, 61]. Since N_{DA} strongly depends on C–C distance formed by the atom [43] the strain-induced stretching of curved graphene bonds evidently causes the value enlarging that is revealed as enhanced brightness of the AFM images. The concept on strain-induced pseudomagnetic field suggested by the authors [58] and further considered in [62] follows just another way to present the bond stretching via the description of their elastic strain in terms of the effective electromagnetic field suggested in [63].

Because N_D is the measure of not only spin contamination but the extent of the relevant molecules radicalization [18–21], locality of chemical reactions can indicate particular places with enhanced N_{DA} values. Thus, a peculiar picture of decoration of graphene reactivity centers with Pd clusters [64] demonstrates possibility to obtain the spatial information about chemical reactivity across the Pd/C system. A particular role of wrinkle as a nanosize gas-inlet for reactions under graphene is shown in [65]. The bright-spot corrugated graphene has been recently observed when covered over gold nanoparticles situated at a substrate [66].

Therefore, computational findings and physicochemical reality do not contradict and sp^2 nanocarbon open-shell molecules (sp^2-OSMs below) do reveal particular properties. Summarizing, the latter can be formulated as following:

1. Singlet ground state of sp^2-OSMs is really spin-contaminated, which means that its singlet spin multiplicity is not exact;
2. Spin contamination is accompanied by the appearance of effectively unpaired electrons that are the measure of the spin-contamination extent;
3. Spin contamination is remarkably strain dependent indicating the crucial role of C–C spacings.

Common for all the sp^2-OSMs and not dependent on chemical content, shape, and size of the latter, the properties evidently have a common nature that has two faces, namely, empirical, connected with peculiar structures of the molecules, and theoretical, implemented in the UHF formalism.

4 What Is the Origin of the sp^2 OSMs UHF Peculiarities?

As shown by the author studies of recent years [28, 29, 31, 32, 35, 37, 40, 43], the variety of C–C bonds length is the cradle of the sp^2-OSMs peculiarities. When the bond lengths exceed a critical value $R_{crit} = 1.395$ Å, RHF-character of the UHF solution is transferred to the spin-contaminated UHF one. Lengths of a considerable part of C–C bonds of sp^2-OSMs, such as polyacenes, fullerenes, CNTs, and graphene, are above the critical value. This characteristic property of the UHF solutions, concerning the RHF/UHF transformation depending on the length of the covalent bond under consideration, is well known [45]. The results of computational experiments [16, 48] over large sets of open-shell molecules, analyzed from this viewpoint, perfectly support this conclusion as well.

Apparently, bond-length concept, so well exhibited by the UHF formalism, could be considered as empirical explanation of the sp^2-OSMs UHF peculiarities. However, the concept does not explain the origin of the main feature of the solutions—their spin contamination pointing to the mixture of pure spin states. Evidently, the reason should be sought among fundamentals that so far have been ignored at the level of the reference quantum molecular theory. The latter should be attributed to the quantum chemical description based on a nonrelativistic Hamiltonian which forms the ground of numerous restricted single-determinant (RSD) computational tools.

It is convenient to represent expected progress in improving calculations from the RSD approach as a gradual inclusion of a series of correction terms into the initial RSD energy E_{RSD}

$$E_{best\,estimate} = E_{RSD} + \Delta E_{corr} + \Delta E_{ScR} + \Delta E_{SO} + \Delta E_{BOC} + \Delta E_{ZPE} + \cdots \quad (4)$$

The main corrections include correlation energy ΔE_{corr}, scalar-relativistic ΔE_{ScR} and vectorial-relativistic spin-orbit ΔE_{SO} contributions, non-Born-Oppenheimer ΔE_{BOC} correction, zero-point vibrational effect ΔE_{ZPE}, and other much less significant corrections intimately connected with nuclear motion. Taking into account each of these terms may considerably change the RSD results, which is confirmed by a profound development of computational ability of quantum chemistry over the last century.

As for spin mixing, two contributions may give birth of this event. The first is connected with the general idea "different orbital for different spins" that, according to Slater, Lowdin, and Pratt (see review [67] and references therein), can correct the correlation errors ΔE_{corr} removing at least part of the defects coming from correlation by letting electrons with different spins occupy different orbitals in space so that they get a chance to avoid each other in accordance with the influence of the Coulomb repulsion. UHF formalism arose just on the way to the implementation of this idea into a computational scheme. Vectorial spin-orbit coupling (SOC) ΔE_{SO} is the second reason. Therefore, spin contamination of the electronic solution is characteristic for two different computational tasks, namely, based on non-relativistic Hamiltonian

$$H^{UHC} + H^{RSD} + \Delta E^{UHF}_{corr}. \quad (5a)$$

and relativistic Hamiltonian

$$H^{rel} + H^{RSD} + \Delta E_{SO}. \tag{5b}$$

Both Hamiltonians, schematically presented above, take into account the correlation of electrons with different spins located in different space. Consequently, a natural question arises: Are the solutions of the two tasks identical (at least, similar) when the electron correlation is significant? As will be shown below, in the case of sp^2 OSMs the answer is positive. For the first time, the question was raised by Deharend and Dive [48] where an analogous view on the spin mixing of ground states of OSMs was expressed. However, the authors limited themselves by the idea only which did not obtain a further development. The second addressing to the issue has appeared just recently [68]. The authors considered the spin-contamination problem in the course of a comparative study exploiting both UHF and Kramers-pairs-symmetry breaking general complex Hartree-Fock (GCHF) approach of the relativistic molecular theory and for the first time showed a full analogy of the data obtained by the two formalisms. This fundamental study tightly connects nonrelativistic UHF formalism with relativistic molecular theory.

The SOC issue classification presents a standard view on the SOC effects for molecules and corresponds to the spin effects consideration in the framework of the competense of Eq. 5b ignoring electron correlation. The latter, once significant, may cause similar empirical features.

5 Empirical Evidences of the Similarity of the SOC and UHF Peculiarities of the sp^2 OSMs

5.1 General Characteristics of Spin-Orbit Coupling in Molecules

The SOC in molecules consisting of light elements has been known quite long ago (for early works see [69] and references therein) as well as a leitmotif of numerous theoretical investigations. Comprehensive reviews [70–74] and the second edition of the well known textbook [75] are a must-have for everyone entering the field. With respect to experimental evidence, SOC in molecules shows itself via the following issues:

1. Splitting of the molecule ground state by removing spatial and/or spin degeneracy;
2. Lifting the ban of spin-forbidden transitions, both radiative and non-radiative;
3. Enhancing the molecule chemical activity caused by their radicalization;
4. Exhibiting a particular molecular 'para- and ferro-diamagnetism' (in terms of [76]).

The first issue is connected with the fact that SO couples the total orbital momentum of molecule L with its total spin S thus depriving them both of the quality of good quantum numbers. Such a role is transferred to the total momentum J with components $(L + S)$; $(L + S)-1$; …$|L - S|$. Consequently, the ground state is split into the relevant J components. Evidently, this causes lowering the ground state energy as a whole with respect to the RSD degenerate case.

The governing role of the total momentum J is manifested in the fact that the same J component can correspond to different combinations of L and S thus providing a mixture of spin states at fixed energy. Such a mixture causes the appearance of issues 2–4, although in different way in each of them. It is necessary to say that in all the studied cases the mixture concerned singlet-triplet one. Thus, issue 2 is usually considered as a result of the mixture of the lowest triplet state either with singlet ground states (optical transitions responsible for the molecule phosphorescence) or with the singlet excited state to provide the radiationless relaxation of the latter, which is consistent with a particular role of triplet states in variable photophysics and photochemistry of molecules. Issue 3 is related to the mixture of singlet ground state with the lowest triplet one, which leads to its spin contamination. Peculiar distribution of spin density, caused by the above mixture, results in a variable response of molecules to the application of magnetic field, which lays the foundation of issue 4.

5.2 *UHF Peculiarities of* sp^2 *OSMs and a Comparative View*

The phenomenology of the UHF peculiarities of sp^2-OSMs comes to the following four issues:

1. Misalignment of the energy of RHF (E^R) and UHF (E^U) solutions $\Delta E^{RU} \geq 0$, where $\Delta E^{RU} = E^R - E^U$;
2. Spin contamination expressed via misalignment of squared spin $\Delta\hat{S}^2 \geq 0$; here $\Delta\hat{S}^2 = \hat{S}_U^2 - S(S+1)$, $\hat{S}_U^2$ is the UHF squared spin while $S(S + 1)$ presents the exact RHF value of $\hat{S}^2$;
3. Effectively unpaired electrons of $N_D \neq 0$ total number (see Eqs. 1 and 2);
4. Molecular magnetism of the sp^2 OSMs.

Similarly to issue 1 of empirical SOC phenomenology outlined in Sect. 5.1, the first issue is a consequence of splitting of the degenerate RHF states. Presented in Fig. 3 are selected sets of energies of HOMO and LUMO orbitals (25 in total) related to fullerene C_{60}, rectangle graphene molecule (5, 5) NGr with five benzenoid units along the armchair and zigzag edges, respectively, and a fragment of a bare (4, 4) single-wall CNT. The data are obtained by using RHF and UHF versions of the CLUSTER-Z1 codes [54].

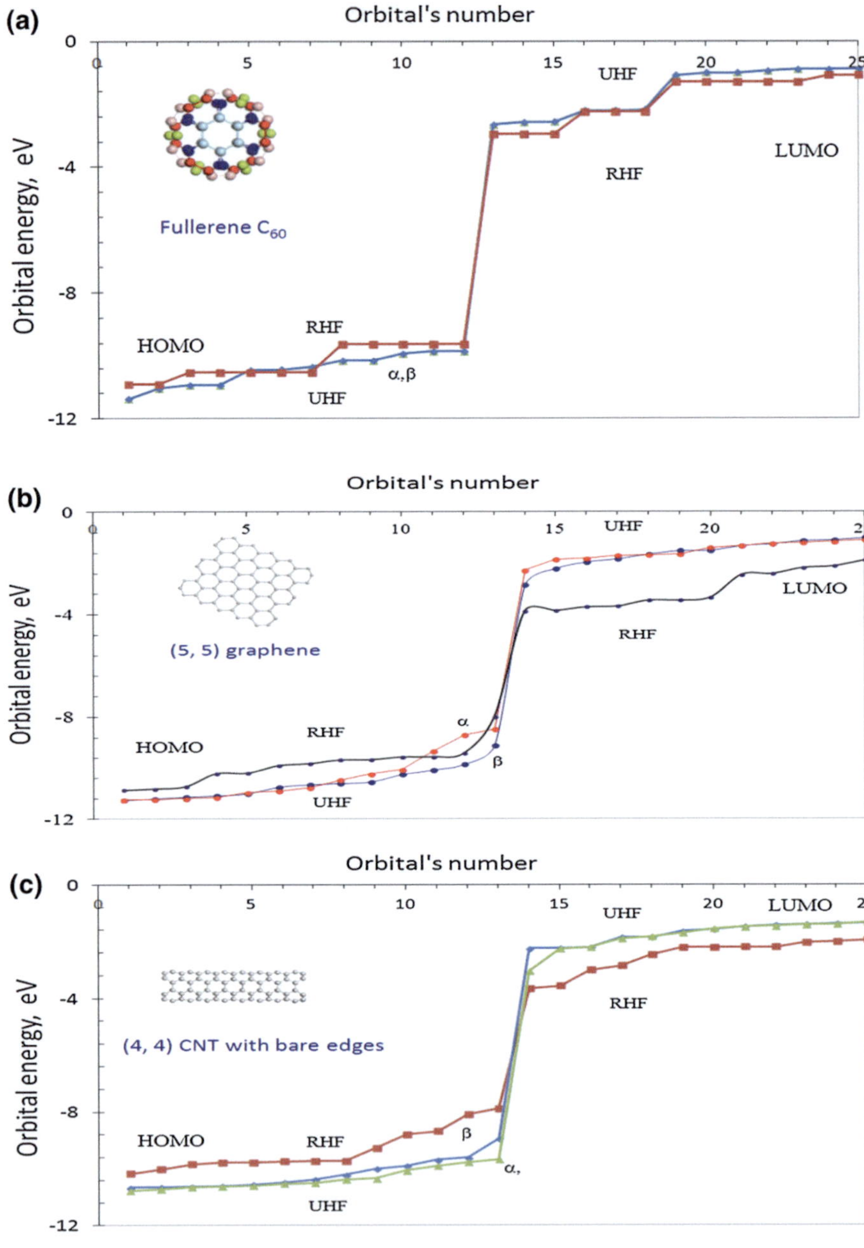

Fig. 3 Energies of 25 spinorbital in the vicinity of HOMO-LUMO gap of fullerene C_{60} **a** (5, 5) NGr molecule **b** and (4, 4) single-walled CNT with bare edges **c** (UHF, current paper)

High degeneracy of the RHF solution of C_{60} (Fig. 3a) is caused by both high spatial (I_h) and spin symmetry. As seen in the figure, the orbitals are clearly split, which causes lowering the molecule spatial symmetry to C_i while conserving the identity of spinorbitals related to α and β spins. The splitting value is different for different orbitals changing from zero to 30 meV.

Data presented in Fig. 3b are related to the (5, 5) NGr molecule. The space symmetry of the molecule (D_{2h}) remains unchanged when going from RHF to UHF formalism. Degeneracy of the RHF orbitals and splitting of the UHF ones exhibits breaking spin symmetry that causes a remarkable distinguishing of orbitals related to α and β spins. As in the case of C_{60}, the splitting is different for different orbital ranging from zero to 1.15 eV.

Fully analogous picture is observed for the (4, 4) SWCNT fragment shown in Fig. 3c. Lowering the molecule space symmetry from C_{4h} to C_{2h} is followed with a considerable decreasing of the HOMO orbitals energies and splitting from zero to 340 meV. Taking as a whole, the data presented in Fig. 3 shows that UHF consideration of the behavior of sp^2-OSMs spinorbitals gives a common picture reflecting the decreasing (increasing) HOMO (LUMO) orbitals energies and degenerate orbitals splitting.

Coming to the second issue, we are facing the problem that for 'singlet-ground-state' fullerene C_{60}, (5, 5) NGr and (4, 4) SWNT the total square spin misalignment $\Delta\hat{S}^2$ is quite considerable and depends on the number of atoms in total and edge atoms with dangling bonds in the two last cases, additionally. Correctly eliminated from edge atoms, $\Delta\hat{S}^2$ can be characterized by an average value attributed to one C–C bond. Such per the bond atom value constitutes ~0.08e for all the cases instead of zero for, say, benzene molecule of D_{6h} symmetry which is a closed-shell molecule. In full consistence with the finding, the phosphorescence of undistorted free benzene molecule has not been observed until now while it is easily fixed for C_{60} under different conditions (see review [77]) and nanosize graphene quantum dots [78]. Benzene is the most neutral solvent that resists to any photophysical event while fullerene C_{60} [22, 79], carbon nanotubes [80] and nanographenes [81, 82] are highly active for photodynamic therapy and so forth.

Issue 3 has a direct connection to that one of molecular SOC phenomenology. However, oppositely to SOC terminology, where the enhancement of chemical activity is described quantitatively, the UHF formalism offers a well determined quantitative description of the issue in terms of the total number of effectively unpaired electrons, N_D. The latter successfully plays the role of molecular chemical susceptibility [20–22] while its fraction N_{DA} on atoms carries the responsibility for the atom chemical activity, atomic chemical susceptibility, distribution of which over the molecule atoms is identical to that of the atom free valence [83] (see Fig. 4 for fullerene C_{60}).

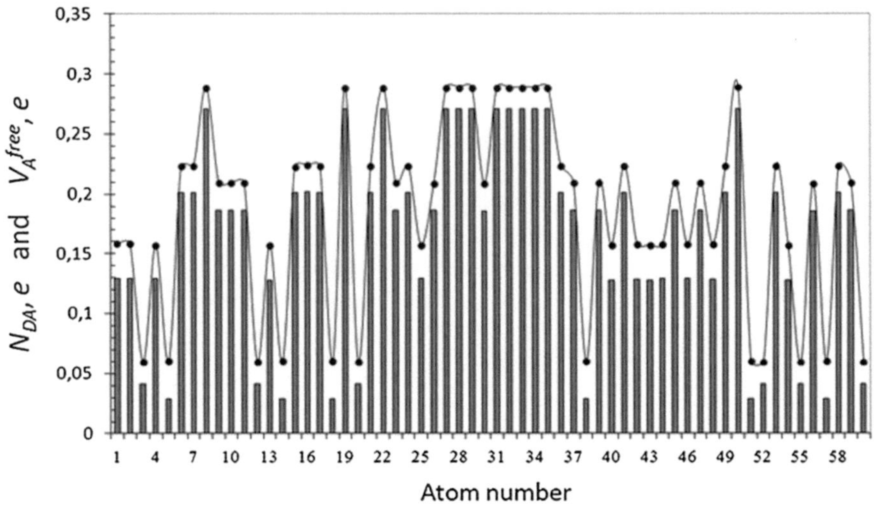

Fig. 4 Atomic chemical susceptibility N_{DA} (histograms) and free valence (*curves with dots*) over atoms of C_{60} fullerene (UHF calculations)

5.3 *Para- and Ferro-Diamagnetism of* sp² *OSMs*

Issue 4 is worth a particular discussion. Empirical observation of para- or ferro-diamagnetism of species with even number of electrons is usually attributed to SOC effects caused by 'local spins' [76]. Contributing to the magnetic effects, UHF formalism suggests a clear vision and quantitative description of the local spins, which can be demonstrated on the example of 'paradiamagnetism' of fullerene C_{60} molecules.

The phenomenon was fixed at low temperature in crystalline state [84, 85]. Providing the kind permission of S.V. Demishev, Fig. 5 presents a confined view of the results obtained. As seen in panel a, the molecules, once exhibiting diamagnetic behavior up to T = 40 K, change the latter for paramagnetic one, 1/T, below the temperature. The observed 'paradiamagnetism' directly evidences spin mixing. Shown in panel b demonstrates that paramagnetic magnetization has a standard dependence on the magnetic field but with the Lande g-factor lying between 1 and 0.3. The strong reduction of the factor from 2 clearly indicates that electrons responsible for the observed magnetization are bound. The most amazing finding is shown in panel c, demonstrating the availability of three g-factor values revealed in the course of magneto-optical study at pulsed magnetic field up to 32 T in the frequency range $\nu = 60 - 90$ GHz at T = 1:8 K. If two first features, namely, paradiamagnetism and deviation of the g-factor values from those related to free electrons, were observed in other cases as well, the third one is quite unique and is intimately connected with the electron structure of fullerene C_{60}.

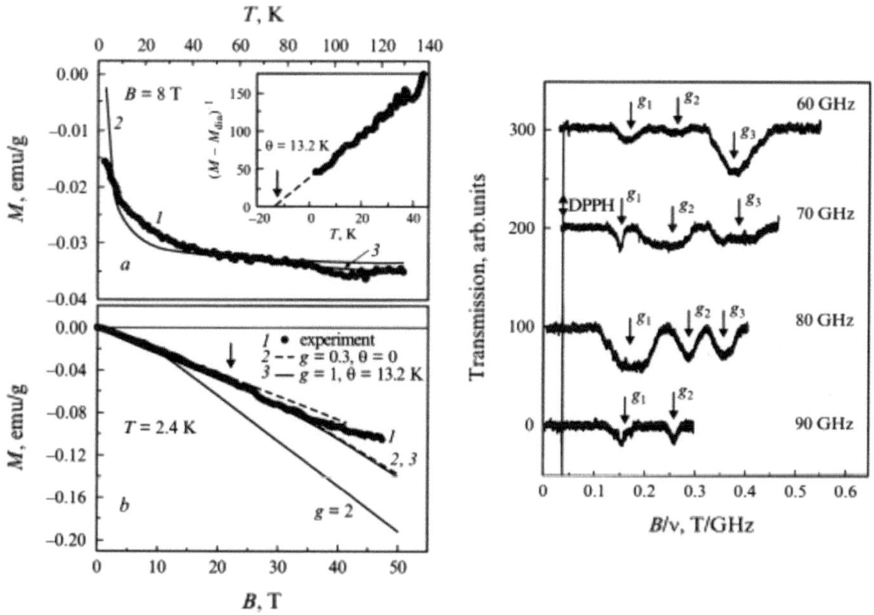

Fig. 5 Magnetization of C_{60} molecule in crystalline state [85]. Temperature (*left top*) and field (*left bottom*) dependence of magnetization for C_{60}-TMTSF-$2CS_2$ molecular complex. *1.* experimental data for *M*(*T*) and *M*(*B*), *2* and *3.* simulations of *M*(*T*) and *M*(*B*) (see details in [85]). Inset in left top shows temperature dependence of magnetization in coordinates $(M - M_{\text{dia}}) - 1 = f\ (T)$. *Right.* ESR absorption lines in (ET)$2C_{60}$ molecular complex at $T = 1{:}8$ K (by kind permission of S.V. Demishev)

Related to the paramagnetic part, the feature should be attributed to a peculiar behavior of the molecule p_z electrons while the other three valence electrons of each carbon atom are sp^2-configured and involved in the formation of spin-saturated σ bonds. According to UHF calculations [20, 21], a part of the p_z electrons, of the total number 9.87*e*, are unpaired and fragmentarily distributed over the molecule atoms. A color image of the molecule shown in Fig. 6a gives a view of this distribution that is presented in terms of the relevant spin density on each of the 60 atoms by histogram. The total spin density equals zero and its negative and positive values are symmetrically (antiferromagnetically) distributed. As seen in the figure, the distribution clearly reveal well configured compositions of the effectively unpaired electrons ('local spins' [86, 87]) with identical positive and negative spin-density values within the latter. Going over the distribution from the highest to the lowest density, one can distinguish two hexagon-packed sets, 12 singles and three sets of pairs covering 12 atoms each. All these compositions, differently colored, are clearly seen in the molecule image. Evidently, the compositions will differently respond to the application of magnetic field thus mimicking the behavior of an atom with different electron orbitals. This allows for exploring the atomic

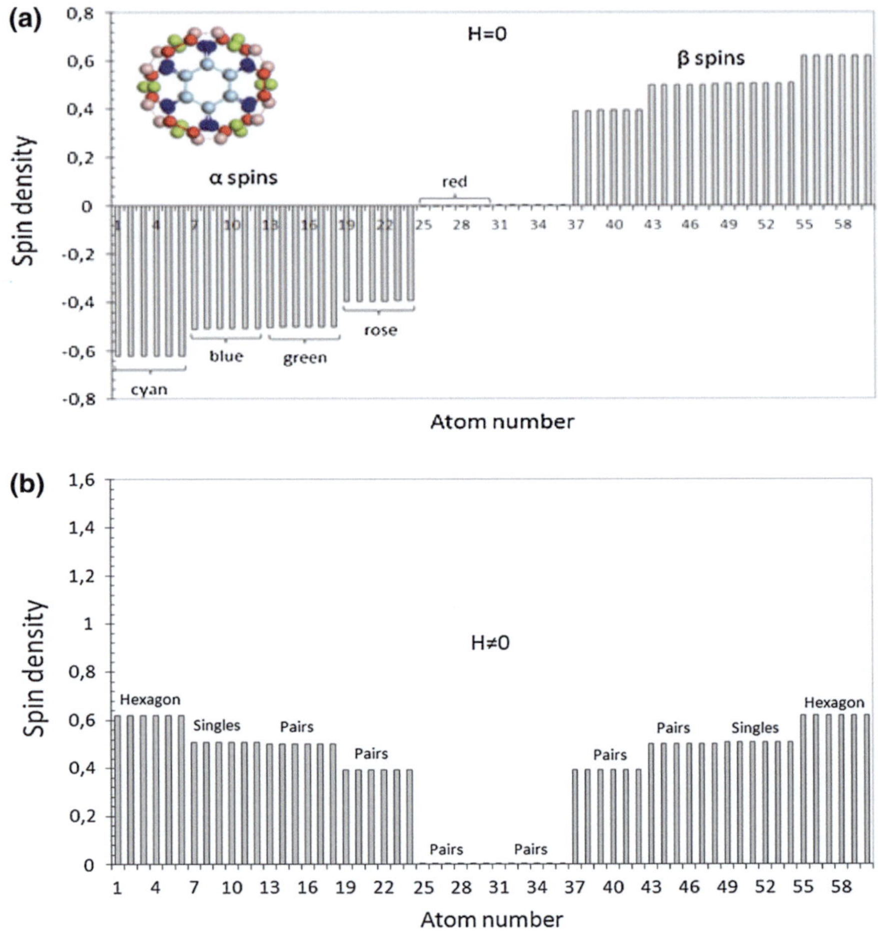

Fig. 6 Spin density distribution over C_{60} atoms in the absence **a** and presence **b** of magnetic field. Insert exhibits the color image of the local spin distribution over the molecule atoms following notations in panels **a** and **b** (UHF calculations)

expression for the evaluation of Lande g-factors related to each set of the local spin configurations of the molecule in the following form [88, 89]

$$g = 1 + \frac{J(J+1) - L(L+1) + S(S+1)}{2J(J+1)}. \quad (6)$$

Application of magnetic field disturbs the antiferromagnetic regularity of the outlined local spins towards ferromagnetic one (see Fig. 6b) that corresponds to the maximum of magnetic ordering.

If experimentally observed, the magnetic properties of fullerene molecules should reflect a variability of g-factor caused by different configuration sets of spins. The corresponding values of the factor can be evaluated by the following way [88, 89].

1. *Hexagon of upright local spins.*

The total spin $S = 1$, the total orbital momentum of the hexagon p_z electrons $L = 6$. Consequently, the total angular momentum J of the hexagon has seven components: (6 + 3), 8, 7, 6, 5, 4, (6 − 3). Usually, the lowest energy state corresponds to the minimum value of the total angular momentum that is $J = 3$ in the case. Following Eq. 6, the relevant Lande factor $g_6 = 0.25$.

2. *Pairs of upright local spin.*

The total spin $S = 1$, the total orbital momentum of pair p_z electrons is $L = 2$ and the total angular momentum J has three components: (1 + 2), 2, (2 − 1). Taking the least value component, obtain the relevant Lande factor $g_2 = 0.5$

3. *Local spin of singles.*

For a single p_z electron, the total spin $\frac{1}{2}$, the orbital angular momentum $L = 1$, while the total angular momentum J has two components: 3/2 and 1/2. According to Eq. 6, the lowest energy term corresponds to $J = ½$ so that the Lande factor $g_1 = 0.66$.

Thus obtained g values are collected in Table 1 alongside with experimental data. As seen in the table, experimental g_1, g_2, and g_3 can be attributed to singles, pairs, and hexagons of local spins, respectively. The calculated and experimental values are in good consent. The absence of the exact coincidence is evident due to a few reasons among which two next are the most important. (1) Eq. 6 is too simplified; its application allowed exhibiting quasi-atomic structure of fullerene molecule related to different configurations of local spins while the exact g values determination is much more complicated problem. (2) The evaluation of g-factors was performed for ferromagnetic ordering of local spins in C_{60}. Experimentally, the studied crystals showed 'paradiamagnetism'. Realization of the 1/T paramagnetic dependence in practice is caused by free rotation of ferromagnetically spin-configured molecules occurred at very low temperatures [90].

Actually, when such a rotation is forbidden, as it is in the case of narrow graphene ribbons (molecules) terminated by hydrogen atoms and rigidly fixed with respect to immobile substrate [91], the ferromagnetic behaviour of the sp2 OSMs has been clearly observed. Graphene webs, presenting a part of the original

Table 1 g-Factors of fullerene C_{60}

Calculated		Experimental [85]	
Attribution	Value	Attribution	Value
Hexagons	0.25	g_3	0.19 ± 0.01
Singles	0.50	g_2	0.27 ± 0.02
Pairs	0.66	g_1	0.43 ± 0.03

graphene sheet in direct contact with the substrate and dividing disks of hydrogenated graphene, may be of different width which occurred to be crucial for the magnetization observed. Thus, the latter is absent when the web width W is zero (bulk graphene) and becomes well observable at W $\approx$ 10 nm. When W is growing, the magnetization gradually falls down. The explanation of the magnetization increase by accompanying size-induced decreasing of exchange integral as well as its decreasing caused by quantifying electron properties towards to crystalline ones at large W is given elsewhere [32]. The observed ferromagnetic magnetization is

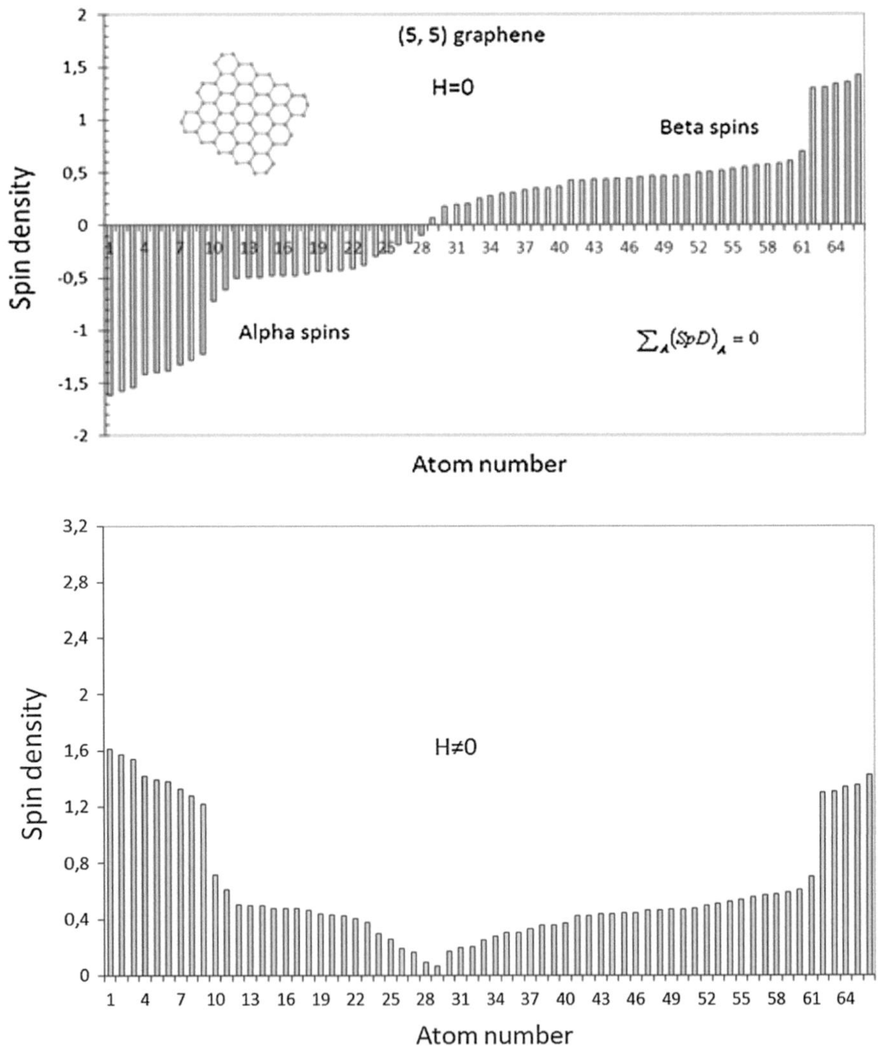

Fig. 7 Spin density distribution over atoms of the (5, 5) NGr molecule (see insert) in the absence **a** and presence **b** of magnetic field (UHF calculations)

consistent with the behavior of local spins of graphene molecules in magnetic field presented in Fig. 7. As seen in the figure, the local spin distribution in zero field is not exactly ordered antiferromagnetically while the total spin density is zero. This is one of many other topological shape-size effects related to condensed compositions of benzenoid rings [61, 92].

6 A Glance at Similarity of SOC and UHF Peculiarities

The smallness of the SOC contribution allows considering H^{SO} as an additive to the total Hamiltonian of a many-electron system in the form

$$H^{tot} = H^{RSD} + H^{SO}. \tag{7}$$

Here H^{RSD} is a spin free nonrelativistic RSD Hamiltonian mentioned earlier. The main computational approach concerns this very Hamiltonian, selecting it within either Hartree-Fock [70, 71, 73] or density functional theory [74] and equation-of-motion approach [74, 75], and treating SOC either perturbationally or variationally. Additional problem consists in a proper choice of the most suitable effective presentation of Hamiltonian H^{SO}. Reviews [70, 71, 73] provide much food for thought on this issue. As suggested in [93–95], SOC appears in the second order of the perturbation theory (PT) applied to $H^{RSD} = H^0$ Hamiltonian. The suggested formalism is widely used for determining governing SOC parameters (see [70] and references therein).

It is the very time to remain that UHF formalism straightly follows from perturbationally considered RHF one as well [50, 95, 97]. In this case Eq. 8 looks like

$$H^{UHF} = H^{RHF} - U. \tag{8}$$

Here U is the spin-polarization operator determined by the difference of H^{RHF} and H^{UHF} operators [95]. The study of Rosski and Karplus [94], undertaken to obtain a proper formalism for describing spin effects in many-electron systems, showed that when the UHF wf is presented as the first PT order to the RHF one, lowering the RHF energy, square spin and spin density are strictly obtained in the second PT order in a perfect consent with the exact UHF solution.

According to the nomenclature suggested by Pople and Nesbet [1], the UHF wf is two-reference once distinguished for α and β electrons located at different spatial orbitals. Caused by the fact, two Fock operators f^{α} and f^{β}, related to electron (1) with either α or β spin, determine the UHF formalism:

$$f^{\alpha,\beta}(1) = h(1) + \sum_a^{N\alpha,\beta} \left[J_a^{\alpha,\beta}(1) - K_a^{\alpha,\beta}(1)\right] + \sum_a^{N\beta,\alpha} J_a^{\beta,\alpha}(1). \tag{9}$$

Here, $h(1)$ presents one-electron part of the operator while $J_a^{\alpha}(1)$ ($J_a^{\beta}(1)$) describes two-electron Coulomb interaction of a selected α electron with all other α (β) electrons and $K_a^{\alpha}(1)$ corresponds to exchange of the selected electron with all other α ones. Similar notations related to a selected β electron compose the bottom Eq. 9. The RHF solution is governed by the Fock operator in the form

$$f(1) = h(1) + \sum_a^{\frac{N}{2}} [J_a(1) - K_a(1). \quad (10)$$

Accordingly, the difference operators $\Delta f^{\alpha}(1) = f^{\alpha}(1) - f(1)$ and $\Delta f^{\beta}(1) = f^{\beta}(1) - f(1)$ in the case when $N^{\alpha} = N^{\beta} = \frac{N}{2}$ can be expressed as $\Delta f^{\alpha} = \sum_a^{N^{\beta}} J_a^{\beta}(1)$ and $\Delta f^{\beta} = \sum_a^{N^{\alpha}} J_a^{\alpha}(1)$, respectively. Consequently, the difference between RHF and UHF formalism is governed by spin-nondiagonal elements of Coulomb interaction (more detailed expressions of the above operators are given in [93, 94]).

In the relativistic theory of many-electron systems the Hamiltonian, $\hat{H}_{DCB}$, is composed of the one-electron Dirac Hamiltonians and two-electron terms describing Coulomb and Breit interactions [70]. This Hamiltonian contains a complete description of the spin-orbit interaction. In computational implementations the operator can be transformed to the Dirac-Fock operator that has the same structure as its nonrelativistic UHF analogue

$$\hat{f}_{DHF}(i)\phi(i) = \varepsilon_i \phi(i). \quad (11)$$

Here, $\phi(i)$ is the four–component-spinor representation of the one–electron state.

As shown recently [68], not only Fock-like Eq. 11 itself is analogous to the UHF one, but its solution can be presented in the UHF terms, such as spin contamination, expressed via ΔS^2, and spin density over molecule atoms. Related to C_6H_5O, ΔS^2 constitutes 0.6024, exactly the same for both UHF and general complex Hartree-Fock (GCHF) method. The GCHF spin contamination was provided with the Kramers pair symmetry breaking. Besides spin contamination, the GCHF solution involves the spin density distribution over the molecule atoms, which is presented in Fig. 8. As seen in the figure, the distribution patterns are fully identical

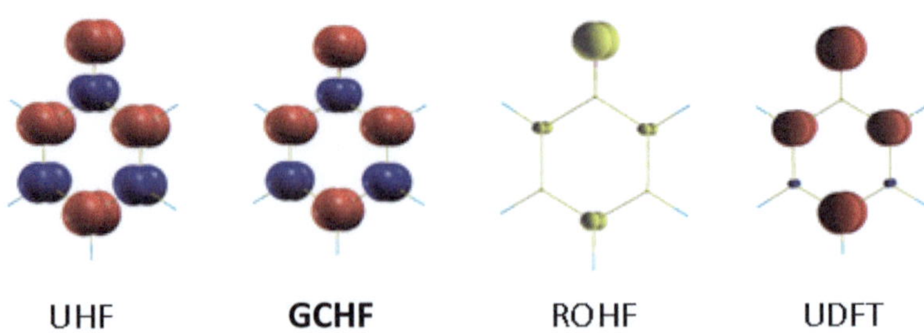

Fig. 8 Spin density distribution over atoms of phenoxyl radical (C_6H_5O) calculated by using different formalisms [68] (by kind permission of L. Buĉinský)

in the case of UHF and GCHF formalisms while considerably differ from both ROHF and UDFT ones [68]. These results alongside with comments given at the end of Sect. 2 give good reason to consider UDFT formalism as a proper technique for treating spin effects in open-shell molecules with great care. At the same time, the results have convincingly shown that the origin of the discussed spin effects is actually common for both relativistic and nonrelativistic description and lies in two-electron spin-nondiagonal elements of the Coulomb interaction.

7 UHF SOC Parameters of sp^2 Nanocarbons

7.1 A Confine Collection of Necessary Relations. Parameters' Formalism

Traditionally, a set of standard SOC parameters has involved two values subjected to experimental verification, namely, energy splitting ΔE^{SO}_{spl} and the rate of inter-system crossing (related to singlet-triplet and vice versa transitions mainly) k_{ISC} as well as theoretically introduced SOC constant a^{SO}. In practice all the three values are determined not by solving either Dirac-Coulomb-Breit or Dirac-Fock, Eq. 11, equations but getting first the relevant nonrelativistic problem solution after which considering H^{SO} Hamiltonian mainly perturbationally, Eq. 7 [70, 74]. An effective Hamiltonian $\hat{H}^{SO}_{eff}$ is commonly considered.

One of the simplest and least demanding approaches is to take the two-electron contributions to the SOC into account through screening the nuclear potential:

$$\widehat{H}^{SO}_{eff} = \frac{1}{2m^2c^2}\sum_I \sum_i \frac{Z^{eff}_{i,l}}{\hat{r}^3_{I_{il}}} \hat{l}_{Iil} \cdot \hat{s}_{il}. \tag{12}$$

In this one-electron one-center spin-orbit operator, I denotes an atom and il an electron occupying an orbital located at center I. Likewise, $\hat{l}_{Iil}$ and $\widehat{s}_{Iil}$ label the angular momentum and spin of electron il with respect to the orbital origin at atom I. The summation over electrons includes only the open shell of an atom I with azimuthal quantum number l.

Supposing that electrons are moving in central field and substituting the fraction under the sum by the relevant static potential U, one gets

$$\widehat{H}^{SO}_{eff} = \frac{1}{2m^2c^2}\sum_I \sum_i \frac{1}{r_{il}} \frac{\partial U_I(r_i)}{\partial r_{il}} \hat{l}_{Iil} \cdot \widehat{s}_{il} \tag{13}$$

Averaging $\widehat{H}_{eff}^{SO}$ over the total angular momentum $\hat{L}_I = \sum_i \hat{l}_i$ and supposing that the self-consistent potential $U_I(r_i)$ of the ith electron in the Ith atomic center is spherically symmetric, we get

$$\widehat{H}_{av}^{SO} \approx \frac{1}{2m^2c^2}\sum\nolimits_I \overline{\left\{\frac{1}{r_I}\left(\frac{d}{dr_I}U_I(r_I)\right)\right\}}.\hat{L}_I.\hat{S}_I \tag{14}$$

Further approximation results in

$$\widehat{H}_{av}^{SO} = \sum\nolimits_I a^{SO}(\hat{L}_I.\hat{S}_I) \tag{15}$$

due to which a^{SO} has the form [89]

$$a^{SO} \approx \overline{\left\{\frac{1}{r_I}\left(\frac{d}{dr_I}U_I(r_I)\right)\right\}}. \tag{16}$$

Important for the following are two more approximate expressions. The first is related to eigenvalues of Eq. 8 that can be written in the form

$$\varepsilon_{SO} = a^{SO}.\frac{1}{2}\{J(J+1) - L(L+1) - S(S+1)\}. \tag{17}$$

The normal ε_{SO} term are related to the minimum projection of the total momentum J. The second presents a well known Lande interval rule

$$\varepsilon_{SO}(J) - \varepsilon_{SO}(J-1) = a^{SO}J. \tag{18}$$

According to Eqs. 16 and 18, the constant a^{SO} can be determined by two ways: the first case concerns the evaluation of the potential gradient while the second addresses the determination of the energy splitting related to particular spinorbitals.

7.2 UHF-Based Determination of ΔE_{spl}^{SO} and a^{SO} SOC Parameters

The best way to illustrate the ability of the UHF formalism in determining a^{SO} constant following the first way is to turn to the bond dissociation and/or breaking. Evidently, if the dissociation concerns the only bond in the molecule r_i in Eq. 16 is just the bond length R and potential $U_I(r_i)$ can be substituted with the molecule total energy $E(R)$. Main UHF characteristics, which accompany stretching the ethylene C–C bond up to 2 Å, are shown in Fig. 9a. Equilibrium C–C distance constitutes 1.326 Å and 1.415 Å in singlet and triplet states of the molecule, respectively. As seen in the figure, as stretching of the bond increases, energies

$E_{sg}(R)$ and $E_{tr}(R)$ approach each other up to quasidegeneracy, which is characteristic for biradicals and which is necessary for an effective SOC [70]. Simultaneously, the number of effectively unpaired electrons N_D, which is zero until C–C distance reaches R_{crit}, starts to grow manifesting a gradual radicalization of the molecule as the bond is stretched as well as exhibiting the transformation of the molecule behavior from closed-shell to open-shell one when R_{crit} is overstepped.

As follows from Eq. 16, a^{SO} is determined by the force acting on the bond under stretching. The deviation of the force under progressive stretching the ethylene C–C bond is presented in Fig. 9b. At the beginning it proceeds linearly starting, however, to slow down when R is approaching R_{crit}. A clearly seen kink is vivid in the region. For comparison, the curve with horizontal bars presents the force caused by stretching a single C–C bond of ethane, R_{crit} for which constitutes 2.11 Å [43], that

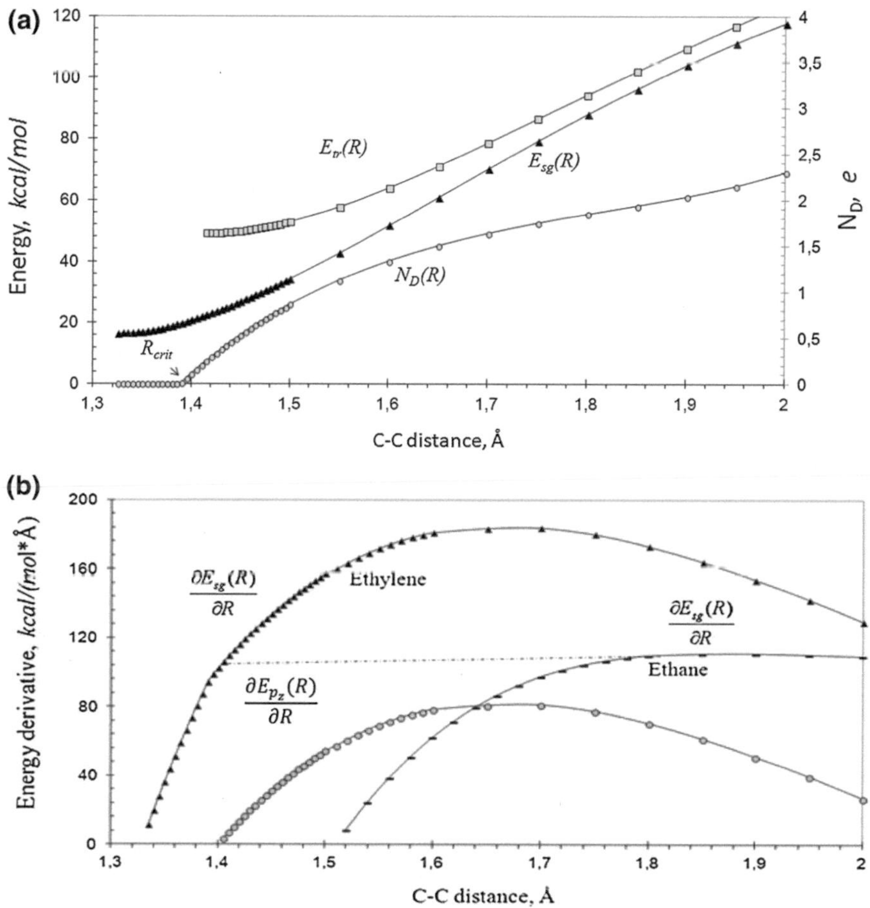

Fig. 9 **a** Energy of singlet and triplet states as well as the total number of effectively unpaired electrons N_D of ethylene versus C–C distance. **b** Energy derivatives versus C–C distance (see text) (UHF calculations)

is why the molecule remains closed-shell one within the interval of C–C distances presented in the figure. As seen, the force is saturated at the level of 110 kcal/(mol * Å) that is close to the kink position on the ethylene curve. It is quite reasonable to suggest that the force excession over this value in the latter case is caused by the closed-open shell transformation of the ethylene molecule over 1.4 Å due to which the force excess, revealed by UHF calculations, can be attributed to the SOC related to p_z electrons. Consequently, this excess can be considered as $\frac{dE_{pz}(R)}{dR}$ that is presented by the gray-ball curve in the figure. Using the curve values in the C–C distance interval from 1.40 to 1.47 Å, which is typical for the C–C bond length dispersion in fullerenes, CNTs, and graphene, and substituting them into Eq. 16, one can obtain the a^{SO} constant laying in the interval from 15 to 110 meV, which is typically expected for molecules of light elements [70]. It is necessary to point out as well that the $\frac{dE_{pz}(R)}{dR}$ force maximum amplitude for ethylene molecule is well consistent with those determined under uniaxial deformation of benzene molecule and graphene [60] due to which the outlined a^{SO} constant values can be considered as typical for the whole family of sp^2 nanocarbons.

8 Conclusion

The paper presents for consideration two new conceptual issues. The first concerns the attribution of the UHF-originated peculiarities of sp^2 nanocarbon open-shell molecules, such as spin contamination of the ground state, depraving the latter from the exact spin multiplicity, and effectively unpaired electrons, to particular spin effects characteristic for the species. The peculiarities are proven to be directly associated with the molecule behavior in reality. The second issue addresses the similarity of these peculiarities consideration that can be on the basis of either non-relativistic unrestricted Hartree-Fock formalism or relativistic Dirac-Fock spin-orbit theory obtained $H^{UHF} = H^{RSD} + \Delta E_{corr}^{UHF}$ and $H^{rel} = H^{RSD} + \Delta E_{SO}$ (H^{RSD} presents a reference restricted single-determinant Hamiltonian) in the form, respectively. This can give a new view of the identity of the discussed peculiarities related to C_6H_5O molecule obtained by the UHF and one of the versions of the Dirac SOC theory [68], as well as the synergistic ability of the SOC theory and UHF to describe molecular magnetism of sp^2 OSMs. Deeply-rooted inherent connection of non-relativistic and relativistic spin effects opens a new vision of complexity of electronic states of many-electron systems.

Acknowledgements The author greatly appreciate fruitful and stimulating discussions with V. Sheka, E. Rashba, J. Karwowski, R. Hoffmann, I. Mayer, E. Orlenko, P. D'yachkov, S. I. Vinitski, A. Gusev, Yu. P. Rybakov, E. Brändas, M. Nacimento, and D. Mukherjee. A particular gratitude to L. Gross, L. Buchinsky and S.V. Demishev for kind permition to use experimental data at own discretion. The work was performed under financial support of the Peoples' Friendship University of Russia, grant: 022,203-0-000.

References

1. Pople JA, Nesbet RK (1954) J Chem Phys 22:571–572
2. Bendikov M, Duong HM, Starkey K, Houk KN, Carter EA, Wudl F (2004) J Am Chem Soc 126:7416–7417
3. Sheka EF (2003) Lecture notes in computer science, vol 2658. In: Sloot PMA, Abramson D, Bogdanov AV, Dongarra J, Zomaya AY, Gorbachev YE (eds) Computational science—ICCS 2003, Part II. Springer, Berlin, pp 386–403
4. Sheka EF (2004) Centr Europ J Phys 2:160–182
5. Sheka EF (2004) Int J Quant Chem 100:375–387
6. Hachmann J, Dorando JJ, Avilés M, Chan GK-L (2007) J Chem Phys 127:134309
7. Jiang D, Dai S (2008) J Phys Chem A 112:332–335
8. Qu Z, Zhang D, Liu C, Jiang Y (2009) J Phys Chem A 113:7909–7914
9. Hajgató B, Szieberth D, Geerlings P, De Proft F, Deleuze MS (2009) J Chem Phys 131:224321
10. Bettinger HF (2010) Pure Appl Chem 82:905–915
11. Rayne S, Forest K (2011) Comput Theor Chem 976:105–112
12. Gao X, Hodgson JL, Jiang D, Zhang SB, Nagase S, Miller GP, Chen Z (2011) Org Lett 13:3316–3319
13. San-Fabián E, Moscardó F (2011) Eur Phys J 64:239–248
14. Rivero P, Jiménez-Hoyos CA, Scuseria GE (2013) J Phys Chem B 117:12750–12758
15. Torres AE, Guadarrama P, Fomine S (2014) J Mol Model 20:2208
16. Blanquart G (2015) Int J Quant Chem 115:796–801
17. Ibeji CU, Ghosh D (2015) Phys Chem Chem Phys 17:9849–9856
18. Sheka EF (2003)Fullerenes as polyradicals. Paper 54 presented at the internet electronic conference of molecular design. http://www.biochempress.com
19. Sheka EF, Zayets VA (2005) Russ J Phys Chem 79(12):2009–2014
20. Sheka EF (2006) J Str Chem 47:593–599
21. Sheka EF (2007) Int J Quant Chem 107:2803–2816
22. Sheka EF (2011) Fullerenes: Nanochemistry, Nanomagnetism, Nanomedicine, Nanophotonics, CRC Press. Taylor and Francis, Boca Raton
23. Sheka EF (2011) In: Bai C, Awadelkarim OO, Kapitza S (eds) Encyclopedia of life support systems(EOLSS): Nanoscience and Nanotechnologies (Kharkin V. UNESCO EOLSS, Oxford, pp 415–444
24. Stück D, Baker TA, Zimmerman Pl, Kurlancheek W (2011) J Chem Phys 135:194306
25. Jiménez-Hoyos CA, Rodríguez-Guzmán R, Scuseria GE (2014) J Phys Chem A 118:9925–9940
26. Sheka EF, Chernozatonskii LA (2007) J Phys Chem C 111:10771–10780
27. Chen Z, Jiang D, Lu X, Bettinger HF, Dai S, Schleyer PR, Houk KN (2007) Org Lett 9:5449–5452
28. Sheka EF, Chernozatonskii LA (2010) Int J Quant Chem 110:1466–1480
29. Sheka EF, Chernozatonskii LA (2009) arXiv:0901.3624v1
30. Jiang D, Sumpter B, Dai S (2007) J Chem Phys 126:134701
31. Sheka EF, Chernozatonskii LA (2010) Int J Quant Chem 110:1938–1946
32. Sheka EF, Chernozatonskii LA (2010) J Exp Theor Phys 110:121–132
33. Nagai H, Nakano M, Yoneda K, Kishi R, Takahashi H, Shimizu A, Kubo T, Kamada K, Ohta K, Botek E, Champagne B (2010) Chem Phys Lett 489:212–218
34. Morita Y, Suzuki S, Sato K, Takui T (2011) Nat Chem 3:197–204
35. Sheka EF (2012) Int J Quant Chem 112:3076–3090
36. Ang LS, Sulaiman S, Mohamed-Ibragim MI (2012) Sains Malaysiana 41:445–452
37. Sheka EF (2013) Progress in theoretical chemistry and physics, vol 27. In: Hotokka M, Brändas EJ, Maruani J, Delgado-Barrio G (eds) Advances in quantum methods and applications in chemistry, physics, and biology. Springer, Berlin, pp 249–284

38. Mizukami W, Kurashige Y, Yanai T (2013) J Chem Theor Comput 9:401–407
39. Plasser F, Pašalić H, Gerzabek MH, Libisch F, Reiter R, Burgdörfer J, Müller T, Shepard R, Lischka H (2013) Ang Chem. Int Ed 52:2581–2584
40. Sheka EF (2014) Int J Quant Chem 114:1079–1095
41. Horn S, Plasser F, Müller T, Libisch F, Burgdörfer J, Lischka H (2014) Theor Chem Acc 133:1511
42. Wu C-S, Chai J-D (2015) J Chem Theory Comput 11:2003–2011
43. Sheka EF (2015) Adv Quant Chem 70:111–161
44. Fradera X, Sola M (2002) J Comput Chem 23:1347–1356
45. Szabo A, Ostlund NS (1996) Modern Quantum Chemistry: Introduction to Advanced Electronic Structure Theory. McGraw-Hill, Toronto, Lancheaster, p 467
46. Kitagawa Y, Saito T, Nakanishi Y, Kataoka Y, Matsui T, Kawakami T, Okumura M, Yamaguchi K (2009) J Phys Chem A 113:15041–15046
47. Lewars E (2011) Computational chemistry. introduction to the theory and applications of molecular and quantum mechanics, 2nd edn. Springer, Heidelberg, pp 535
48. Dehareng D, Dive G (2000) Int J Compt Chem 21:483–504
49. Cizek J, Paldus J (1967) J Chem Phys 47:3976–3985
50. Hubač I, Čàrski P (1983) Int J Quant Chem 24:141–148
51. Kaplan I (2007) Int J Quant Chem 107:2595–603
52. Jacob CR, Reiher M (2012) Int J Quant Chem 112:3661–3684
53. K, Fueno T, Yamaguchi K (1978) Theor Chim Acta 48:175–83
54. Zayets VA (1990) CLUSTER-Z1: quantum-chemical software for calculations in the s, p-basis. Institute of Surface Chemistry National Academy of Sciences, Ukraine, Kiev
55. Staroverov VN, Davidson ER (2000) Chem Phys Lett 330:161–168
56. Gross L, Mohn F, Moll N, Liljeroth P, Meyer G (2009) Science 325:1110–1114
57. Smallest possible five-ringed structure made: 'Olympicene' molecule built using clever synthetic organic chemistry.www.sciencedaily.com/releases/2012/05/120528100253.htm
58. Levy N, Burke SA, Meaker KL, Panlasigui M, Zettl A, Guinea F, Castro Neto AH, Crommie MF (2010) Science, 329(5991):544–547
59. Georgiou T, Britnell L, Blake P, Gorbachev RV, Gholinia A, Geim AK, Casiraghi C, Novoselov KS (2011) Appl Phys Let 99:093103
60. Sheka EF, Popova NA, Popova VA, Nikitina EA, Shaymardanova LKh (2011) J Mol Mod 17:1121–1131
61. Sheka EF, Popova VA, Popova NA (2013) Progress in theoretical chemistry and physics, vol 27. In: Hotokka M, Brändas EJ, Maruani J, Delgado-Barrio G (eds) Advances in quantum methods and applications in chemistry, physics, and biology. Springer, Berlin, pp 285–299
62. Kim K-J, Blanter YM, Ahn K-H (2011) Phys Rev B 84:081401(R)
63. Mañes JL (2007) Phys Rev B 76:045430
64. Pentsak EO, Kashin AS, Polynski MV, Kvashnina KO, Glatzel P, Ananikov VP (2015) Chem Sci 6:3302–3313
65. Zhang Y, Fu Q, Cui Y, Mu R, Jin L, Bao X (2013) Phys Chem Chem Phys 15:19042–19048
66. Osváth Z, Deák A, Kertész K, Molnár Gy, Vértesy G, Zámbó D, Hwang C, Biró LP (2015) Nanoscale 7:5503–5509
67. Löwdin P-O (1958) Adv Chem Phys 2:209–322
68. Bučinský L, Malček M, Biskupič S, Jayatilaka D, Büchel GE, Arion VB (2015) Comp Theor Chem 1065:27–41
69. Marian CM (2001) In: Lipkowitz KB, Boyd DB (eds) Reviews in computational chemistry, vol 17, pp 99–204
70. Iliaš M, Kellö V, Urban M (2010) Acta Phys Slov 60:259–391
71. Mück LA (2012) Highly accurate quantum chemistry: spin-orbit splittings via multireference coupled-cluster methods and applications in heavy-atom main-group chemistry. DN dissertation at Johannes Gutenberg-Universität in Mainz
72. Möck LA, Gauss J (2012) J Chem Phys 136:111103
73. Marian CM (2012) WIREs Comput Mol Sci 2:187–203

74. Reiher M, Wolf A (2014) Relativistic quantum chemistry: the fundamental theory of molecular science, 2nd edn, Willey
75. Epifanovsky E, Klein K, Stopkowicz S, Gauss J, Krylov AI (2015) J Phys Chem 143:064102
76. Magarill LI, Chaplik AV (1999) J Exp Theor Phys 88:815–818
77. Orlandi G, Negri F (2002) Photochem Photobiol Sci 1:289–308
78. Razbirin BS, Rozhkova NN, Sheka EF, Nelson DK, Starukhin AN (2014) J Exp Theor Phys 118:735–746
79. Sheka EF (2011) Nanosci Nanotech Let 3:28–33
80. SáFar GAM, Gontijo RN, Fantini C, Martins DCS (2015) Idemori im, Pinheiro MVB, Krambrock K. J Phys Chem C 119:4344–4350
81. Liu H, Ryu S, Chen Z, Steigerwald ML, Nuckolls C, Brus LE (2009) J Am Chem Soc 131:17099–17101
82. Belousova IM, Videnichev DA, Kislyakov IM, Krisko TM, Rozhkova NN, Rozhkov S (2015) Opt Mat Express 5:169–175
83. Mayer I (1986) Int. J. Quant. Chem 29:73–84
84. Demishev SV, Pronin AA, Sluchanko NE, Weckhuysen L, Moshchalkov VV, Spitsina NG, Yagubskii ÉB (1999) JETP Lett 69:785–791
85. Demishev SV, Sluchanko NE, Weckhuysen L, Moshchalkov VV, Ohta H, Okubo S, Oshima Y, Spitsina NG (2002) Phys Sol State 44:441–443
86. Mayer I (2007) Chem Phys Let 440:357–359
87. Mayer I (2012) Chem Phys Let 539–540:172–174
88. Orlenko EV, Sheka EF, Orlenko FE (2016) Eur Phys J D 70:59
89. Sheka EF, Orlenko EV (2015) arXiv 1511.04327 [cond-mat. mes-hall]
90. Johnson RD, Yannoni CS, Dorn HC, Salem JR, Bethune DS (1992) Science 255:1235–1238
91. Tada K, Haruyama J, Yang HX, Chshiev M, Matsui T, Fukuyama H (2011) Phys Rev Lett 107:217203
92. Sheka EF (2013) Carbon materials: chemistry and physics, vol 7. In: Ashrafi AR, Cataldo F, Iranmanesh A, Ori O (eds) Topological modelling of nanostructures and extended systems. Springer, Berlin, pp 137–197
93. Rossky PJ, Karplus M (1980) J Chem Phys 73:6196–6214
94. Hubač I, Čàrski P (1980) Phys Rev A 22:2392–2399
95. Havriliak SJ, Yarkony DR (1985) J Chem Phys 83:1168–1172
96. I.Hubač, P. Čàrski Int J Quant Chem 24:141–148
97. Steele GA, Pei F, Laird EA, Jol JM, Meerwaldt HB, Kouwenhoven LP (2013) Nat Commun 4:1573
98. Marchenko D, Varykhalov A, Scholz MR, Bihlmayer G, Rashba EI, Rybkin A, Shikin AM, Rader O (2012) Nature Commun 3:1232

Population Analyses Based on Ionic Partition of Overlap Distributions

Ya Kun Chen and Yan Alexander Wang

Abstract Borrowing ideas from Hess's law in thermodynamics and the holographic electron density theorem, we have proposed two novel population analyses based on ionic partition of overlap distributions (IPOD). The IPOD population analyses combine the conceptual simplicity of the Mulliken and Löwdin population analyses with low computational cost and mild basis-set variation effect. More promisingly, the resultant IPOD charges are of the similar quality to the natural population analysis.

Keywords Population analysis · Partial charge · Mulliken population analysis · Löwdin population analysis · Christoffersen-Baker population analysis · Hirshfeld population analysis · Natural population analysis

1 Introduction

According to the basic theorems of the density functional theory (DFT) [1], the electron density, $\rho(\mathbf{r})$, a function of spatial coordinates, determines the wave function and therefore all the properties of an entire molecule. On the one hand, the explicit knowledge of the electron density provides complete information to understand the molecular system. On the other hand, such complete information is often unnecessary for calculating many properties that can be described simply by atomic charges. For example, evaluation of the multipoles of a molecular system only requires a set of compact information including proper atomic charges and nuclear locations. For such purposes, several population analyses [2–21] have been invented to estimate partial atomic charges in chemical systems.

One class of population analyses partition the entire molecular wave function into atomic contributions and regard the total atomic charge as the difference between the nuclear charge and the sum of the electron numbers over all basis functions

Y.K. Chen · Y.A. Wang (✉)
Department of Chemistry, University of British Columbia,
Vancouver, BC V6T 1Z1, Canada
e-mail: yawang@chem.ubc.ca

A. Tadjer et al. (eds.), *Quantum Systems in Physics, Chemistry, and Biology*,
Progress in Theoretical Chemistry and Physics 30,
DOI 10.1007/978-3-319-50255-7_4

centered on each constituent atom in a molecule. The most famous of all is the Mulliken population analysis (MPA), which directly assigns atomic charges according to the original nonorthogonal atomic orbitals [2, 3]. In so doing, some of the orbital populations inevitably exceed two or become negative [2–9]. To avoid such a deficiency, the Löwdin population analysis (LPA) first transforms the original nonorthogonal atomic orbitals into a set of orthogonal atomic orbitals through symmetrical orthogonalization and then performs the MPA on this new set of orthogonal atomic orbitals [4, 5].

In both of these two population analyses, the two-center (off-diagonal) overlap contributions to the electron density are divided equally between the two atoms involved. Obviously, such an equal partition of the off-diagonal overlap distributions is unphysical because non-equivalent atoms, either different atoms or the same atom at different electronic states, do have different electronegativities to attract the shared electrons in chemical bonds, thus resulting in uneven shares of the off-diagonal contributions to the electron density.

Though with many drawbacks, the MPA and LPA, for their conceptual simplicity and superior computational speed, are still extremely popular and widely used especially for large molecular systems. In many quantum chemistry software packages that use atomic centered basis functions, the MPA and LPA are usually calculated by default. However, the unphysical partition of the off-diagonal electron distributions and the strong basis-set dependence prohibit these two population analyses from predicting accurate atomic charges. In addition, the non-pre-orthogonalized LPA suffers serious unphysical rotational variance when pure spherical harmonic basis functions are not used [5].

In 1971, an early attempt to remedy some of the major conceptual deficiencies of the MPA and LPA was published by Christoffersen and Baker [6], who devised a simple scheme of only using the squares of the molecular orbital (MO) expansion coefficients of the nonorthogonal atomic basis functions to proportionally partition the total number of electrons into individual atomic centers. Edgecombe and Boyd later further modified the original Christoffersen-Baker population analysis (CBPA), but the quality of the resulting atomic charges is still similar to that of the MPA [6, 7].

From a different direction, in the pursuit of improving the MPA and LPA, Reed et al. utilized the concept of the natural orbitals and developed the natural population analysis (NPA) [8], in which occupancy-weighted symmetrical orthogonalization technique is employed to construct a set of orthonormal natural atomic orbitals and their corresponding occupation numbers are then used to compute atomic charges. Over the years, the NPA has gradually gained popularity and has been widely applied to analyze the intra- and inter-molecular interactions and the aromaticity of molecules [9, 10]. Despite that the NPA circumvents many shortcomings of the MPA and LPA and often demonstrates good convergence behavior upon basis-set change, it still confronts severe basis-set effects in the valence/Rydberg orbital space for transition-metal complexes [11].

In comparison, the topology of the electron density can also be used to partition the density into atomic contributions. The atoms in molecules (AIM) scheme developed by Bader uses the zero-flux surfaces, whose normal vectors are perpendicular to the gradient of the electron density, to divide space into rigid atomic domains [12, 13]. Such a method offers a unique, unambiguous way to partition the entire space for constituent atoms in a molecule.

Alternatively, based on the aspherical atom model, which has been proposed to minimize the error in reproducing the data obtained from X-ray diffraction [18–20], the Hirshfeld population analysis (HPA) partitions the electron density at each point in space in such a way that the superimposed atomic densities yield zero net charges [13–18]. The merit of the HPA is that it retains more information when atoms come together to form a molecule [18, 21]. Specifically, it can ensure the condensed Fukui function indices non-negative, which is requisite for a realistic system [21].

However, both the AIM and HPA have to explicitly manipulate the electron density at every grid point in space and hence are much more time-consuming and expensive in terms of computational cost, in comparison with other population analyses (e.g., MPA, LPA, CBPA, and NPA) without employing a numerical grid.

Hereafter, based on ideas of ionic partition of overlap distributions (IPOD), we will develop two novel population analyses, which possess the conceptual simplicity of the MPA, CBPA, and LPA and rival the quality of the NPA partial charges.

2 IPOD Population Analyses

The development of the IPOD schemes is based on the analysis of the decomposition of the molecular electron density, similar to the MPA, CBPA, and LPA.

Let us first introduce a set of atomic-centered basis functions $\{\phi_\mu^A(\mathbf{r})\}$, where the Greek subscript μ and the Roman superscript A are the indices for atomic basis functions and constituent atoms, respectively. After a self-consistent field calculation is finished, the ith molecular orbital (MO), ψ_i^{MO}, can be expressed as a linear combination of atomic basis functions:

$$\psi_i^{\mathrm{MO}}(\mathbf{r}) = \sum_{\mu}^{\mathrm{AO}} \sum_{A}^{\mathrm{atom}} C_{i\mu}^A \phi_\mu^A(\mathbf{r}), \tag{1}$$

where $\{C_{i\mu}^A\}$ are the MO expansion coefficients. Then, the electron density of ψ_i^{MO} is simply written as

$$\rho_i^{\mathrm{MO}}(\mathbf{r}) = f_i \left|\psi_i^{\mathrm{MO}}(\mathbf{r})\right|^2, \tag{2}$$

where f_i is the occupation number of ψ_i^{MO}. Summing up the electron densities of all molecular orbitals, we obtain the total electron density $\rho_{\mathrm{tot}}(\mathbf{r})$,

$$\rho_{\text{tot}}(\mathbf{r}) = \sum_{i}^{\text{MO}} \rho_i^{\text{MO}}(\mathbf{r}), \tag{3}$$

which is normalized to the total number of electrons in the system, $n_{\text{tot}} = \langle \rho_{\text{tot}}(\mathbf{r}) \rangle$. Plugging Eqs. (1) and (2) into Eq. (3), one arrives at

$$\rho_{\text{tot}}(\mathbf{r}) = \sum_{\mu,\nu}^{\text{AO}} \sum_{A,B}^{\text{atom}} d_{\mu\nu}^{AB} \phi_{\mu}^{A*} \phi_{\nu}^{B}, \tag{4}$$

where $d_{\mu\nu}^{AB} = \sum_i f_i C_{i\mu}^{A*} C_{i\nu}^{B}$ is the density matrix element and the asterisk indicates the complex conjugate.

In Eq. (4), the double summation over the atomic centers can be regrouped into the squared terms on the same center and the cross terms between two different centers:

$$\rho_{\text{tot}}(\mathbf{r}) = \underbrace{\sum_{\mu,\nu}^{\text{AO}} \sum_{A}^{\text{atom}} d_{\mu\nu}^{AA} \phi_{\mu}^{A*} \phi_{\nu}^{A}}_{\rho_{\text{tot}}^{\text{ionic}}(\mathbf{r})} + \underbrace{2 \sum_{\mu,\nu}^{\text{AO}} \sum_{A>B}^{\text{atom}} d_{\mu\nu}^{AB} \phi_{\mu}^{A*} \phi_{\nu}^{B}}_{\rho_{\text{tot}}^{\text{covalent}}(\mathbf{r})}. \tag{5}$$

On the right-hand side of the above equation, the electron density of the first squared terms clearly originates from single center contributions, characterizing the *ionic* component of the total electron density. The electron density of the second cross terms is due to the overlap distributions between distinct atomic pairs, defining the *covalent* component of the total electron density.

Up to this point, our derivation is still in line with the MPA: after assuming the covalent electron density of every atomic pair to be equally partitioned between the two atomic centers involved, we can again recover the MPA. As already criticized above, such an unphysical partition scheme, however, is one of the major weakness of the MPA. To overcome this conceptual difficulty, we thus dig deeper into the expression of the electron density and seek other strategies to partition the covalent electron density instead.

The first term on the right-hand side of Eq. (5), the total ionic density $\rho_{\text{tot}}^{\text{ionic}}(\mathbf{r})$, can be written as a sum over the ionic densities from different atomic centers:

$$\rho_{\text{tot}}^{\text{ionic}}(\mathbf{r}) = \sum_{A}^{\text{atom}} \rho_A^{\text{ionic}}(\mathbf{r}), \tag{6}$$

where $\rho_A^{\text{ionic}}(\mathbf{r})$ is the ionic electron density on atom A:

$$\rho_A^{\text{ionic}}(\mathbf{r}) = \sum_{\mu,\nu}^{\text{AO}} d_{\mu\nu}^{AA} \phi_{\mu}^{A*} \phi_{\nu}^{A}. \tag{7}$$

Then, the number of ionic electrons on atom A is the integral of $\rho_A^{\text{ionic}}(\mathbf{r})$ over the entire space:

$$n_A^{\text{ionic}} = \sum_{\mu,\nu}^{\text{AO}} d_{\mu\nu}^{AA} S_{\mu\nu}^{AA}, \tag{8}$$

where $S_{\mu\nu}^{AA} = \langle \phi_\mu^A | \phi_\nu^A \rangle$ is the overlap matrix element confined for atom A. The total number of ionic electrons is a simple sum of n_A^{ionic} over all atomic centers in the molecule,

$$n_{\text{tot}}^{\text{ionic}} = \sum_{A}^{\text{atom}} n_A^{\text{ionic}}. \tag{9}$$

In retrospective, the CBPA [6, 7] simply excludes the cross terms in Eq. (7), such that the ionic electron density on atom A is solely determined by the sum of the *pure* squared terms, $d_{\mu\mu}^{AA} |\phi_\mu^A|^2$. Evidently, such a drastic simplification is theoretically less sound.

Before partitioning the total covalent electron density, $\rho_{\text{tot}}^{\text{covalent}}(\mathbf{r})$, we can recast it in terms of distinct atomic pair covalent densities:

$$\rho_{\text{tot}}^{\text{covalent}}(\mathbf{r}) = \sum_{A>B}^{\text{atom}} \rho_{AB}^{\text{covalent}}(\mathbf{r}), \tag{10}$$

where $\rho_{AB}^{\text{covalent}}(\mathbf{r})$ is the covalent density for atomic pair $\{AB\}$ $(A \neq B)$,

$$\rho_{AB}^{\text{covalent}}(\mathbf{r}) = 2 \sum_{\mu,\nu}^{\text{AO}} d_{\mu\nu}^{AB} \phi_\mu^{A*} \phi_\nu^B, \tag{11}$$

whose integral over the entire space yields the number of covalent electrons for atomic pair $\{AB\}$,

$$n_{AB}^{\text{covalent}} = 2 \sum_{\mu,\nu}^{\text{AO}} d_{\mu\nu}^{AB} S_{\mu\nu}^{AB}, \tag{12}$$

with $S_{\mu\nu}^{AB} = \langle \phi_\mu^A | \phi_\nu^B \rangle$. The total number of covalent electrons is then a sum over distinct atomic pairs,

$$n_{\text{tot}}^{\text{covalent}} = \sum_{A>B}^{\text{atom}} n_{AB}^{\text{covalent}}. \tag{13}$$

To this point, we have successfully decomposed the molecular density into the ionic and covalent contributions and are ready to invoke the IPOD procedure to partition the covalent density according to the ionic distribution. Conceptually speaking, the task of partitioning the covalent electron density between any pair of atomic centers must consider electronegativities of constituent atoms within a molecular environment. This central issue has not been handled properly in the MPA and HPA.

We thus decide to take a different route to tackle this problem by envisioning a mind experiment of filling electrons into a molecule according to Hess's law in thermodynamics. First, we only allow ionic electrons to fall into their places at various atomic centers within a molecule. Of course, how many ionic electrons will reside at which atomic center must be a direct measure of the *transient* electronegativity of that atom within the particular molecular environment before the chemical potential equilibrates throughout the entire molecular structure. Second, we permit the rest covalent electrons to occupy the remaining space in chemical bonds. Through this two-step electron filling process, we obtain a reasonable estimate of the relative transient electronegativities of constituent atoms within a molecular environment, which will enable us to partition the covalent density distributions in a more meaningful way.

In the end, the IPOD population analyses should not come as a total surprise, because they somehow root in the holographic electron density theorem (HEDT) [22]. The HEDT states that given a known electron density function in a finite region of a Coulomb system, the density functions in other subdomains and so as to the entire electron density can be uniquely determined [22]. Even the CBPA and HPA can also be developed by applying this theorem [23]. However, the HPA has its own limitations: it uses the densities of the free ground-state atoms as references [13–18], which are apparently unphysical for charged and excited atoms in molecular systems. In the CBPA [6, 7], the total ionic electron density is underestimated (*vide ante*).

According to the HEDT, the ionic density of each constituent atom, which retains some atomic feature for each atom in a molecule, can be used to recover the overall electron density in a molecule. Consequently, the covalent density distribution should follow the ionic density distribution. The IPOD methods, therefore, use the distribution of the ionic density to determine the distribution of the entire electron density among all constituent atoms.

Particularly, instead of the equal partition scheme exerted by the MPA, the IPOD scheme of the first kind (IPOD1) partitions the total covalent electron number $n_{\text{tot}}^{\text{covalent}}$ in proportion to the ionic atomic charge distribution. As a result, the electron number on atom A is also proportional to the ionic electron number on the same atom, leading to the IPOD1 value for atomic charge q_A:

$$q_A^{\text{IPOD1}} = Z_A - \left(\frac{n_A^{\text{ionic}}}{n_{\text{tot}}^{\text{ionic}}}\right) n_{\text{tot}}, \tag{14}$$

where Z_A is the nuclear charge of atom A. Equation (14) is virtually identical to that used in the CBPA [6, 7], except that the ionic electron number was estimated differently (*vide supra*).

Different from IPOD1, which globally partitions the total electron number directly proportional to the ionic electron number on each atom, the IPOD scheme of the second kind (IPOD2) partitions the covalent electron number of every distinct atomic pair proportional to the ionic electron numbers of the two atomic centers involved. Specifically, the covalent electron number on atom A within an atomic pair $\{AB\}$, $n_{A(B)}^{\text{covalent}}$, can be evaluated as

$$n_{A(B)}^{\text{covalent}} = \left(\frac{n_A^{\text{ionic}}}{n_A^{\text{ionic}} + n_B^{\text{ionic}}} \right) n_{AB}^{\text{covalent}}. \tag{15}$$

Summing over all other atomic centers, we obtain the IPOD2 atomic charge on atom A:

$$q_A^{\text{IPOD2}} = Z_A - \left(n_A^{\text{ionic}} + \sum_{B \neq A}^{\text{atom}} n_{A(B)}^{\text{covalent}} \right). \tag{16}$$

3 Computational Details

To benchmark our IPOD analyses against other population analyses, we implemented the two IPOD1 and IPOD2 methods in the NWChem 6.0 source code [24], which already have the MPA and LPA built in. We also calculated the NPA charges for each molecule using Gaussian 09 package [25]. The B3LYP DFT calculations were carried out to obtain the wave functions [26, 27]. A suite of Pople basis sets were employed to test the basis-set variation effect.

These population analyses were applied to some representative molecules with different bonding characters. First, we chose several molecules with different bond polarities, from nearly non-polar to ionic. The perfect tetrahedral methane molecule (CH_4) with C–H bond lengths of 1.093 Å was adopted to represent the molecules containing slightly polarized single bonds. Hydrogen fluoride (HF) with the H–F bond length of 0.920 Å is a good example of highly polarized covalent molecule. Lithium fluoride (LiF) with the Li–F bond length of 1.552 Å was chosen to exemplify ionic molecules.

Other than the above concern over bond polarity, there was also a set of molecules that have been believed to be good test cases for population analyses because these molecules have counter-intuitive charge distributions or unusual electronic structures. For example, carbon monoxide (CO) with the C≡O bond of 1.137 Å has a counter-intuitive dipole moment direction. The triangular boron trifluoride (BF_3) with the B–F bond length of 1.318 Å represents typical electron-deficient molecules. The planar benzene molecule (C_6H_6) is an excellent prototype of aromatic systems. In C_6H_6, all C–H and aromatic C=C bond lengths and H–C–C bond angles are 1.396 Å, 1.097 Å, and 120.0°, respectively.

4 Results and Discussions

For all systems studied here (see Fig. 1), the most visible feature of the IPOD analyses is their milder basis-set fluctuation effects, much better than the MPA and LPA. For basis sets larger than 4–31G, the IPOD charges often exhibit fast convergence. This is quite remarkable, given the fact that the original nonorthogonal atomic basis functions are used, without the elaborate orthogonalization process employed in the LPA and NPA.

For the molecules with weakly polar bonds, such as CH_4 (Fig. 1a) and C_6H_6 (Fig. 1f), both IPOD1 and IPOD2, similar to the NPA, yield relatively larger atomic charges compared to the MPA and LPA. Such a behavior is intrinsic to the IPOD methods, because they partition the covalent distributions according to the ionic electron numbers, which might put more emphasis on the ionic feature. To some extent, the IPOD schemes improve the MPA and LPA that ignore the uneven partition nature of the covalent electrons among non-equivalent atomic centers. For such molecules, basis sets play significant roles in the MPA and LPA, whereas the IPOD1 and IPOD2 methods are relatively less sensitive to the basis-set change.

For the molecules with polar bonds, such as HF (Fig. 1b), the IPOD1 and IPOD2 methods are in good agreement with the NPA. The MPA and LPA underestimate the bond polarity because they divide the overlap electron densities evenly among the participating atomic centers. The basis-set effect in this case is still dramatic for the MPA and LPA while it is significantly less so for the IPOD schemes.

For ionic molecular systems, such as LiF (Fig. 1c), both the MPA and LPA greatly underestimate the ionic nature of the molecule by giving very small atomic charges to ionic centers. In comparison, the IPOD methods and the NPA predict large magnitude of atomic charges.

For more challenging cases, both IPOD methods outperform the MPA and LPA. For BF_3 (Fig. 1e), the atomic charges of the F atom predicted by the IPOD methods are closer to the NPA predictions and exhibit less sensitivity to the basis-set change than the MPA and LPA. In the case of CO (Fig. 1d), the NPA predicts a large negative charge on the O atom, contradicting the experimental observation in both magnitude and sign: there is a small dipole moment of 0.122 D pointing from the C atom to the O atom [28]. The MPA and LPA yield small atomic charges but show drastic undulance upon basis-set change. The IPOD methods, for basis sets larger than 3–21G, persistently predict a small negative charge on the O atom, being closer to the experimental observations.

As expected, the IPOD1 and IPOD2 methods always predict atomic charges in relatively larger magnitudes than the MPA and LPA. Using the NPA charges as guide, a closer observation of the data shown in Fig. 1 reveals that IPOD1 is more suitable for non-polar systems, whereas IPOD2 is ideal for polar and ionic systems. In terms of computational cost, the IPOD methods are only marginally more expensive than the MPA and LPA but are much cheaper than the NPA, AIM, and HPA. Therefore, the IPOD analyses are immaculate for fast computing high-quality atomic charges for large molecules.

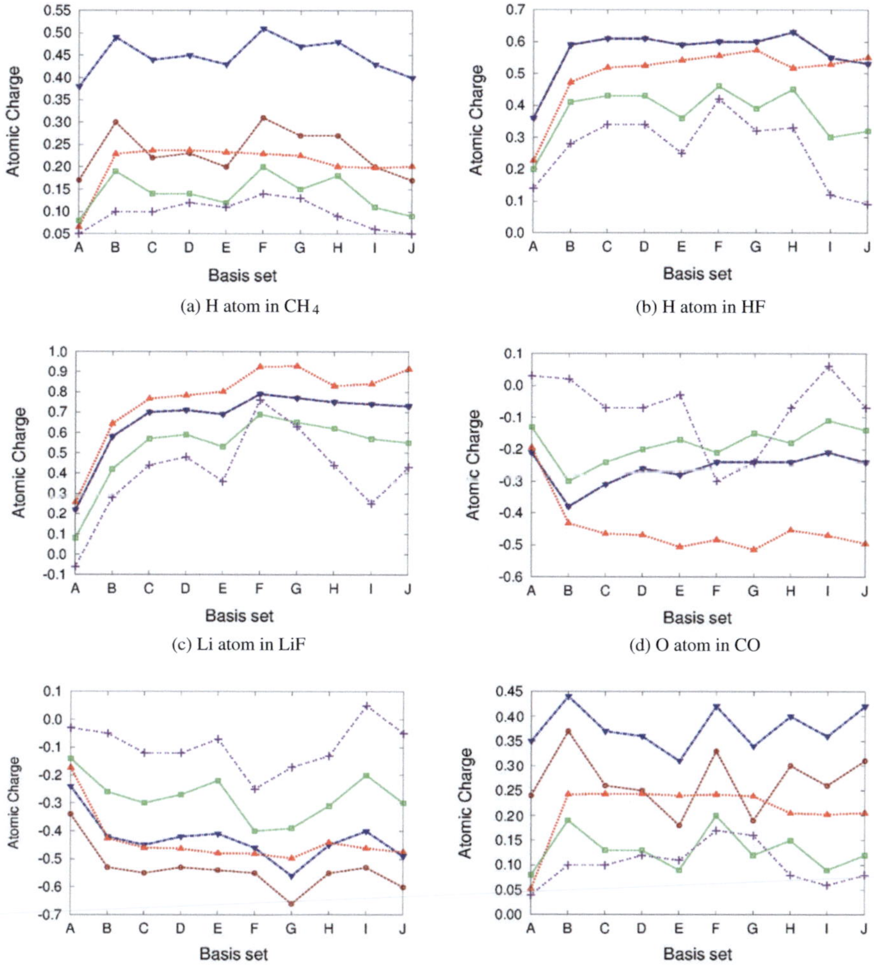

(a) H atom in CH_4

(b) H atom in HF

(c) Li atom in LiF

(d) O atom in CO

(e) F atom in BF_3

(f) H atom in C_6H_6

Fig. 1 Atomic charges from the IPOD1 (*circles*), IPOD2 (*down triangles*), NPA (*up triangles*), MPA (*squares*), and LPA (*crosses*) using the B3LYP functional with different Pople basis sets: A for STO-3G, B for 3–21G, C for 4–31G, D for 6–31G, E for 6–31G**, F for 6–31++G, G for 6–31++G**, H for 6–311G, I for 6–311G**, and J for 6–311++G(2d,2p). For diatomic molecules (HF, LiF, and CO), IPOD1 and IPOD2 atomic charges are the same

5 Conclusions

In summary, we have advanced two new population analyses, IPOD1 and IPOD2, based on ionic partition of overlap distributions. Numerical tests show remarkable stability and physical meaningfulness of the resulting atomic partial charges. More significantly, our new population analyses not only correct the conceptual flaws of other commonly used Mulliken, Löwdin, Christoffersen-Baker, and Hirshfeld population analyses, but also avoid the numerical complexity of the well-known natural population analysis. Because of their conceptual simplicity, easy implementation, low computational cost, and high-quality resultant charges, the IPOD methods should replace the Mulliken and Löwdin population analyses in routine calculations.

Acknowledgements Financial support for this study was provided by a grant from the Natural Sciences and Engineering Research Council (NSERC) of Canada.

References

1. Honhenberg P, Kohn W (1964) Phys Rev B 136:864–871
2. Mulliken RS (1955) J Chem Phys 23:1833–1840
3. Mulliken RS (1955) J Chem Phys 23:1841–1846
4. Löwdin P-O (1950) J Chem Phys 18:365–375
5. Bruhn G, Davidson ER, Mayer I, Clark AE (2006) Int J Quantum Chem 106:2065–2072
6. Christoffersen RE, Baker KA (1971) Chem Phys Lett 8:4–9
7. Edgecombe KE, Boyd RJ (1987) J Chem Soc Faraday Trans 2(83):1307–1315
8. Reed AE, Weinstock RB, Weinhold F (1985) J Chem Phys 83:735–746
9. Reed AE, Curtiss LA, Weinhold F (1988) Chem Rev 88:899–926
10. Zubarev DY, Boldyrev AI (2008) Phys Chem Chem Phys 10:5207–5217
11. Maseras F, Morokuma K (1992) Chem Phys Lett 195:500–504
12. Bader RFW (2005) Monatsh Chem 136:819–854
13. Clark AE, Davidson ER (2003) Int J Quantum Chem 93:384–394
14. Davidson ER, Chakravorty S (1992) Theor Chim Acta 83:319–330
15. Becke AD (1988) J Chem Phys 88:2547–2553
16. Yang W (1991) Phys Rev Lett 66:1438–1441
17. Hirshfeld FL (1977) Theor Chim Acta 44:129–138
18. Roy RK (2003) J Phys Chem A 107:10428–10434
19. Hirshfeld FL (1971) Acta Cryst 18:769–781
20. Coppens P, Volkov A (2004) Acta Cryst A60:357–364
21. Roy RK, Hirao K, Pal S (2000) J Chem Phys 113:1372–1379
22. Riess J, Münch W (1981) Theor Chim Acta 58:295–300
23. Geerings P, Boon G, van Alsenoy C, de Proft F (2005) Int J Quantum Chem 101:722–732
24. Valiev M, Bylaska EJ, Govind N, Kowalski K, Straatsma TP, van Dam HJJ, Wang D, Nieplocha J, Apra E, Windus TL, de Jong WA (2010) Comput Phys Commun 181:1477–1489
25. Frisch MJ, Trucks GW, Schlegel HB, Scuseria GE, Robb MA, Cheeseman JR, Scalmani G, Barone V, Mennucci B, Petersson GA, Nakatsuji H, Caricato M, Li X, Hratchian HP, Izmaylov AF, Bloino J, Zheng G, Sonnenberg JL, Hada M, Ehara M, Toyota K, Fukuda R, Hasegawa J, Ishida M, Nakajima T, Honda Y, Kitao O, Nakai H, Vreven T, Montgomery JA Jr, Peralta JE, Ogliaro F, Bearpark M, Heyd JJ, Brothers E, Kudin KN, Staroverov VN, Kobayashi R, Normand J, Raghavachari K, Rendell A, Burant JC, Iyengar SS, Tomasi J, Cossi M, Rega N, Millam NJ, Klene M, Knox JE, Cross JB, Bakken V, Adamo C, Jaramillo J, Gomperts R, Stratmann RE, Yazyev O, Austin AJ, Cammi R, Pomelli C, Ochterski JW, Martin RL, Morokuma K, Zakrzewski VG, Voth GA, Salvador P, Dannenberg JJ, Dapprich S, Daniels AD, Farkas Ö, Foresman JB, Ortiz JV, Cioslowski J, Fox DJ (2009) Gaussian 09, Revision A.1. Gaussian, Inc., Wallingford CT
26. Becke AD (1993) J Chem Phys 98:5648–5672
27. Lee C, Yang W, Parr RG (1988) Phys Rev B 37:785–789
28. Scuseria GE, Miller MD, Jensen F, Geertsen J (1991) J Chem Phys 94:6660–6663

Topological Quantum Computation with Non-Abelian Anyons in Fractional Quantum Hall States

Lachezar S. Georgiev

Abstract We review the general strategy of topologically protected quantum information processing based on non-Abelian anyons, in which quantum information is encoded into the fusion channels of pairs of anyons and in fusion paths for multi-anyon states, realized in two-dimensional fractional quantum Hall systems. The quantum gates which are needed for the quantum information processing in these multi-qubit registers are implemented by exchange or braiding of the non-Abelian anyons that are at fixed positions in two-dimensional coordinate space. As an example we consider the Pfaffian topological quantum computer based on the fractional quantum Hall state with filling factor $\nu_H = 5/2$. The elementary qubits are constructed by localizing Ising anyons on fractional quantum Hall antidots and various quantum gates, such as the Hadamard gate, phase gates and CNOT, are explicitly realized by braiding. We also discuss the appropriate experimental signatures which could eventually be used to detect non-Abelian anyons in Coulomb blockaded quantum Hall islands.

Keywords Quantum computation · Non-Abelian anyons · Fractional quantum Hall states · Conformal field theory · Topological protection · Braid group

1 Introduction: Quantum Computation in General

Quantum Computation (QC) is a relatively new field [1] of computational research in which information is encoded in two-level quantum systems called *qubits*, or *qu*antum *bits*, in analogy with the classical bits. In contrast with the classical bits, which can have only two states '0' and '1', the qubit is an arbitrary linear combination of the two basic quantum states $|0\rangle$ and $|1\rangle$

L.S. Georgiev (✉)
Institute for Nuclear Research and Nuclear Energy, Bulgarian Academy of Sciences, 72 Tsarigradsko Chaussee, Sofia, Bulgaria
e-mail: lgeorg@inrne.bas.bg

A. Tadjer et al. (eds.), *Quantum Systems in Physics, Chemistry, and Biology*, Progress in Theoretical Chemistry and Physics 30,
DOI 10.1007/978-3-319-50255-7_5

$$|\psi\rangle = \alpha|0\rangle + \beta|1\rangle, \quad \text{with} \quad \alpha, \beta \in \mathbb{C} \quad \text{and} \quad |\alpha|^2 + |\beta|^2 = 1, \tag{1}$$

i.e., a qubit $|\psi\rangle$ belongs to the projective $\mathbb{C}^2$ space which is called the Bloch sphere [1]. In order to prepare one qubit state for QC we have to define first an orthonormal basis $|0\rangle$, $|1\rangle$, which could be for example the two spin-projection states of a spin 1/2 particle, and then construct physically the state $|\psi\rangle = |0\rangle$ or $|\psi\rangle = |1\rangle$.

Next, just like in classical computation, where we need a register of bits which belong to the direct sums of bit spaces, QC requires the construction of a register of qubits, or, multiple-qubit, which however belong to the tensor product of the single-qubit spaces, e.g., the n-qubit register belongs to the space

$$\mathcal{H}^n = \underbrace{\mathbb{C}^2 \otimes \mathbb{C}^2 \cdots \mathbb{C}^2}_{n} \simeq \left(\mathbb{C}^2\right)^n \simeq \mathbb{C}^{2^n},$$

that should be projective by definition, thus expressing the required normalization of the n-qubit states [1]. This difference in the register spaces dimension in the classical (direct sum space has dimension $2n$) and quantum computation (tensor product space has dimension 2^n) is one of the reasons for the significantly bigger computational power of the quantum computers as compared to the classical ones [1].

The first step in a concrete QC is the initialization of the n-qubit register, which is done in analogy with the initialization of a classical one, i.e.,

$$010011000\ldots01 \quad \stackrel{\text{QC}}{\to} \quad |010011000\ldots01\rangle,$$

where we used the standard notation for the tensor-product basis states, e.g., $|01\rangle = |0\rangle \otimes |1\rangle$ and $|101\rangle = |1\rangle \otimes |0\rangle \otimes |1\rangle$, etc.

The second step in QC is to process the initial information, which is done by applying a quantum operation, called *quantum gate*, on the n-qubit register [1]. The quantum gates are realized by unitary quantum operators acting over the space $\mathcal{H}^n$. Sequential application of two quantum gates is a quantum gate again and any quantum gate possesses its inverse. In other words, quantum gates G should realize by appropriate physical processes all unitary $2^n \times 2^n$ matrices and therefore belong to the unitary group $G \in SU(2^n)$.

The third step in QC is the measurement of a given observable, as described above, of the n-qubit register after its processing with a number of quantum gates and this measurement is the result of the QC.

It is important to remember, that universal quantum computation requires physical realization of all unitary matrices in the unitary group $SU(2^n)$ for an arbitrary finite number n of qubits and their measurements.

After the work of Peter Shor [2], unveiling an algorithm for factorization of large numbers N into primes on a quantum computer, which requires time that is polynomial in the size $\log N$, while classical factorization algorithms require exponential time, it has become obvious that quantum computers, if they can be constructed, would be much faster than classical ones. This anticipated exponential speed-up with respect to classical computing is due to the fundamental properties of the quantum systems, such as quantum parallelism and entanglement [1]. The difficulties to factorize large numbers on classical computers is the security foundation of some public-key cryptographic algorithms, such as the RSA, which explains the wide interest of the banks, intelligence services and military services in the physical realization of quantum computers.

Another important theoretical observation in the field of QC is that any unitary operator, i.e., any quantum gate, can be approximated by products (sequential application, or concatenation) [1] of only 3 universal gates: H, T and CNOT, that can be applied to any qubit in the n-qubit register, with arbitrary precision. The first two quantum gates, H and T act on single qubits, while CNOT is a quantum gate acting on two qubits in the n-qubit register. These three basic quantum gates can be written explicitly, in the basis $\{|0\rangle, |1\rangle\}$ for H and T and in the two-qubit basis $\{|00\rangle, |01\rangle, |10\rangle, |11\rangle\}$ for CNOT as [1]

$$H = \frac{1}{\sqrt{2}}\begin{bmatrix} 1 & 1 \\ 1 & -1 \end{bmatrix}, \quad T = \begin{bmatrix} 1 & 0 \\ 0 & e^{i\pi/4} \end{bmatrix}, \quad \text{CNOT} = \begin{bmatrix} 1 & 0 & 0 & 0 \\ 0 & 1 & 0 & 0 \\ 0 & 0 & 0 & 1 \\ 0 & 0 & 1 & 0 \end{bmatrix}. \tag{2}$$

However, there are huge difficulties on the way of constructing a stable quantum computer due to the unavoidable decoherence and noise [1] resulting from the local interactions of the qubits with their environment, destroying in this way all coherent phenomena and flipping uncontrollably $|0\rangle \leftrightarrow |1\rangle$.

One way to make the fragile quantum information more robust is to use the so-called quantum error correcting algorithms [1].

Another possible way out might be the topological quantum computation (TQC) [3, 4], whose strategy is to improve QC hardware by using intrinsic topological protection instead of compensating hardware deficiency by clever circuit design. This new idea requires a fundamentally new concept: the non-Abelian exchange statistics of quasiparticles which are believed to exist in the fractional quantum Hall states [4].

2 Non-Abelian Anyons and Topological QC

In this section we will explain the new concept of non-Abelian exchange statistics of particles and will try to give an idea of what it can be used for. Consider a system of indistinguishable particles at fixed positions in space. One of their important statistical characteristics is their statistical angle θ_A/π, which can be defined in the following way: the quantum state of the system with many indistinguishable particles of type 'A' can be expressed as a correlation function (or a vacuum expectation value) of some quantum filed operators $\psi_A(z)$ which represent the act of creation of a particle of type 'A' at position z in the space. Let us focus only on two particles of type 'A', at positions z_1 and z_2 in space, although the system may contain more particles. If we exchange the two particles adiabatically, then the state after the exchange differs from that before it by the statistical phase $\mathrm{e}^{i\pi(\theta_A/\pi)}$, i.e.,

$$\langle\cdots\psi_A(z_1)\psi_A(z_2)\cdots\rangle \to \langle\cdots\psi_A(z_2)\psi_A(z_1)\cdots\rangle = \mathrm{e}^{i\pi(\theta_A/\pi)}\langle\cdots\psi_A(z_1)\psi_A(z_2)\cdots\rangle,$$

where θ_A/π is by definition the statistical angle of the particles of type 'A'.

In three-dimensional space (or four-dimensional space-time) only two type of particles could exist, as long as the statistical angle is concerned[1]: bosons, which correspond to $\theta_A/\pi = 0$ and fermions which correspond to $\theta_A/\pi = 1$. However, in two-dimensional space the variety of statistical angles is much richer than in three dimensions. For example, the so-called Laughlin anyons correspond to $\theta_L/\pi = 1/3$, and actually all values of θ_A/π between 0 and 1 are admissible. That is why such particles, which could exist only in two-dimensional (and one-dimensional) space, are called any-ons, in analogy with the bosons and fermions.

2.1 Construction of n-Particle States: The Braid Group

The many-particle quantum states in three-dimensional space are built as representations of the symmetric group $\mathscr{S}_n$, which are symmetric for bosons and antisymmetric for fermions. The symmetric group $\mathscr{S}_n$, which is the group of permutation of n objects, is a finite group generated by $n-1$ elementary transpositions of neighboring objects i.e., $\sigma_i : i \leftrightarrow i+1$ and obviously satisfy $\sigma_i = \sigma_i^{-1}$.

On the other hand the many-particle quantum states in two dimensional space are constructed as representations of the braid group [5] $\mathscr{B}_n$. The braid group $\mathscr{B}_n$ is an infinite group, which is an extension of the symmetric group $\mathscr{S}_n$ whose generators, unlike those for $\mathscr{S}_n$, do not satisfy $(B_i)^2 = \mathbb{I}$. The genetic code of the braid group $\mathscr{B}_n$ is know as the *Artin relations* [5]

$$\begin{aligned} B_iB_j &= B_jB_i, \qquad &\text{for} \quad |i-j| \geq 2 \\ B_iB_{i+1}B_i &= B_{i+1}B_iB_{i+1}, \qquad &i = 1, \dots, n-1. \end{aligned}$$

The very concept of the non-Abelian anyons requires the existence of degenerate multiplets of n-particle states, with fixed coordinate positions of the anyons, and the exchanges of the coordinates of these anyons generate statistical phases $\mathrm{e}^{i\theta_A}$ which

[1]For simplicity we disregard here the case of relativistic parafermions and parabosons.

might be non-trivial matrices acting on those multiplets. That is why this statistics is called non-Abelian (see the example below).

2.2 *Fusion Paths: Labeling Anyonic States of Matter*

As mentioned above, the states with many non-Abelian anyons at fixed coordinate positions form degenerate multiplets which means that specifying the positions of the anyons and their quantum numbers, such as the electric charge, single-particle energies and angular momenta, are not sufficient to specify a concrete n-particle state. More information is needed and this information is a non-local characteristics of the n-particle state as a whole.

This additional information is provided by the so-called *fusion channels* [6]. Consider the fusion of two particles of type 'a' and 'b', i.e., consider the process when the two particles come to the same coordinate position and form a new particle of type 'c'. This can be written formally as

$$\Psi_a \times \Psi_b = \sum_{c=1}^{g} N_{ab}{}^c \Psi_c,$$

where the fusion coefficients $(N_{ab})^c$ are integers, specifying the different possible channels of the fusion process, which are symmetric and associative [6]. Two particles Ψ_a and Ψ_b, where b could be the same as a, are called non-Abelian anyons if the fusion coefficients $N_{ab}{}^c \neq 0$ for more than one c. The most popular example of non-Abelian anyons are the Ising anyons realized in two-dimensional conformal field theory with $\widehat{u(1)} \times$ Ising symmetry by the primary field [6]

$$\Psi_I(z) = \sigma(z) \ : \mathrm{e}^{i\frac{1}{2\sqrt{2}}\phi(z)} :,$$

where $\phi(z)$ is a normalized $\widehat{u(1)}$ boson [6] and $\sigma(z)$ is the chiral spin field of the two-dimensional Ising model CFT [6]. The fusion rules of the Ising model are non-Abelian since the fusion process of two Ising anyons σ could be realized in two different fusion channels: that of the identity operator $\mathbb{I}$ and that of the Majorana fermion ψ [6], i.e.,

$$\sigma \times \sigma = \mathbb{I} + \psi, \quad \sigma \times \psi = \sigma. \tag{3}$$

The quantum information is then encoded into the fusion channel and the computational basis is defined by pairs of Ising anyons whose fusion channel is fixed

$$\begin{aligned} |0\rangle = (\sigma, \sigma)_{\mathbb{I}} \quad &\longleftrightarrow \quad \sigma \times \sigma \to \mathbb{I} \\ |1\rangle = (\sigma, \sigma)_{\psi} \quad &\longleftrightarrow \quad \sigma \times \sigma \to \psi, \end{aligned} \tag{4}$$

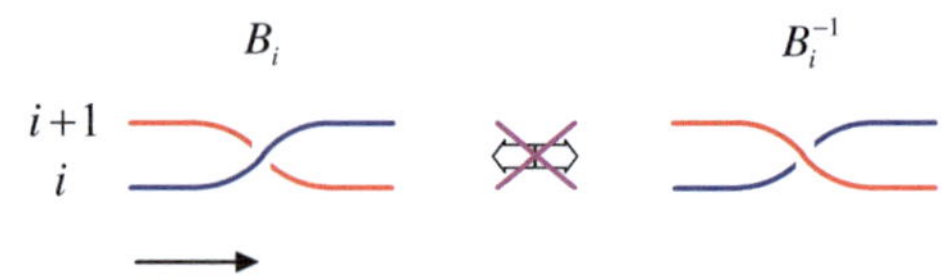

Fig. 1 Braid diagrams: exchange of particles with number i and $i+1$ in counter-clockwise and clockwise directions are distinct and inverse to each other

i.e. the state is $|0\rangle$ if the two Ising anyons σ fuse to $\mathbb{I}$ and $|1\rangle$ if they fuse to ψ. The fusion channel is a topological quantity–it is independent of the fusion process details and depends only on the topology of the coordinate space with positions of the anyons removed. This quantity is non-local: the fusion channel is independent of the anyon separation and is preserved after separation. It is also robust and persistent: if we fuse two particles and then split them again, their fusion channel does not change.

The concatenation of several fusion channels of neighboring pairs of anyons is called a *fusion paths* and can be displayed in Bratteli diagrams [4].

In other words, quantum states with many non-Abelian anyons at fixed positions in two-dimensional space are specified/labeled by fusion paths and can be plotted in Bratteli diagrams.

2.3 *Braiding of Anyons: Topologically Protected Quantum Gates*

Quantum operations in TQC, needed for processing of quantum information, are implemented by braiding non-Abelian anyons [4, 7]. Braiding of two anyons is the adiabatic exchange of the coordinate positions of the two anyons in the counter-clockwise direction, without crossing any other coordinates. The clockwise exchanges represent the inverse of the braids described above and they are not equal to the counter-clockwise exchanges as illustrated in Fig. 1.

For the purpose of illustration of the non-Abelian statistics acting on a degenerate multiplet of multyanyon states we consider the wave functions representing 8 Ising anyons σ at fixed positions with coordinates $\eta_1, \ldots, \eta_8$, which correspond to three-qubit states. Each pair of Ising anyons σ is characterized by the fermion parity of their fusion channel in Eq. (3), i.e., it is '+' for the fusion channel of the identity operator $\mathbb{I}$ and '−' for the fusion channel of the Majorana fermion ψ. Taking into account the information encoding (4) in the quantum information language we note that '+' corresponds to the computational state $|0\rangle$ while '−' corresponds to $|1\rangle$ and therefore the multiplet can be explicitly written in the three-qubit computational basis as

$$
\begin{aligned}
|000\rangle &\equiv \langle[\sigma(\eta_1)\sigma(\eta_2)]_+[\sigma(\eta_3)\sigma(\eta_4)]_+[\sigma(\eta_5)\sigma(\eta_6)]_+[\sigma(\eta_7)\sigma(\eta_8)]_+\rangle \\
|001\rangle &\equiv \langle[\sigma(\eta_1)\sigma(\eta_2)]_+[\sigma(\eta_3)\sigma(\eta_4)]_+[\sigma(\eta_5)\sigma(\eta_6)]_-[\sigma(\eta_7)\sigma(\eta_8)]_-\rangle \\
|010\rangle &\equiv \langle[\sigma(\eta_1)\sigma(\eta_2)]_+[\sigma(\eta_3)\sigma(\eta_4)]_-[\sigma(\eta_5)\sigma(\eta_6)]_+[\sigma(\eta_7)\sigma(\eta_8)]_-\rangle \\
|011\rangle &\equiv \langle[\sigma(\eta_1)\sigma(\eta_2)]_+[\sigma(\eta_3)\sigma(\eta_4)]_-[\sigma(\eta_5)\sigma(\eta_6)]_-[\sigma(\eta_7)\sigma(\eta_8)]_+\rangle \\
|100\rangle &\equiv \langle[\sigma(\eta_1)\sigma(\eta_2)]_-[\sigma(\eta_3)\sigma(\eta_4)]_+[\sigma(\eta_5)\sigma(\eta_6)]_+[\sigma(\eta_7)\sigma(\eta_8)]_-\rangle \\
|101\rangle &\equiv \langle[\sigma(\eta_1)\sigma(\eta_2)]_-[\sigma(\eta_3)\sigma(\eta_4)]_+[\sigma(\eta_5)\sigma(\eta_6)]_-[\sigma(\eta_7)\sigma(\eta_8)]_+\rangle \\
|110\rangle &\equiv \langle[\sigma(\eta_1)\sigma(\eta_2)]_-(\sigma(\eta_3)\sigma(\eta_4)]_-[\sigma(\eta_5)\sigma(\eta_6)]_+[\sigma(\eta_7)\sigma(\eta_8)]_+\rangle \\
|111\rangle &\equiv \langle[\sigma(\eta_1)\sigma(\eta_2)]_-[\sigma(\eta_3)\sigma(\eta_4)]_-[\sigma(\eta_5)\sigma(\eta_6)]_-[\sigma(\eta_7)\sigma(\eta_8)]_-\rangle,
\end{aligned} \tag{5}
$$

where the subscript $\pm$ of the pair $[\sigma(\eta_i)\sigma(\eta_{i+1})]_\pm$ denotes the fermion parity of the fusion channel and is $+$ if the state of the pair is $|0\rangle$ and $-$ if the state is $|1\rangle$. If we now transport adiabatically the Ising anyon with coordinate η_7 along a complete loop around that with coordinate η_6 this is equivalent to two subsequent applications of two braid generators $B_6^{(8,+)}$ (both in the counter-clockwise direction) over the multiplet defining the basis (5). Taking the explicit expression for the matrix $B_6^{(8,+)}$, where we have chosen for concreteness the positive parity representation [8] from Ref. [8], we obtain the following monodromy matrix acting over the 8-fold degenerate multiplet (5)

$$
\left(B_6^{(8,+)}\right)^2 = \begin{bmatrix} 0&1&0&0&0&0&0&0 \\ 1&0&0&0&0&0&0&0 \\ 0&0&0&1&0&0&0&0 \\ 0&0&1&0&0&0&0&0 \\ 0&0&0&0&0&1&0&0 \\ 0&0&0&0&1&0&0&0 \\ 0&0&0&0&0&0&0&1 \\ 0&0&0&0&0&0&1&0 \end{bmatrix}. \tag{6}
$$

This result illustrates why non-Abelian statistics is so interesting-by simply exchanging two Ising anyons we obtained a statistical 'phase' which is not diagonal-it is a non-trivial non-diagonal statistical matrix. Other exchanges generate other non-diagonal matrices acting on the same multiplet (5) and in general these non-diagonal statistical matrices do not commute, hence the name non-Abelian statistics.

The non-Abelian statistics is definitely a new fundamental concept in two-dimensional particle physics, which is interesting on its own. However, it might also have a promising application in quantum computation. Notice that in quantum information language the matrix (6) is an implementation of the quantum NOT gate X on the third qubit [1], i.e.,

$$
\left(B_6^{(8,+)}\right)^2 = X_3 = \mathbb{1}_2 \otimes \mathbb{1}_2 \otimes X.
$$

Because information is encoded globally in degenerate multiplets of quantum states with n-anyons at fixed positions no local interaction can change or corrupt this information. In other words, quantum information is hidden from its enemies (noise and decoherence) which are due to local interactions, however, it is hidden even from us. In order to read this information non-local topologically nontrivial operations are needed [4, 9]. This leads to the so-called *topological protection* of the encoded information and its processing. For example, for Ising anyons, the unprecedented precision of quantum information processing is due to the exponentially small probability for accidental creation of quasiparticle-quasihole pairs which is expressed in terms of the temperature and the experimentally estimated energy gap $\Delta \approx 500$ mK [9–11] resulting in

$$\text{Error rate} \simeq \left(\frac{k_B T}{\Delta}\right) \exp\left(-\frac{\Delta}{k_B T}\right) < 10^{-30}$$

for temperatures below 5 mK [9].

Given that the new concept of non-Abelian statistics is so interesting a natural question arises how could it be discovered. Experiments with Fractional Quantum Hall (FQH) states, such as the $\nu_H = 5/2$ state in which non-Abelian anyons are expected to exist, are conducted in extreme conditions and are difficult and expensive. On top of that it appeared that there are other candidate FQH states, such as the 331 state, having the same electric properties and identical patterns of Coulomb blockaded conductance peaks but without non-Abelian anyons [12]. Therefore the conductance spectrometry [13–15] of single-electron transistors, which was expected to detect non-Abelian statistics experimentally, is not sufficient to do that.

A possible resolution of this problem could be to measure the thermoelectric characteristics of the Coulomb-blockaded FQH islands. The thermoelectric conductance, thermopower and especially the thermoelectric power factor [16–18] considered here could be the appropriate tools for detecting non-Abelian anyons, should they exist in Nature.

3 The Pfaffian Quantum Hall State and TQC with Ising Anyons

The Pfaffian FQH state, also known as the Moore–Read state [19] is the most promising candidate to describe the Hall state with filling factor $\nu_H = 5/2$, which is routinely observed in ultrahigh-mobility samples [11, 20, 21]. It is believed to that the observed state at $\nu_H = 5/2$ is in the universality class of the Moore–Read state [19], whose CFT is $\widehat{u(1)} \times \text{Ising}$.

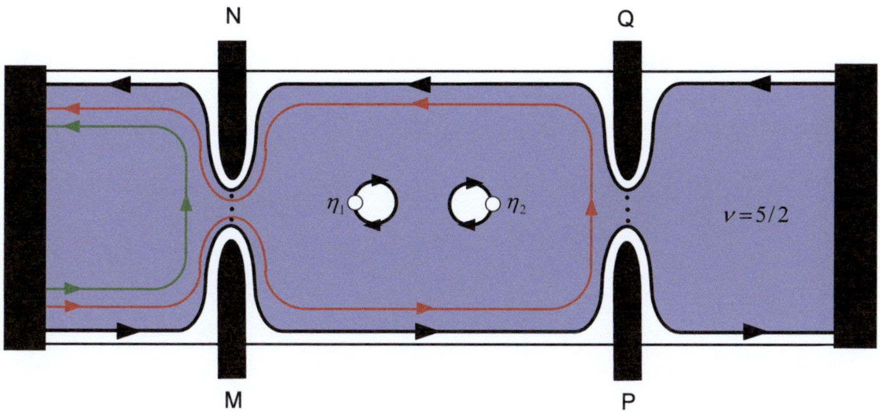

Fig. 2 The qubit scheme of Das Sarma et al. Only two antidots are shown and there are two anyons with coordinates η_1 and η_2 localized on the antidots

3.1 TQC Scheme with Ising Anyons: Single Qubit Construction

The TQC scheme proposed by Das Sarma et al. [9] is based on a $\nu_H = 5/2$ FQHE sample with 4 non-Abelian quasiparticles at fixed positions η_a localized on 4 quantum antidots as shown in Fig. 2. The N-electron wave function describing the sample with 4 anyons with coordinates $\eta_1, \ldots, \eta_4$ can be expressed as a CFT correlation function [19]

$$\Psi_{\mathrm{qubit}}(\eta_1, \eta_2, \eta_3, \eta_4, \{z_i\}) = \langle \sigma(\eta_1)\sigma(\eta_2)\sigma(\eta_3)\sigma(\eta_4) \prod_{i=1}^{N} \psi_{\mathrm{el}}(z_i) \rangle,$$

where $\psi_{\mathrm{el}}(z)$ is the field operator of the physical electron at position z in the coordinate plane and $\sigma(\eta)$ is the non-Abelian Ising anyon at position η. Following Eq. (4) the qubit is in the state $|0\rangle$ if the two anyons $\sigma(\eta_1)$ and $\sigma(\eta_1)$ fuse together to the unit operator $\mathbb{I}$ while it is in the state $|1\rangle$ if they fuse together to the Majorana fermion ψ and the state could be measured by interferometric measurement of the conductance. For more details on the TQC scheme of Das Sarma et al. see Refs. [4, 9, 22].

The quantum gates in the TQC scheme with Ising anyons are implemented by braiding, i.e., by adiabatic exchange of the antidots on which the Ising anyons are localized. The braiding of the 4 anyons generate two finite two-dimensional representations of the braid group $\mathscr{B}_4$, which are characterized by the positive or negative total fermion parity of the 4 anyon fields σ [8]. Denoting the braid matrices in the positive-parity representations as $B_1^{(4,+)} = R_{12}^{(4)}$, $B_2^{(4,+)} = R_{23}^{(4)}$ and $B_3^{(4,+)} = R_{34}^{(4)}$, where e.g. $R_{23}^{(4)}$ denotes the matrix representing the exchange of the anyons with coordinates η_2 and η_3, we can write them explicitly as [8]

$$B_1^{(4,+)} = \begin{bmatrix} 1 & 0 \\ 0 & i \end{bmatrix}, \quad B_2^{(4,+)} = \frac{e^{i\frac{\pi}{4}}}{\sqrt{2}} \begin{bmatrix} 1 & -i \\ -i & 1 \end{bmatrix}, \quad B_3^{(4,+)} = \begin{bmatrix} 1 & 0 \\ 0 & i \end{bmatrix} \tag{7}$$

and similarly for the negative-parity representation (with the corresponding notation)

$$B_1^{(4,-)} = \begin{bmatrix} 1 & 0 \\ 0 & i \end{bmatrix}, \quad B_2^{(4,-)} = \frac{e^{i\frac{\pi}{4}}}{\sqrt{2}} \begin{bmatrix} 1 & -i \\ -i & 1 \end{bmatrix}, \quad B_3^{(4,-)} = \begin{bmatrix} i & 0 \\ 0 & 1 \end{bmatrix}. \tag{8}$$

Multi-qubit states are realized by adding more pairs of Ising anyons. Since each qubit state is encoded into one pair of anyons $(\sigma\sigma)_\pm$, where the subscript $\pm$ denotes the total fermion parity of the pair,

> an n-qubit state can be realized by $2n+2$ Ising anyons σ localized on $2n+2$ antidots. Therefore, the exchanges of the $2n+2$ Ising anyons in the n-qubit register generate representations of the braid group $\mathscr{B}_{2n+2}$ which are again characterized by the total fermion parity of the σ fields.

The last 2 anyons (or any other chosen pair) among the $2n+2$ anyons are inert because they carry no information–their only purpose is to compensate the total fermion parity so that the CFT correlation function is non-zero.

Interestingly enough, it was found in Ref. [8], that the generators $B_j^{(2n+2,\pm)}$, $j = 0, \ldots, 2n+1$, of the braid group $\mathscr{B}_{2n+2}$ can be expressed in terms of the generators $B_j^{(2n,\pm)}$, $j = 0, \ldots, 2n-1$, for $\mathscr{B}_{2n}$ due to the following *recursive relations*

$$\begin{aligned} B_j^{(2n+2,+)} &= B_j^{(2n+2,-)} && \text{for} \quad 1 \le j \le 2n \\ B_j^{(2n+2,\pm)} &= B_j^{(2n,\pm)} \otimes \mathbb{I}_2 && \text{for} \quad 1 \le j \le 2n-3 \\ B_j^{(2n+2,\pm)} &= B_{j-2}^{(2n,\pm)} \oplus B_{j-2}^{(2n,\mp)} && \text{for} \quad 3 \le j \le 2n+1. \end{aligned} \tag{9}$$

Using the recursion relations (9) together with Eqs. (7) and (8) we can find explicitly all braid generators. This will allow us to build almost all quantum gates as products of the braid generators and implement them by subsequent braiding of Ising anyons. We emphasize here that all quantum gates which can be implemented by braiding of non-Abelian anyons are topologically protected hardware for topological quantum computers [4].

3.2 *Single-Qubit Gates: The Pauli X Gate*

The first gate which has been implemented by braiding of Ising anyons [9] is the NOT gate [1] which is usually denoted as the Pauli X matrix

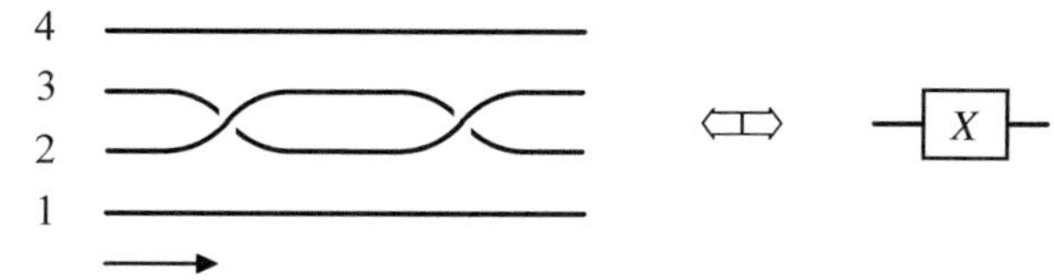

Fig. 3 Braiding diagram for the Pauli X gate and its quantum computation symbol

$$X \equiv (R_{23})^2 = \left(B_2^{(4,\pm)}\right)^2 = \begin{bmatrix} 0 & 1 \\ 1 & 0 \end{bmatrix}.$$

It can be realized by taking the anyon with coordinate η_2 along a complete loop around the anyon with coordinate η_3. Using Eqs. (7) and (8) we can easily check that the square of the generator $B_2^{(4,\pm)}$ indeed implements the NOT gate and this process corresponds to the braid diagram given in Fig. 3.

3.3 The Hadamard Gate

Another important single-qubit gate is the Hadamard gate [1]. It can be implemented by braiding [22, 23] as follows: first exchange the anyons with coordinates η_1 and η_2, then exchange the anyons with coordinates η_2 and η_3 and finally exchange again the anyons with coordinates η_1 and η_2, as shown in the braid diagram in Fig. 4 which is equivalent to exchanging counter-clockwise first the anyons with coordinates η_1 and η_3 and then taking the anyon with coordinate η_2 along a complete loop around that with coordinate η_1, i.e.

$$H \simeq \left(R_{12}\right)^2 R_{13} = R_{12}R_{23}R_{12} = B_1^{(4,\pm)}B_2^{(4,\pm)}B_1^{(4,\pm)} = \frac{e^{i\frac{\pi}{4}}}{\sqrt{2}}\begin{bmatrix} 1 & 1 \\ 1 & -1 \end{bmatrix}, \tag{10}$$

where R_{ij} in the notation of Refs. [22, 23] is the operation representing the counter-clockwise exchange of anyons with coordinates η_i and η_j.

3.4 Two-Qubits Construction

Taking into account the quantum information encoding into the Ising anyon pairs fusion channel, specified in Eq. (4), we consider the following 6 Ising anyons realization of the two-qubit computational basis in the Ising TQC

Fig. 4 Hadamard gate implemented by braiding and its quantum computation symbol

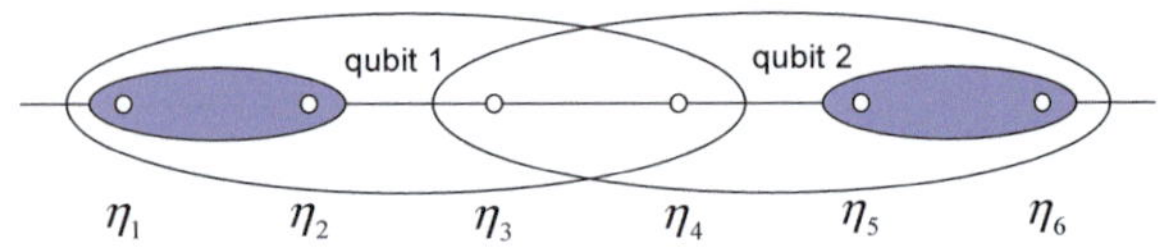

Fig. 5 Two qubits constructed from 6 Ising anyons $\sigma(\eta_i)$ at fixed positions η_i

$$|00\rangle \equiv \langle \sigma_+\sigma_+\sigma_+\sigma_+\sigma_+\sigma_+\rangle, \quad |01\rangle \equiv \langle \sigma_+\sigma_+\sigma_+\sigma_-\sigma_+\sigma_-\rangle$$
$$|10\rangle \equiv \langle \sigma_+\sigma_-\sigma_+\sigma_-\sigma_+\sigma_+\rangle, \quad |11\rangle \equiv \langle \sigma_+\sigma_-\sigma_+\sigma_+\sigma_+\sigma_-\rangle,$$

as shown in Fig. 5, where the first pair of Ising anyons (with coordinates η_1 and η_2) correspond to the first qubit, the third pair of Ising anyons (with coordinates η_5 and η_6) represents the second qubit and the two Ising anyons between them (with coordinates η_3 and η_4) is an inert pair which, on one side compensates total fermion parity so that the correlation function is non-zero, while on the other side creates topological entanglement between the two Ising qubits which can be used to construct by braiding some entangling two-qubit gates, such as the CNOT gate.

3.5 *Two-Qubit Gates: The Controlled-NOT Gate*

The two-qubit CNOT gate is one of the most important resource for quantum computation [1] because this entangling gate can transfer information from one qubit to another. Moreover, due to the CNOT gate, one can express any n-qubit gate, which can be written as a unitary matrix from the group $SU(2^n)$, as a product of single-qubit gates (two-dimensional unitary matrices in tensor product with $n-1$ unit matrices $\mathbb{I}_2$ completing the dimension of the matrix to 2^n) with two-qubit gates (four-dimensional unitary matrices in tensor product with $n-2$ unit matrices $\mathbb{I}_2$ unit matrices completing the dimension of the matrix to 2^n), see Ref. [1]. In this subsection we will demonstrate that the CNOT gate can be constructed explicitly in terms of 6-anyon elementary braidings using the notations of Ref. [22] in which $R_{i,i+1} = B_i^{(6,+)}$ are the representations of the generators of the braid group $\mathscr{B}_6$ corresponding to the elementary braids of 6 Ising anyons.

Our strategy [22, 23] is to use first the well known connection [1] between the CNOT and the Controlled-Z (CZ) gate, which is the diagonal matrix with the following elements on the diagonal $\mathrm{CZ} = \mathrm{diag}(1, 1, 1, -1)$, in terms of the Hadamard gate[2] acting on the second qubit $H_2 = \mathbb{I}_2 \otimes H$ and then to try to construct the diagonal gate CZ in terms of the diagonal braid generators. It is not difficult to check [22] that $R_{12}R_{34}^{-1}R_{56} = \mathrm{diag}(1, 1, 1, -1)$ and $R_{56}R_{45}R_{56} = H_2$ so that one realization of CNOT is

[2]recall that the single-qubit Hadamard gate has been constructed in Eq. (10).

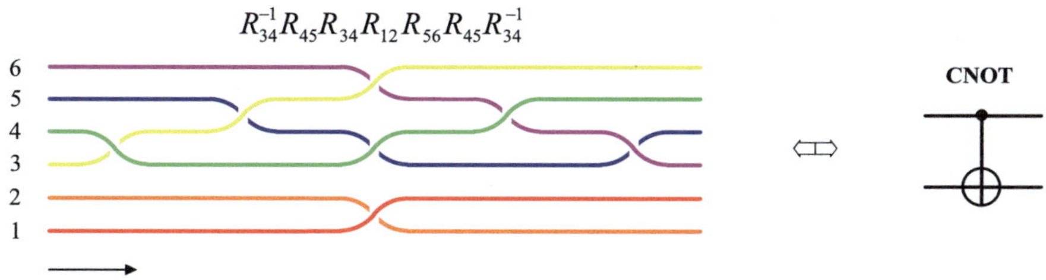

Fig. 6 The CNOT gate realized by braiding of 6 Ising anyons and its quantum computation symbol

$$\text{CNOT} = H_2 \,\text{CZ}\, H_2 = R_{56}R_{45}R_{56}^{-1}R_{34}^{-1}R_{12}R_{45}R_{56} \simeq \begin{bmatrix} 1 & 0 & 0 & 0 \\ 0 & 1 & 0 & 0 \\ 0 & 0 & 0 & 1 \\ 0 & 0 & 1 & 0 \end{bmatrix}.$$

Alternatively the CNOT can be constructed by another combination of 6-anyon braids, namely

$$\text{CNOT} = R_{34}^{-1}R_{45}R_{34}R_{12}R_{56}R_{45}R_{34}^{-1},$$

which is graphically shown on Fig. 6

What is remarkable in this construction of the CNOT gate by braiding 6 Ising anyons is that it is completely topologically protected and consist only 7 elementary braids. Unfortunately, the construction of the embedding of the two-qubit CNOT into Ising systems with more than 2 qubits is not possible by braiding Ising anyons only [24] which is a clear limitation of the Ising TQC.

In the rest of this paper we will focus on how non-Abelian Ising anyons could be detected experimentally if they exist. We start by the description of the Coulomb blockaded quantum Hall islands which are equivalent to single-electron transistors and review their conductance spectroscopy and then we will consider the thermoelectric characteristics of these islands.

4 Coulomb-Blockaded Quantum Hall Islands: QD and SET

The experiments, which are expected to shed more light on the nature of the quasiparticle excitations in FQH states, are performed in single-electron transistors (SET) constructed as Coulomb blockaded islands, or quantum dots, equipped with drain, source and side gates as shown in Fig. 7. The quantum dot (QD) is realized by splitting a larger FQH bar with the help of two quantum point contacts (QPC) and it is assumed that the left- and right- FQH liquids are much larger in size than the QD,

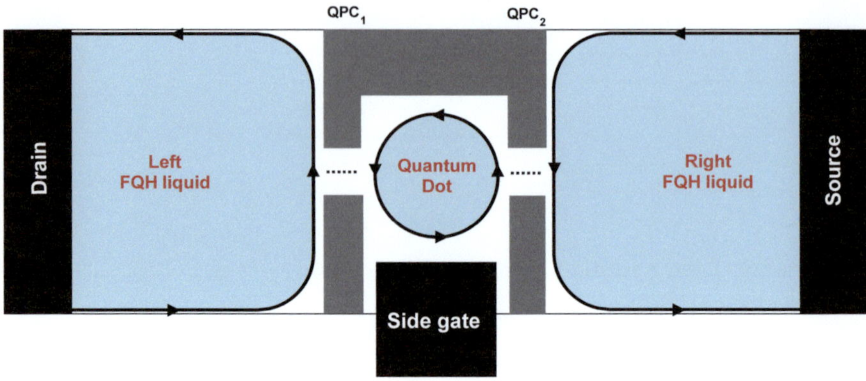

Fig. 7 Single-electron transistor

so that the energy spacing in the left- and right- FQH liquids is much smaller [25] than the energy spacing in the QD, which is proportional to $\hbar 2\pi v_F/L$, where v_F is the Fermi velocity of the electrons on the QD's edge and L is the circumference of the edge. When L is small enough the energy spacing is big and the QD's energy levels are discrete. In this setup, for low temperatures, electron can tunnel from the left FQH liquid to the QD and then to the right FQH liquids only when the chemical potentials of the left and right FQH liquids are aligned with some of the discrete energy levels of the QD. On the other hand, changing the potential of the side gate, which is capacitively coupled to the QD, continuously shifts up or down the energy levels of the QD. Therefore, measuring the electric conductance as a function of the gate voltage can be viewed as precise energy level spectroscopy of the QD and this can be used to distinguish FQH systems.

4.1 Coulomb Island's Conductance–CFT Approach

We are going to use the Grand canonical partition function for a FQH disk, whose edge could be described by an effective rational unitary CFT [6, 26, 27]. The Grand partition function for the edge of the FQH disk can be written as

$$Z(\tau,\zeta) = \mathrm{tr}_{\mathscr{H}_{\mathrm{edge}}}\, \mathrm{e}^{-\beta(H-\mu N)} = \mathrm{tr}_{\mathscr{H}_{\mathrm{edge}}}\, \mathrm{e}^{2\pi i\tau(L_0-c/24)} e^{2\pi i\zeta J_0}, \tag{11}$$

where the Hamiltonian $H = \hbar\frac{2\pi v_F}{L}\left(L_0 - \frac{c}{24}\right)$ is expressed in terms of the zero mode L_0 of the Virasoro stress-energy tensor [6] with central charge c, the electron number operator $N = -\sqrt{\nu_H} J_0$ is expressed in terms of the zero mode of the $\widehat{u(1)}$ current and ν_H is the (quantum) Hall filling factor. The trace is taken over the Hilbert space $\mathscr{H}_{\mathrm{edge}}$ for the edge states and might depend on the type and number of the quasiparticles which are localized in the bulk. The modular parameters ζ and τ of the rational CFT [6], which appear in Eq. (11) are related to the temperature T and chemical potential μ as follows

$$\tau = i\pi\frac{T_0}{T}, \quad T_0 = \frac{\hbar v_F}{\pi k_B L}, \quad \zeta = -i\frac{\mu}{2\pi k_B T} = \frac{\mu}{\Delta\varepsilon}\tau, \quad \Delta\varepsilon = \hbar\frac{2\pi v_F}{L}, \tag{12}$$

where v_F is the Fermi velocity on the edge, L is the edge's circumference and k_B is the Boltzmann constant.

When the FQH disk is threaded by perpendicular magnetic field B the edge is affected by the flux of this field BA, where A is the area of the disk and the concrete type of the field B is not important. Therefore we can assume that the disk is threaded by $B_{\text{total}} = B_0 + B$, where B_0 is a constant homogeneous magnetic field, corresponding to the center of a FQH plateau, while B is magnetic field of the Aharonov–Bohm (AB) type. This is very convenient because the AB flux can be treated analytically [28] and the disk CFT partition function $Z_\phi(\tau, \zeta)$ in presence of (dimensionless) AB flux ϕ is simply obtained by a shift in the modular parameter $\zeta \to \zeta + \phi\tau$ [28], i.e.

$$Z_\phi(\tau, \zeta) = Z(\tau, \zeta + \phi\tau), \quad \phi = \frac{e}{h}(B_{\text{total}} - B_0)A, \tag{13}$$

where h is the Plank constant. Interestingly enough, the variation of the side-gate voltage is affecting the QD in the same way [17] as the AB flux ϕ through the *externally induced electric charge* Q_{ext} on QD, which changes continuously with the gate voltage V_g

$$-\frac{C_g V_g}{e} \equiv \nu_H \phi = Q_{\text{ext}}, \tag{14}$$

where C_g is the capacitance of the gate. Therefore we can use the partition function (13) to compute various thermodynamic quantities as functions of the gate voltage V_g.

The Grand potential on the edge, in presence of AB flux ϕ or gate voltage V_g defined in Eq. (14), can be expressed as

$$\Omega_\phi(T, \mu) = -k_B T \ln Z_\phi(\tau, \zeta), \tag{15}$$

where Z_ϕ is defined in Eq. (13) and the thermal average of the electron number can be computed by [15]

$$\begin{aligned} \langle N_{\text{el}}(\phi)\rangle_{\beta,\mu_N} &= -\frac{\partial\Omega_\phi(\beta, \mu_N)}{\partial\phi} + \nu_H\phi + \nu_H\left(\frac{\mu_N}{\Delta\varepsilon}\right) \\ &= \nu_H\left(\phi + \frac{\mu_N}{\Delta\varepsilon}\right) + \frac{1}{2\pi^2}\left(\frac{T}{T_0}\right)\frac{\partial}{\partial\phi}\ln Z_\phi(T, \mu_N), \end{aligned} \tag{16}$$

where μ_N is the chemical potential of a QD with N electrons. Similarly, the edge conductance G_{is} of the Coulomb blockade island, in presence of AB flux ϕ or gate voltage V_g defined in Eq. (14), can be computed by [15]

$$G_{\text{is}}(\phi) = \frac{e^2}{h}\left(\nu_H + \frac{1}{2\pi^2}\left(\frac{T}{T_0}\right)\frac{\partial^2}{\partial\phi^2}\ln Z_\phi(T,0)\right). \tag{17}$$

As an example the profile of the conductance for the Pfaffian FQH disk is given in Sect. 5.

4.2 Thermopower: A Finer Spectroscopic Tool

The thermopower, or the Seebeck coefficient [25], is defined as the potential difference V generated between the two leads of the SET when the temperatures T_L and T_R of the two leads is different and $\varDelta T = T_R - T_L \ll T_L$, under the condition that $I = 0$. Usually thermopower is expressed [25] as the ratio G_T/G, where G_T and G are the thermal and electric conductances respectively, however for a SET, both G and G_T are 0 in large intervals of gate voltages, called the Coulomb valleys, so it is more convenient to use another expression for thermopower [25]

$$S \equiv -\lim_{\varDelta T\to 0}\frac{V}{\varDelta T}\bigg|_{I=0} = -\frac{\langle\varepsilon\rangle}{eT}, \tag{18}$$

where $\langle\varepsilon\rangle$ is the average energy of the tunneling electrons through the SET. It is intuitively clear that the average tunneling energy can be expressed in terms of the total energies of the QD with $N+1$ electrons and of the QD with N electrons as

$$\langle\varepsilon\rangle^{\phi}_{\beta,\mu_N} = E_{\text{QD}}^{\beta,\mu_{N+1}}(\phi) - E_{\text{QD}}^{\beta,\mu_N}(\phi),$$

where the total QD energy (with N electrons on the QD) can be written (in the Grand canonical ensemble) as

$$E_{\text{QD}}^{\beta,\mu_N}(\phi) = \sum_{i=1}^{N_0} E_i + \langle H_{\text{CFT}}(\phi)\rangle_{\beta,\mu_N}.$$

Here E_i, $i = 1,\ldots,N_0$ are the occupied single-electron states in the bulk of the QD, and $\langle\cdots\rangle_{\beta,\mu}$ is the Grand canonical average of H_{CFT} on the edge at inverse temperature $\beta = (k_BT)^{-1}$ and chemical potential μ. However, because we are working with Grand canonical partition functions the difference of the thermal averages $\langle N(\phi)\rangle_{\beta,\mu_{N+1}} - \langle N(\phi)\rangle_{\beta,\mu_N}$ of the electron numbers of the QDs with $N+1$ and N electrons is not 1 for all values of V_g, as can be seen if we plot this difference using Eq. (16). Therefore, it is more appropriate to express the average tunneling energy as [17]

$$\langle \varepsilon \rangle^{\phi}_{\beta,\mu_N} = \frac{\langle H_{\mathrm{CFT}}(\phi)\rangle_{\beta,\mu_{N+1}} - \langle H_{\mathrm{CFT}}(\phi)\rangle_{\beta,\mu_N}}{\langle N_{\mathrm{el}}(\phi)\rangle_{\beta,\mu_{N+1}} - \langle N_{\mathrm{el}}(\phi)\rangle_{\beta,\mu_N}}. \tag{19}$$

All thermal averages in Eq. (19) can be computed within our CFT approach from the partition function (13) for the FQH edge in presence of AB flux ϕ or gate voltage V_g. For example, the electron number average can be computed from Eq. (16), while the edge energy average can be computed from the standard Grand canonical ensemble expression

$$\langle H_{\mathrm{CFT}}(\phi)\rangle_{\beta,\mu_N} = \Omega_{\phi}(T,\mu_N) - T\frac{\partial \Omega_{\phi}(T,\mu_N)}{\partial T} - \mu_N \frac{\partial \Omega_{\phi}(T,\mu_N)}{\partial \mu} \tag{20}$$

where $\Omega_{\phi}(T,\mu_N)$ is defined in Eq. (15) and the chemical potentials μ_N and μ_{N+1} corresponding to QD with N and $N+1$ electrons respectively are given by [17]

$$\mu_N = -\frac{1}{2}\Delta\varepsilon, \quad \mu_{N+1} = \frac{1}{2}\Delta\varepsilon,$$

where $\Delta\varepsilon$ is defined in Eq. (12). One very interesting thermoelectric quantity is the *thermoelectric power factor* $\mathscr{P}_T$, which is defined as the electric power P generated by the temperature difference ΔT and can be expressed from Eq. (18) in terms of the thermopower S and the electric conductance G as [17]

$$P = V^2/R = \mathscr{P}_T(\Delta T)^2, \quad \mathscr{P}_T = S^2 G, \tag{21}$$

where $R = 1/G$ is the electric resistance of the CB island. In the next section we will calculate the power factor for SET in the Pfaffian FQH state.

5 The Pfaffian Quantum Hall State

Here we will consider the case of the Pfaffian SET with even number of quasiparticles localized in the bulk, which is promising from the experimental point of view. The disk partition function for this case is given by [17]

$$Z^{\mathrm{even}}(\tau,\zeta) = K_0(\tau,2\zeta;8)\mathrm{ch}_0(\tau') + K_4(\tau,2\zeta;8)\mathrm{ch}_{1/2}(\tau') \tag{22}$$

where $q = \mathrm{e}^{2\pi i \tau} = \mathrm{e}^{-\beta\Delta\varepsilon}$, $\beta = (k_B T)^{-1}$ and $\tau' = (v_n/v_c)\tau$ takes into account the anticipated difference in the Fermi velocities v_n and v_c of the neutral and charged modes respectively. The $u(1)$ partition functions K are those of the Luttinger liquid and are defined (upto an unimportant ζ-independent multiplicative factor) as [6, 17]

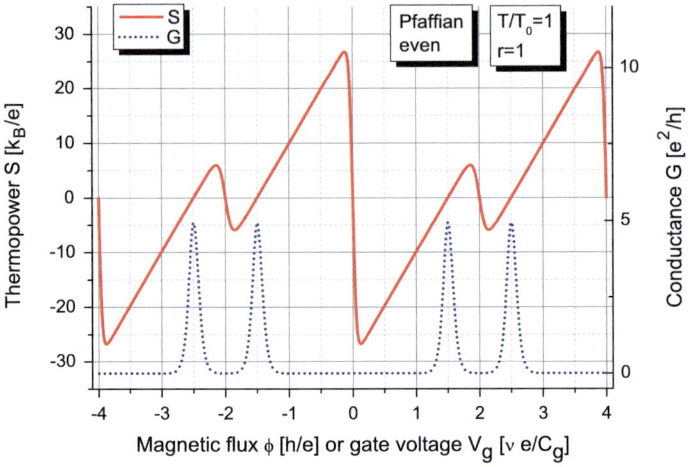

Fig. 8 Thermopower (*left*) and conductance (*right*) of the Pfaffian state with even number of bulk quasiparticles

$$K_l(\tau,\zeta;m) \propto \sum_{n=-\infty}^{\infty} q^{\frac{m}{2}\left(n+\frac{l}{m}\right)^2} e^{2\pi i\zeta\left(n+\frac{l}{m}\right)}, \tag{23}$$

where l corresponds to static quasiparticles localized in the bulk with the electric charge $l/4$. The neutral-sector partition functions in Eq. (22) are defined as [17]

$$\mathrm{ch}_{0,1/2}(\tau) = \frac{q^{-1/48}}{2}\left(\prod_{n=1}^{\infty}(1+q^{n-1/2}) \pm \prod_{n=1}^{\infty}(1-q^{n-1/2})\right),$$

where the subscript denotes the vacuum or the Majorana fermion as a neutral topological charge of the bulk quasiparticles. Using the explicit partition function (22) and substituting again in Eqs. (17), (19), (16), and (20) we can compute the conductance and the thermopower for the Pfaffian state with even number of bulk quasiparticles and plot them for $T = T_0$ in Fig. 8. The centers of the conductance peaks correspond to the zeros of the thermopower, however, the conductance peaks are grouped in pairs and the sawtooth curve of the thermopower is modulated in a corresponding way. This modulation is due to the neutral sector of the corresponding CFT and its vanishes when the ratio $r = v_n/v_c$ becomes much smaller than 1 since then the role of the neutral part is decreased.

Next, using the computed thermopower for the Pfaffian FQH state with even number of bulk quasiparticles and Eq. (21) we plot in Fig. 9 the power factor profile, together with the conductance [17] for $T = T_0$ with $r = 1$. This modulation of power factor's dips could in general be used to estimate experimentally the ratio $r = v_n/v_c$ by measuring the two periods $\Delta\phi_1$ and $\Delta\phi_2$ between the dips because [17] $\Delta\phi_1 = 2 - r$ and $\Delta\phi_2 = 2 + r$, see Fig. 9.

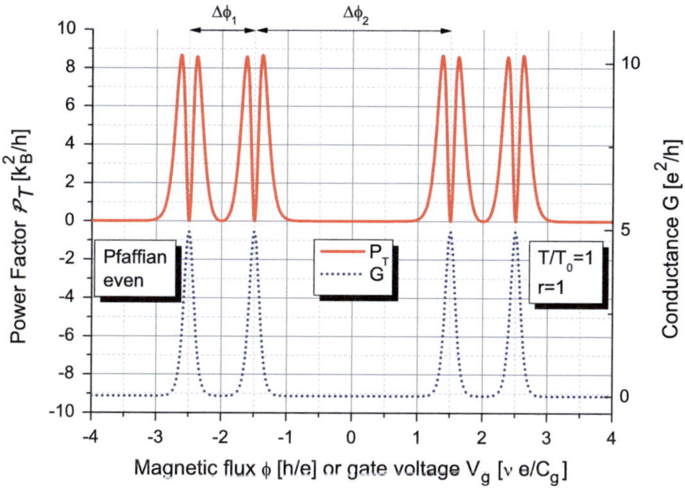

Fig. 9 Thermoelectric power factor (*left*) and conductance (*right*) of the Pfaffian state with even number of bulk quasiparticles

Acknowledgements This work has been partially supported by the Alexander von Humboldt Foundation under the Return Fellowship and Equipment Subsidies Programs and by the Bulgarian Science Fund under Contract No. DFNI-E 01/2 and DFNI-T 02/6.

References

1. Nielsen M, Chuang I (2000) *Quantum Computation and Quantum Information*. Cambridge University Press
2. Shor PW (1997) SIAM. J. Sci. Statist. Comput. 26:1484
3. Kitaev A (2003) Ann. Phys. 303:2. (N.Y.)
4. Sarma SD, Freedman M, Nayak C, Simon SH, Stern A (2008) Rev. Mod. Phys. 80:1083
5. Birman JS (1974) *Braids, links and mapping class groups*, 82nd edn. Princeton University Press. (Ann. Math. Stud.)
6. Di Francesco P, Mathieu P, Sénéchal D (1997) Conformal field theory. Springer, New York
7. Stern A (2008) Ann. Phys. 323:204
8. Georgiev LS (2009) J. Phys. A: Math. Theor. 42:225203
9. Sarma SD, Freedman M, Nayak C (2005) Phys. Rev. Lett. 94:166802
10. Pan W, Xia JS, Shvarts V, Adams DE, Störmer HL, Tsui DC, Pfeiffer LN, Baldwin KW, West KW (1999) Phys. Rev. Lett. 83:3530
11. Pan W, Xia JS, Stormer HL, Tsui DC, Vicente C, Adams ED, Sullivan NS, Pfeiffer LN, Baldwin KW, West KW (2008) Phys. Rev. B 77:075307
12. Bonderson P, Nayak C, Shtengel K (2010) Phys. Rev. B 81:165308
13. Ilan R, Grosfeld E, Stern A (2008) Phys. Rev. Lett. 100:086803
14. Cappelli A, Georgiev LS, Zemba GR (2009) J. Phys. A: Math. Theor. 42:222001
15. Georgiev LS (2010) EPL 91:41001
16. Viola G, Das S, Grosfeld E, Stern A (2012) Phys. Rev. Lett. 109:146801
17. Georgiev LS (2015) Nucl. Phys. B 894:284

18. Georgiev LS (2015) Nucl. Phys. B 899:289
19. Moore G, Read N (1991) Nucl. Phys. B 360:362
20. Eisenstein J, Cooper K, Pfeiffer L, West K (2002) Phys. Rev. Lett. 88:076801
21. Choi H, Kang W, Sarma SD, Pfeiffer L, West K (2008) Phys. Rev. B 77:081301
22. Georgiev LS (2008) Nucl. Phys. B 789:552
23. Georgiev LS (2006) Phys. Rev. B 74:235112
24. Ahlbrecht A, Georgiev LS, Werner RF (2009) Phys. Rev. A 79:032311
25. Matveev K (1999) Lect. Notes Phys. LNP 547:3
26. Fröhlich J, Studer UM, Thiran E (1997) J. Stat. Phys. 86:821
27. Cappelli A, Zemba GR (1997) Nucl. Phys. B 490:595
28. Georgiev LS (2005) Nucl. Phys. B 707:347

Part II
Structure and Dynamics

$Rb_2(^3\Sigma_u^+)$ Dimer Interacting with He Atoms: Quantum Structures of Small Clusters and Reactive Scattering Calculations $Rb + RbHe \rightarrow Rb_2(^3\Sigma_u^+, v) + He$ at Ultralow and Intermediate Energies

Gerardo Delgado-Barrio and Pablo Villarreal

Abstract We review in this work some studies done in our group on complexes involving a rubidium dimer in its lowest triplet state and several helium atoms. By freezing the Rb_2 bond length, a suitable analytical form was fitted to an accurate ab initio 2D potential energy surface (Rodríguez-Cantano et al., J Phys Chem A 116:2394 (2012)) for the three-body complex He-$Rb_2(^3\Sigma_u)$, and energetic and geometric properties of small $(^4He)_N$-$Rb_2(^3\Sigma_u)$ clusters, $1 \leq N \leq 4$, were determined by using a variational treatment. The calculations show a transition in the arrangement of the Helium atoms from $N = 2$, where they tend to be separated across the diatom bound, to $N = 4$, in which a closer packing of the rare gas particles is reached. The deepest well of the He-Rb_2 interaction is placed at the T-shape configuration, a feature which makes the dopant to be parallel to the Helium "minidroplet". In turn, by using the best available full dimensional potential energy surface (Guillon et al., J Chem Phys 136:174307 (2012)) atom-diatom reactive collisions were investigated. A simple model (Rodríguez-Cantano et al., J Chem Phys 142: 164304 (2015)) was applied to estimate reactive probabilities for different values of the total angular momentum. Noticeable values of the probabilities in the ultracold regime, fulfilling numerically the Wigner threshold law, support the feasibility of the reaction process. The rubidium dimer is mainly produced at high vibrational states, and the reactivity is more efficient for a bosonic helium partner than for the fermion one.

PACS number: 34.50.Lf · 34.50.Cx · 34.10.+x · 34.20.-b

1 Introduction

Helium nanodroplets provide an ideal matrix for aggregation and stabilization of a variety of weakly bound complexes and for trapping atoms and molecules in a weakly

---

G. Delgado-Barrio (✉) · P. Villarreal
Instituto de Física Fundamental, IFF-CSIC, Serrano 123, 28006 Madrid, Spain
e-mail: g.delgado@csic.es

A. Tadjer et al. (eds.), *Quantum Systems in Physics, Chemistry, and Biology*,
Progress in Theoretical Chemistry and Physics 30,
DOI 10.1007/978-3-319-50255-7_6

interacting environment. The helium droplet technique combines the benefits of both the gas phase and the classical matrix-isolation techniques [1]. Nanodroplets of ^{4}He and/or ^{3}He of different sizes constitute a good spectroscopic environment where non-homogeneous quantum systems can be studied in detail. Understanding of complex structures of atoms and molecules requires to start with well-defined structures and well-defined distributions of populated states. This calls for spectroscopic studies at very low temperatures. In this regard, spectroscopic experiments involving helium droplet beams (Helium nanodroplet isolation (HENDI)) proved, since their introduction in 1992 [1], to be a versatile method that obtains temperatures below 1 K and offers the possibility of studying well-defined and complex structures of atoms and molecules [2].

Superfluid helium creates a gentle matrix around the impurities and due to the very weak interactions of helium atoms with other atoms or molecules allows measurements of the spectra with high precision. Consequently, helium-nanodroplet spectroscopy enables very accurate probing of molecules or clusters which cannot be investigated in the gas phase due to their instability. Today, the field of helium droplets is well established and several review articles have appeared during the last decade highlighting a variety of experimental as well as theoretical aspects [1, 3–12].

The solvation process of an atomic or molecular impurity in an helium drop depends on the competition between the different interactions He-He and He-impurity. If the stronger interaction is the He-He the impurity will be located outside of the Helium drop, if the situation is the opposite, the dopant will be in the inner part of the drop. The attraction between two helium atoms is very weak (only $\sim$8 cm^{-1}) and thus weaker than most of the He-dopant interactions, nevertheless there are He-impurity potentials smaller than He-He ones, this is the case of He with alkali atoms [13].

Their dimers behave in the same way and have been extensively studied by the groups of Ernst and Stienkemeier: predictions for the observation of KRb spectra under cold conditions and (HENDI) experiments [14, 15] and also with Na_2 [16]. More recent papers of these groups, in this field, are the reinvestigation of the Rb_2 $(2)^3\Pi_g$-$a^3\Sigma_u^+$ band [17], and a wave packet dynamics study of alkali dimers [18]. The conclusion is that the diatom is placed in a dimple on the surface of the helium nanodroplet, which produce small modifications in the spectroscopic properties of the impurity [19].

Several theoretical studies have appeared the last years on alkaline dimers in helium clusters. Bovino et al. [20] have presented Diffusion Monte Carlo (DMC) calculations on Li_2 dimers, singlet and triplet species. They shown that the dimers remain on the surface of small helium clusters, up to 30 particles, in a perpendicular or in a parallel arrangement with respect to the rare gas depending on the $^1\Sigma/^3\Sigma$ spin state. Later our group published a new Potential Energy Surface (PES) for the complex He-$Cs_2(^3\Sigma_u)$ [21], that allowed a Path Integral Monte Carlo (PIMC) [22] and DMC [23] calculations for $(^4He)_N$-$Cs_2(^3\Sigma_u)$, $1 \leq N \leq 30$, (see also Refs. [24–28] for other studies on clusters of dimer with Helium, with Br_2 and ICl).

In a recent paper Guillon et al. have presented a PES for the He-$Rb_2(^3\Sigma_u)$ system and using this PES they have performed DMC and PIMC calculations on

$(^4He)_N$-$Rb_2(^3\Sigma_u)$ clusters [29]. They have interpolated their ab initio points by using the RKHS method [30]. To continue the analysis of this weakly bound many-particle complexes, we have reported a new He-$Rb_2(^3\Sigma_u)$ PES [31] and after that, we did some variational and DMC studies of the $(^4He)_N$-$Rb_2(^3\Sigma_u)$ clusters, $1 \leq N \leq 4$; finally we have compared our results with those of Guillon et al. [29]. Both PESs are very similar, but our model makes the dynamical calculations very fast, reducing the computational effort. Due to the similarities between the He-$Cs_2(^3\Sigma_u)$ and He-$Rb_2(^3\Sigma_u)$ PES's, one can expect many similar results in both systems, assumption that was confirmed [22–24, 26–29].

High-spin alkali dimers have attracted during the last decade considerable attention mainly due to their role in producing samples of cold molecules [32]. Translationally cold alkali-metal molecules have been produced [33–36]. The molecules are usually produced in weakly bound states corresponding to high vibrational excitations [37, 38]. In the earlier studies they have focused on heteronuclear and homonuclear mixtures of the lighter species, Li, Na, and K, more recently on the heavier Rb and Cs molecules [39], while currently have extended to ultracold polar alkali-alkali earth molecules [40, 41].

Rubidium dimers, in their lowest triplet state, have been formed on the surface of cold helium droplets [42, 43], as it was expected due to the very weak high-spin alkali dimers and helium atoms interaction, the alkali metal atoms remain on the helium surface, where they can *skate*, forming cold molecules through collisions [43, 44], These collisions lead preferentially to high spin (triplet) state alkali dimers [17, 43], as the binding energy that is released during the collision is larger for the low spin (singlet) state for which the subsequent evaporation of the droplet could be significant. In fact, singlet states, with a stronger bond, may cause either the desorption of the alkali molecule or the complete evaporation of small helium clusters, and thus, triplet-state alkali dimers were selectively produced [15]. Later on the dynamics of these complexes have been extensively studied (see Refs. [45–47]).

We have addressed to the collisional formation of rubidium dimers in the presence of just one helium atom as a preliminary step for analyzing the feasibility of the *skating* mechanism. We have calculated reactive probabilities for the collisional process $^{87}Rb + {}^{87}Rb{}^{n}He \rightarrow Rb_2(^3\Sigma_u^+, v) + {}^{n}He$, n = 3, 4, by means of a simplified model [48] which is based on the Rosen treatment to study the lifetime of a linear triatomic molecule [49]. It is clear that a full scattering treatment would be better by solving the corresponding close-coupling equations in the energy domain. Nevertheless, in this case we have very low collision energies, together with the moderate well-depths leading up a long spatial range of integration with small steps. Even more if several total angular momentum values are taken, all of these lead to an extremely difficult computational problem.

Guillon et al. have calculated an accurate three-dimensional (PES) of He-Rb_2 [50], with the proper long range behavior. They have also reported in that paper a study on the vibrational relaxation of Rb_2 molecules by collisions with He for a total angular momentum $J = 0$ at low vibrational states of the rubidium dimer and after that they have extended to higher ($J \leq 20$) angular momenta [51]. In our work [48] the coupling of final vibrational states of rubidium dimer was neglected, as well as the hyperfine, spin-spin, and spin-rotation terms in the Rb_2 molecule.

The structure of the article is as follows: in Sect. 2 we describe the interaction potentials employed in the present work. In Sect. 3 we provide details on the variational and reactive theoretical treatments. In Sect. 4 we present and discuss the main approximations involved in our calculations and the results obtained. Finally, our conclusions and some prospectives are presented in Sect. 5.

2 The Interaction Potentials

2.1 2D Model

For describing the PES of He-$Rb_2(^3\Sigma_u)$ complex $W(r, R, \theta)$, Jacobi coordinates (r, R, θ) were used, where $\boldsymbol{r}$ is the vector joining the two atoms of the dimer ($r \equiv |\boldsymbol{r}|$), $\boldsymbol{R}$ is the vector of the He atom with respect to the center of mass of the Rb_2 ($R \equiv |\boldsymbol{R}|$), and θ is the angle between $\boldsymbol{R}$ and $\boldsymbol{r}$.

All ab initio calculations were performed with the Gaussian03 program [52] and all the details are shown in Ref. [31]. These calculations to determine $W(r, R, \theta)$ were done with the spin restricted single and double excitations coupled cluster method with perturbative triples [RCCSD(T)] correlating only the valence electrons.

In the first step, we have calculated, with the CCSD(T) method, the potential curve of the Rb_2, by using different basis sets and several effective core potentials (ECP). In Table 1 we present for the isolated dimer, the values of r_{eq}, and of D_e (the well-depth). The agreement between recent experimental data, on the dimer Expt. Beser et al. [53], and the theoretical results Soldán [54] are quite good. To determine the three body PES, keeping on the computational effort within aceptable limits, we selected the def2-TZVP [55] basis set, which uses the ECP28MWB pseudopotential [56].

The first two bound levels held by this potential are: $E_0 = -188.289$ and $E_1 = -176.372$ cm^{-1}. If we use the corresponding wave-functions we can obtain the averaged values $\bar{r}_v = \langle v\,|r|\,v\rangle$, $v = 0, 1$, and the $\bar{r}_{01} = \langle v = 0\,|r|\,v = 1\rangle$ non-diagonal term. The values in (Å) are: $\bar{r}_0 = 6.39$, $\bar{r}_1 = 6.46$, and $\bar{r}_{01} = 0.18$. The small

Table 1 Theoretical values of equilibrium distance, r_{eq}, and well-depth D_e, of the rubidium dimer, together with some experimental data

Ref. [31]	r_{eq} (Å)	D_e (cm^{-1})
def2-TZVP [55]	6.35	194.37
def2-TZVPP [55]	6.35	196.80
def2-QZVP [55]	6.35	196.14
Expt. Beser et al. [53]	6.069	241.45
Soldán [54]	6.208	229.6

values of the difference $\bar{r}_0 - r_{eq} = 0.04$ Å and $\bar{r}_{01}$, together with the weakness of the He-Rb_2 interaction as compared with the vibrational diatomic spacing, make almost negligible the difference of the results when we take the rubidium dimer as a rigid rotor, with bond-length fixed at r_{eq} within the complex, or to take into account its vibrational content at $v = 0$.

In order to decide the basis set for the He atom we have calculated the interaction energies, using the supermolecular approach including the BSSE correction [57], with several basis sets in [RCCSD(T)]. Finally we have used the aug-cc-pVQZ that produces small differences with the results provided by a better, but which leads to greater computational effort, aug-cc-pV5Z basis [29].

To represent the PES for the $HeRb_2$, we have used an analytical function to fit the CCSD(T) ab initio points, where the diatomic bond-length remains at its equilibrium distance $r_{eq} = 6.35$ Å. The 2D function $W(r_{eq}, R, \theta)$, was written as an addition of the two angle-dependent He-Rb interactions, each of them represented by a Lennard-Jones function, $W(r_{eq}, R, \theta) \equiv V(\widetilde{R}_1, \theta) + V(\widetilde{R}_2, \theta)$, with $V(x, \theta) = d(\theta)\left[\left(\frac{\bar{x}(\theta)}{x}\right)^{12} - 2\left(\frac{\bar{x}(\theta)}{x}\right)^{6}\right]$. where $\widetilde{R}_i$ is the distance from the helium to the i-th rubidium atom, and $d(\theta)$ and $\bar{x}(\theta)$ are, at a fixed orientation θ, the well-depth and the equilibrium distance, respectively.

By using a cubic spline fitting to the calculated ab initio points one gets values of d and $\bar{x}$ for each θ angle and then a two-parameter expression of these coefficients on the orientation angle, $d(\theta) = d(0) + \left[d\left(\frac{\pi}{2}\right) - d(0)\right] \sin^{2\alpha}(\theta)$, and a similar expression for $\bar{x}(\theta)$ are assumed. The final fit to the ab initio points produces values of $\alpha = 15.214$ for $d(\theta)$ and $\alpha = 6.173$ for $\bar{x}(\theta)$. If we plot $W(r_{eq}, R, \theta)$ obtained using the procedure above described together with the ab initio data, the agreement is really good.

2.2 3D Model

The 3D PES describing the He-$Rb_2(^3\Sigma_u)$ complex is decomposed as the sum of three two-body (2B) terms, plus a three-body (3B) residual part:

$$V_{\mathrm{TOT}}(R_1, R_2, R_3) = V^{\mathrm{2B}}_{\mathrm{Rb-Rb}}(R_1) + V^{\mathrm{2B}}_{\mathrm{Rb-He}}(R_2) + V^{\mathrm{2B}}_{\mathrm{Rb-He}}(R_3) + V^{\mathrm{3B}}(R_1, R_2, R_3), \quad (1)$$

where R_1 and R_3 are the distances from the incoming Rb atom to the Rb and He atoms, in the RbHe complex, respectively, while R_2 is the diatomic Rb-He distance.

We have analytically represented the 2B terms, as:

$$V^{\mathrm{2B}}(R) = V_M(R) \times f(R; R_m, \beta) + \left(\frac{C_6}{R^6} + \frac{C_8}{R^8} + \frac{C_{10}}{R^{10}}\right) \times [f(R; R_m, \beta) - 1]. \quad (2)$$

Table 2 Parameters of the 2B interactions

	D (cm^{-1})	α (Å^{-1})	R_e (Å)	β (Å^{-1})	R_m (Å)	C_6 ($E_h a_0^6$)	C_8 ($E_h a_0^8$)	C_{10} ($E_h a_0^{10}$)
Rb-Rb	223.6471	0.6928	6.1067	1.3280	10.8116	4.707×10^3	5.730×10^5	7.665×10^7
Rb-He	0.9363	0.8130	7.4497	1.6820	9.5921	4.469×10^1	3.145×10^3	3.038×10^5

In this equation $V_M(R)$ is a Morse potential describing the well region, C_n, $n = 6, 8, 10$, are the long-range coefficients, and $f(R; R_m, \beta) = \{1 - \tanh[\beta(R - R_m)]\}/2$ is a matching function connecting the two ranges.

All the parameters of the 2B part were obtained by using non-linear least-square fittings, and are shown in Table 2. Concerning the 3B part, it was fitted by a reproducing kernel Hilbert space (RKHS) method [30] following a recent paper Ref. [50] where the ab initio points were obtained.

3 Theoretical Approachs

3.1 *The Variational Treatment*

The Hamiltonian for the $HeRb_2$ system, where the diatomic is fixed at the equilibrium distance, can be written in Jacobi coordinates as [58, 59]:

$$H = -\frac{\hbar^2}{2\mu_{Rb_2}}\nabla_{\boldsymbol{r}}^2 - \frac{\hbar^2}{2\mu}\nabla_{\boldsymbol{R}}^2 + W(\boldsymbol{R}), \tag{3}$$

where μ_{Rb_2} and μ are the Rb_2 and the He-Rb_2 reduced masses, respectively. $W(\boldsymbol{R})$ is the $HeRb_2$ interaction, $W(r_{eq}, R, \theta)$, where $\boldsymbol{R} \equiv (R, \theta)$. In the first part of this work the dimer is treated, as a $^1\Sigma$ rigid rotor at its equilibrium distance, r_{eq}, and therefore the spin effects are neglected [21, 60] and its Hamiltonian has only the kinetic term. The angular momentum $\boldsymbol{j}$ is associated to the vector $\boldsymbol{r}$ and the angular momentum $\boldsymbol{l}$ to $\boldsymbol{R}$, yielding the total angular momentum vector $\boldsymbol{J} = \boldsymbol{j} + \boldsymbol{l}$.

In a body-fixed (BF) frame, where the Z^{BF} axis is parallel to $\boldsymbol{R}$, we have used the basis set of wave-functions:

$$\Phi_{nj\Omega}^{JM}(\hat{\boldsymbol{r}}, \boldsymbol{R}) = f_n(R) W_{j\Omega}^{JM}(\hat{\boldsymbol{r}}, \hat{\boldsymbol{R}}). \tag{4}$$

In this basis set f_n are radial functions associated with the He-Rb_2 stretching motion. In order to get these f_n, in a first step we have numerically solved, for the ground state energy level, E_k, a set of Schrdinger equations at different orientations $\{\theta_k\}_{k=1,K}$: $\left\{-\frac{\hbar^2}{2\mu}\frac{\partial^2}{\partial R^2} + W(R;\theta_k)\right\}\phi(R;\theta_k) = E_k\phi(R;\theta_k)$. After that, the eigenfunctions $\phi(R;\theta_k)$ are orthogonalized, given rise to an orthonormal set $\{f(R)\}_{n=1,K}$. The angular functions $W_{j\Omega}^{JM}$ depend on unitary vectors $\hat{\boldsymbol{r}}$, $\hat{\boldsymbol{R}}$, with $\hat{\boldsymbol{r}} \equiv (\theta, \phi)$ being the orientation in the BF frame and $\hat{\boldsymbol{R}} \equiv (\theta_R, \phi_R)$ the orientation with respect to a space-fixed (SF) frame. A symmetry-adapted basis set composed of functions of Eq. (4) is used to expand the total wave-function, taking into account the relevant symmetry operators of the system [61, 62], and finally, the Hamiltonian matrix is diagonalized [63].

For the complex He_2Rb_2, its Hamiltonian, in satellite coordinates and also at r = r*eq*, can be expressed as [61]:

$$H = -\frac{\hbar^2}{2\mu_{Rb_2}}\nabla_r^2 + \sum_{i=1}^{2}\left(-\frac{\hbar^2}{2\mu}\nabla_i^2 + W(\boldsymbol{R}_i)\right) - \frac{\hbar^2}{2m_{Rb}}\nabla_1 \cdot \nabla_2 + V\left(|\boldsymbol{R}_1 - \boldsymbol{R}_2|\right), \quad (5)$$

where m_{Rb} is the rubidium mass, $\boldsymbol{R}_i$ are the coordinates of the two rare gas atoms with respect to the center of mass of the Rb_2, and $V\left(|\boldsymbol{R}_1 - \boldsymbol{R}_2|\right)$ is the potential between the two Helium atoms [61].

In this case the total wave-function in the Z^{BF} axis was written as a linear combination of the basis set [61]:

$$\Phi^{JK}_{q_1 q_2 L\Omega}(\hat{\boldsymbol{r}}, \hat{\boldsymbol{R}}_1, \hat{\boldsymbol{R}}_2) = f_{n_1}(R_1) f_{n_2}(R_2) W^{JK}_{l_1 l_2 L\Omega}(\hat{\boldsymbol{r}}, \hat{\boldsymbol{R}}_1, \hat{\boldsymbol{R}}_2), \quad (6)$$

where we have collected, for each helium atom, in a general index: $\{q_i\} \equiv \{l_i, n_i\}$, the corresponding orbital quantum number and the He vibration; with L the quantum number corresponding to the total orbital angular momentum $\boldsymbol{L} = \boldsymbol{l}_1 + \boldsymbol{l}_2$ and $\hat{\boldsymbol{R}}_k \equiv (\theta_k, \phi_k)$ the orientations of the particles 1 and 2 in the BF frame. The f_k's are the radial functions obtained as before, and $W^{JK}_{l_1 l_2 L\Omega}(\hat{\boldsymbol{r}}, \hat{\boldsymbol{R}}_1, \hat{\boldsymbol{R}}_2)$ are the angular functions. All the details are shown in [61].

In the case of He_NRb_2 with $N \geq 3$ we have used an approximate method [22], where the Hamiltonian was written as:

$$H = -\frac{\hbar^2}{2\mu_{Rb_2}}\nabla_r^2 + \frac{N}{2}\sum_{i=1}^{2}\left(-\frac{\hbar^2}{2\mu}\nabla_i^2 + W(\boldsymbol{R}_i)\right) + \binom{N}{2} V\left(|\boldsymbol{R}_1 - \boldsymbol{R}_2|\right). \quad (7)$$

3.2 The Reaction Model

In this subsection we describe an approximate method, related with an old idea proposed by Rosen in 1933 [49] to study the lifetimes of unstable linear triatomic molecules. The main idea, in the Rosen approach, was to choose a set of coordinates which produces an almost separable problem. It would be separable if one neglects some of the smaller terms in the Hamiltonian, which are later introduced as a perturbation that contains the couplings among the zero-order states via a Golden Rule expression. In a previous paper [48] we extended the Rosen treatment to study a full-collision process instead a half-collision (dissociation) one. In that paper we were studying the reaction of a rubidium atom that collides with a HeRb dimer to produce $Rb_2(^3\Sigma_u^+) + He$. We need to calculate the reaction probability of a rubidium atom that collides at an energy ϵ_1 with a HeRb dimer to produce $Rb_2(^3\Sigma_u^+, \ell'_1, v)$,where the He atom is ejected with an energy ϵ_2 and an orbital angular momentum ℓ'_2.

The Hamiltonian describing the Rb-Rb-He system can be written in satellite coordinates $\{\boldsymbol{R}_k\}$, after separation of the center of mass motion, as [61]:

$$H = -\frac{\hbar^2}{2\mu_1}\left(\frac{\partial^2}{\partial R_1^2} + \frac{2}{R_1}\frac{\partial}{\partial R_1}\right) + \frac{\boldsymbol{\ell}_1^2}{2\mu_1 R_1^2} + V_{Rb_2}^{2B}(R_1) - \frac{\hbar^2}{2\mu_2}\left(\frac{\partial^2}{\partial R_2^2} + \frac{2}{R_2}\frac{\partial}{\partial R_2}\right)$$
$$+\frac{\boldsymbol{\ell}_2^2}{2\mu_2 R_2^2} + V_{RbHe}^{2B}(R_2) + V_{RbHe}^{2B}(R_3) + V^{3B}(R_1, R_2, R_3) - (\hbar^2/m_{Rb})\boldsymbol{\nabla}_{\boldsymbol{R}_1}\cdot\boldsymbol{\nabla}_{\boldsymbol{R}_2}, \tag{8}$$

where, μ_1 is the reduced mass of the Rb_2 and μ_2 is the reduced mass of the RbHe complex; $\boldsymbol{R}_k$ are the vectors going from the central Rb atom to the other Rb atom ($k = 1$) or to the He atom ($k = 2$) and $\boldsymbol{\ell}_k$ are the angular momenta associated with $\boldsymbol{R}_k$. Denoting the angle formed by $\boldsymbol{R}_1$ and $\boldsymbol{R}_2$ as γ the R_3 distance is defined as $R_3 = \sqrt{R_1^2 + R_2^2 - 2R_1R_2\cos\gamma}$. In a space-fixed coordinate system we have used the following zero-order states describing the reactants (*in*), and the products (*out*):

$$\Psi^{(in)}(\boldsymbol{R}_1, \boldsymbol{R}_2) = \frac{\phi_{\epsilon_1}(R_1)\varphi_{v_0}(R_2)\mathcal{Y}_{\ell_1\ell_2}^{JM}(\hat{\boldsymbol{R}}_1, \hat{\boldsymbol{R}}_2)}{R_1R_2}$$
$$\Psi^{(out)}(\boldsymbol{R}_1, \boldsymbol{R}_2) = \frac{\phi_v(R_1)\varphi_{\epsilon_2}(R_2)\mathcal{Y}_{\ell'_1\ell'_2}^{JM}(\hat{\boldsymbol{R}}_1, \hat{\boldsymbol{R}}_2)}{R_1R_2}, \tag{9}$$

where the ϕ_{ϵ_1} and ϕ_v radial functions correspond to the Rb_2 molecule in a continuum (incoming) or a discrete (outgoing) state, respectively, while the φ radial functions describe similar states of the HeRb subunit. In Eq. (9), the $\mathcal{Y}_{\ell_1\ell_2}^{JM}$ are angular functions in the coupled representation,

$$\mathcal{Y}_{\ell_1\ell_2}^{JM}(\hat{\boldsymbol{R}}_1, \hat{\boldsymbol{R}}_2) = (-1)^J\sqrt{2J+1}\sum_{\omega}\begin{pmatrix}\ell_1 & \ell_2 & J\\ -\omega & \omega - M & M\end{pmatrix} \times Y_{\ell_1\omega}(\theta_1, \phi_1)Y_{\ell_2 M-\omega}(\theta_2, \phi_2), \tag{10}$$

with $\hat{\boldsymbol{R}}_k = \boldsymbol{R}_k/R_k = (\theta_k, \phi_k)$ are unit vectors, $\begin{pmatrix}\cdot & \cdot & \cdot\end{pmatrix}$ are 3-j symbols, and $Y_{\ell_k\omega}$ are spherical harmonics. J is the quantum number associated with the total angular momentum $\boldsymbol{J} = \boldsymbol{\ell}_1 + \boldsymbol{\ell}_2$ with third component M.

Denoting by $W(R_1, R_2, cos\gamma) = V_{RbHe}^{2B}(R_3) + V^{3B}(R_1, R_2, R_3)$ the two last potential terms in Eq. (8), and taking into account that average values of $cos\gamma$ become zero

[61, 64], $\langle \mathcal{Y}_{\ell_1\ell_2}^{LM} | cos\gamma | \mathcal{Y}_{\ell_1\ell_2}^{LM} \rangle = 0$, we consider a Taylor expansion around $cos\gamma = 0$ up to the first order,

$$W(R_1, R_2, cos\gamma) \approx W_0(R_1, R_2) + cos\gamma\, \Delta W(R_1, R_2) \tag{11}$$

where $W_0 = W(R_1, R_2, cos\gamma = 0)$ and $\Delta W = \partial W / \partial cos\gamma|_{cos\gamma=0}$. For this later contribution the portion coming from the 2B interaction is obtained by multiplying the analytic derivative of Eq. (2) with $\partial R_3 / \partial cos\gamma = -R_1 R_2 / R_3$. Regarding the 3B contribution, we have used the RKHS fitting method followed in Ref. [50].

The ϕ and φ functions in Eq. (9) are solutions of the following Schrödinger equations:

$$\begin{aligned}
&\left[-\frac{\hbar^2}{2\mu_2}\frac{\partial^2}{\partial R_2^2} + V_{\text{RbHe}}^{2B}(R_2) - E_0\right]\varphi_{v_0}(R_2) = 0\\
&\left[\frac{\hbar^2}{2\mu_1}\left(-\frac{\partial^2}{\partial R_1^2} + \frac{\ell_1(\ell_1+1)}{R_1^2}\right) + V_{\text{Rb}_2}^{2B}(R_1) + \langle \varphi_{v_0}(R_2)|W_0|\varphi_{v_0}(R_2)\rangle - \epsilon_1\right]\phi_{\epsilon_1}(R_1) = 0\\
&\left[\frac{\hbar^2}{2\mu_1}\left(-\frac{\partial^2}{\partial R_1^2} + \frac{\ell_1'(\ell_1'+1)}{R_1^2}\right) + V_{\text{Rb}_2}^{2B}(R_1) - \varepsilon_v^{(\ell_1')}\right]\phi_v^{(\ell_1')}(R_1) = 0\\
&\left[\frac{\hbar^2}{2\mu_2}\left(-\frac{\partial^2}{\partial R_2^2} + \frac{\ell_2'(\ell_2'+1)}{R_2^2}\right) + V_{\text{RbHe}}^{2B}(R_2) + \left\langle \phi_v^{(\ell_1')}(R_1)|W_0|\phi_v^{(\ell_1')}(R_1)\right\rangle - \epsilon_2\right]\varphi_{\epsilon_2}^{(\ell_2')}(R_2) = 0.
\end{aligned} \tag{12}$$

The first equation corresponds to the unique existing eigenstate $\varphi_{v_0=0}$ for $v_0 = \ell_2 = 0$ of the HeRb diatomics, with eigenvalue E_0, and the second equation to a continuum function ϕ_{ϵ_1} of the incoming Rb atom with kinetic energy ϵ_1, respectively. In turn, the two last equations describe an exiting Rb_2 molecule in a ro-vibrational (ℓ_1', v) state, and an ejected He atom with kinetic energy ϵ_2 respectively, Eq. (12). All the calculations are performed *on shell* energy, that is, $\epsilon_1 + E_0 = \varepsilon_v^{(\ell_1')} + \epsilon_2$.

Extending now the Rosen calculation of discrete-continuum transition rates for unstable linear triatomic molecules [49] towards the present full-collision process, for a given total angular momentum J, one estimates the reaction probability of a rubidium atom colliding at an energy ϵ_1 with a HeRb dimer to produce $Rb_2(^3\Sigma_u^+, \ell_1', v)$ while ejecting the He atom with an energy ϵ_2 and an orbital angular momentum ℓ_2' as

$$P_{\ell_1', v}^J(\epsilon_1) = \left|\langle \Psi^{(in)} | (cos\gamma\, \Delta W + T_{12}) | \Psi^{(out)} \rangle\right|^2. \tag{13}$$

Summing over the vibrational states of Rb_2 gives a total probability of $P_{\ell_1'}^J(\epsilon_1) = \sum_v P_{\ell_1', v}^J(\epsilon_1)$. The two terms in Eq. (13) are small enough to apply first order perturbation theory and they allow for a partially analytical evaluation. The first term, involves a double quadrature, [61, 64] and the second contribution, involves single quadratures [61].

The expression of the partial reactive cross sections is: $\sigma_r^J(\epsilon_1) = \frac{\pi\hbar^2}{2\mu_1\epsilon_1}\sum_{\ell'_1,v} P^J_{\ell'_1 v}(\epsilon_1)$, and finally the total cross section is given by:

$$\sigma_r(\epsilon_1) = \sum_J (2J+1)\sigma_r^J(\epsilon_1) \tag{14}$$

4 Results

4.1 Bound States

All the numerical details in the variational calculations of bound states are shown in a previous paper [31], in order to get the radial wave-functions we have used a Truhlar and Numerov-Cooley procedure [65]. In the full process the energy was converged to within 10^{-3} cm^{-1}.

In Fig. 1 we plot the cuts of the computed PES of the T-shaped and the collinear configurations, together with the He-He interaction (V at Eq. 5) which was described by the Aziz-Slaman semiempirical potential [66]. The cuts in the figure show that the T-shape complex, with an equilibrium distance of $R_{min} \sim 6$ Å, is stronger bounded than the collinear one, with an equilibrium distance of $R_{min} \sim 11$ Å. On the other hand we can see that the He-He interaction is deeper than the T-shaped complex He-Rb_2, and as it was mentioned before, so in a big cluster of He_N-Rb_2 a packing of the rare gas atoms is most likely to happen, with the ensuing isolation of the dimer. Nevertheless, it constitutes only a possibility, and a quantum mechanics calculation would be necessary to confirm this point.

In Table 3 one can see how, our variational and DMC energies for HeRb_2, match quite well, with the results in Ref. [29]. In the following, we have used the

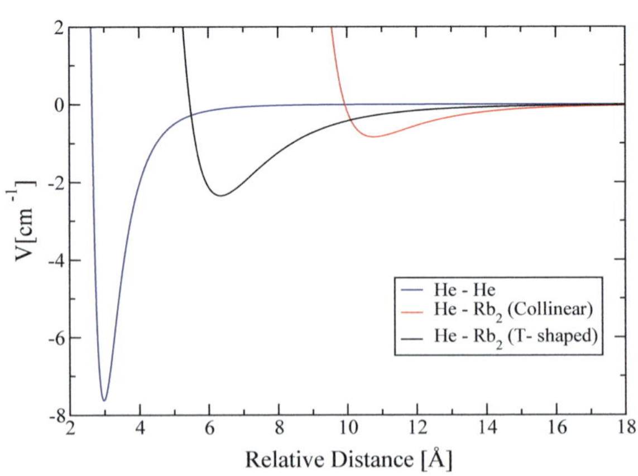

Fig. 1 A comparison of the computed PES for the He-Rb_2 complex at extreme orientations together with the He-He interaction from Ref. [66]

three-body complex: (1) to asses the quality of our fit to the calculated ab initio points, (2) to analyze the relevance of the dopant vibration, and (3) to estimate the temperature effect by including the excited rotational states. Regarding the first issue, we applied a RKHS interpolation method [30] in a similar way to what was done in Ref. [29]. The variational ground energy of $HeRb_2$, with the bond length fixed at $r_{eq} = 6.35$ Å, becomes -0.1150 cm^{-1}, which only differs 0.004 cm^{-1} from the energy calculated with the analytical fit. Concerning the second point, and using again the RKHS procedure, we have repeated the calculation with the bond length fixed at $r_0 = 6.39$ Å, that is the averaged diatomic distance in the ground vibrational state. The result is now -0.1146 cm^{-1}, which confirms the small contribution of the Rb_2 vibration. As regards to the third issue, we found, together with the $J = 0$ state at -0.1195 cm^{-1}, three additional bound levels at -0.1003, -0.062, and -0.006 cm^{-1} for $J = 1$, 2, and 3, respectively. Taking into account that the whole rotation of the complex really corresponds to a rotation of the rubidium diatomics, and by using a Boltzmann averaging,

$$\langle E(T)\rangle = \frac{\sum_J (2J+1)\, E_j\, e^{-\frac{E_J}{KT}}}{\sum_J (2J+1)\, e^{-\frac{E_J}{KT}}},$$

we are able to estimate that for very low temperatures the system remains bound. In particular, at a temperature of $T = \frac{1}{25.6}$ K one gets an averaged energy of $\langle E(T)\rangle = -0.096$ cm^{-1}, very different from the $+0.09$ cm^{-1} value reported in Ref. [29] (belonging to the continuum), and slightly above our results at $T = 0$ K.

The results obtained for the ground state energies for the rest of complexes, He_NRb_2, with $N \in \{2, 3, 4\}$ are also presented in Table 3, where a comparison with the values of Ref. [29] and for pure He aggregates showed in Lewerenz [67] are also included. These results indicate that all the doped species are bound, because their energies are below those for pure ^4He complexes, with both sets of results close to each other and the variational treatment yielding more deeply bound clusters than the results provided by the DMC method.

In general, the results provided by Guillon et al. on DMC binding energies are larger than ours (except He_4Rb_2). In the case $N = 1$ one possible explanation is that the PES used by them exhibits a slightly deeper well, with a ratio of the potential well minima of $2.59/2.35 = 1.1$. To understand this point, we have repeated the calculations using our PES but multiplied by a factor of 1.1 and obtained the DMC result for $N = 1$ that appears in the table with $E^{DMC}(W \times 1.1) < E^{DMC}$ [29] $< E^{DMC}(W)$.

For the cluster with $N = 2$ our results almost follow the independent particles model (IPM) and $E(He_2Rb_2) \simeq 2 \times E(HeRb_2)$, because the correlation energy of the rare gas particles $\simeq 0$, which make sense for this case since the two He adatoms of the complex are placed so far from one another that their interaction energy is negligible, (as we will see in the next Figures that display their spatial distributions). For bigger clusters, He_3Rb_2 and He_4Rb_2, the situation are very different and they no

Table 3 Variational and DMC energies (cm^{-1}) of the He_NRb_2 complexes, with $N \in \{1, 2, 3, 4\}$, together with those from Guillon et al. in Ref. [29]. For $N \geq 2$, energies of pure clusters reported by Lewerenz [67]

	$N = 1$	$N = 2$	$N = 3$	$N = 4$
Variational [31]	−0.119	−0.218	−0.494	−1.671
DMC [31]	-0.106 ± 0.005	-0.197 ± 0.024	-0.481 ± 0.022	-1.409 ± 0.011
DMC [29]	-0.1341 ± 0.0003	-0.3908 ± 0.0006	-0.850 ± 0.004	-1.52 ± 0.01
PIMC [29]	$+0.09 \pm 0.04$	-0.18 ± 0.01	-0.54 ± 0.03	-1.15 ± 0.03
Estimate at $T = 1/25.6$ K [31]	−0.096			
VAR this work × 1.1	−0.166			
DMC this work × 1.1	-0.151 ± 0.005			
Pure He clusters [67]		-8.6×10^{-4}	−0.0872	−0.388

longer follow the IPM model with an important interaction energy among the rare gas atoms.

Figure 2 displays the variational angular distribution $D_1(\cos\theta)$ for the four studied clusters. We can observe that, by putting more helium atoms, it follows the initial tendency exhibited by the three-body complex, i.e., the Helium atoms are placed in a region around the T-shaped geometry with respect to the diatomic dimer. The Fig. 3 shows the radial one-particle distribution $D_1(R)$ for these clusters. The calculations also evidence that, the larger the cluster the shorter the distance of the solvent atoms to the diatom: from $R \sim 10 - 11$ Å for $N = 2$ to $R \sim 8.5 - 10$ Å for $N = 4$. One can also see that, all distributions become more localyzed with narrower regions of max-

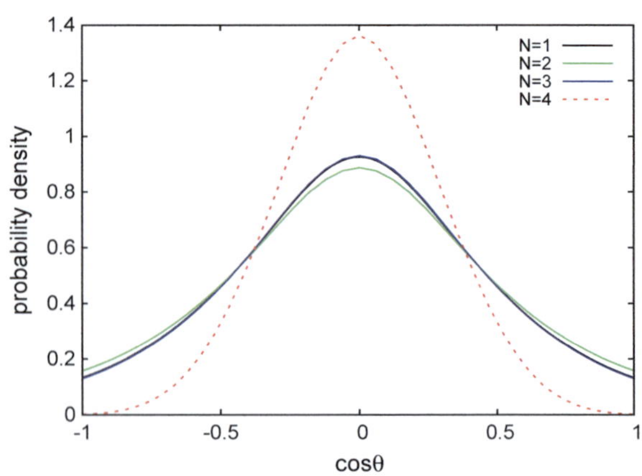

Fig. 2 Computed angular distributions, $D_1(\cos\theta)$, for the clusters $^4He_NRb_2$, $N \in \{1, 2, 3, 4\}$

Fig. 3 Radial distributions, $D_1(R)$, for the clusters $^4He_NRb_2$, $N \in \{1, 2, 3, 4\}$

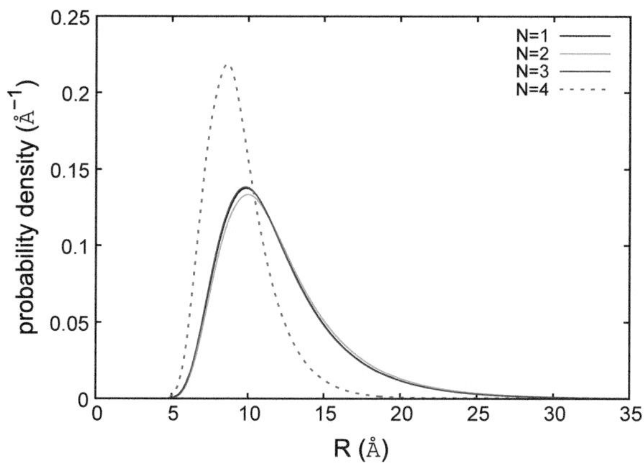

Fig. 4 Computed two-particle angular distributions, $D_2(\cos\gamma)$, for the clusters $^4He_NRb_2$, $2 \leq N \leq 4$

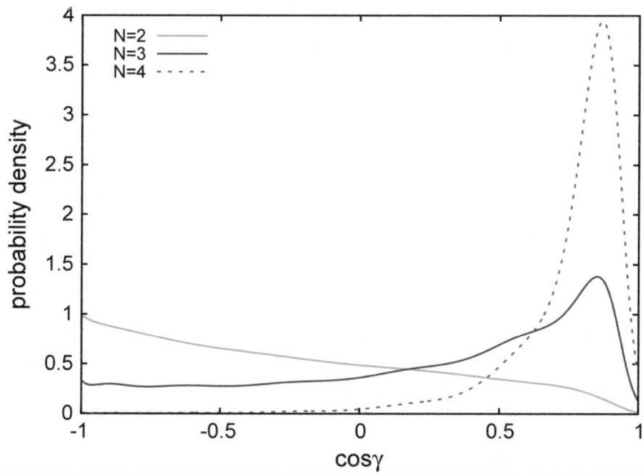

ima as N increases, indicating that the systems with the stronger binding energies exhibit a more compact positioning of the He adatoms around the weakly bound dimer.

In Fig. 4 we show the two-particle angular distributions. These distributions are presented as function of the angle γ formed between any two He atoms with respect to the dopant's center of mass, taken as the vertex of such angle. It is worthy to note the dramatic change of the distributions when one goes from $N = 2$ to the $N = 3$ and $N = 4$ cases.

The four-body cluster appears as a cross-like structure with the two lighter adatoms placed, away from one another, almost along a straight line that cuts the dimer at its midpoint, this cluster is also shows a strong delocalization of the He atoms radial positions, as indicated by the two-particle radial distribution that appears in Fig. 5.

When one or two more He atoms are added to He_2Rb_2, one sees that the angular distributions are markedly modified (see Fig. 4); the subregion of the solvent atoms

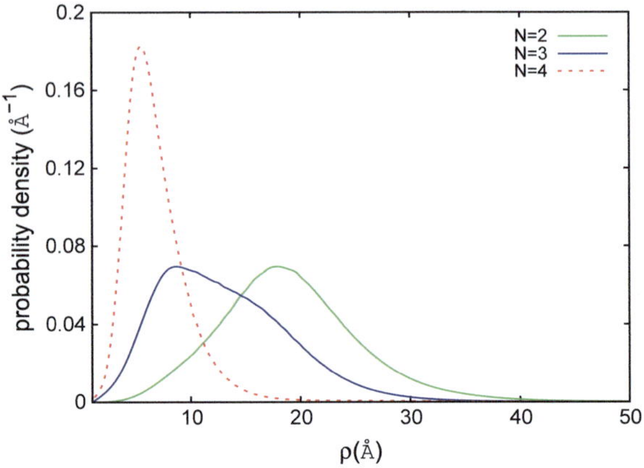

Fig. 5 Two-particle radial distributions, $D_2(\rho)$, for the clusters $^4He_NRb_2$, $2 \le N \le 4$

acquires greater compactness in space and all atoms are now "piled up" in the interaction region where the strongest He-Rb_2 attaction exists, i.e., around the T-shaped configuration for the complex. As a consequence of this effect, therefore, we see in the Figs. 4 and 5 that all the He adatoms of the clusters with $N > 2$ exhibit two-particle angle values close to $\gamma \sim 0°$ and relative He-He distances which move down to $\sim$5 Å on the same side of the dopant for $N = 4$, because the He-He interactions dominates over the He-dimer one for $N \gtrsim 3$.

4.2 Reactive Process

As in the vibrational relaxation studies all the mass parameters and the conversion factors appear in [50, 51]. We have used 40,000 points of R_1 and R_2 in the interval $[2, 750]$ Å. see e.g. Ref. [68]. The radial discrete equations in Eq. (12), were solved by a mixed Truhlar and Numerov-Cooley procedure [65], while continuum energy normalized solutions in Eq. (12) were calculated by outward Numerov propagation of the wave function and by matching it to a sine behavior at large distances to determine the phase-shift [69].

For energies below 100 cm^{-1}, the analytical fits of the 2B interactions, present root mean squares (rms) errors $\approx 10^{-3}$ cm^{-1} for Rb_2 and HeRb, when compared with the RKHS values [50]. At the minimum configuration, the relative error was also very small, less than 0.5% for Rb_2, and 0.2% for HeRb.

The calculations were realized for the reactive channels $(J, \ell_1 = J, \ell_2 = 0) \rightarrow (J, \ell'_1 = J \pm 1, \ell'_2 = 1)$, with $J \le 3$, where the $^3He^{87}Rb$ ($^4He^{87}Rb$) dimer only has a rotationless ($\ell_2 = 0$) level at -5.26×10^{-3} cm^{-1} (-3.029×10^{-2} cm^{-1}) in such a way that $\ell_1 = J$ and $\ell'_2 = 1$.

The rubidium $^{87}Rb_2$ dimer in its lowest triplet electronic state presents 39 vibrational levels for angular momenta $1 \le \ell'_1 \le 4$, and 40 for $\ell'_1 = 0$ (41 levels were reported for the model potential obtained from high-resolution two-photon dark-state spectroscopy in the presence of a magnetic field [70]). Concerning the He-Rb, with a different He-Rb interaction [71] They did not find any bound level for the $^3He^{87}Rb$ dimer, in spite of the value of the long range coefficient (atomic units) C_6, 46.48, is greater than the present one of 44.69, (an accurate value of 43.4 [72] is widely recognized [50, 71]).

The reaction probabilities present at $J = 1$, and for the possible values $\ell_1' = 0$, 2, a maximum located at low energy of 1.69×10^{-5} cm^{-1}, and reaching a value of $\approx$0.22 for $\ell_1' = 2$ [48]. For all the channels, a secondary maximum near $\epsilon_1 = 0.036$ cm^{-1} does appear, reaching a maximum value of $\approx 1.5 \times 10^{-2}$ for $J = 0$. For energies beyond the cold regime, $\epsilon_1 > 1$ cm^{-1}, the reaction probabilities decay quickly and become negligible ($\leq 1 \times 10^{-3}$). Actually both maxima are two orbiting resonances: the first one is supported by $\ell_1 = 1$ in the entrance channel, i.e. $J = 1$, and the second, common to all the channels, corresponds to the exiting value $\ell_2' = 1$. The values attained by these reaction probabilities are close to the vibrational relaxation of Rb_2 by collision with He that appear in [50], although a narrower interval of energy.

The total cross section, Eq. (14), and the corresponding J-dependent contributions, $\sigma_r^J(\epsilon_1)$, already multiplied by the $(2J + 1)$ factor are shown in a logarithmic scale in Fig. 6, where the maxima reached in probabilities appear at almost same energies. For the peak at lower energy (see inset plot in Fig. 6), the width at half maximum is of $\sim 1.1 \times 10^{-5}$ cm^{-1}, corresponding to a long lifetime resonance of about 0.48 μs. It is interesting to note that the different partial cross sections follow almost perfectly the Wigner threshold law [73] at energies below 10^{-6} cm^{-1}, $\sigma^J(\epsilon_1) \sim \epsilon_1^{\ell_1 - 1/2}$. Total reaction cross section is dominated by s-waves in the ultra-cold regime, $\epsilon_1 \leq 10^{-7}$ cm^{-1}, and is fully converged up to an energy of 10^{-4} cm^{-1}. Hereupon, higher values of the total angular momentum ($J > 3$) should be considered to attain convergence as it occurs in the vibrational relaxation [51], but

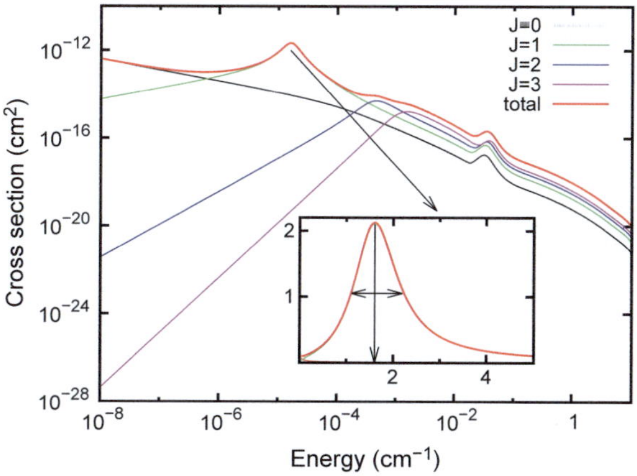

Fig. 6 For the $(J, \ell_1 = J, \ell_2 = 0) \rightarrow (J, \ell_1' = J \pm 1, \ell_2' = 1)$ reactive channels, cross sections depending on the kinetic energy of the incoming Rb atom, ϵ_1, colliding with ^{4}HeRb. The scales are logarithmic. An enlargement of the main peak is displayed in the *inset*, where in linear scales the units are 10^{-5} cm^{-1} and 10^{-12} cm^2 for the x and y axis, respectively

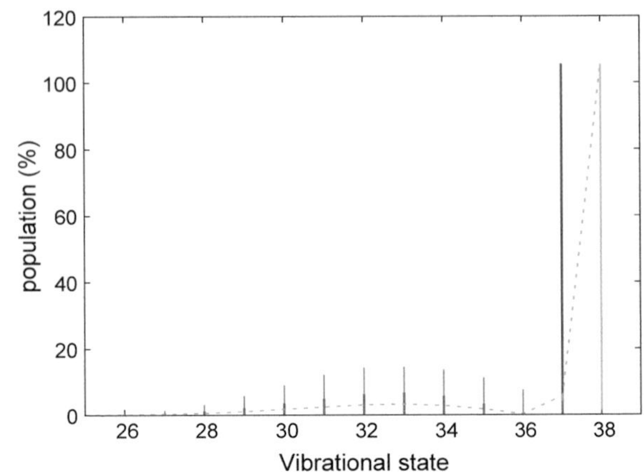

Fig. 7 Vibrational population of the formed Rb_2 molecule at three energies of the incoming Rb atom for the $(J = \ell_1 = 1, \ell_2 = 0) \rightarrow (J = 1, \ell_1' = 2, \ell_2' = 1)$ reactive channel. *Red* $\epsilon_1 = 1$ cm^{-1}; *green* $\epsilon_1 = 3.57 \times 10^{-2}$ cm^{-1}; *blue* $\epsilon_1 = 1.69 \times 10^{-5}$ cm^{-1}. The *green dashed line* is depicted just to guide the eye

we do not expect that these results would be qualitatively affected by the addition of extra partial waves.

In Fig. 7 one can see vibrational populations of the Rb_2 product for the $(\ell_1 = 1, \ell_1' = 2)$ channel at three energies: 1.69×10^{-5}, 3.57×10^{-2} cm^{-1}, and at a higher energy of 1 cm^{-1}. The two first correspond to the maxima previously described.

As it was mentioned above, for the rotational $\ell_1' = 2$ state, the diatom has 39 bound levels. For the two lowest collision energies considered, the rubidium dimer is preferently formed at very high vibrational levels, but at the highest energy, the vibrational population moves towards intermediate excitations, with a maximum at $v = 32 - 33$, and shows a Gaussian-like behavior. In any case at all energies, the population of states below $v \leq 25$ becomes negligible. If we take into account the coupling among vibrational states of rubidium induced by the presence of the helium atom, and in particular for highly excited states, would lead to a redistribution, and therefore to a modification of the individual populations, although the general behavior would remain essentially unchanged. These results are similar to the findings reported in the Knoop et al. in this work they have studied the competition of processes of exchange/relaxation in ultracold mixtures of atom/diatom caesium species in the presence of magnetic fields [74].

In Fig. 8 one can see reaction probabilities for channels with $J = 0, 1$ as a function of the kinetic energy of the incoming Rb atom, for ^{3}He and ^{4}He. Both isotopes ^{3}He and ^{4}He, show a main peak $(J = 1, \ell_1' = 0, 2)$ at almost the same energy of 1.69×10^{-5} cm^{-1} mentioned above, although the probability corresponding to the lightest isotope reaches only 1/2 of the value attained for the heaviest one. In general, the probabilities for ^{3}He are lower than for ^{4}He. This results suggest that by increasing the mass of the third body (or by considering the presence of additional helium atoms), the formation of molecular rubidium would be enhanced.

Using the energy conservation and the detailed balance principle, one can consider the inverse reaction He+$Rb_2(v, \ell_1')$ →HeRb+Rb for a vibrational state of Rb_2

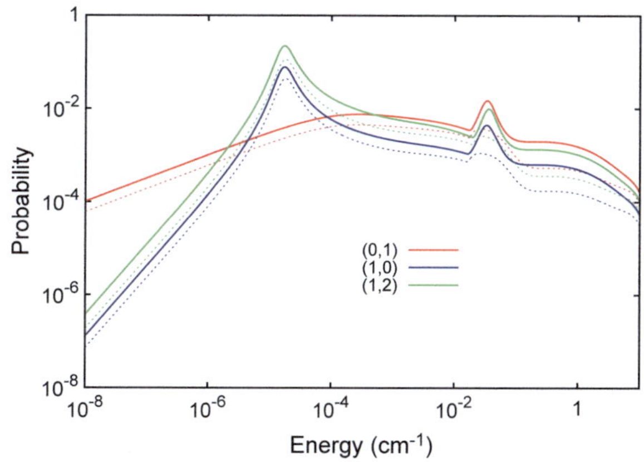

Fig. 8 Comparison of reaction probabilities for the ^{4}He (*solid lines*) or ^{3}He (*dotted lines*) partners for the leading channels $J = 0, 1$, as functions of the incoming energy of the Rb projectile. The scales are logarithmic. The *labels* denote ($\ell_1 = J, \ell_1' = J \pm 1$) channels

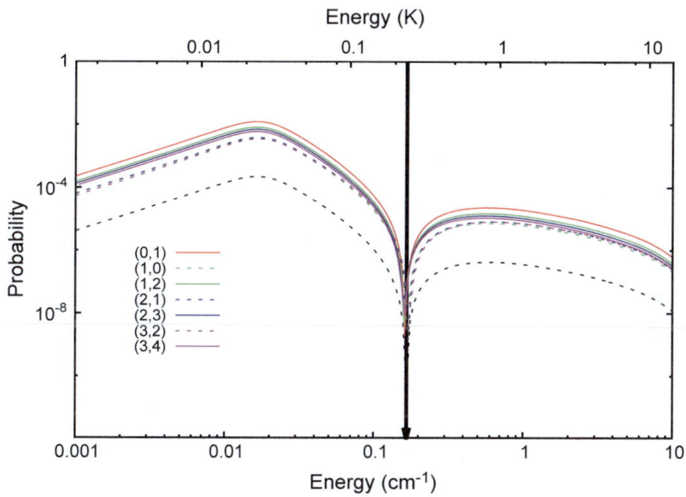

Fig. 9 Reaction probabilities for the inverse reaction He+$Rb_2(v, \ell_1')$ →HeRb+Rb as function of the kinetic energy of the helium atom, ϵ_2 (in units of cm^{-1} and K). The *labels* denote (ℓ_1, ℓ_1'). They correspond to $v = 38$, although $v = 39$ (*dashed black line*) is also included

at an energy of the He atom that is its departure energy in the direct reaction. Reaction probabilities $P^J_{\ell_1',v}$ are calculated as in Eq. (13) by replacing ϵ_1 by ϵ_2.

In Fig. 9 we display these reaction probabilities for the ^{4}He isotope, using logarithmic scale, for the different channels for $v = 38$ (as well as $v = 39$ for the unique channel in which it exist, that is, $J = \ell_1 = 1 , \ell_1' = 0$). One can see a fulfillment of Wigner threshold law for $\epsilon_2 \leq 0.01$ cm^{-1}, that is for reactive probabilities,

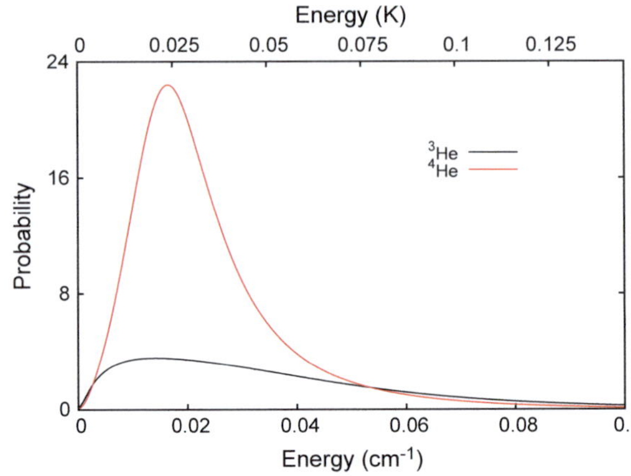

Fig. 10 For the two helium isotopes, reaction probabilities, in units of 10^{-5}, for the $J = 1$ inverse reaction He+Rb$_2$($v = 39, \ell_1' = 0$) →HeRb+Rb in terms of the kinetic energy of the incoming helium atom, ϵ_2 (in cm^{-1} and K)

$P(\epsilon_2) \propto \epsilon_2^{\ell_2'+1/2} = \epsilon_2^{3/2}$. We can also see that whatever be the channel analyzed and for the two initial vibrational states of rubidium dimer considered, a sink in the reactive probabilities at an energy of ≈0.165 cm^{-1} (0.23 K) is obtained, indicating that the reaction is practically forbidden at this energy and non-reactive processes should be enhanced there. Actually, for all initial vibrational states of the rubidium dimer considered, $1 \leq v_i \leq 4$, a clear increase of the vibrational relaxation probability is reported at 0.145 K in Ref. [50], that is close to our results.

In Fig. 10 we plot, for both helium isotopes, the reaction probabilities for the inverse reaction He+Rb$_2$($v = 39, \ell_1' = 0$) →HeRb+Rb, for the only reactive channel $J = \ell_1 = 1, \ell_1' = 0$ in which the formation of the rubidium dimer in an excited vibrational state is possible through the direct reaction, as function of the kinetic energy of the incoming helium atom. The reaction probability for ^{4}He has at 0.017 cm^{-1} (24 mK) a maximum of 2.2×10^{-4}, while for ^{3}He the peak is located at lower energy, 0.0084 cm^{-1} (12 mK) with a value of ∼3 ×10^{-5}. A similar behaviour has been presented in [51] for the $J = 1$ main peaks in the rate coefficient for vibrational relaxation of Rb$_2$($v = 1, j = 0$) by collisions with ^{4}He or ^{3}He.

It is interesting to note that the potentials employed in this paper have been obtained after several ab initio calculations, it is clear that we still have a remaining inaccuracy in the PES that we have used, and this could lead to an uncertainty in the results. Previous collision calculations at ultralow kinetic energies have analized the sensitivity of their results to the used PES by introducing a multiplicative λ (scaling) parameter in the potential form. The results become very sensitive to details of the potential used in the calculations [75, 76]. To explore this effect and for the channel ($\ell_1 = J = 1, \ell_1' = 2$) leading to the highest reaction probability, we have performed several calculations for λ varying in the range [0.95–1.05], and, some variations of the probability profile were obtained. Nevertheless, the average profile shows a similar behavior to the original one ($\lambda = 1$). Moreover, the population of high vibrational states of the formed Rb_2 molecule maintains this trend at all the energies and all the λ values considered. Therefore, within a variation of the full PES of the order of $\pm 5\%$, we may conclude that our semi-quantitative results are reliable.

5 Concluding Remarks

In the first part we have presented a 2D PES of the He-$Rb_2(^3\Sigma_u)$ complex, where the dimer was taken as a rigid rotor at the equilibrium distance. We have used a simple analytical fit that makes possible to carry out bound state calculations with a reasonable computational effort. We have shown variational procedures to yield energy and geometric results for $(^4He)_N$-$Rb_2(^3\Sigma_u)$ clusters $N \in [1, 4]$. The helium-impurity potential was obtained from ab initio calculations and was written as a pair of Lennard-Jones functions whose terms take into account the anisotropy of the problem. For the He-He potential we have used the PES described by Aziz-Slaman [66]. The overall PES was modelled as sum of pair-wise functions which give the interactions between the diatom and each helium atom and those of the helium atoms among themselves. The study of such small systems has allowed us to attain a high level of numerical accuracy and a detailed description of both the energy and structure parameters as they evolve within a limited set of aggregates.

The main conclusions in this context are:

(1) The vibrational degree of freedom of the Rb_2 can be neglected in the present calculations without loss of accuracy since the interaction energy $W_0(R, \theta)$ is almost identical to the 2D surface $W(r_{eq}, R, \theta)$.

(2) There is a transition in the species arrangements: from the cross-like structure in He_2Rb_2 to the gathering of the rare gas atoms in a pure sub-complex of helium in He_4Rb_2, in agreement with previous works [20, 22, 29].

(3) All clusters are found to be bound with respect to the $(^4He)_N$+Rb_2 asymptotes, even more the results suggest that the dopant dimer will be located outside the complex and parallel to it, this fact was found before in the case of other triplet diatoms [20, 22, 23].

In the second part we have addressed, as a preliminary attempt, to investigate the formation of rubidium dimers in their lowest triplet electronic state by atomic collisions on the surface of helium droplets. We have studied reactive collisions $^{87}Rb + {}^{87}Rb^{n}He \rightarrow Rb_2(^3\Sigma_u^+, v) + {}^{n}He$ ($n = 3, 4$) in a range from ultracold to moderate collision energies, 10^{-8}–100 cm^{-1}. Using the best PES available [50], the model to treat photodissociation of linear triatomic molecules [49] was extended to an approximate three-dimensional collisional treatment which accounts for the main features of the process. This model leads to a simplification of significant reactive channels and it almost reduces to analytic evaluations.

The conclusions in this topic can be summarized as follows:

(4) The reactive process shows a high efficiency at ultralow energies for $J = 1$, that is similar to what has been recently reported on vibrational relaxation of the rubidium dimer by collisions with atomic helium [51]. Reaction probabilities, and

cross-sections fulfill numerically the Wigner threshold law, and become negligible at collision energies greater than 1 cm^{-1}. The rubidium dimer is preferently produced at highly excited vibrational levels, i.e. $v > 25$ for the collision energies of interest.

(5) Fermionic and bosonic helium as partners show similar results although with quantitative differences. The two species display clear maxima in the reaction probabilities for $J = 1$ (orbiting resonances in the entrance channel) at the same ultralow energy (1.69×10^{-5} cm^{-1}), as well as a secondary peak near 0.04 cm^{-1}, corresponding to orbiting resonances in the exiting channel, for all the J values. The efficiency of the reaction is clearly greater for the ^4He than for the ^3He for all the energies and in all the channels.

(6) A complementary study of the inverse reaction reveals the presence, for all the reactive channels, of a sinking energy at which non-reactive processes should be dominant, close to the findings in Ref. [50]. For $J = 1$, and $Rb_2(\ell'_1 = 0,\ v = 39)$, fermionic as well as bosonic helium give rise to peaks in probabilities of the inverse reaction. The maximum in the case of the ^3He appears at a lower energy than it does for the boson, as it has also been found in the vibrational relaxation process [51]. However, the efficiency of the reactive process is clearly lower for the fermion than that for the boson, in contrast to the findings on the vibrational relaxation results [51].

Finally, since the presence and further ejection of just one helium atom enables the formation of the rubidium dimer, the reactive model can be extended to Rb_2-$(He)_N$ systems containing a number $N \geq 2$ of helium atoms by using some spectator model [77–79]. According to the present model, which involves the breaking and formation of two bonds, only one helium atom participates in the reaction, while the other $N - 1$ helium atoms would be frozen all along the reactive process. With this approach, mimicking the scenario of rubidium atoms on the surface of helium droplets, one could estimate the efficiency of the formation of molecular rubidium in terms of the number of spectators.

Acknowledgements We thank Centro de Cálculo (IFF, CSIC), Centro Técnico de Informática (CTI, CSIC) and Centro de Supercomputación de Galicia (CESGA) for the allocation of computer time. This work has been supported by MICINN Grants FIS2011-29596-C02-01 and FIS2014-51933-P. The aid of COST Action CM1405 (MOLIM) is also appreciated.

References

1. Toennies JP, Vilesov AF (2004) Angew Chem Int Ed 43:2622
2. Callegari C, Lehmann KK, Schmied R, Scoles G (2001) J Chem Phys 115:10090
3. Toennies JP, Vilesov A (1988) Annu Rev Phys Chem 49:1
4. Grebenev S, Toennies JP, Vilesov AF (1998) Science 279:2083
5. Toennies JP, Vilesov AF, Whaley KB (2001) Phys Today 54:31
6. Stienkemeier F, Vilesov AF (2001) J Chem Phys 115:10119
7. Choi MY, Douberly GE, Falconer TM, Lewis WK, Lindsay CM, Merritt JM, Stiles PL, Miller RE (2006) Int Rev Phys Chem 25:15
8. Stienkemeier F, Lehmann KK (2006) J Phys B: Atom Mol Opt Phys 39:R127
9. Moroni S, Baroni S (2005) Comput Phys Commun 169:404

10. Paesani F, Gianturco FA, Whaley KB (2001) J Chem Phys 115:10225
11. Goyal S, Schutt DL, Scoles G (1992) Phys Rev Lett 69:933
12. Ceperley DM (1995) Rev Mod Phys 67:279
13. Hernando A, Barranco M, Mayol R, Pi M, Ancilotto F, Bünermann O, Stienkemeier F (2010) J Low Temp Phys 158:105
14. Allard O, Nagl J, Auböck G, Callegari C, Ernst WE (2006) J Phys B 39:s1169
15. Higgins J, Callegari C, Reho J, Stienkemeier F, Ernst WE, Gutowski M, Scoles G (1998) J Phys Chem A 102:4952
16. Class P, Droppelmann G, Schulz CP, Mudrich M, Stienkemeier F (2007) J Phys Chem A 111:7537
17. Auböck G, Aymar M, Dulieu O, Ernst WE (2010) J Chem Phys 132:054304
18. Schlesinger M, Mudrich M, Stienkemeier F, Strunz WT (2010) Chem Phys Lett 490:245
19. Auböck G, Nagl J, Callegari C, Ernst WE (2007) J Phys Chem A 111:7404
20. Bovino S, Coccia E, Bodo E, López-Durán D, Gianturco FA (2009) J Chem Phys 130:224903
21. Prosmiti R, Delgado-Barrio G, Villarreal P, Yurtsever E, Coccia E, Gianturco FA (2009) J Phys Chem A 113:14718
22. Pérez de Tudela R, López-Durán D, González-Lezana T, Delgado-Barrio G, Villarreal P, Gianturco FA, Yurtsever E (2011) J Phys Chem A 115:6892
23. López-Durán D, Pérez de Tudela R (2011) R, Rodríguez-Cantano T, González-Lezana MP, Lara-Castells G, Delgado-Barrio P. Villarreal Phys Scr 84:028107
24. López-Durán D, de Lara-Castells MP, Delgado-Barrio G, Villarreal P, Di Paola C, Gianturco FA, Jellinek J (2004a) Phys Rev Lett 93:053401
25. López-Durán D, de Lara-Castells MP, Delgado-Barrio G, Villarreal P, Di Paola C, Gianturco FA, Jellinek J (2004b) J Chem Phys 121:2975
26. de Lara-Castells MP, López-Durán D, Delgado-Barrio G, Villarreal P, Di Paola C, Gianturco FA, Jellinek J (2005) Phys Rev A 71:033203
27. Di Paola C, Gianturco FA, López-Durán D, de Lara-Castells MP, Delgado-Barrio G, Villarreal P, Jellinek J (2005) Chem Phys Chem 6:1348
28. de Lara-Castells MP, Prosmiti R, Delgado-Barrio G, López-Durán D, Villarreal P, Gianturco FA, Jellinek J (2006) Phys Rev A 74:053201
29. Guillon G, Zanchet A, Leino M, Viel A, Zillich RE (2011) J Phys Chem A 115:6918
30. Ho TS, Rabitz H (1996) J Chem Phys 104:2584
31. Rodríguez-Cantano R, López-Durán D, González-Lezana T, Delgado-Barrio G, Villarreal P, Yurtsever E, Gianturco FA (2012) J Phys Chem A 116:2394
32. Krems R, Friedrich B, Stwalley W (2009) Cold molecules: theory, experiment, applications. CRC Press, Boca Raton, FL
33. Fioretti A, Comparat D, Crubellier A, Dulieu O, Masnou-Seeuws F, Pillet P (1998) Phys Rev Lett 80:4402
34. Jones KM, Tiesinga E, Lett PD, Julienne PS (2006) Rev Mod Phys 78:483
35. Donley EA, Claussen NR, Thompson ST, Wieman CE (2002) Nature (London) 417:529
36. Köhler T, Góral K, Julienne PS (2006) Rev Mod Phys 78:1311
37. Deiglmayr J, Grochola A, Repp M, Mörtlbauer K, Glück C, Lange J, Dulieu O, Wester R, Weidemüller M (2008) Phys Rev Lett 101:133004
38. Lang F, Winkler K, Strauss C, Grimm R, Hecker J (2008) Denschlag. Phys Rev Lett 101:133005
39. Quemener G, Julienne PS (2012) Chem Rev 112:4949
40. Krois G, Pototschnig JV, Lackner F, Ernst WE (2013) J Phys Chem A 117:13719
41. Chen T, Zhu S, Li X, Qian J, Wang Y (2014) Phys Rev A 89:063402
42. Stienkemeier F, Higgins J, Ernst WE, Scoles G (1995) Phys Rev Lett 74:3592
43. Brühl FR, Miron RA, Ernst WE (2001) J Chem Phys 115:10275
44. Ernst WE, Huber R, Jiang S, Beuc R, Movre M, Pichler G (2006) J Chem Phys 124:024313
45. Theisen M, Lackner F, Ernst W (2011) J Phys Chem A 115:7005
46. Fechner L, Gruner B, Sieg A, Callegari C, Ancilotto F, Stienkemeier F, Mudrich M (2012) Phys Chem Chem Phys 14:3843

47. von Vangerow J, Sieg A, Stienkemeier F, Mudrich M, Leal A, Mateo D, Hernando A, Barranco M, Pi M (2014) J Phys Chem A 118:6604
48. Rodríguez-Cantano R, González-Lezana T, Prosmiti R, Delgado-Barrio G, Villarreal P, Jellinek J (2015) J Chem Phys 142:164304
49. Rosen N (1933) J Chem Phys 1:319
50. Guillon G, Viel A, Launay J-M (2012) J Chem Phys 136:174307
51. Viel A, Launay J-M (2014) J Phys Chem A 118:6529
52. Frisch MJ, Trucks GW, Schlegel HB, Scuseria GE, Robb MA, Cheeseman JR, Montgomery Jr JA, Vreven T et al (2004) Gaussian 03. Gaussian Inc., Wallingford, CT
53. Beser B, Sovkov VB, Bai J, Ahmed EH, Tsai CC, Xie F, Li L, Ivanov VS, Lyyra AM (2009) J Chem Phys 131:094505
54. Soldán P (2010) J Chem Phys 132:234308
55. Weigend F, Ahlrichs R (2005) Phys Chem Chem Phys 7:3297
56. Leininger T, Nicklass A, Küchle W, Stoll H, Dolg M, Bergner A (1996) Chem Phys Lett 255:274
57. Boys SF, Bernardi F (1970) Mol Phys 19:553
58. Beswick JA, Delgado-Barrio G (1980) J Chem Phys 73:3653
59. Gianturco FA (1979) The transfer of molecular energy by collisions. Springer, Berlin
60. Wernli M, Bodo E, Gianturco FA (2007) Eur Phys J D 45:267
61. Villarreal P, Roncero O, Delgado-Barrio G (1994) J Chem Phys 101:2217
62. Hernández MI, Halberstadt N, Sands WD, Janda KC (2000) J Chem Phys 113:7252
63. Linear algebra package 3.2.1 (2009) http://www.netlib.org/lapack/
64. López-Durán D, Rodríguez-Cantano R, González-Lezana T, Delgado-Barrio G, Villarreal P, Gianturco FA (2012) Phys Rev A 86:022501
65. Delgado-Barrio G, Cortina AM, Varadé A, Mareca P, Villarreal P, Miret-Artés S (1986) J Comput Chem 7:208
66. Aziz RA, Slaman MJ (1991) J Chem Phys 94:8047
67. Lewerenz M (1997) J Chem Phys 106:4596
68. Gianturco FA, Palma A, Villarreal P, Delgado-Barrio G, Roncero O (1987) J Chem Phys 87:1054
69. Roncero O, Miret-Artés S, Delgado-Barrio G, Villarreal P (1986) J Chem Phys 85:2084
70. Strauss C, Takekoshi T, Lang F, Winkler K, Grimm R, Hecker Denschlag J, Tiemann E (2010) Phys Rev A 82:052514
71. Tscherbul TV, Zhang P, Sadeghpour HR, Dalgarno A (2009) Phys Rev A 79:062707
72. Zhu C, Dalgarno A, Porsev SG, Derevianko A (2004) Phys Rev A 70:032722
73. Bell MT, Softley TP (2009) Mol Phys 107:99
74. Knoop S, Ferlaino F, Berninger M, Mark M, Nägerl H-C, Grimm R, D'Incao JP, Esry BD (2010) Phys Rev Lett 104:053201
75. Żuchowski PS, Hutson JM (2009) Phys Rev A 79:062708
76. Janssen LMC, van der Avoid A, Groenenboom GC (2013) Phys Rev Lett 110:063201
77. Zhang DH, Yang M, Lee S (2002) Phys Rev Lett 89:103201
78. Wrede E, Schnieder L, Seekamp-Schnieder K, Niederjohann B, Welge KH (2005) Phys Chem Chem Phys 7:1577
79. Donald SB, Harrison I (2012) Phys Chem Chem Phys 14:1784

Water Structuring at Non-Polar Fluid Interfaces

Yana Tsoneva and Alia Tadjer

Abstract The structuring of water molecules at the water/vapour interface is an object of scientific interest for decades. After the first successful attempts to explore liquid water with the help of theoretical chemistry, the number of studies on this topic grows progressively. Most of them are focused on bulk water but there is still need of a more detailed research on surface water. In addition, interfaces with alkanes are interesting as being instructive from both biological and industrial perspectives. Since in both bio- and industrial applications water/air and water/oil interfaces are mediated by amphiphiles, the role of a surfactant monolayer on surface water structuring deserves more attention as well. In the present study several atomistic water models were chosen—non-polarisable (SPC, TIP3P, and TIP4P) and polarisable (SW-RIGID-ISO, SWM4-NDP, and COS/G2) and classical molecular dynamics simulations were carried out on bulk water, water/vapour and water/alkane (from pentane to nonane) systems, as well as on water/DLPC/vapour and water/DLPC/octane models. In all cases the temperature was kept at 298 K. Several structural properties of bulk and surface layers were examined by means of radial distribution functions and Voronoi analysis. Dipole moments, surface tension and hydrogen bonding were addressed too. The objective was to estimate the impact of accounting for polarisability on the water properties of interest and to select a cost-efficient water model for describing them, as well as to add new data to the existing knowledge about interfacial water structuring.

Keywords Molecular dynamics · Polarisable water models · Voronoi analysis · Asphericity parameter · Alkanes · DLPC monolayer

---

Electronic supplementary material The online version of this chapter (doi:10.1007/978-3-319-50255-7_7) contains supplementary material, which is available to authorized users.

Y. Tsoneva · A. Tadjer (✉)
Faculty of Chemistry and Pharmacy, University of Sofia, 1 James Bourchier Ave, 1164 Sofia, Bulgaria
e-mail: tadjer@chem.uni-sofia.bg

A. Tadjer et al. (eds.), *Quantum Systems in Physics, Chemistry, and Biology*, Progress in Theoretical Chemistry and Physics 30,
DOI 10.1007/978-3-319-50255-7_7

1 Introduction

Water is the substance with the largest number of anomalies, most of which are crucial for life. Although the water molecule has a simple structure, its behaviour is quite intricate [1–3] and water assemblies exhibit numerous interesting properties [4]. The *instantaneous* (time periods less than 1 fs) water structuring can be determined by means of X-ray absorption spectroscopy (XAS). XAS results allowed Wernet et al. to promote the hypothesis that 80% of the molecules in liquid water have one tightly and one loosely bound O–H group organized in ring- or chain-like structures and the remaining 20% are in tetrahedral coordination [5]. An outline of the existing experimental techniques for investigation of water samples is presented in a review by Clark et al. [6], concluding that the predominating scientific evidence supports the conventional vision of tetrahedral coordination. Yet, disparate opinions are published too, such as the paradigm of Nilsson and Pettersson [7] endorsing the concept of high density and low density liquids. Rationalising experimental results, they contend that at room temperature water is a high-density liquid with low-density fluctuations becoming more intense and frequent with temperature decrease [8].

The water structure averaged over longer time periods is termed *vibrational structure* and can be studied by utilizing infrared or Raman spectroscopy [9–13], and X-ray [14–18] or neutron [19–25] scattering. A Fourier transformation of the spectral data gives the radial distribution functions (RDFs), whereof the number of closest neighbours has been estimated ranging from 4.4 to 4.9 with temperature raise from 1.5 to 83 °C [5].

Surface water plays a key role in a variety of physical, chemical and biological processes [26–28]. Its structuring is sensitive to temperature changes, gas molecules penetration, ions flux, and electric gradients [29]. Being one of the most significant boundaries, the water/vapour interface is among the most scrupulously investigated systems [27, 30–46]. Almost all properties of bulk water are modified at the surface. Therefore, the solvating capacity and a number of other features differ in the surfacial layer and attract unremitting scientific attention. For example, the molecules at the surface of both liquid water and ice are supposed to coordinate lower number of neighbours, thus forming low-density phases which are hydrophobic, non-elastic, with higher fluidity and thermal stability than bulk water [47, 48]. The hydrogen bonding at the surface is deemed stronger than in the bulk [49] but some H-bonds are lost which invokes enhanced reactivity [30]. Another controversial question is the molecular orientation at the surface, dating back to the first computational treatments of the system [50], reporting dominating orientation of the interfacial water molecules with the oxygen towards the vapour and the hydrogens towards the bulk. These findings have been corroborated by other authors [51], but soon after opposite statements were published affirming that the hydrogen atoms are pointing to the vapour [52, 53]. Nowadays, it is considered that one of the hydrogens sticks out of the liquid [36], whereas the other one is inside. This was confirmed by ab initio calculations on water clusters [54] and ab initio path integral

MD (PIMD) [55]. Nevertheless, there is evidence for occurrences of structuring patterns with both hydrogens staying out of the liquid [42]. The latter was demonstrated by the results of molecular dynamics (MD) simulations using various force fields [35, 54–59], revealing that the H–O–H plane is perpendicular to the liquid/vapour interface. About ¼ of the water molecules have one O–H group at a small angle above the surface [49, 60, 61], while the majority of waters have electron accepting-centres spread out [62] thus creating a weak negative charge.

The water structuring close to hydrophobic surfaces is of primary importance for the description of numerous processes, such as: wetting, protein folding, oil extraction, micelle formation, etc. Among those, the water/oil interface is most frequently addressed. A model system providing insight into the water/oil interaction, related directly to a number of biological and technological surface phenomena, is the water/alkane liquid interface. Differences in the structuring at the water/vapour and water/non-polar phase are demonstrated in Richmond's review [63]. Diminished coupling of the water molecules is established, attributed to the reduced coordination number of the surfacial water molecules or/and H-bond weakening. Molecular modelling of liquid/liquid interfaces was performed by Benjamin [27, 64]. A detailed up-to-date summary of the achievements in the investigation of water/organic liquid interfaces from computational viewpoint was published by Pratt and Pohorille [65].

Surfactant monolayers have been studied persistently both as models of half-membranes and as targets of independent interest and application potential. The introduction of a lipid monolayer at the water/non-polar fluid interface presents a new element which modifies the structure of the two-phase systems. Knowledge about the phospholipid-specific water structuring would allow the determination of surfactants heads effect on the membrane structure [66].

The literature addressing water structuring at various interfaces is extremely rich and the present introduction has no ambition to cover it in detail. Yet, to our knowledge, there exists no methodical investigation of the change in some properties of liquid water at various interfaces using an assortment of classical atomistic water models and estimating the effect of polarisability. In the present work the most widely-used rigid and some polarisable water models are employed and their performance in the bulk and at the interface with vapour is compared. The polarisable models employed are chosen to belong to one family—the one utilising Drude oscillators. This selection is based on their compatibility with other force fields and the comparatively good and intuitively equitable representation of the polarisability impact at a low cost. This type of description has proven itself as very effective, including its latest implementation, namely the QDO [62] (quantum Drude oscillators) approach. Drude oscillators reproduce very accurately the water-vapour coexistence curve and water structuring for a variety of conditions [67] at a reasonable computational price. In addition, the systems water/liquid alkanes, water/lipid monolayer/vapour and water/lipid monolayer/octane are studied aiming at generalising the water structuring at soft non-polar interfaces and at appreciating the necessity of employing polarisable water models.

2 Computational Protocol

Molecular dynamics simulations were carried out for the aforementioned systems. The periodic boxes of the initial models are presented in Fig. 1; all of them feature mirror symmetry and therefore only ½ boxes are shown. Bulk water in an approximately 50 × 50 × 50 Å periodic box, and a water/vapour system in a ~50 × 50 × 100 Å one were initially studied with various water models. The latter box was prepared by symmetrically extending the former one with vacuum layers along the Z axis. Two groups of water models were used—non-polarisable (SPC, TIP3P, and TIP4P) and polarisable (SW-RIGID-ISO, SWM4-NDP, and COS/G2). Next, the vacuum was filled with alkanes (C5 to C9) for the TIP4P water model. The nonpolar molecules were described with the AMBER99 [68] force field. Then, the systems TIP4P/vapour and TIP4P/octane were enriched with monolayers of the phospholipid dilauroyl phosphatidylcholine (DLPC), where both non-water compounds were described by the CHARMM27 [69] force field. In each case the simulations were performed in NVT ensemble and the temperature was kept at 298 K; the GROMACS software package [70] was utilised. The hydrogen-containing bonds were restrained by means of the LINCS algorithm [71] while the water geometry was rigidised by the SETTLE scheme [72]. The Lennard-Jones potential for non-bonded interactions was applied with a cutoff set to

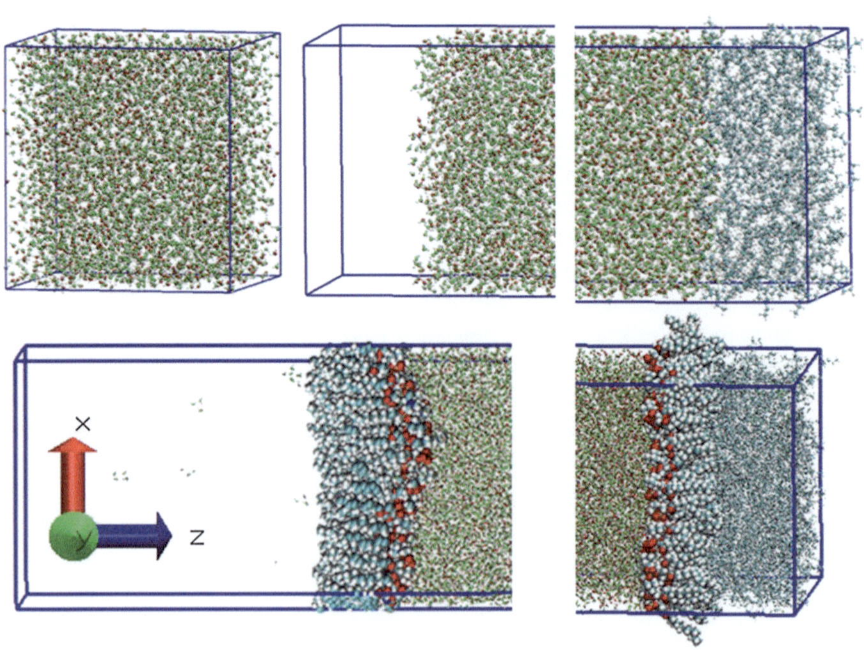

Fig. 1 Initial ½ models of the studied systems: bulk water (*top left*), water/vapour (*top centre*), water/alkane (*top right*), water/DLPC/vapour (*bottom left*), water/DLPC/octane (*bottom right*)

12 Å and a switch function activated at 10 Å. The electrostatic interactions were assessed in the monopole approximation with a cutoff of 14 Å and a switch function launched at 12 Å, making use of the PME method for quantification of the long-range electrostatics. The time step was set to 2 fs and integration of the equations of motion was done with the leap-frog algorithm.

The systems were equilibrated following a standard protocol [73]. Before the equilibration in NVT ensemble, relaxation in NPT (1 ns) was carried out for bulk water letting the box size to adjust itself, which is the reason the box dimensions are not exactly 50 Å in each direction. A 25 ns production trajectory was subject to statistical analysis with frames extracted at intervals of 2 fs, except for the systems that include phospholipids, where the analysis is over 10 ns trajectory. For visualization and calculation of the radial distribution functions the VMD program was employed [74]. The Voronoi analysis was done with the voro++ software package [75].

3 Results and Discussion

3.1 Density and Density Profiles

The average values of the mass density in the interior of the studied systems: bulk water, water/vapour, water/alkane, water/DLPC/vapour and water/DLPC/octane, are presented in Table 1. The mean values differ insignificantly from the reference ones (RD, reported by the developers). In the water/vapour systems the values are averaged from 34 to 66 Å along the Z axis (Fig. 2). The density profiles reveal that the bulk density (BD) is reached at about 10 Å away from the interface (Fig. 2). One of the techniques for quantification of the surface layer width is based on the density variation from 5 to 95% of the bulk value, which is how the breadth of the surface water layer (interfacial slab, IS) was determined (Table 1) and used for further analyses. 10 Å thick water layers from the interior are selected being representative of the bulk (bulk slab, BS).

Overall, it may be stated that polarisable force fields give more accurate results for density—two of them reproduce remarkably well the experiment. TIP4P performs best among the non-polarisable force fields—at the level of the less intricate polarisable models. Most models yield concerted results. The inset in Fig. 2 demonstrates that COS/G2 gives enhanced density at the interface but since the increment is within the standard deviation, there should be another property proving the superiority of the polarisable model.

In the water/alkanes models water has noticeably higher density compared to bulk water or water/vapour (Table 1), particularly at the interface. There are "horns"—density peaks in the interfacial slabs (Fig. 3, centre), which are not observed in the absence of alkanes, since the latter suppress evaporation and their hydrophobicity rather repels surface water.

Table 1 Average bulk (B) or bulk slab (BS) densities (BD) [g/cm^3], reference density (RD), width [Å] of the interfacial slab (IS), and number of first neighbours (N_{fn}) obtained from the RDFs of the studies systems. ***All data refer to water only***

Model	BD ± 0.00	RD[a] [77–80]	IS Width	N_{fn}	
				B/BS	IS
SPC	0.976	0.971	–	5.0	–
/vapour	0.974		6.9	4.9	5.8
TIP3P	0.985	0.982	–	4.7	–
/vapour	0.982		6.9	5.5	6.1
TIP4P	0.992	0.999	–	4.7	–
/vapour	0.990		6.5	4.5	5.2
/DLPC/vapour	0.997		22.2	4.6	4.4
/pentane	1.003		6.3	4.7	4.5
/hexane	1.001		6.4	4.5	5.2
/heptane	1.002		5.3	4.6	4.9
/octane	1.003		5.8	4.6	4.8
/DLPC/octane	0.988		19.7	4.4	4.9
/nonane	1.001		5.8	4.7	4.9
SW	0.991	0.996	–	6.4	–
/vapour	0.990		6.7	6.3	5.7
SWM4	0.996	0.996	–	4.5	–
/vapour	0.997		6.2	4.5	5.7
COS/G2	0.998	0.997	–	4.2	–
/vapour	0.998		5.7	4.4	4.9
Experiment [76]	0.997		–		

[a]The reference values are taken from the publications of the developers of the respective models

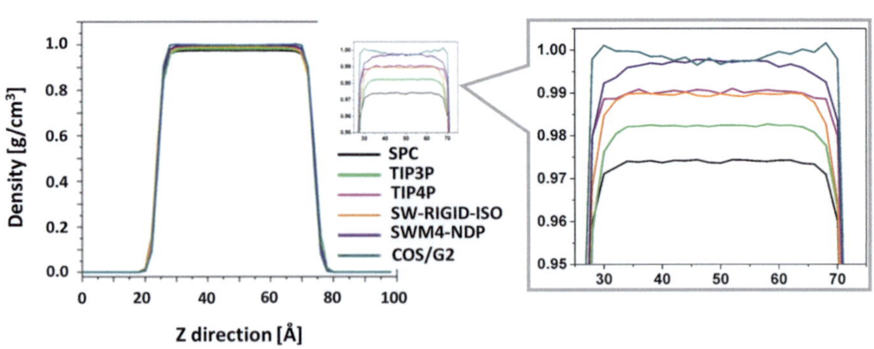

Fig. 2 Density profiles along the Z-axis of the water/vapour system using different water models. The inset is a zoom of the plateau along the ordinate

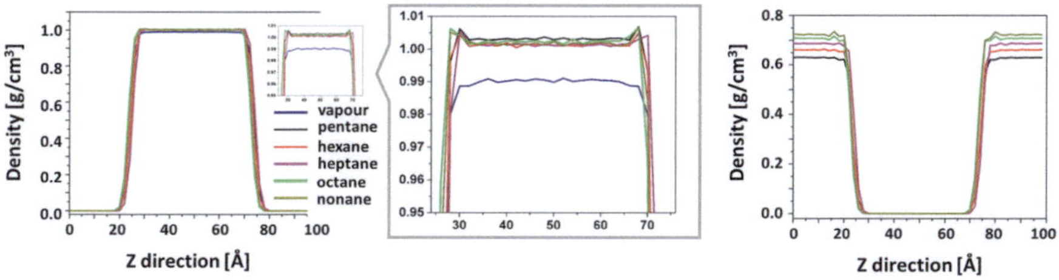

Fig. 3 Density plots of water (*left*) and alkanes (*right*) along the Z-axis in the water/alkane systems. The inset (*centre*) is a zoom of the plateau along the ordinate for water

No density/chain-length correlation is identified. The same holds for the IS width—no definite tendency in the change of the IS width with alkane chain extension, as established with other methods [100], can be outlined, which may be due to the approximate assessment of the interface position, the classical description of the systems, and the non-polarisable water model. The IS widths determined from the density plots (5–95% of bulk density) are shown in Table 1. On the whole, the IS width decreases with chain length, converging to a constant value.

From the density profiles of the DLPC-containing systems (ESM: Fig. S1) can be seen that the contributions from the different components overlap more markedly compared to the direct water/vapour and water/alkane interfaces. This is owed to the fact that water hydrates the heads and penetrates in the surfactant layer, while tails mix with octane which results in broader and lower DLPC density profiles in the water/DLPC/octane system. The hydrophilicity of surfactants' heads is the reason for the much wider IS—22.2 Å for water/DLPC/vapour and 19.7 Å for water/DLPC/octane. The data in Table 1 attest that the BS density of the water/vapour systems with and without DLPC is the same, while in water/DLPC/octane the bulk value is not reached.

3.2 Radial Distribution Functions

Short-range structuring data for liquid water can be extracted from the radial distribution functions. Therefore, RDFs of bulk and interfacial slabs are plotted for all investigated systems. Figure 4 contains the oxygen–oxygen RDFs (g_{OO}) for bulk water and water/vapour systems. No sizeable differences between the different layers are seen except that in the interfacial one the slope is less slanting which should be attributed to a less expressed structuring. The polarisable models are expected to produce more accurate results. The presented data show that the SW-RIGID-ISO model does not reproduce correctly the RDF profile. COS/G2 gives higher maxima and lower minima and can be considered as the best representation of liquid water structuring among the water models in focus.

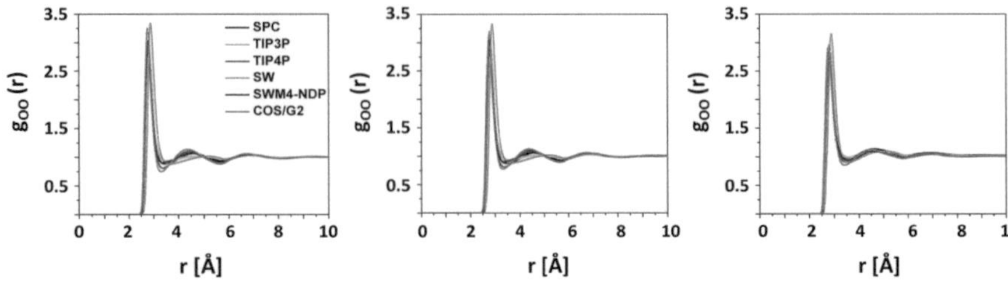

Fig. 4 Oxygen–oxygen radial distribution functions for bulk water (*left*), and for BS (*centre*) and IS (*right*) of the water/vapour models

The first maxima of all bulk RDFs are at 2.77 ± 0.01 Å (non-polarisable models) up to 2.89 Å (polarisable ones). For the ISs of the water/vapour and water/alkane systems this value is negligibly higher, while in the DLPC mediated interfaces it is immaterially lower (deviations up to 0.02 Å). The second g_{OO} peaks are at 4.53 ± 0.02 Å for the non-polarisable models, down to 4.39 Å for some polarisable ones. At the surface this value grows up to 4.71 Å for water/vapour and water/alkane interfaces and decreases down to 4.28 Å for the DLPC mediated ones. The numerical values of the maxima reproduce well the experimental estimates for bulk water: 2.875 Å and 4.475–4.525 Å, obtained from neutron diffraction experiments [19], which validates the computational protocol applied.

Using the area beneath the first maximum of g_{OO} yields the number of first neighbours N_{fn} (Eq. 1, where ρ is the number density and $r_{\min 1}$ the position of the first minimum). The minimum experimental value for liquid water is 4.4 [18, 24]. The results obtained in the current study are collected in Table 1. The average number of first neighbours in all models agrees with that reported by other authors. Notably, the larger values obtained for the surfacial layer support the hypothesis that in this region a more compact organisation exists.

$$N_{fn} = \int_0^{r_{\min 1}} 4\pi r^2 \rho g_{OO}(r) dr \tag{1}$$

RDFs for the bulk and surfacial layers of the water/alkane models are built as well. g_{OO} can be found in the ESM: Fig. S2. The only difference registered is that the RDFs of all surfacial layers decay faster in the presence of alkanes. This may be due to suppressed perturbations in this region due to the presence of a denser phase compared to vapour. The values for the number of first neighbours are also somewhat different—those for the surfacial layers are again larger than the bulk ones but are smaller than in the water/vapour models (Table 1). The presence of an alkane phase enhances the similarity between surfacial and bulk layers.

The plots for the systems enriched with phospholipid monolayers look very similar (ESM: Fig. S2) to those of the other models. Yet, the surfacial layers of the DLPC mediated interfaces are discernible for their faster decay.

3.3 Voronoi Diagrams

Voronoi cells (Voronoi polyhedra, VP) [81] are constructed in such a way that their faces are equidistant to neighbouring centres and normal to the line connecting them. In our case, the centres are the oxygen atoms of water. The VP parameters can be related to structural properties of the studied systems. Here the VP volume, the number of shared VP faces and their area, the number of VP vertices as well as the VP asphericity parameter are discussed. Dominating and mean values of these parameters are presented in Table S1.

The VP volume defines the space belonging to a centre; thus, its reciprocal value is a measure of the local particle density around a centre. Bearing in mind the close density values obtained with the different water models, the VP volumes are expected to be model insensitive as well. The distributions for bulk water and water/vapour systems are shown in Fig. 5. The values for bulk water and bulk slabs in the water/vapour systems are identical, in accord with the findings of other authors [82–88]. The surfacial layers profiles are substantially different—no values smaller than 20 Å^3 are noticeable and 30 Å^3 is scarcely populated, while this is the highest probability volume in the bulk. In addition, the profile shape is different—it is non-symmetrical, proving that the liquid is inhomogeneous in this region.

The VP volume of the surfacial molecules should exclude the vapour space but due to the dynamical process of evaporation and condensation it is hard to avoid the partial inclusion of vapour phase which renders the VP volume of the surface centres excessively large. High values of the latter may be also due to the presence of capillary waves, since a water molecule from the lateral area of such waves would have fewer neighbours. No dependence on the type of water model is registered in the bulk, while mild differences are observed for the interfacial slab (Table S1). The above reasons cannot be the explanation for the presence of large VP volumes in the surfacial slabs of the water/alkane systems (Fig. 6). One may speculate that the alkanes at the borderline are organised in a less compact manner leaving substantial cavities but this needs to be investigated in a separate study. Unambiguous conclusions about water VP volume sensitivity to the alkane chain length cannot be drawn.

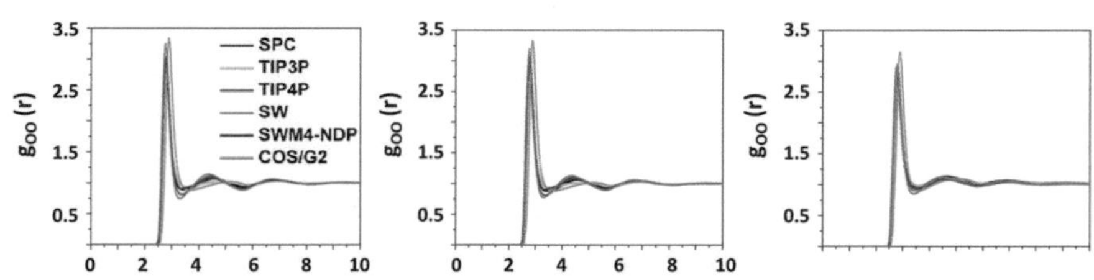

Fig. 5 Distribution of VP volume of water molecules in bulk water (*left*), and in bulk slab (*centre*) and interfacial slab (*right*) of the water/vapour models

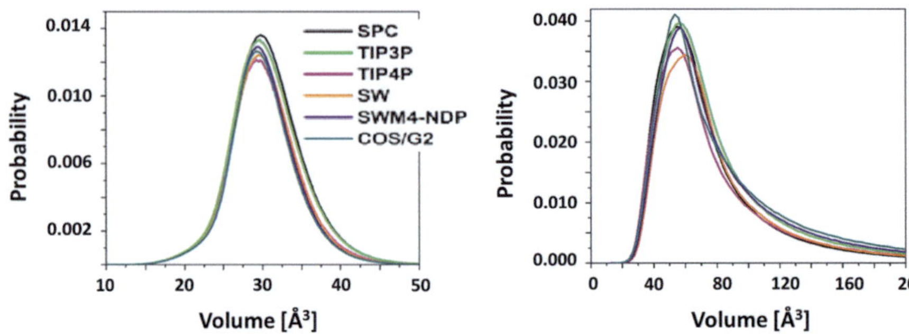

Fig. 6 Distribution of VP volume of water in the bulk slab (*left*) and interfacial slab (*right*) of the TIP4P/alkanes models

In order to compare the results for the DLPC mediated interfaces with the direct-contact ones, simulations of water/vapour and water/octane systems using periodic boxes of the size utilized for the surfactant-containing ones, namely, 70 × 70 × 140 Å, were carried out and all comparisons are made between equal-size systems. The change of box size had essentially no impact on the results. The VP volume distribution of DLPC mediated interfaces differ materially from the direct interfaces, particularly the TIP4P/DLPC/octane profile (Fig. 7). Both in the bulk and in the interfacial slab plots two peaks are visible, evidencing the presence of two VP sizes in this system. In the bulk, the less populated size is half the size of the preferred one, both maxima being at lower value than the single symmetric peak of direct interfaces. At the interface, the distribution is similar—a less populated size at approximately half the volume of the more populated one. Most probably, since the water molecules hydrating the heads experience stronger attraction, they pack closer. The maxima for the DLPC enriched systems are at lower values compared with those of the direct water/vapour and water/octane interfaces which reveals that the presence of a surfactant invokes the occurrence of a more compact structuring even more condensed than that due to alkanes.

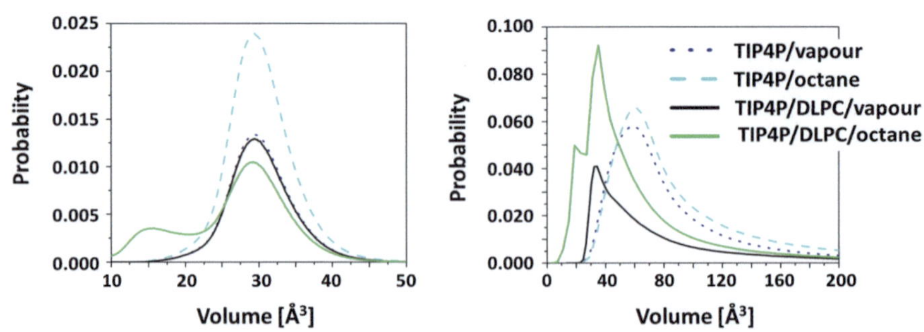

Fig. 7 Distribution of VP volume of water in the bulk slab (*left*) and interfacial slab (*right*) of the DLPC containing models

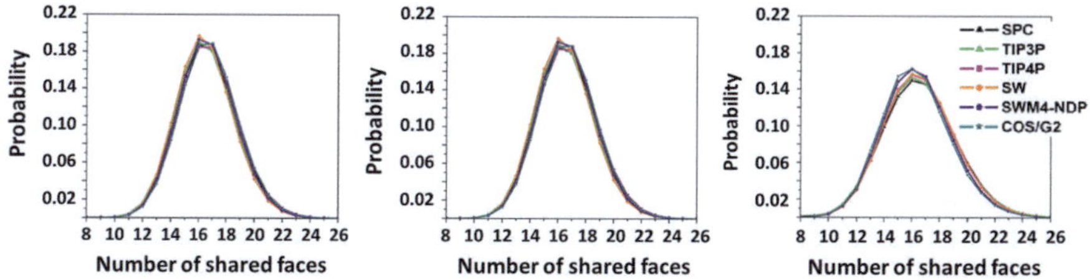

Fig. 8 Distribution of the number of shared VP faces for bulk water (*left*), and for BS (*centre*), and IS (*right*) in the water/vapour models

As each Voronoi cell contains solely one molecule, two adjacent molecules share a VP face. The number of shared faces of a VP should equal the number of neighbours a molecule has. In principle, these should be only the closest neighbours but for liquids forming hydrogen bonds the second and even the third coordination shell is often included. The closest neighbours, however, should feature large common areas of the shared faces and based on the portion of the shared faces the number of molecules in the first coordination shell can be assessed [88]. The number of shared faces varies from 10 to 24 with a bell-shaped distribution profiles (Fig. 8 for systems containing only water and ESM: Figs. S3, S4 for the others), the average value in all cases being around 15.5 (Table S1). Nevertheless, it is visible that 16 and 17 are the most probable in the bulk, while at the surface all three values: 15, 16 and 17 are equally populated. In addition, the distribution is broader in the IS not because of the larger proportion of the most probable number of neighbours but due to the presence of more small faces. The distribution of the individual faces areas is shown in Fig. 9. The sharp peak at close to zero areas should be ignored as it originates from the massive amount of very small faces shared by relatively distant neighbours. It is not possible to make the same calculation for the interfacial layers as for the purpose the VPs with artificially large volumes have to be excluded but we did not succeed in doing that. The second

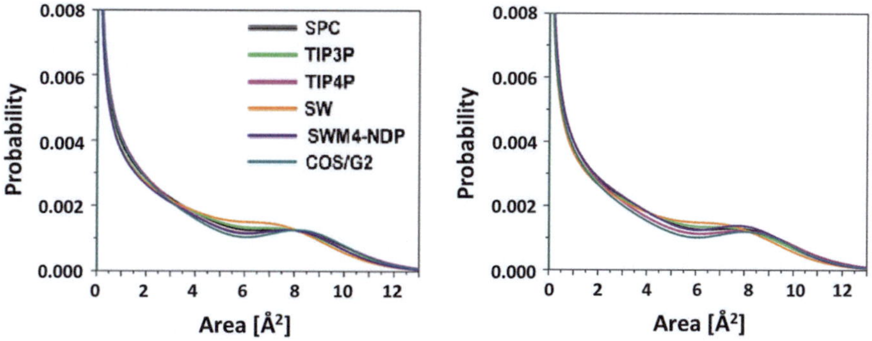

Fig. 9 Distribution of the individual areas of the shared VP faces in bulk water (*left*) and bulk slab (*right*) obtained with different water models

Table 2 Fraction of the larger faces from the individual areas distribution [%] and average number of neighbours with larger shared faces (N)

		SPC	TIP3P	TIP4P	SW-RIGID-ISO	SWM4-NDP	COS/G2
Fraction [%]	*B*	22.5	22.4	24.0	21.6	23.9	24.4
	BS	22.9	23.3	22.9	22.8	24.6	24.6
N	*B*	3.5	3.7	3.9	3.3	3.7	3.8
	BS	3.6	3.6	3.6	3.4	3.8	3.9

maximum at above 6 Å^2 is assigned to the closest neighbours. Their fraction is represented in Table 2. The large faces are about 23% of the total number for bulk water and using the data in Table S1 one can come up with a value for the number of neighbours in the first shell. The average number of neighbours obtained with SWM4-NDP and COS/G2 are the most proximate to the expected ones gathered by other authors (3.9 [88]). The coordination numbers obtained for each system are close to the ones extracted from the RDFs.

Another VP characteristic considered informative for the structuring of a liquid is the asphericity parameter, η, defined [88] as a ratio comprising the total area (A) and the volume (V) of the Voronoi cell:

$$\eta = \frac{A^3}{36\pi V^2} \tag{2}$$

$\eta = 1$, 1.33, 1.35, 1.91 and 3.33 correspond to sphere, truncated octahedron, rhombic dodecahedron, cube and tetrahedron, respectively, i.e., the larger the value, the more compactly packed the system. η of liquid water at room temperature is in the range 1.3–2.0 with a bell-like distribution [88]; the same is observed in this study (Fig. 10 for the systems containing only water and ESM: Figs. S5, S6 for the others). There are some differences in the mean values obtained with the different models but in all cases they are within the error margins. The most populated mean values for the surfacial layers (Table S1) are shifted to the higher values (>1.7), indicating that the liquid there is more compactly organised.

The values of the asphericity parameter found here depart completely from the idea of tetrahedral ($\eta = 3.33$) coordination of water. However, they are not so

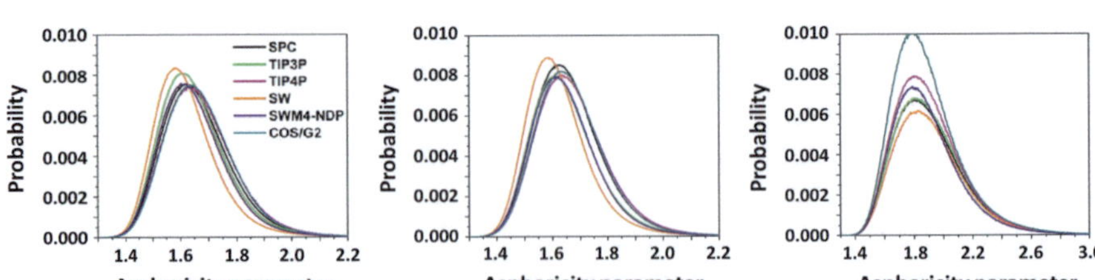

Fig. 10 Distribution of the VP asphericity parameter of bulk water (*left*), and of bulk slab (*centre*) and interfacial slab (*right*) in the water/vapour systems

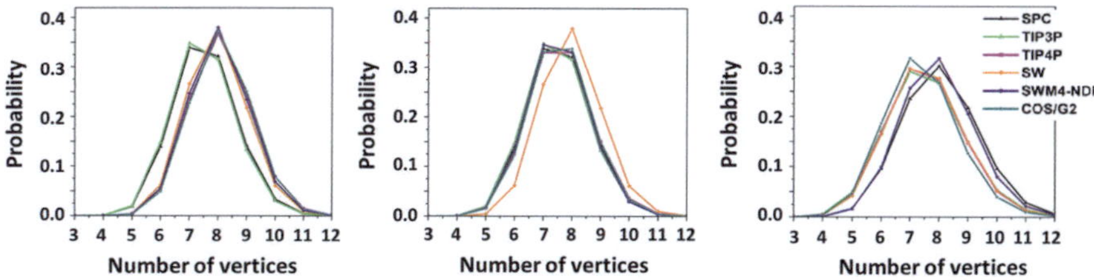

Fig. 11 Distribution of the vertices of individual VP faces in the bulk (*left*), bulk slab (*centre*) and surfacial slab (*right*) of water described with different models

distant from the value for the icosahedron ($\eta = 1.13$), considered as one of the possible ordering pattern in water clusters [29]. Such a fluctuating repetitive network of water molecules with icosahedron-type symmetry has been suggested first in 1998 [89]. Experimentally, it has been observed in water nanoparticles by means of X-ray diffraction [90, 91]. These polyhedral structures are idealised and in reality they are significantly fragmented and ill-ordered due to thermal effects but the existence of long-lived ring fragments has been proven [92]. The formation of a cluster with approximately icosahedral shape does not exclude the tetrahedral coordination inside and is in accord with the large number of shared faces found.

The icosahedron's faces are 12 regular pentagons, 30 rectangles, and 20 regular triangles. Analysis of the number of vertices of the VP faces could be used to probe this shape hypothesis. The distribution of VP vertices is given in Fig. 11, Figs. S7, S8, and Table S1 contains their mean values. It can be seen that the polygons building a VP have more vertices but this can be due to thermal fluctuations, overlapping of different polygons (e.g., a triangle and a pentagon or two rectangles) or to merging of two or more faces in the framework of this analysis. No sizeable difference is found between the results obtained with different water models or interfaces.

3.4 Water Dipole Moment—Size and Direction

Major feature emphasising the difference between polarisable and non-polarisable force fields is the dipole moment. Table 3 contains the mean values obtained for bulk water with different water models and the reference values reported by their developers. The average dipole moments of the water/vapour system are 2.60 ± 0.15 D, 2.45 ± 0.17 D and 2.59 ± 0.18 D for SW-RIGID-ISO, SWM4-NDP and COS/G2 respectively. The slightly lower values are due to the presence of evaporated water molecules which are not polarised by their environment but their number participates in the quotient.

Table 3 Mean and reference values of the dipole moment per molecule of bulk water obtained using various water models

	SPC	TIP3P	TIP4P	SW-RIGID-ISO	SWM4-NDP	COS/G2	Experiment
μ [D]	2.27	2.35	2.18	2.61 ± 0.15	2.46 ± 0.16	2.61 ± 0.17	2.95 [93]
Ref. μ [D]	2.27 [94]	2.35 [95]	2.18 [77]	2.61 [78]	2.46 [94]	2.59 [79]	

A specific characteristic rendering additional information about the structuring and the electric properties is the average dipole moment component along the Z axis. It is related to the local polarisation of the system:

$$P_z(z) = \langle \sum_m \delta(z - z_m) \left(\sum_j q_{jm} z_{jm} \right) \rangle \tag{3}$$

where $\delta(z - z_m)$ is the Dirac delta-function, m and j label the molecules and atoms, and q_{jm} and z_{jm} are atomic net charges and Z-coordinates, respectively. Dipole orientation with respect to the Z-axis is presented in Fig. 12, revealing that the residual μ_z values are essentially zero in the bulk with pronounced variation at the surface. In all cases the outermost water molecules are oriented with the hydrogens towards the vapour, while the more inner species are directed oppositely. Depending on the water model, three types of behaviour is demonstrated at the water/vapour interface: SPC and TIP3P water molecules are more oriented in the outer than in the inner sublayer which seems unphysical; the SW-RIGID-ISO waters have the same degree of orientation of the inner and outer molecules; the TIP4P, SWM4-NDP and COS/G2 models display more intense 'Z-ward' orientation of the subsurfacial waters while the outer dipoles are more parallel to the surface having at least one hydrogen directed to the non-polar phase. This result indicates that there is a region of predominating positive charges in the "evaporating" part of the interfacial layer and more marked negative charges at the inner "condensed" sublayer.

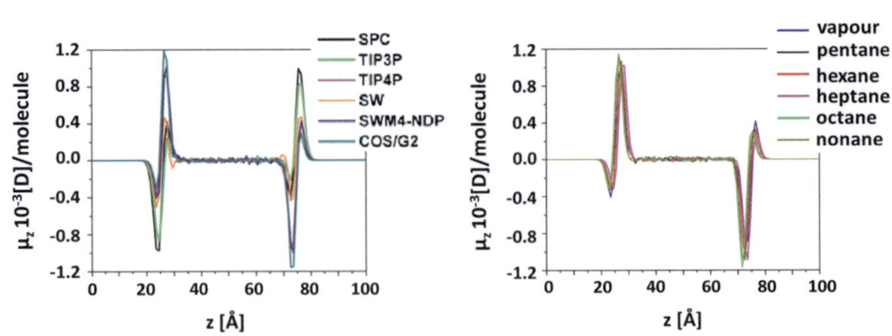

Fig. 12 Residual Z-component of the dipole moment at the water/vapour (*left*) and at TIP4P/alkanes (*right*) interfaces

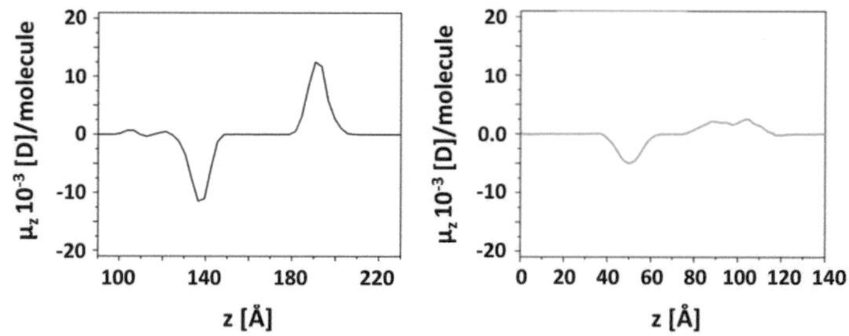

Fig. 13 Residual Z-component of the dipole moment at the water/DLPC/vapour (*left*) and at water/DLPC/alkanes (*right*) interfaces

The presence of alkanes does not change the profile and no particular dependence on the hydrocarbon chain length can be identified. Overall, the peaks in both interfacial sublayers intensify with chain extension.

The dipole moment orientation in the DLPC containing systems is shown in Fig. 13. The values are larger by an order of magnitude revealing much more pronounced orientation of the surfacial water molecules in a direction normal to the interface. The profiles are completely different: the water molecules are oriented with the hydrogen atoms towards the hydrophilic phospholipid heads and there is no compensating water counterpart—the compensatory role is played by the heads. Yet in the TIP4P/DLPC/vapour case water molecules stuck among the tails (as evidenced in the density plots, ESM: Fig. S1) are oriented in a way balancing the IS polarity, while in the TIP4P/DLPC/octane this contribution is almost missing. The dissimilarity of the two DLPC films is clearly visible.

In brief, the presence of surfactant monolayer causes marked water reorganization at both the water/DLPC/vapour and water/DLPC/alkanes interfaces, more sizeable in the former system, resulting in inverted polarity of these layers.

3.5 *Surface/Interface Tension*

MD simulations data allow the assessment of surface tension according to the following equation:

$$\sigma = \frac{1}{2} L_z \left[P_{zz} - \frac{1}{2} \left(P_{xx} + P_{yy} \right) \right] \tag{4}$$

where P_{qq} are the statistical averages of elements of the pressure tensor and L_z is the periodic box length along the axis normal to the surface, the Z-axis [96]. The calculated values are presented in Table 4 and they fully agree with those obtained by other authors—53.4, 49.5 and 54.7 mN/m for the SPC, TIP3P and TIP4P models respectively [97]. Apparently, the values obtained with the polarisable models are much closer to the experimental findings.

Table 4 Surface/Interface tension (σ) obtained for different water models and interfaces

Model		σ [mN/m]	σ_{exp} [mN/m]
SPC		51.99 ± 0.08	71.99 ± 0.05 [98]
TIP3P		48.28 ± 0.04	
SW-RIGID-ISO		76.57 ± 0.03	
SWM4-NDP		63.26 ± 0.03	
COS/G2		65.82 ± 0.01	
TIP4P	/vapour	53.72 ± 0.02	
	/C5	48.57 ± 0.01	48.74–50.80 [99]
	/C6	48.97 ± 0.01	50.38–51.25 [99]
	/C7	49.92 ± 0.07	50.71–51.64 [99]
	/C8	51.19 ± 0.04	51.30–52.30 [100]
	/C9	51.61 ± 0.06	52.20–53.00 [100]

The surface tension is a feature which is highly sensitive to interface modifications. The presence of alkanes should reduce its value which is confirmed by the results obtained and the experimental values provided (Table 4). The surface tension grows monotonously with alkane chain extension as it does at alkane/air interfaces.

The magnitude of σ is not influenced by changes of the box size or of the force fields employed for description of the alkanes (results not shown here), yet in all cases where polarisation is ignored the computed values are well below the experimental ones. The use of polarisable water models produces much more reliable results for both kinds of direct interfaces.

3.6 Hydrogen Bonds

Hydrogen bonds are the main reason for the peculiar structure and properties of water. In this study, the number of hydrogen bonds during the simulation was monitored and averaged per molecule at each step of the simulation (mean values presented in Table 5) aiming at the evaluation of H-bonding robustness. In the bulk, water molecules participate in 4 H-bonds on the average, which matches the conventional observations. However, using the same criteria (bond length $R_{HB} \leq 2.5$ Å [101] and angle H–O … H $\alpha \leq 70°$, as these are the values recommended by IUPAC [102]), and employing the polarisable COS/G2 water model, we found only 2 H-bonds per molecule, which agrees with another theory of water structuring suggesting the formation of chains and rings [7]. For the interfacial water/vapour layer, half of this number is registered (~2 with non-polarisable fields and ~1 with COS/G2), which is also in accord with literature data, disclosing reduced number of hydrogen bonds at the interface invoking higher reactivity of surfacial water. Replacement of the vapour with alkanes results in a minor increase both in the bulk and at the surface with no distinct correlation with the alkane chain length.

Table 5 Average number of hydrogen bonds per molecule during the simulation of bulk water and of water/vapour and water/alkanes systems using different water models

Model		*B*	*BS*	*IS*
SPC		4.06 ± 0.07	4.05 ± 0.07	2.33 ± 0.13
TIP3P		4.01 ± 0.07	4.01 ± 0.07	2.18 ± 0.15
TIP4P	/vapour	3.87 ± 0.07	3.86 ± 0.07	2.10 ± 0.14
	/C5		3.93 ± 0.07	2.01 ± 0.17
	/C6		3.92 ± 0.07	2.10 ± 1.17
	/C7		3.92 ± 0.07	1.99 ± 0.17
	/C8		3.93 ± 0.07	1.86 ± 0.18
	/C9		3.92 ± 0.07	1.93 ± 0.17
SW-RIGID-ISO		4.05 ± 0.07	4.04 ± 0.08	2.05 ± 0.14
SWM4-NDP		4.01 ± 0.07	4.01 ± 0.07	1.95 ± 0.16
COS/G2		2.01 ± 0.05	2.01 ± 0.05	1.05 ± 0.12

It is obvious for both interfaces that the average number of neighbours in the bulk is smaller than that of H-bonds, indicating that part of the neighbours form more than one H-bonds, whereas at the interface, irrespective of the more compact packing, the lack of neighbours in one of the half-spaces reduces the hydrogen bonding capacity. It should be noted that enhanced compactness does not necessarily mean better organisation or long-range ordering. Overall, the interfacial water/vapour and water/alkanes layers feature tighter packing and deteriorated ordering corroborated by all types of analyses presented in this study.

4 Conclusion

The study addresses the water structuring in bulk and surfacial water at the interface with water vapour or alkanes of different chain length. One of the aims of the work was to assess the reproducibility of experimental data using an assortment of polarisable and non-polarisable water models and to check for which properties the utilisation of polarisable models is critical. Simple polarisable models based on Drude oscillators were tested in order to keep the computational costs low. For bulk water and water/vapour systems the models TIP4P, SWM4-NDP and COS/G2 performed best. The superiority of the polarisable water models stood out most markedly in the evaluation of dipole moments and surface tensions. Since the TIP4P model produced results commeasurable with the polarisable ones, it was used predominantly further on to simulate water/alkane (C5–C9) interfaces and to quantify the structural parameters of water obtained from RDFs and Voronoi analyses. It was found that the surface layer width is in the range 5.7–6.9 Å in accordance with literature data. The molecules in this layer are organised in a more compact and less ordered manner. The ordering is owed mainly to hydrogen bonds which are twice as many in the bulk compared to the surface. The number of nearest neighbours, the asphericity parameter and analysis of the characteristics of the Voronoi cells showed that the tetrahedral coordination was blurred and more complex polyhedra were formed. The surface layer was found to consist of two sublayers, inner and outer, with oppositely oriented unequal polarity, defining areas of residual charges at the interface.

In addition to the systems with direct contact between water and non-polar fluids, interfaces mediated by lipid (DLPC) monolayers were modelled. The polar and non-polar fragments of the mediating surfactant are covalently bound and the monolayer was meant to seam together the two phases. The width of the surface water layer tripled upon DLPC inset. The compactness of the surfacial water, which was enhanced by the presence of alkanes, was tightened further by the lipid introduction. However, the water orientation was changed and the surfacial polarity was inverted, balanced by the lipid heads instead of the diffuse outer sublayer.

References

1. Finney JL (2001) J Mol Liq 90:303–312
2. Finney JL (2004) Philos Trans R Soc London 359:1471–2970
3. Ball P (2001) Life's matrix. A biography of water. University of California Press, Berkeley
4. Maréchal Y Hydrogen (2007) Bond and the water molecule: the physics and chemistry of water, aqueous and bio-media. Elsevier, Amsterdam
5. Wernet P, Nordlund D, Bergmann U, Cavalleri M, Odelius M, Ogasawara H, Naslund LA, Hirsch TK, Ojamae L, Glatzel P, Pettersson LGM, Nilsson A (2004) Science 304:995–999
6. Clark GNI, Cappa CD, Smith JD, Saykally RJ, Head-Gordon T (2010) Mol Phys 108:1415–1433
7. Nilsson A, Pettersson LGM (2011) Chem Phys 389:1–34
8. Huang C, Wikfeldt KT, Tokushima T, Nordlund D, Harada Y, Bergmann U, Niebuhr M, Weiss TM, Horikawa Y, Leetmaa M, Ljungberg MP, Takahashi O, Lenz A, Ojamae L, Lyubartsev AP, Shin S, Pettersson LGM, Nilsson A (2008) Proc Natl Acad Sci USA 106:15214–15218
9. Walrafen GE (1964) J Chem Phys 40:3249–3256
10. Walrafen GE (1967) J Chem Phys 47:114–126
11. Monosmith B, Walrafen GE (1984) J Chem Phys 81:669–674
12. Mizoguchi K, Hori Y, Tominaga Y (1992) J Chem Phys 97:1961–1968
13. Lock AJ, Bakker HJ (2002) J Chem Phys 117:1708–1713
14. Narten AH, Levy HA (1972) Water a comprehensive treatise, Liquid water: scattering of X-rays, USA
15. Page DI, Water a comprehensive treatise, USA, 1972
16. Narten AH, Thiessen WE, Blum L (1982) Science 217:1033–1034
17. Yamanaka K, Yamaguchi T, Wakita H (1994) J Chem Phys 101:9830–9836
18. Hura G, Sorenson JM, Glaeser RM, Head-Gordon T (2000) J Chem Phys 113:9140–9148
19. Soper AK, Phillips MG (1986) Chem Phys 107:47–60
20. Dore JC (1991) J Mol Struct 250:193–211
21. Dore JC, Blakey DM (1995) J Mol Liq 65–66:85–90
22. Soper AK (1997) J Phys: Condens Matter 9:2717–2730
23. Jedlovsky P, Brodholt JP, Bruni F, Ricci MA, Soper AK, Vallauri R (1998) J Chem Phys 108:8528–8540

24. Soper AK (2000) ChemPhys 258:121–137
25. Nakahara M, Matubayasi N, Wakai C (2001) J Mol Liq 90:75–83
26. McConnell HM (1991) Annu Rev Phys Chem 42:171–195
27. Benjamin I (1996) Chem Rev 96:1449–1476
28. Chandler D (2005) Nature 437:640–647
29. http://www1.lsbu.ac.uk/water/
30. KuoI-FW, Mundy CJ (2004) Science 303: 658–660
31. Miranda PB, Shen YR (1999) J Phys Chem B 103:3292–3307
32. Braslau A, Deutsch M, Pershan PS, Weiss AH, Als-Nielsen J, Bohr J (1985) Phys Rev Lett 54:114–117
33. Braslau A, Pershan PS, Swislow G, Ocko BM, Als-Nielsen J (1988) Phys Rev A 38:2457–2470
34. Goh MC, Hicks JM, Kemnitz K, Pinto GR, Bhattacharyya K, Eisenthal KB, Heinz TF (1988) J Phys Chem 92:5074–5075
35. Townsend RM, Rice SA (1991) J Chem Phys 94:2207–2218
36. Du Q, Superfine R, Freysz E, Shen YR (1993) Phys Rev Lett 70:2313–2316
37. Benjamin I (1994) Phys Rev Lett 73:2083–2086
38. Morita A, Hynes JT (2000) Chem Phys 258:371–390
39. Morita A, Hynes JT (2002) J Phys Chem B 106:673–685
40. Brown MG, Raymond EA, Allen HC, Scatena LF, Richmond GL (2000) J Phys Chem A 104:10220–10226
41. Wei X, Shen YR (2001) Phys Rev Lett 86:4799–4802
42. Wilson KR, Cavalleri M, Rude BS, Schaller RD, Nilsson A, Pettersson LGM, Goldman N, Catalano T, Bozek JD, Saykally RJ (2002) J Phys: Condens Matter 14:L221–L226
43. Perry A, Ahlborn H, Space B, Moore PB (2003) J Chem Phys 118:8411–8419
44. Fecko CJ, Eaves JD, Loparo JJ, Tokmakoff A, Geissler PL (2003) Science 301:1698–1702
45. Paul S, Chandra A (2003) Chem Phys Lett 373:87–93
46. Paul S, Chandra A (2004) Chem Phys Lett 386:218–224
47. Sun CQ, Zhang X, Zhou J, Huang Y, Zhou Y, Zheng W (2013) J Phys Chem Lett 4:2565–2570
48. Zhang X, Huang Y, Ma Z, Zhou Y, Zheng W, Zhou J, Sun CQ (2014) Phys Chem Chem Phys 16:22987–22994
49. Gan W, Wu D, Zhang Z, Guo Y, Wan H (2006) Chinese J Chem Phys 19:20–24
50. Weyl WA (1951) J Colloid Sci 6:389–405
51. Stillinger FH, Ben-Nairn A (1967) J Chem Phys 47:4431–4437
52. Fletcher NH (1968) Philos Mag 18:1287–1300
53. Croxton CA (1981) Phys A 106:239–259
54. Kuo IFW, Mundy CJ, Eggimann BL, McGrath MJ, Siepmann JI, Chen B, Vieceli J, Tobias DJ (2006) J Phys Chem B 110:3738–3746
55. Kessler J, Elgabarty H, Spura T, Karhan K, Partovi-Azar P, Hassanali AA, Kühne TD (2015) J Phys Chem B 119:10079–10086
56. Lee CY, McCammon JA, Rossky P (1984) J Chem Phys 80:4448–4455
57. Wilson MA, Pohorille A, Pratt LR (1987) J Phys Chem 91:4873–4878
58. Matsumoto M, Kataoka Y (1988) J Chem Phys 88:3233–3245
59. Walker DS, Hore DK, Richmond GL (2006) J Phys Chem B 110:20451–20459
60. Wilson KR, Schaller RD, Co DT, Saykally RJ, Rude BS, Catalano T, Bozek JD (2002) J Chem Phys 117:7738–7744
61. Fan Y, Chen X, Yang L, Cremer PlS, Gao YQ (2009) J Phys Chem B 113:11672–11679
62. Cipcigan FS, Sokhan VP, Jones AP, Crain J, Martyna GJ (2015) Phys Chem Chem Phys 17:8660–8669
63. Richmond G (2002) Chem Rev 102:2693–2724
64. Benjamin I (1997) Annu Rev Phys Chem 48:407–451
65. Pratt L, Pohorille A (2002) Chem Rev 102:2671–2692
66. Watry MR, Tarbuck TL, Richmond GL (2003) J Phys Chem B 107:512–518

67. Sokhan VP, Jones AP, Cipcigan FS, Crain J, Martyn GJ (2015) PNAS 112:6341–6346
68. Weiner SJ, Kollman PA, Case DA, Singh UC, Ghio C, Profetajr S, Wiener P (1984) J Am Chem Soc 106(106):765–784
69. Brooks CL, Brucoleri RE, Olafson BD, Slater DJ, Swaminathan S, Karplus M (1983) J Comput Chem 4(4):187–217
70. van der Spoel D, Lindahl E, Hess B, Groenhof G, Mark AE, Berendsen HJC (2005) J Comput Chem 26:1701–1718
71. Hess B, Bekker H, Berendsen HJC, Fraaije JGEM (1997) J Comput Chem 18:1463–1472
72. Miyamoto S, Kollman PA (1992) J Comput Chem 13:952–962
73. Duan Y, Wu C, Chowdhury S, Lee MC, Xiong G, Zhang W, Yang R, Cieplak P, Luo R, Lee T, Caldwell J, Wang J, Kollman P (2003) J Comput Chem 24:1999–2012
74. Humphrey W, Dalke A, Schulten K, Molec J (1996) Graphics 14:33–38
75. Rycroft CH (2007) Multiscale modeling in granular flow. Ph.D. thesis, Massachusetts Institute of Technology
76. http://www.engineeringtoolbox.com/water-density-specific-weight-d_595.html
77. Jorgensen WL, Chandrasekhar J, Madura JD, Impey RW, Klein ML (1983) J Chem Phys 79:926–935
78. van Maaren P, van der Spoel D (2001) J Phys Chem B 105:2618–2626
79. Yu H, van Gunsteren WF (2004) J Chem Phys 121:9549–9564
80. Bret C, Field MJ, Hemmingsen L (2000) Mol Phys 95:751–763
81. Voronoi GF (1908) J Reine Angew Math 134:198–287
82. Ruocco G, Sampoli M, Vallauri R (1992) J Chem Phys 96:6167–6176
83. Ruocco G, Sampoli M, Torcini A, Vallauri R (1993) J Chem Phys 99:8095–8104
84. Shih JP, Sheu SY, Mou CY (1994) J Chem Phys 100:2202–2212
85. Yeh Y, Mou C-Y (1999) J Phys Chem B 103:3699–3705
86. Jedlovszky P (1999) J Chem Phys 111:5975–5985
87. Jhon YI, No KT, Jhon JS (2006) Fluid Phase Equilibria 244:160–166
88. Jedlovszky P, Pártay LB, Bartók AP, Voloshin VP, Medvedev NN, Garberoglio G, Vallauri R (2008) J Chem Phys 128:244503–244512
89. Chaplin MF (2000) Biophys Chem 83:211–221
90. Müller A, Bögge H, Diemann E (2003) Inorg Chem Commun 6:52–53
91. Garcia-Ratés M, Miró P, Poblet JM, Bo C, Avalo JB (2011) J Phys Chem B 115:5980–5992
92. Bednyakov AS, Stepanov NF, Novakovskaya YV (2014) Russian J Phys Chem A88:287–294
93. Gubskaya AV, Kusalik PG (2002) J Chem Phys 117:5290–5302
94. Kiyohara K, Gubbins KE, Panagiotopoulos AZ (1998) Mol Phys 94:803–808
95. Mahoney MW, Jorgensen WL (2000) J Chem Phys 112:8910–8922
96. Harris JG (1992) J Phys Chem 96:5077–5086
97. Chen F, Smith PE (2007) J Chem Phys 126:221101–221103
98. Pallas NR, Harrison Y (1990) Colloid Surf 43:169–194
99. Matsubara H, Murase M, Mori YH, Nagashima A (1988) Int J Thermophys 9:409–424
100. Mitrinovic DM, Tikhonov AM, Li M, Huang ZQ, Schlossman ML (2000) Phys Rev Lett 85:582–585
101. Xenides D, Randolf BR, Rode BM (2006) J Mol Liq 123:61–67
102. Arunan E, Desiraju GR, Klein RA, Sadlej J, Scheiner S, Alkorta I, Clary DC, Crabtree RH, Dannenberg JJ, Hobza P, Kjaergaard HG, Legon AC, Mennucci B, Nesbitt DJ (2011) Pure Appl Chem 83:1637–1641

Molecular Design of Organometallic Materials: Effect of the Metallophilic Interactions, Ligand, Metal, and Oxidation State

Julia Romanova, M.R. Ranga Prabhath, Yousif Sadik and Peter D. Jarowski

Abstract The present book chapter is devoted to metal (**M**) homoleptic complexes with pyridyl–triazolate ligands (**L**) and their importance for the molecular design of optical and conducting materials. It summarizes our recently published, as well as new DFT/TD-DFT results on the role of the ligand, metal and oxidation state for tuning of molecular and supramolecular properties in $\mathbf{ML_2}$ complexes. The intermolecular interactions are taken into account by comparison between monomer ($\mathbf{ML_2}$) and dimer organometallic complexes ($\mathbf{ML_2 \ldots ML_2}$). The ligand effect is demonstrated in Pt(II)-complexes by modifying the electronic structure of the pyridyl-triazolate ligands. The role of the metal is estimated in the case of Pd(II) and Pt(II)-complexes with the same pyridyl-triazolate ligand. The effect of the oxidation state is revealed by simulations on Pt(II)-dimers in mixed valence $\mathbf{[ML_2 \ldots ML_2]^{+\bullet}}$ and doubly oxidized $\mathbf{[ML_2 \ldots ML_2]^{2+}}$ states. The obtained structure-property relationships are interpreted with respect to available experimental data and are used for the formulation of molecular design strategies.

Keywords Metallophilic interaction · Excimer · Ligand effect · Oxidation (p-doping) · TD-DFT

1 Introduction

Organometallic complexes represent powerful building blocks for advanced materials with applications in catalysis [1], molecular electronics and photonics [2–5], magnetic devices [6], solar cells [7, 8], anticancer therapy [9], biomolecular and

Electronic supplementary material The online version of this chapter (doi:10.1007/978-3-319-50255-7_8) contains supplementary material, which is available to authorized users.

J. Romanova · M.R. Ranga Prabhath · Y. Sadik · P.D. Jarowski (✉)
Advanced Technology Institute, University of Surrey, Guildford GU2 7XH, UK
e-mail: peterjarowski@gmail.com

A. Tadjer et al. (eds.), *Quantum Systems in Physics, Chemistry, and Biology*,
Progress in Theoretical Chemistry and Physics 30,
DOI 10.1007/978-3-319-50255-7_8

cellular probes [10–13], and many more. The wide area application of these molecules originates from their hybrid structural theme, which successfully combines the functionalities of purely organic and inorganic compounds. Moreover, the organic/inorganic fusion promotes a rich response to external stimuli, such as light, voltage, pressure, pH, solvent, temperature, as well as electric and magnetic fields. As the molecular response to stimuli is the main mechanism to transfer information at the material level, the hybrid or nanohybrid approach targets this scale directly through molecular design [14, 15]. The hybrid approach can be also applied to supramolecular interactions, though with less precision, when aiming to design multi-stimuli (smart) materials.

However, even armed with organometallic building blocks [16, 17], there are many challenges in the molecular design of advanced materials. At the molecular level, the challenges of material design are associated with the need for a better fundamental understanding of structural concepts toward the creation of organic-inorganic motifs, i.e. what is the best combination between metal and ligand for a given functionality. At the supramolecular level, the material design challenges are related to the intermolecular interactions and self-assembly of the constituent molecules [18–22]. For the latter, the challenges are even greater since, in addition to the structural diversity of metal-ligand combinations, different intermolecular interactions can occur and strongly affect the material structure and response. Moreover, these interactions are not only limited to the organometallic building blocks but also include molecules in their environment (solvent and counter-ions).

Metallophilic interaction (metal-metal attraction) represents an intriguing type of intermolecular interactions between organometallic complexes [23–25]. It has been demonstrated that external stimuli such as light and voltage can modulate intermolecular metallophilic interactions, which in turn give rise to smart materials with unique luminescent and conducting properties [14]. It has also been shown that the metal-metal attraction can be the driving force for self-assembly and formation of discrete dimer structures or even metallostrings [26]. Therefore, the ability to modulate the metallophilic interactions represents a promising approach to tune the material properties. However, in order to take advantage of the metallophilic interaction in the design of new materials, complexes of metals with low coordination number and relatively high relativistic effects are required. Such metals are Pt(II), Au(I) and Hg(II) [23]. In addition, due to the complexity of the organic-inorganic theme, the metallophilic interaction can be regulated by ligand architecture, π-π stacking strength and solubility [27].

Good examples for luminescent materials working through light-driven intermolecular metallophilic interactions can be found in the family of Pt(II)-complexes [28–31]. The luminescent properties of such complexes vary substantially going from monomer (low concentrated solutions) to supramolecular level (high concentrated solutions and solid state). In most of the cases, the emission maxima measured in the solid state are strongly red-shifted with respect to the isolated monomers, and this behaviour has been explained by an excited state aggregation, as well as by the formation of excimers with covalent metal-metal bonds [32–34]. It has also been shown that such excimer formation, i.e. the metallophilic interactions in the excited state, can be suppressed by introducing bulky substituents in the ligand [35].

The family of Pt(II)-complexes also contains many representatives with good conducting properties originating from metallophilic interactions. There are many examples of one-dimensional (1D) structures composed of square-planar Pt(II)-complexes, where the metallophilic bonding is strengthened by partial oxidation [25, 36–39]. Due to the partial oxidation and short metal-metal contacts, such materials have a mixed valance (MV) nature [40, 41] and enhanced conducting properties. In addition, recently metallopolymers with direct Pt–Pt bonds have shown to be strong near-infrared (NIR) emitters with emissions above 1000 nm, due to the formation of MV species [42]. In some cases the metallophilic interactions in the 1D mixed-valence structures occur only within discrete dimers, because the latter are split by an organic or halide ligand (X). Such structures are called 1D MMX complexes and also possess good semiconducting or metallic properties [43–46].

The present DFT/TD-DFT investigation gives insights into molecular design of new luminescent and conducting materials through the exploration of *different metal-ligand combinations*. It aims to demonstrate how the ligand, metal and the oxidation state can be used in order to tune molecular and supramolecular properties in organometallic $\mathbf{ML_2}$ complexes. In particular, our molecular set is composed of organometallic complexes, which contain four different pyridyl–triazolate ligands as coordination sites and Pd(II) or Pt(II) as metal centers (Scheme 1). In order to contribute to the decoding of the supramolecular puzzle in organometallics, in addition to the *monomer complexes*, we simulated *metallophilic dimers* (Scheme 1). Since, the investigated organometallic complexes and their dimers can be regarded as promising building blocks for luminescent and conducting materials, we simulated their response to *external stimuli, such as light and voltage*. The interactions with light and the effect of an applied voltage were simulated by excited state calculations and ground state computations on oxidized mixed valent forms, respectively.

The chapter is organized as follows. Section 2 represents the computational methodology. Section 3 presents our previous results on the role of the ligand [32–34], as well as new results on the effect of the metal type and oxidation state, and it contains discussions on the molecular and supramolecular structure and properties. Section 4 summarizes the main conclusions on the theoretically derived structure-property relationships and defines molecular design guidelines that can be applied in the laboratory synthesis of new functional organometallic materials.

2 Computational Procedure

The ground-state (S_0) and excited-state (S_n, T_n) structures of the organometallic complexes and their dimers (Scheme 1) were obtained with the PBE0 [47] and PBE0-D3 [48] functionals by using the density functional theory (DFT) and its time dependent formulation (TD-DFT). The electronic shells of the metal atoms are described by Stuttgart/Dresden (SDD) pseudopotential and basis set, while Pople 6-31G* split-valence basis set is used for the ligands. In order to take into account the electrostatic interactions with the solvent ethanol ($\varepsilon_0 = 24.85$, $\varepsilon_\infty = 1.85$), the polarizable continuum model (PCM) in its integral equation formalism variant was applied [49–51]. The frequency analysis confirms that the optimized ground-state structures represent true minima on the potential energy surface. In the instances where the convergence in the excited triplet states were not achieved, the T_1 excited state geometry was taken from DFT optimization. The TD-DFT optimizations were performed without symmetry constraints, which allows symmetry breaking in excited state. In case the symmetry is preserved, the equivalent bond lengths from all four ligands are averaged and only the symmetrically unique bond lengths are listed. The vertical absorption and emission energies were obtained with linear response solvation. All quantum-chemical calculations were performed with the Gaussian 09 program package [52]. The molecular orbitals were visualized with GaussView 5 [53] with an isosurface value of 0.02.

Dispersion correction was used only for the ground states of the dimers, because it was previously demonstrated that its inclusion has a strong impact on the Pt–Pt intermolecular distance and hence on the calculated dimerization energies [34]. On the other hand, due to the strong Pt–Pt intermolecular interaction in excited states it was found that the PBE0 and PBE-D3 methods predict identical optimized geometries at the TD-DFT level [34].

3 Results and Discussion

3.1 *Effect of the Ligand*

Our recent experimental and theoretical investigations demonstrate the ligand effect on the metallophilic interaction in luminescent organometallic complexes [32–34]: **Pt(LD)**$_2$, **Pt(LH)**$_2$, **Pt(LA)**$_2$ and **Pt(LA2)**$_2$. These studies were focused on homoleptic Pt complexes with pyridyl-triazolate coordination sites (Scheme 1), where the electronic structure of the ligand is systematically modified by an attachment of a substituent to the pyridine cycle. The substituents $–N(CH_3)_2$, –H, –CHO or $–CH{=}C(CN)_2$ were chosen due to the variation of their π-donor/acceptor properties [54]. The numerical results were obtained by the computational protocol described in this chapter and herein we briefly present the main results on the ligand effect.

The TD-DFT/PBE0/PCM calculations for the monomers reveal that the $S_1 \rightarrow S_0/T_1 \rightarrow S_0$ emission energies lie between 375/473 nm for **Pt(LH)$_2$** and 461/956 nm for **Pt(LA2)$_2$**. The photoluminescence spectra of the complexes in diluted solutions show maxima in the range between 397 nm for **Pt(LD)$_2$** and 408 nm for **Pt(LA2)$_2$**. The simulations for the dimers predict $S_1 \rightarrow S_0/T_1 \rightarrow S_0$ emission energies at 524/584 nm for **Pt(LD)$_2$**, 554/591 nm for **Pt(LH)$_2$**, 587/613 nm for **Pt(LA)$_2$** and 745/795 nm for **Pt(LA2)$_2$**. The experimentally observed emission maxima at high concentrations/thin films demonstrate 459/487 nm for **Pt(LD)$_2$**, 540/541 nm for **Pt(LH)$_2$**, 575/602 nm for **Pt(LA)$_2$** and 599/625 nm for **Pt(LA2)$_2$**. Based on the experimental data and the PBE0 results for the excited-state structure of the monomer and dimer complexes it was concluded that [32–34]:

(1) in the monomers the emitting states are with metal-to-ligand charge transfer (MLCT) character, while in the dimers the emission originates from metal-to-metal-to-ligand charge transfer (MMLCT) states
(2) in the monomers the excitation to the first MLCT state mainly affect the Pt-triazole bond and the triazolate cycle, while in the dimers the excitation to the first MMLCT state induces strong structural changes to the Pt–Pt intermolecular bond
(3) in excited states the metallophilic interaction is strong and causes an excimer formation (Pt–Pt bond lengths < 2.9 Å); in ground states the metal-metal attraction is weak (Pt–Pt bond lengths > 3.5 Å) but slightly improves when the π-acceptor strength of the substituent/ligand increases
(4) the tunability in the emission properties at the monomer level is weak but at the dimer level becomes strong and dependent on the π-donor/acceptor strength of the ligand; the stronger the π-acceptor character of the substituent (ligand) the larger the red shift in emission energy
(5) S_1 and T_1 in the dimers lie closely with the $\Delta E_{S_1-T_1}$ energy difference decreasing from 0.24 to 0.10 eV when the π-acceptor strength of the ligands increases
(6) the possibility to simultaneously tune the emission wavelength and $\Delta E_{S_1-T_1}$ difference by ligand modification in Pt(II) complexes was suggested as a promising approach for the discovery of new highly efficient luminescent [55] and singlet fission materials [56].

The effect of the ligand on metallophilic interaction in the ground and excited states is qualitatively explained in Scheme 2. The metallophilic interaction in ground state occurs due to overlap between the $5d_z{}^2$ orbitals of adjacent Pt-atoms. This leads to the formation of bonding (σ) and antibonding (σ*) sigma molecular orbitals within the dimers. The dimer molecular orbitals are doubly occupied and therefore in the ground-state, the metallophilic intermolecular interaction is weak. The ligand field determines the energy of the monomer $5d_z^2$ orbitals. When the π-acceptor strength of the ligand (substituent) increases, the energy of the monomer $5d_z^2$ orbital decreases [34]. The lower the energy of the $5d_z^2$, the higher $\Delta E_{\sigma\text{-}\sigma^*}$ splitting and the shorter Pt–Pt bond length is [34]. Therefore, the metal-metal attraction in ground-state slightly increases in line with the π-acceptor strength of the substituent (ligand). The excitation in the dimers causes one-electron extraction from $\sigma^*(d_z^2\text{-}d_z^2)$, which reduces the number of electrons in the antibonding molecular orbital and stabilizes the Pt–Pt interaction. Therefore, the Pt–Pt bonds in the excited state are shorter and their length approaches the covalent region ~2.7–2.8 Å. On the other hand, the $\Delta E_{\sigma\text{-}\sigma^*}$ splitting in excited-state remains affected by the ligand field.

3.2 *Effect of the Metal*

Here, we present new results on the effect of metal (Pd or Pt) on the metallophilic interaction in ground and excited states of homoleptic **M(LA2)$_2$** complexes. For this purpose, we focus on complexes with the **LA2** ligand, where the pyridyl-triazolate coordination sites are functionalized with $-CH{=}C(CN)_2$ groups (Scheme 1). We chose this ligand as the **Pt(LA2)$_2$** complex was part of the molecular set in the investigation of the ligand effect, which allows us to compare different aspects of the metallophilic interaction. Additionally, the comparison between **Pt(LA2)$_2$** and **Pd(LA2)$_2$** is motivated by the fact that both metals have d^8-electronic configuration and form square-planar complexes but Pd has a lower atomic number than Pt, which allows us to track the impact of the relativistic effects on the metallophilic interaction [23].

The optimized ground state geometries of **Pd(LA2)$_2$** and **Pt(LA2)$_2$** monomers are presented in Table 1. The structural differences are very small. For the C–C, C–N and N–N bonds, the absolute bond length differences between the monomers are below 0.003 Å, while for the M-N1$_{Py}$ and M-N2$_{Tr}$ bonds they amount to 0.006 Å. It should be noted that the Pt-N1$_{Py}$ bond is shorter than the Pd-N1$_{Py}$ one, which indicates slightly weaker interaction between the Pd and pyridine ring. However, an opposite behavior is observed for M-N2$_{Tr}$ bonds and hence slightly stronger interaction between Pd and the triazolate cycles. The results on the absorption properties of **Pd(LA2)$_2$** and **Pt(LA2)$_2$** monomers are also very similar (Table 2). The $S_0 \rightarrow S_1/S_0 \rightarrow T_1$ transition energies for **Pd(LA2)$_2$** are predicted at 385/597 nm, while for **Pt(LA2)$_2$** they are at 415/600 nm. Therefore, our simulations reveal rather weak tunability in absorption properties by metal ion modification. Both $S_0 \rightarrow S_1$ transitions are predicted with relatively high oscillator strength, however, a hypochromic shift is observed for the Pt-complex (0.4189) with respect to its Pd-equivalent (1.1888). The molecular orbital analysis indicates that S_1/T_1 excitations in **Pt(LA2)$_2$** and S_1 in **Pd(LA2)$_2$** are mainly associated with

Table 1 Optimized PBE0/PCM bond lengths [Å] in the ground- and excited-states of **Pd(LA2)$_2$** and **Pt(LA2)$_2$**. The atom labelling is according to Scheme 3 and L1/2 denote ligand 1/2

	M = Pd			M = Pt		
	S_0	S_1	T_1	$^aS_0$	$^aS_1$	T_1
	L1	L1	L1	L1	L1	L1
M-N1$_{Py}$	2.061	2.055	2.059	2.055	2.063	2.051
N1$_{Py}$-C1$_{Py}$	1.363	1.392	1.376	1.366	1.377	1.380
C1$_{Py}$-C2$_{Py}$	1.397	1.406	1.408	1.396	1.397	1.407
C2$_{Py}$-C3$_{Py}$	1.380	1.376	1.370	1.380	1.377	1.370
C3$_{Py}$-C4$_{Py}$	1.409	1.428	1.434	1.409	1.420	1.434
C4$_{Py}$-C5$_{Py}$	1.403	1.434	1.444	1.402	1.407	1.442
C5$_{Py}$-N1$_{Py}$	1.330	1.310	1.315	1.333	1.327	1.318
M-N2$_{Tr}$	2.003	2.002	2.002	2.009	1.964	2.007
N2$_{Tr}$-N3$_{Tr}$	1.310	1.299	1.316	1.313	1.328	1.318
N3$_{Tr}$-N4$_{Tr}$	1.327	1.350	1.325	1.327	1.309	1.324
N4$_{Tr}$-C6$_{Tr}$	1.344	1.326	1.345	1.343	1.359	1.345
C6$_{Tr}$-C7$_{Tr}$	1.389	1.413	1.393	1.389	1.377	1.392
C7$_{Tr}$-N2$_{Tr}$	1.358	1.376	1.358	1.358	1.382	1.358
C7$_{Tr}$-C1$_{Py}$	1.435	1.413	1.429	1.433	1.427	1.428
C4$_{Py}$-C8$_R$	1.444	1.412	1.389	1.445	1.430	1.414
C8$_R$-C9$_R$	1.363	1.407	1.487	1.363	1.379	1.412
C9$_R$-C10$_R$	1.427	1.413	1.401	1.430	1.422	1.415
C10$_R$-N5$_R$	1.162	1.168	1.171	1.162	1.165	1.166
		L2	L2			L2
M-N1$_{Py}$		2.063	2.063			2.055
N1$_{Py}$-C1$_{Py}$		1.364	1.363			1.366
C1$_{Py}$-C2$_{Py}$		1.397	1.397			1.396
C2$_{Py}$-C3$_{Py}$		1.380	1.380			1.379
C3$_{Py}$-C4$_{Py}$		1.409	1.409			1.409
C4$_{Py}$-C5$_{Py}$		1.403	1.403			1.402
C5$_{Py}$-N1$_{Py}$		1.330	1.330			1.333
M-N2$_{Tr}$		1.999	2.008			2.013
N2$_{Tr}$-N3$_{Tr}$		1.311	1.310			1.312
N3$_{Tr}$-N4$_{Tr}$		1.326	1.328			1.327
N4$_{Tr}$-C6$_{Tr}$		1.345	1.344			1.343
C6$_{Tr}$-C7$_{Tr}$		1.388	1.389			1.389
C7$_{Tr}$-N2$_{Tr}$		1.359	1.358			1.358
C7$_{Tr}$-C1$_{Py}$		1.434	1.435			1.433
C4$_{Py}$-C8$_R$		1.444	1.444			1.414
C8$_R$-C9$_R$		1.364	1.364			1.412
C9$_R$-C10$_R$		1.427	1.427			1.415
C10$_R$-N5$_R$		1.162	1.162			1.166

^{a}Results are from Ref. [34]

Table 2 Vertical absorption and emission energies E [eV], wavelengths λ [nm] and oscillator strength *f* computed for Pd and Pt complexes with LA2 ligands. The results are obtained at the TD-DFT/PBE0/PCM level with linear response solvation in ethanol

	Absorption				Emission			
	ES	E	λ	*f*	ES	E	λ	*f*
$Pd(LA2)_2$	S_1	3.22	385	1.1888	S_1	2.81	441	1.3781
	T_1	2.08	597	–	T_1	1.31	945	–
$Pd(LA2)_2$ … Pd $(LA2)_2$	S_1	3.00	413	0.0239	S_1	2.42	512	0.4187
	S_2	3.07	404	0.6103				
	T_1	2.06	602	–	T_1	NA*	NA*	–
[a]$Pt(LA2)_2$	S_1	2.99	415	0.4189	S_1	2.69	461	0.4771
	T_1	2.07	600	–	T_1	1.30	956	–
[a]$Pt(LA2)_2$ … Pt $(LA2)_2$	S_1	2.73	453	0.0180	S_1	1.66	745	0.0309
	S_3	2.93	423	0.4998				
	T_1	2.05	603	–	T_1	1.56**	796**	–
$[Pt(LA2)_2$ … Pt $(LA2)_2]^+$	D_1	1.26	986	0.0026				

[a]Results are from Ref. [34]

*The excitation energy is negative due to the multireference character of the ground state caused by rotation around $–CH=C(CN)_2$ in T_1

**Estimated on DFT optimized T_1 geometry

HOMO to LUMO transitions (**Figure S1, SI**). These excitations are assigned as MLCT transitions. However, T_1 in $\mathbf{Pd(LA2)_2}$ has mixed character and, in addition to HOMO to LUMO transition, involves HOMO-1 to LUMO+1 transition—both configurations with equal weights (**Figure S1, SI**). Therefore, the T_1 excitation in $\mathbf{Pd(LA2)_2}$ is assigned as mixed MLCT and intra-ligand (IL) transition. It is important to note that in $\mathbf{Pd(LA2)_2}$ complex, the metal coefficient in the HOMO orbital is much smaller than in the Pt-analogue and hence Pd(II) is less involved in the MLCT excitation than Pt(II).

Despite the similar S_0 geometries of $\mathbf{Pd(LA2)_2}$ and $\mathbf{Pt(LA2)_2}$ monomers, the optimized S_1 structures are quite different (Table 1). An excitation from S_0 to S_1 in $\mathbf{Pd(LA2)_2}$ slightly affects the M-N bonds and only one of the ligands is subject to structural modifications. In this case, the bond length differences (BLD) between the ground and excited states are 0.005 Å for Pd-$N1_{Py}$ and 0.001 Å for Pt-$N2_{Tr}$. The most pronounced BLD is observed for the $C8_R$-$C9_R$ bond, which elongates with 0.043 Å. Such elongation in $\mathbf{Pd(LA2)_2}$ is also associated with an increased quinoid character of the affected ligand upon excitation. The larger quinoid character of the ligand in the excited state is also obvious from the increased 'double' character of $C8_R$-$C4_{Py}$, $N1_{Py}$-$C5_{Py}$, $C2_{Py}$-$C3_{Py}$, $C7_{Tr}$-$C1_{Py}$, $N2_{Tr}$-$N3_{Tr}$ and $N4_{Tr}$-$C6_{Tr}$ bonds in S_1. When $\mathbf{Pd(LA2)_2}$ is excited to T_1, again only one of the ligands is affected. Comparison between S_0 and T_1 optimized geometries shows an increased quinoid character of the pyridine cycle but slightly affected triazolate ring (BLDs $\leq$ 0.006 Å) and metal-nitrogen bonds (BLDs = 0.002 Å). Indeed, the first triplet excitation is localized in the $C8_R$-$C9_R$ region and this bond elongates substantially with 0.123 Å. Such strong elongation is associated with a change in the $C8_R$-$C9_R$ bond character from 'double' to 'single'. According to the molecular orbital analysis, the optimized S_1 and T_1 excited states of **Pd(LA2)$_2$** can be classified as HOMO to LUMO transition. In these excitations, HOMO spans one of the ligands and possess very small metal contribution, while LUMO is entirely ligand based (**Figure S2, SI**). The results on the Pt-complex show that when **Pt(LA2)$_2$** is excited to S_1, both ligands are equally involved. Moreover, in this case the $S_1 \rightarrow S_0$ transition induces strongest structural changes in the Pt-$N2_{Tr}$ bonds, which decrease by 0.045 Å. The better coupling between Pt(II) and 5-membered ring in S_1 leads to stronger geometrical redistribution in the triazolate than in pyridine moieties. The 'double' character of $N2_{Tr}$-$N3_{Tr}$ and $N4_{Tr}$-$C6_{Tr}$ bonds decreases when going from S_0 to S_1, while it increases for the other bonds in the triazolate cycle. Therefore, the triazolate ring shows a different conjugation pattern in the S_1 excited state of Pd- and of Pt-complexes. On the other hand, in both complexes the pyridine ring and its substituents have qualitatively similar structures. In **Pt(LA2)$_2$**, the $C8_R$-$C9_R$ bond is also elongated but only by 0.016 Å. Therefore, in a quantitative aspect the S_1 excitation in **Pt(LA2)$_2$** causes much weaker modifications in the aromatic structure of the pyridine moiety than in **Pd (LA2)$_2$**. The molecular orbital analysis indicates that the optimized S_1/T_1 excited states of **Pt(LA2)$_2$** possess MLCT character, involving HOMO and LUMO orbitals (**Figure S2, SI**).

The $S_1 \rightarrow S_0$/$T_1 \rightarrow S_0$ emission wavelengths for **Pd(LA2)$_2$** are predicted at 441/945 nm, while for **Pt(LA2)$_2$** they are at 461/956 nm. This result is similar to the absorption properties and indicates weak tunability in emission energies going from **Pd(LA2)$_2$** to **Pt(LA2)$_2$**. However, despite the weak tunability, there is a difference between the electronic structure of the S_1/T_1 excited states in Pd- and Pt-complexes. This difference is associated with much smaller contribution of Pd (II) in the excitations with respect to Pt(II).

The optimized ground state geometries of **Pd(LA2)$_2$…Pd(LA2)$_2$** and **Pt(LA2)$_2$ …Pt(LA2)$_2$** are presented in Tables 3 and 4. The results indicate that in both dimers, the molecular structure is almost unaffected by aggregation. The calculated free energies of dimerization are slightly positive at PBE0 level and suggest weak intermolecular interaction (5.5 kcal/mol for Pd and 7.2 kcal/mol for Pt). When the dispersion correction is taken into account stronger interaction between the monomers is predicted (−17.5 kcal/mol for Pd and −19.7 kcal/mol for Pt). In addition, the intermolecular metal-metal distances in both complexes are very close to each other. At PBE0/PBE0-D3 levels they amount to 3.582/3.087 Å for Pd and 3.572/3.058 Å for Pt. Therefore, at first sight it can be concluded that weak metallophilic interaction is the driving force for aggregation in both types of complexes. However, in the case of **Pd(LA2)$_2$…Pd(LA2)$_2$** HOMO is ligand based and it represents π-π interaction between the monomers (**Figure S1, SI**). For the **Pd (LA2)$_2$** dimer, the first occupied MO with $d_z{}^2$ atomic coefficients is HOMO-4 and it shows $\sigma^*(d_z{}^2\text{-}d_z{}^2)$ overlap between Pd ions (**Figure S1, SI**). On the other hand, in the case of **Pt(LA2)$_2$…Pt(LA2)$_2$**, HOMO represents the $\sigma^*(d_z{}^2\text{-}d_z{}^2)$ overlap

Table 3 Optimized PBE0/PCM bond lengths [Å] in the ground- and excited-states of **Pd(LA2)$_2$...Pd(LA2)$_2$**. The atom labelling is according to Scheme 3, where M1/2 and L1/2 denote monomer 1/2 and ligand 1/2, respectively

	S_0	S_0^{D3}	S_1		T_1^{DFT}	
	M1/L1	M1/L1	M1/L1	M2/L1	M1/L1	M2/L1
Pd-N1$_{Py}$	2.060	2.054	2.065	2.059	2.061	2.060
N1$_{Py}$-C1$_{Py}$	1.363	1.363	1.370	1.376	1.368	1.363
C1$_{Py}$-C2$_{Py}$	1.397	1.397	1.400	1.403	1.401	1.397
C2$_{Py}$-C3$_{Py}$	1.379	1.379	1.382	1.380	1.376	1.380
C3$_{Py}$-C4$_{Py}$	1.409	1.410	1.408	1.415	1.420	1.408
C4$_{Py}$-C5$_{Py}$	1.403	1.403	1.412	1.421	1.419	1.403
C5$_{Py}$-N1$_{Py}$	1.330	1.329	1.323	1.319	1.325	1.330
M-N2$_{Tr}$	2.003	2.001	2.017	2.010	2.004	2.005
N2$_{Tr}$-N3$_{Tr}$	1.310	1.310	1.306	1.309	1.313	1.310
N3$_{Tr}$-N4$_{Tr}$	1.328	1.329	1.340	1.336	1.326	1.328
N4$_{Tr}$-C6$_{Tr}$	1.344	1.344	1.331	1.334	1.346	1.345
C6$_{Tr}$-C7$_{Tr}$	1.389	1.389	1.412	1.411	1.389	1.389
C7$_{Tr}$-N2$_{Tr}$	1.357	1.358	1.358	1.361	1.357	1.357
C7$_{Tr}$-C1$_{Py}$	1.435	1.434	1.426	1.424	1.435	1.435
C4$_{Py}$-C8$_{R}$	1.444	1.443	1.443	1.431	1.407	1.445
C8$_{R}$-C9$_{R}$	1.364	1.364	1.366	1.384	1.474	1.362
C9$_{R}$-C10$_{R}$	1.427	1.427	1.427	1.420	1.405	1.427
C10$_{R}$-N5$_{R}$	1.162	1.162	1.162	1.165	1.169	1.162
			M1/L2	M2/L2	M1/L2	M2/L2
Pd-N1$_{Py}$			2.050	2.059	2.058	2.056
N1$_{Py}$-C1$_{Py}$			1.367	1.364	1.363	1.363
C1$_{Py}$-C2$_{Py}$			1.399	1.398	1.398	1.397
C2$_{Py}$-C3$_{Py}$			1.379	1.379	1.380	1.380
C3$_{Py}$-C4$_{Py}$			1.417	1.410	1.409	1.409
C4$_{Py}$-C5$_{Py}$			1.409	1.403	1.403	1.403
C5$_{Py}$-N1$_{Py}$			1.327	1.329	1.329	1.329
M-N2$_{Tr}$			1.990	1.997	2.006	2.003
N2$_{Tr}$-N3$_{Tr}$			1.314	1.310	1.310	1.310
N3$_{Tr}$-N4$_{Tr}$			1.325	1.328	1.328	1.328
N4$_{Tr}$-C6$_{Tr}$			1.346	1.344	1.344	1.344
C6$_{Tr}$-C7$_{Tr}$			1.389	1.389	1.389	1.388
C7$_{Tr}$-N2$_{Tr}$			1.358	1.358	1.358	1.357
C7$_{Tr}$-C1$_{Py}$			1.435	1.433	1.435	1.435
C4$_{Py}$-C8$_{R}$			1.433	1.443	1.444	1.444
C8$_{R}$-C9$_{R}$			1.382	1.364	1.364	1.364
C9$_{R}$-C10$_{R}$			1.421	1.427	1.427	1.427
C10$_{R}$-N5$_{R}$			1.165	1.162	1.162	1.162
Pd-Pd	3.582	3.087	3.237		4.253	

^{D3}PBE0 results with D3 dispersion correction

^{DFT}Geometry obtained by ground state DFT optimization of T_1

Table 4 Optimized PBE0/PCM bond lengths [Å] in the ground- and excited-states of **Pt(LA2)$_2$…Pt(LA2)$_2$**, as well as in its oxidized form **[Pt(LA2)$_2$…Pt(LA2)$_2$]$^{+\bullet}$**. The atom labelling is according to Scheme 3, where M1 and L1 denote monomer 1 and ligand 1

	$^aS_0$	$^aS_0^{D3}$	$^aS_1$	$^aS_1^{D3}$	$^aT_1^{DFT}$	D_1
	M1/L1	M1/L1	M1/L1	M1/L1	M1/L1	M1/L1
M-N1$_{Py}$	2.054	2.052	2.064	2.058	2.066	2.070
N1$_{Py}$-C1$_{Py}$	1.366	1.366	1.369	1.370	1.368	1.365
C1$_{Py}$-C2$_{Py}$	1.396	1.396	1.398	1.398	1.397	1.395
C2$_{Py}$-C3$_{Py}$	1.379	1.379	1.377	1.377	1.378	1.380
C3$_{Py}$-C4$_{Py}$	1.409	1.410	1.416	1.416	1.415	1.408
C4$_{Py}$-C5$_{Py}$	1.402	1.401	1.403	1.402	1.404	1.403
C5$_{Py}$-N1$_{Py}$	1.332	1.332	1.331	1.330	1.331	1.333
M-N2$_{Tr}$	2.009	2.007	2.014	2.014	2.016	2.016
N2$_{Tr}$-N3$_{Tr}$	1.312	1.312	1.316	1.316	1.316	1.313
N3$_{Tr}$-N4$_{Tr}$	1.327	1.329	1.323	1.323	1.322	1.323
N4$_{Tr}$-C6$_{Tr}$	1.343	1.343	1.346	1.346	1.346	1.346
C6$_{Tr}$-C7$_{Tr}$	1.389	1.389	1.389	1.389	1.388	1.388
C7$_{Tr}$-N2$_{Tr}$	1.357	1.358	1.358	1.359	1.357	1.356
C7$_{Tr}$-C1$_{Py}$	1.433	1.432	1.432	1.431	1.433	1.435
C4$_{Py}$-C8$_R$	1.444	1.443	1.438	1.438		1.447
C8$_R$-C9$_R$	1.364	1.364	1.371	1.371		1.362
C9$_R$-C10$_R$	1.427	1.427	1.425	1.424		1.428
C10$_R$-N5$_R$	1.162	1.162	1.163	1.163		1.162
Pt–Pt	3.573	3.058	2.845	2.837	2.839	2.853

^{a}Results are from Ref. [34]
^{D3}PBE0 results with dispersion correction
^{DFT}Geometry obtained by ground state DFT optimization of T_1

between the $d_z{}^2$ atomic orbitals of Pt centers (**Figure S1, SI**). Comparison between the complexes reveals that the $\Delta E_{\sigma\text{-}\sigma^*}$ splitting for Pd/Pt complexes are 0.83/1.13 eV at PBE0 level and 1.54/2.10 at PBE0-D3 level. Due to the smaller $\Delta E_{\sigma\text{-}\sigma^*}$ splitting in the case of the Pd-dimer, the metal-metal attraction in the ground state is weaker than in the Pt-dimer. The latter can be associated with the lower atomic number and smaller relativistic effects of Pd with respect to Pt.

The calculated excitation energies for the dimers of **Pd(LA2)$_2$** and **Pt(LA2)$_2$** are presented in Table 2. Similar to the case of monomers, weak tunability in the absorption properties is observed and the aggregation induces a slight bathochromic shift of the $S_0 \rightarrow S_n$ transitions. Both dimers are characterized with two closely lying singlet excited states S_1 and $S_{2/3}$. For these $S_0 \rightarrow S_1/S_0 \rightarrow S_{2/3}$ transitions, TD-DFT predicts 413/404 nm in **Pd(LA2)$_2$…Pd(LA2)$_2$** and 453/423 nm in **Pt(LA2)$_2$…Pt(LA2)$_2$**. In addition, the excitations to the first triplet state are estimated as 602 nm in **Pd(LA2)$_2$…Pd(LA2)$_2$** and 603 nm in **Pt(LA2)$_2$…Pt(LA2)$_2$**.

Despite the weak tunability in absorption energies by metal ion modification, different excited state characters are observed for Pd- and Pt-dimers. The MO analysis indicates that S_1 of **Pd(LA2)$_2$...Pd(LA2)$_2$** is mainly represented by one-electron excitation from HOMO to LUMO, where both orbitals are ligand based and the transition is of IL character (**Figure S1, SI**). On the other hand, the T_1 state in the Pd-dimer has high multi-reference character and is associated with four one-electron configurations: HOMO to LUMO, HOMO-1 to LUMO+3, HOMO-2 to LUMO+1 and HOMO-2 to LUMO+2 (**Figure S1, SI**). All these configurations have a comparable contribution with MO coefficients between 0.25030 and 0.36909. Therefore, the T_1 excited state in **Pd(LA2)$_2$...Pd(LA2)$_2$** has mixed IL and MLCT character. In Pt-dimer, however, S_1 and T_1 are associated with HOMO to LUMO transition and represent MMLCT excitations (**Figure S1, SI**), where HOMO is $\sigma^*(d_z{}^2\text{-}d_z{}^2)$ and LUMO is ligand based.

The optimized singlet excited state geometry of **Pd(LA2)$_2$...Pd(LA2)$_2$** is presented in Table 3. The results indicate an increase in the monomer-monomer interaction when the dimer is excited to S_1. The Pd-Pd intermolecular distance decreases from 3.582 Å (S_0) to 3.237 Å (S_1). However, the Pd-Pd bond length in the excited state is relatively long and does not suggest covalent metal-metal interaction. Indeed, the reduction in intermolecular distances can be associated with π-π interaction between the ligands (**Figure S2, SI**). The monomers in S_1 of **Pd (LA2)$_2$...Pd(LA2)$_2$** are characterized with an asymmetrical molecular geometry with respect to the ligands. Regarding the BLDs, one can see that in M1/L1 the excitation mainly affects the triazolate cycle and Pd-N2$_{Tr}$ bond, while in M1/L2, Pd-N2$_{Tr}$, Pd-N1$_{Py}$, C4$_{Py}$-C8$_R$ and C8$_R$-C9$_R$ bonds are subject to structural modifications. The results also show that the second monomer in **Pd(LA2)$_2$...Pd(LA2)$_2$** has geometry in which M2/L2 is almost unaffected upon excitation, while the main structural modification in M2/L1 represents the increased C8$_R$-C9$_R$ bond length and ligand quinoid character. Indeed, monomer M2 has a geometry very similar to the excited singlet state of the isolated complex. However, in the dimer, the BLDs with respect to the ground state are smaller than in the isolated complex and therefore the excitation induces weaker geometrical modifications. For example, in the dimer when going from S_0 to S_1, C8$_R$-C9$_R$ elongates by 0.021 Å (M2/L2), while in the isolated complex the same bond length increases by 0.043 Å.

It is important to note that the TD-DFT optimization of T_1 excited state of **Pd (LA2)$_2$...Pd(LA2)$_2$** was not converged. The reason for the un-optimized geometry is that during the TD-DFT optimization, rotation around the C8$_R$-C9$_R$ bond in one of the ligands occurs, which causes the formation of a biradical structure and RHF/UHF instability in the ground state wave function. This result is not surprising since the MO analysis of the vertical $S_0 \rightarrow T_1$ excitation suggests a multi-reference character of the first triplet excited state.

An idea of how the dimer in a triplet state looks like can be obtained from the DFT geometry optimization, T_1^{DFT} (Table 3). The DFT results for the triplet state show that the C8$_R$-C9$_R$ bond in one of the ligands (M1/L1) increases from 1.364 Å (S_0) to 1.474 Å (T_1^{DFT}) and that there is a photo-induced twist around this bond (**Figure S2, SI**). Except the twist, the excited state geometry of monomer M1 is very similar to the excited triplet state of the isolated complex. On the other hand, the excitation does not induce geometrical changes in M2. Therefore, despite the aggregation, the excitation can be regarded as monomolecular. It is important to note that, due to the rotation around the $C8_R$-$C9_R$ bond the calculated excitation energy on T_1^{DFT} geometry are negative.

Our results on the Pd-complex show no evidence for strong metallophilic interaction in the excited state. Indeed, the Pd-Pd bond length increases in the excited state by 0.680 Å with respect to the DFT optimized high spin (T_1^{DFT}) structure and this is due to the observed photo-induced twist (Table 3, **Figure S2, SI**). The lack of metallophlic interaction in the excited state is additionally confirmed by the MO analysis (**Figure S2, SI**), which shows ligand based HOMO and LUMO orbitals. Due to the lack of metallophilic interaction in excited states of the **Pd(LA2)$_2$** complex, one can not exclude the possibility of photo-induced twisting around the $C8_R$-$C9_R$ region even in the isolated monomer. Moreover, previous experimental and theoretical calculations on purely organic compounds have demonstrated photo-induced twisting in –CH=C(CN)$_2$ groups associated with S_0 S_1 transition [57–59]. However, further experimental investigations on **Pd(LA2)$_2$** complex would be required to confirm our theoretical findings. If confirmed, the **Pd (LA2)$_2$** complex cannot be regarded as a promising phosphorescent material due to this 'dark' triplet state. In such a case, **Pd(LA2)$_2$** complex would represent a promising luminescent molecular rotor, which would emit only when the excited state rotation is constrained. Such luminescent molecular rotors are extremely usable as microenvironmental restriction and viscosity-sensitive bioprobes [60–64]. Indeed, molecular rotors with photo-induced twist in –CH=C(CN)$_2$ groups have been already reported but only in the cases of for purely organic push-pull chromophores, which fluoresce [65]. Our results reveal that the same photo-induced twist is also possible for organometallic complexes.

Although the same ligand is involved, the character of the excited states of Pt-dimers differs from the case of Pd. The S_1 and T_1 excited state geometries for the **Pt(LA2)$_2$…Pt(LA2)$_2$** dimer are presented in Table 4. The results indicate that S_1 and T_1 have almost identical molecular structures. Moreover, these excited states possess geometries very similar to the ground state. The main structural difference between the ground and excited states is in the Pt–Pt bond, which when going from $S_0 \rightarrow S_1/S_0 \rightarrow T_1$ decreases by 0.728 Å/0.734 Å. As discussed earlier, such bond length shortening is associated with strong metallophilic interaction in the excited states and an excimer formation. The molecular orbital analysis demonstrate that these excitations result from one electron transition from HOMO, which is $\sigma^*(d_z^2$-$d_z^2)$ to ligand based LUMO orbital and represent MMLCT states (**Figure S2, SI**).

The calculated emission energies for **Pd(LA2)$_2$…Pd(LA2)$_2$** and **Pt(LA2)$_2$…Pt (LA2)$_2$** are represented in Table 2. In the case of the Pd-dimer there is a relatively small bathochromic shift (0.41 eV) in the fluorescent energy with respect to the monomer, while for the Pt-dimer the red-shift associated with the aggregation is substantial (1.03 eV). Moreover, tunability in the emission energies is high and depends on the type of the metal. The $S_1 \rightarrow S_0$ transitions are predicted at 512 nm for Pd and 745 nm for Pt dimer. This observation can be explained by the excited state metallophilic interaction, which occurs in Pt-complex and persists in the Pd-analogue.

3.3 Effect of the Oxidation State

In the last section of the book chapter we will present new results on the effect of oxidation state on the metallophilic interaction in **Pd(LA2)$_2$** complex. In this case, the focus is only on the ground state of the dimers, in which none or one electrons were removed: **Pt(LA2)$_2$...Pt(LA2)$_2$** and **[Pt(LA2)$_2$...Pt(LA2)$_2$]$^{+\bullet}$**. The choice of complexes with the same **LA2** ligand (Scheme 1) is driven by the possibility to compare with the results on the ligand and metal effects.

The optimized bond lengths for the neutral and oxidized **Pt(LA2)$_2$...Pt(LA2)$_2$** dimers, are presented in Table 4. The results clearly demonstrate that the metallophilic interaction is enhanced by chemical or electrochemical oxidation (p-doping). Moreover, the intermolecular attraction induced by oxidation occurs mainly due to the metallophilicity and the ligands are fairly involved. The latter is obvious from the relatively small bond length differences for the Pt–N, C–C, C–N and N–N bonds ($<|0.020|$ Å), as well as from the substantial reduction in the Pt–Pt distance by 0.720 Å going from neutral to positively charged dimer. Due to the strong decrease in the Pt–Pt bond length upon oxidation, the metal-ligand interaction in the oxidized dimers decreases. This is obvious from the elongation of Pt-N1$_{Py}$ and Pt-N2$_{Tr}$ bonds in **[Pt(LA2)$_2$...Pt(LA2)$_2$]$^{+\bullet}$**. Moreover, the Pt–Pt bond length in the oxidized dimer is within the covalent region (2.853 Å). Therefore, the oxidized dimers can be regarded as mixed valence compounds, where there is no bridging ligand between the metal centres. The optimized geometry of **[Pt(LA2)$_2$... Pt(LA2)$_2$]$^{+\bullet}$** is symmetrical with respect to all four ligands and corresponds to delocalized $Pt^{2.5+}$–$Pt^{2.5+}$ mixed valence situation.

TD-DFT results for the monooxidized dimer show that the first electronic transition $D_0 \rightarrow D_1$ lies in the near infrared region ~986 nm and it is predicted with a relatively low oscillator strength (0.0026). As a mixed valence compound, the first excited state (D_1) in **[Pt(LA2)$_2$...Pt(LA2)$_2$]$^{+\bullet}$** has inter-valence charge transfer character. It is associated with one-electron transition from HOMO-4$^{\beta}$ to LUMO$^{\beta}$, where both orbitals represent metallophilic interaction (**Figure S1, SI**). The donating HOMO-4$^{\beta}$ represents $\sigma^*(d_{xz}\text{-}d_{xz})$ overlap between the metal ions and has also contribution from triazolate cycles, while the accepting LUMO$^{\beta}$ is associated with $\sigma^*(d_{z^2}\text{-}d_{z^2})$ overlap between the Pt centres. The first excited state for the neutral dimer is at 453 nm and therefore, the appearance of near infrared absorption for the monooxidized form reveals substantial changes in the optical properties of the dimer. The low energy electronic transition also suggests an increase in the conductivity going from neutral to monooxidized dimer. Moreover, due to the short Pt–Pt contacts in the oxidized dimer, one can expect strong metal-metal coupling and stabilization of **[Pt(LA2)$_2$...Pt(LA2)$_2$]$^{+\bullet}$** with respect to **Pt(LA2)$_2$...Pt (LA2)$_2$**. Such effects are not common in mixed valence compounds, where the metal centres are bridged via a molecular linker, but are highly desirable when aiming to design advanced conducting materials. Therefore, our theoretical results suggest that materials, where **$Pt(LA2)_2...Pt(LA2)_2$** dimers exist (thin films for example) will possess interesting redox behaviour and improved conducting properties upon oxidation (p-doping). And finally, the enhanced metallophilic interaction in **$[Pt(LA2)_2...Pt(LA2)_2]^{+\bullet}$** can be understood from the MO analysis. The HOMO orbital in **$Pt(LA2)_2...Pt(LA2)_2$** represents $\sigma^*(d_z{}^2\text{-}d_z{}^2)$ overlap between the $d_z{}^2$ atomic orbitals of Pt ions. Electrochemical or chemical one-electron oxidation of **$Pt(LA2)_2...Pt(LA2)_2$** will remove an electron from this antibonding orbital and stabilize the Pt–Pt interaction. This stabilization is reminiscent of the case of photoexcitation. Furthermore, our results on the molecular structure show that the geometry of the **$[Pt(LA2)_2...Pt(LA2)_2]^{+\bullet}$** dimer is very similar to the S_1/T_1 excited state geometries in **$Pt(LA2)_2...Pt(LA2)_2$**.

The calculations on the doubly oxidized dimer **$[Pt(LA2)_2...Pt(LA2)_2]^{2+}$** show that the antibonding $\sigma^*(d_z{}^2\text{-}d_z{}^2)$ is empty, which also results in strong metal-metal interaction (~2.9 Å). The optimized geometries for **$[Pt(LA2)_2...Pt(LA2)_2]^{2+}$** in different spin states suggest Pt^{3+}–Pt^{3+} non-mixed valence situation. The broken symmetry singlet ($\langle S^2 \rangle = 1.04$) and triplet geometries are almost identical and the energy difference between the spin states is negligible ~0.01 kcal/mol. These results also indicate that the **$[Pt(LA2)_2...Pt(LA2)_2]^{2+}$** dimer possess high diradical character with respect to the neutral form. Due to the multireference character of **$[Pt(LA2)_2...Pt(LA2)_2]^{2+}$** and the need to go beyond the DFT/TD-DFT approaches, we do not present and discuss here its absorption and emission properties. However, our results suggest that upon successive oxidation, the system goes from closed-shell **$[Pt(LA2)_2...Pt(LA2)_2]$**, open-shell doublet **$[Pt(LA2)_2...Pt(LA2)_2]^{+\bullet}$** to diradical **$[Pt(LA2)_2...Pt(LA2)_2]^{2+}$**, which as shown previously would lead to a very intriguing behavior of the second hyperpolarizability [66]. In particular, open-shell singlet molecules with intermediate diradical characters exhibit larger values of the second hyperpolarizability than closed-shell and pure open-shell molecules [67]. In this respect, our Pt-complexes can be regarded also as a promising material for nonlinear optics.

4 Summary

The present theoretical investigation on **ML_2** complexes demonstrates the power of the organometallic structural theme in the molecular design of new luminescent and conducting materials. Starting from d^8 metal ions, such as Pd(II) and Pt(II), and pyridyl–triazolate ligands, analogous molecular and supramolecular (dimer) structures can be designed. Despite the similarity in the organometallic structures, different molecular functionalities can be obtained (Fig. 1). Moreover, such functional variety originates not only from the specific metal-ligand combination but also from specific intermolecular interactions.

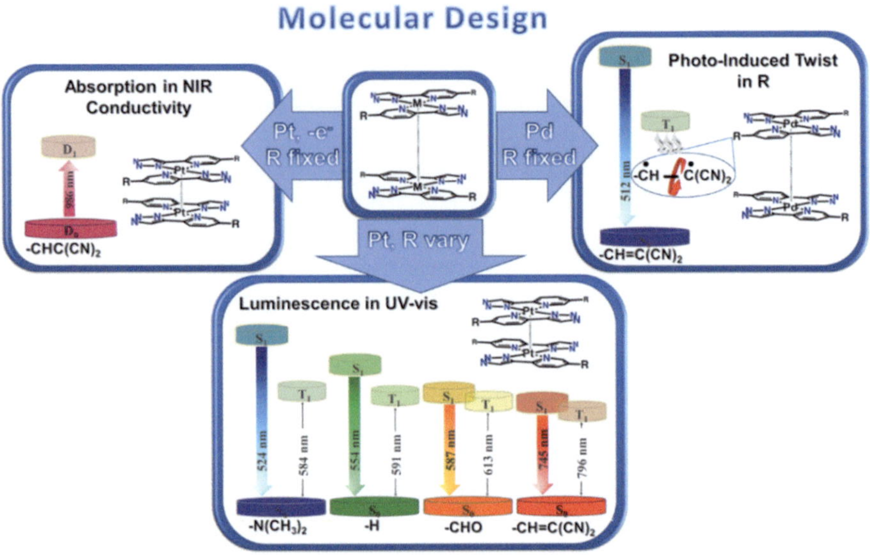

Fig. 1 Molecular design strategies by using organometallic theme with pyridyl-triazolate ligands, Pt(II) and Pd(II)

Scheme 1 Chemical structure of the investigated monomer and dimer complexes

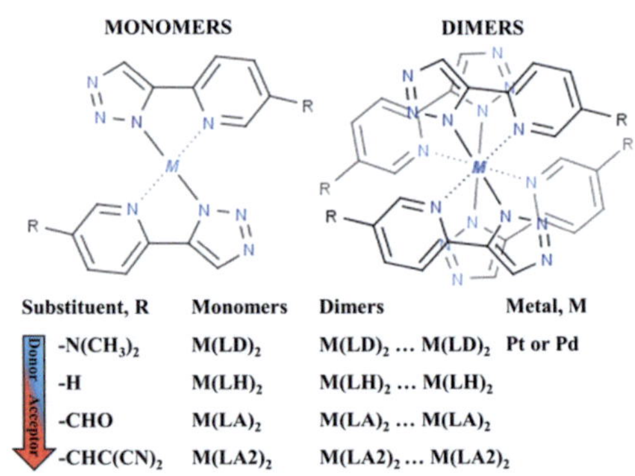

Scheme 2 Qualitative explanation of the ligand effect (acceptor strength of the substituent) on the formation of ground and excited state dimers in Pt-complexes due to metallophilic interaction

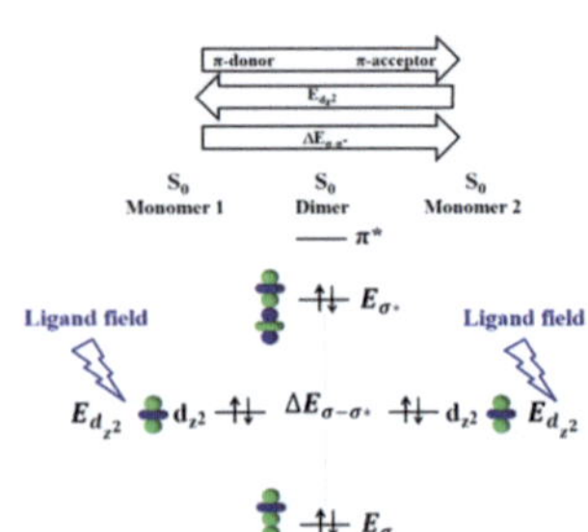

Scheme 3 Atom labelling in the complexes with **LA2** ligand, where M = Pt or Pd

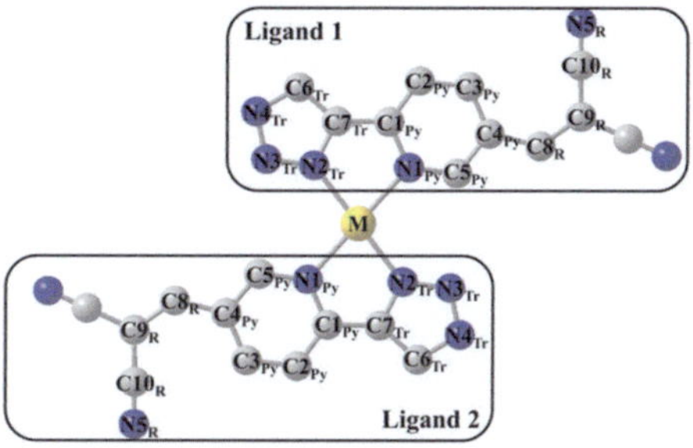

At a molecular level, rather week tunability in absorption and emission energies of $\mathbf{ML_2}$ complexes are observed by the alteration of Pd(II) with Pt(II) and by the modification of π-donor/acceptor properties of the pyridine substituents of the ligands (–$N(CH_3)_2$, –H, –CHO or –CH=$C(CN)_2$).

At the dimer level, weak metallophilic interactions between the monomers in the ground state are observed. Although generally week, the ground state metal-metal attraction in $\mathbf{PtL_2}$ complexes improves when the π-acceptor strength of the ligand increases and when Pd (II) is replaced by Pt(II). The low-energy absorption transitions in the dimers are slightly affected by ligand or metal ion modification and are close to those for the isolated monomers. On the other hand, the low energy emission wavelengths and excited state character are both very sensitive to ligand and metal ion alteration (Fig. 1). In all Pt-dimers, the metallophilic interaction is enhanced upon excitation and this leads to the formation of excimers with covalent metal-metal bonds (Fig. 1). As a result, the excimers possess MMLCT character, and the emission wavelengths are strongly red-shifted with respect to the monomers. Moreover, when the π-acceptor strength of the ligands in the $\mathbf{PtL_2}$ series increases, the red-shift in the emission wavelength becomes stronger and the $\Delta E_{S1\text{-}T1}$ gap decreases. However, when the dimer of Pd-complexes with –CH=C$(CN)_2$ substituents is excited, metallophilic interaction is not observed. The excited states in the Pd-dimer have IL character and monomer origin. Moreover, due to the IL excitations in T_1, the quinoid character of one of the ligands increases and this causes rotation of the –$C(CN)_2$ group. As a result, the excited triplet state in the Pd-dimer is expected to be non-emitting (**Figure S1, SI**).

It was also demonstrated that the first oxidation of Pt-dimer improves the metallophilic interaction and leads to the formation of a mixed valent compound with a direct link between the metal centers. The results predict NIR absorption for the oxidized dimer and suggest improved conductivity upon p-doping. The doubly oxidized Pt-dimer is characterized also with strong metallophilic interaction and possesses diradical character, which is attractive for non-linear optical applications.

The proposed hybrid theme is demonstrated as a promising approach for the molecular design of optical and conducting materials with different functionalities.

Acknowledgements The authors would like to thank Dr. David J. Carey for the helpful discussions, the Leverhulme Trust (RPG-2014-006) for funding and the National Service for

Computational Chemistry Software (NSCCS) at Imperial College London. M. R. R. P. would like to thank the Southeast Physics Network (SEPnet) for student funding.

References

1. Lee J, Farha OK, Roberts J, Scheidt KA, Nguyen ST, Hupp JT (2009) Chem Soc Rev 38:1450
2. Kalinowski J, Fattori V, Cocchi M, Williams JAG (2011) Coord Chem Rev 255:2401
3. Yersin H, Rausch AF, Czerwieniec R (2012) Organometallic emitters for OLEDs: triplet harvesting, singlet harvesting, case structures, and trends. In: Physics of organic semiconductors, vol 371. Wiley-VCH Verlag
4. Liao C, Shelton AH, Kim K-Y, Schanze KS, Appl ACS (2011) Mater Interf 3:3225
5. Ho C-L, Wong W-Y (2011) Coord Chem Rev 255:2469
6. Miller JS (2014) Mater Today 17:207
7. Pashaei B, Shahroosvand H, Abbasi P (2015) RSC Adv 5:94814
8. Hagfeldt A, Grätzel M (2000) Acc Chem Res 33:269
9. (a) Martins P, Marques M, Coito L, Pombeiro AJ, Baptista PV, Fernandes AR (2014) Anticancer Agents Med Chem 14:1199; (b) Alberto ME, Butera V, Russo N, Inorg Chem 50:6965; (c) Alberto ME, Russo N (2011) Chem Commun 47:887; (d) Pucci D, Bellusci A, Bernardini S, Bloise R, Crispini A, Federici G, Liguori P, Lucas MF, Russo N, Valentini A (2008) Dalton Trans 5897; (e) Alberto ME, Lucas MF, Pavelka M, Russo N (2008) J Phys Chem B 112:10765
10. Hemmert C, Gornitzka H (2016) Dalton Trans 45:440
11. Arrowsmith RL, Pascu SI, Smugowski H (2012) Organomet Chem 38:1
12. Zhang KY, Lo KKW (2014) Metal complexes for cell and organism imaging. In: Inorganic chemical biology: principles, techniques and applications. Wiley, UK, p 99
13. Stephenson GR (2005) Organometallic bioprobes. In: Bioorganometallics: biomolecules, labeling, medicine. Wiley-VCH Verlag GmbH & Co. KGaA, Weinheim, p 215
14. McConnell AJ, Wood CS, Neelakandan PP, Nitschke JR (2015) Chem Rev 115:7729
15. Zhuang J, Gordon MR, Ventura J, Li L, Thayumanavan S (2013) Chem Soc Rev 42:7421
16. Foscato M, Occhipinti G, Venkatraman V, Alsberg BK, Jensen VR (2014) J Chem Inf Model 54:767
17. Foscato M, Venkatraman V, Occhipinti G, Alsberg BK, Jensen VR (2014) J Chem Inf Model 54:1919
18. Lehn J-M (2002) Proc Natl Acad Sci USA 99:4763
19. Machado VG, Baxter PN, Lehn J-M (2001) J Braz Chem Soc 12:431
20. Lanigan N, Wang X (2013) Chem Commun 49:8133
21. Haiduc I, Edelmann FT (2013) Chem Commun 49:8133
22. Alvarez S, Ruiz E (2012) Self-assembly of coordination compounds: design principles. In: Supramolecular chemistry: from molecules to nanomaterials, vol 5. Wiley, Chichester, UK, 1993
23. Pyykkö P (1997) Chem Rev 97:597
24. (a) Schmidbaur H, Schier A (2012) Chem Soc Rev 41:370; (b) Krogmann K (1969) Angew Chem Int Ed 8:35
25. Georgieva I, Trendafilova N, Dodoff NI (2013) J Photochem Photobiol A: Chem 267:35
26. Campbell MG, Powers DC, Raynaud J, Graham MJ, Xie P, Lee E, Ritter T (2011) Nat Chem 3:949
27. Po C, Ke ZH, Tam AYY, Chow HF, Yam VWW (2013) Chem Eur J 19:15735
28. Tsai JL-L, Zou T, Liu J, Chen T, Chan AO-Y, Yang C, Lok C-N, Che C-M (2015) Chem Sci 6:3823
29. Kim D, Brédas J-L (2009) J Am Chem Soc 131:11371
30. Wan KT, Che CM, Cho KC (1991) J Chem Soc Dalton Trans 1077

31. Ma BW, Djurovich PI, Thompson ME (2005) Coord Chem Rev 249:1501
32. Ranga Prabhath MR, Romanova J, Curry RJ, Silva SRP, Jarowski PD (2015) Angew Chem Int Ed 54:7949
33. Ranga Prabhath MR, Romanova J, Curry RJ, Silva SRP, Jarowski PD (2014) Abstracts of Papers of the American Chemical Society 248. 248th National Meeting of the American-Chemical-Society (ACS), San Francisco, CA
34. Romanova J, Ranga Prabhath MR, Jarowski PD (2016) J Phys Chem C 120:2002
35. (a) Amar A, Meghezzi H, Boixel J, Le Bozec H, Guerchais V, Jacquemin D, Boucekkine A (2014) J Phys Chem A 118:6278
36. Underhill AE, Watkins DM (1980) Chem Soc Rev 9:429
37. Mitsumi M, Ueda H, Furukawa K, Ozawa Y, Toriumi K, Kurmoo M (2008) J Am Chem Soc 130:14102
38. Sakai K, Ishigami E, Konno Y, Kajiwara T, Ito T (2002) J Am Chem Soc 124:12088
39. Minot MJ, Perlstein JH (1971) Phys Rev Lett 26:371
40. Day P, Hush NS, Clark RJ (2008) Philos Trans A Math Phys Eng Sci 366:5
41. Parthey M, Kaupp M (2014) Chem Soc Rev 43:5067
42. Klaus DR, Keene M, Silchenko S, Berezin M, Gerasimchuk N (2015) Inorg Chem 54:1890
43. Matsumoto K, Arai S, Ochiai M, Chen W, Nakata A, Nakai H, Kinoshita S (2005) Inorg Chem 44:8552
44. Kimizuka N, Oda N, Kunitake T (2000) Inorg Chem 39:2684
45. Iguchi H, Nafady A, Takaishi S, Yamashita M, Bond AM (2014) Inorg Chem 53:4022
46. Ozawa Y, Kim M, Toriumi K (2013) Acta Crystallogr Sect C Cryst Struct Commun 69:146
47. Adamo C, Barone V (1999) J Chem Phys 110:6158
48. Grimme S, Antony J, Ehrlich S, Krieg H (2010) J Chem Phys 132:154104
49. Miertus S, Scrocco E, Tomasi J (1981) Chem Phys 55:117
50. Cammi R, Mennucci B (1999) J Chem Phys 110:9877
51. Tomasi J, Mennucci B, Cammi R (2005) Chem Rev 105:2999
52. Gaussian 09, Revision A1, Frisch MJ, Trucks GW, Schlegel HB, Scuseria GE, Robb MA, Cheeseman JR, Scalmani G, Barone V, Mennucci B, Petersson GA, Nakatsuji H, Caricato M, Li X, Hratchian HP, Izmaylov AF, Bloino J, Zheng G, Sonnenberg JL, Hada M, Ehara M, Toyota K, Fukuda R, Hasegawa J, Ishida M, Nakajima T, Honda Y, Kitao O, Nakai H, Vreven T, Montgomery JA, Peralta Jr JE, Ogliaro F, Bearpark M, Heyd JJ, Brothers E, Kudin KN, Staroverov VN, Kobayashi R, Normand J, Raghavachari K, Rendell A, Burant JC, Iyengar SS, Tomasi J, Cossi M, Rega N, Millam JM, Klene M, Knox JE, Cross JB, Bakken V, Adamo C, Jaramillo J, Gomperts R, Stratmann RE, Yazyev O, Austin AJ, Cammi R, Pomelli C, Ochterski JW, Martin RL, Morokuma K, Zakrzewski VG, Voth GA, Salvador P, Dannenberg JJ, Dapprich S, Daniels AD, Farkas Ö, Foresman JB, Ortiz JV, Cioslowski J, Fox DJ (2009) Gaussian Inc., Wallingford CT
53. Dennington R, Keith T, Millam J (2009) GaussView, Version 5. Semichem Inc., Shawnee Mission, KS
54. Hansch C, Leo A, Taft RW (1991) Chem Rev 91:165
55. Wang X, Gong SL, Song D, Lu Z-H, Wang S (2014) Adv Funct Mater 24:7257
56. Smith MB, Michl J (2013) Ann Rev Phys Chem 64:361
57. Louant O, Champagne B, Liégeois V (2015) Chem Phys Lett 634:249
58. Allen BD, Benniston AC, Harriman A, Rostron SA, Yu C (2005) Phys Chem Chem Phys 7:3035
59. Zhang W, Lan Z, Sun Z, Gaffney KJ (2012) J Phys Chem B 116:11527
60. Haidekker MA, Theodorakis EA (2010) J Biol Eng 4:11
61. Kuimova MK (2012) Phys Chem Chem Phys 14:12671
62. Kottas GS, Clarke LI, Horinek D, Michl J (2005) Chem Rev 105:1281

63. Haidekker MA, Nipper ME, Mustafic A, Lichlyter D, Dakanali M, Theodorakis EA (2010) Dyes with segmental mobility: molecular rotors. In: Advanced fluorescence reporters in chemistry and biology, vol I8. Springer, Berlin, p 267
64. Kuimova MK, Botchway SW, Parker AW, Balaz M, Collins HA, Anderson HL, Suhling K, Ogilby PR (2009) Nat Chem 1:69
65. Goh WL, Lee MY, Joseph TL, Quah ST, Brown CJ, Verma C, Brenner S, Ghadessy FJ, Teo YN (2014) J Am Chem Soc 136:6159
66. Inoue Y, Yamada T, Champagne B, Nakano M (2013) Chem Phys Lett 570:75
67. Nakano M, Kishi R, Nitta T, Kubo T, Nakasuji K, Kamada K, Ohta K, Champagne B, Botek E, Yamaguchi K (2005) J Phys Chem A 109:885

Electrodynamical and Quantum Chemical Modelling of Electrochemical and Catalytic Processes on Metals and Semiconductors: A Review

A.V. Glushkov, A.A. Svinarenko, O.Yu. Khetselius, Yu.V. Dubrovskaya, A.S. Kvasikova, A.A. Kuznetsova and E.L. Ponomarenko

Abstract The advanced quantum-mechanical and electrodynamical approaches in the electron theory of catalysis have been generalized to determine quantitatively the catalytic activity for metals, binary metallic alloys and semiconductor materials. The solutions of some model tasks associated with stabilization of ionic states of atomic hydrogen and molecular oxygen in effective electron gas are given. Within the approach the quantitative correlation between the electron structure parameters of the materials and their catalytic activity is found on example of simple model reactions.

Keywords Electron theory of catalysis · Quantum and electrodynamics models · Metals and binary metallic alloys · Semiconductor materials

1 Introduction

At present time a great interest attracts studying an electrochemical and catalytic activity properties of the different materials, including a development of advanced chemical industry technologies, search and construction of the electrochemical solid-state energy sources, etc. [1–13]. A great number of the attempts to develop a quantitative theory of electron catalysis for the metal-like and semiconductor systems including the description of processes on electrodes' surfaces of the electrochemical solid-state energy sources have been undertaken (see [1–8]). One could mention quite good understanding of a mechanism of heterogeneous catalytic process under obligatorily treating intermediate stages, namely, stages of adsorption and desorption [1, 2]. Naturally, any heterogeneous reaction can be treated as the

A.V. Glushkov (✉) · A.A. Svinarenko · O.Yu. Khetselius · Yu.V. Dubrovskaya · A.S. Kvasikova · A.A. Kuznetsova · E.L. Ponomarenko
Odessa State Environmental University, L'vovskaya str., 15, Odessa-9 65016, Ukraine
e-mail: glushkovav@gmail.com; dirac13@mail.ru

A. Tadjer et al. (eds.), *Quantum Systems in Physics, Chemistry, and Biology*, Progress in Theoretical Chemistry and Physics 30,
DOI 10.1007/978-3-319-50255-7_9

process based on a radical mechanism. The matter is in the fact that the radicals and ion-radicals appear on the surface under chemosorption and provide the corresponding mechanism of these reactions. It is important to note that the reaction path through one-electron charged intermediates (ion-radicals) is not the main mechanism in a heterogeneous catalysis. Nevertheless, it can be possible in some special cases. The detailed review of different approaches to adsorption and catalysis, including the key aspects of different conceptual models (such as the density functional method, ab initio quantum chemistry methods) are presented in [9–26]. The cited methods allowed to get very useful information about the electrochemical and catalytic activity properties of the different materials, nevertheless, different fundamental physical and chemical aspects of the problem remain up to known degree veiled [1–15]. In this chapter we present an advanced quantum-mechanical and electrodynamical approach to description of the catalytic activity for metals, binary metallic alloys and semiconductor materials. Within the approach the quantitative correlation between the electron structure parameters of the materials and their catalytic activity is found on example of simple model reactions.

2 Electrodynamical and Quantum-Mechanical Models for Metals, Alloys and Semiconductors

It is well known that the catalytic properties of different metallic and semiconductors materials are linked with the electronic processes inside and on the surface of the materials. The Fermi level position determines the adsorption and catalytic activity of the surface in relation to molecules of the given kind under other equal conditions. The chemosorption ability of the surface, a degree of its charging, a reactive ability of the chemisorbed particles etc. are dependent upon the Fermi level position on the surface of crystal (speech is about distance between the Fermi level and the conductivity band bottom: E_F^s). The position of the Fermi level inside the crystal can be defined as E_F^ν. The corresponding expression linking the values E_F^s and E_F^ν is as follows:

$$\sigma + \int_0^\infty \rho(x)dx = 0 \tag{1}$$

where ρ is a density of the volume charge in the plane, σ is a density of the surface charge. The value of σ can be expressed as function of a pressure P, temperature T: $\sigma = \sigma(P, T, E_F)$ [2, 6, 7]. The second item in (1) is the function of E_F^s and E_F^ν:

$$\int_0^\infty \rho(x)dx = R\left(T; E_F^s; E_F^\nu\right) \tag{2}$$

The relationship between E_F^s and E_F^ν (defining the correlation between the surface and bulky properties of the material) can be written as follows:

$$\sigma\left(P,T;E_F^s\right)+R\left(T;E_F^s;E_F^\nu\right)=0 \tag{3}$$

or

$$E_F^s=f\left(P;T;E_F^\nu\right) \tag{4}$$

Let us further to consider in details the electrodynamical approach to structure of a metallic system. Within the simple approximation [9] with using a set of isotropic s-d energy bands, one could write the next obvious formula for a static dielectric permeability:

$$\varepsilon=1+\varepsilon_{ss}+\varepsilon_{dd}+\varepsilon_{sd}+\varepsilon_{ds}, \tag{5}$$

In Eq. (5) ε *(ij)* means the contribution into ε due to the i-j transitions. If one accept here the approximation of the free electrons, then it is easily to write the expression for ε_{ss} component:

$$\varepsilon_{ss}=2\pi\nu_s(E_F)k^{-2}\left\{1+\left[4\left(k_F^s\right)^2-k^2\right]\ln\left|\left(2k_F^s+k\right)/2k_F^s-k\right)\left|/4k_F^sk\right\}. \tag{6}$$

$$\nu_i(E_F)=N_i(E_F)a_B^2e$$

Here the following notations are used: a_B is the Bohr radius, $k = q\cdot a_B$, q is the wave number, $q_F=(3\pi^2z_i/\Omega)^{1/3}$; z_i is a number of electrons in "*i*" band; $N_i(E_F)$ is a density of states on the Fermi surface in "i" band.

This corresponding ε_{ds} contribution can be represented as follows:

$$\varepsilon_{ds}=\left[2m_sk_de^2f_c/\pi\pi^2k^2\right]\left\{1+\left[4(k_d)^2-k^2\right]\ln|(2k_d+k)/2k_d-k)|/4k_dk\right\} \tag{7}$$

where m_s is the effective mass of electron in the conductivity band; k_d, f_c are the numeral parameters [7, 8, 11]. At last, the corresponding formula for ε_{dd} looks as:

$$\varepsilon_{dd}=2\pi\nu_d(E_F)k^{-2}|M_{dd}|^2\left\{1+\left[4\left(k_F^d\right)^2-k^2\right]\ln\left|\left(2k_F^d+k\right)/2k_F^d-k\right)\left|/4k_F^dk\right\} \tag{8}$$

where M_{dd} is the matrix element, determined by the superposition of the wave functions for d electrons.

The next key step is in supposing that we can work using the spherical symmetry and the crystal potential is fully screened by the conductivity electrons. Then one could introduce an effective potential, which imitates an influence of the metallic field on the hydrogen atom (for example, for process of the following type: $H = H^+ + e^-$). This effective potential can be determined in the following form:

$$\Phi(r) = -\frac{2e^2}{\pi r}\int_0^{\infty}\frac{\sin kr}{k\varepsilon(k)}dk \tag{9}$$

or after substitution of (5) to (7) resulted in the expression:

$$\Phi(r) = -\left(e^2 a/r\right) exp[-\alpha R] cos[\alpha R] \tag{10}$$

where

$$\alpha = \left[\pi^{-1}12\left(k_F^s\right)^2\right]^{-1/4}\left\{\nu_s(E_F) + \left(k_F^d/k_F^s\right)\nu_d(E_F) + f_c\left(k_F^d/k_F^s\right)^2\nu_s(E_F)\right\}^{1/4}$$

$$R = 2q_F^s r, \tag{11}$$

$$a = \left(k_F^s\right)^{-1}$$

According to Ref. [8, 23] further key idea in the following. One could find the numerical solution of the Schrödinger equation for the hydrogen atom in a field $\Phi(r)$ and obtain a spectrum of states, which could be continual or discrete in dependence upon the parameter $\zeta^{-1} = \alpha/a$ [8]. The spectrum is continual, if $\zeta < \zeta_0$ ($\zeta_0 = 0,362$) and the corresponding material is a catalyst for the hydrogen ionization reaction. If $\zeta > \zeta o$, the spectrum is discrete (metal or alloy does not demonstrate catalytic activity for cited reaction). In Refs. [8, 23–25] such an approach was successfully applied to studying the metal catalytic properties in relation to reaction of the H ionization and obtained excellent agreement with experiment [2, 5–7]. In Ref. [8] there are listed the corresponding parameters α and a for a number of metals.

The analogous consideration can be performed for the binary metallic alloy. Obviously, one should take into account a dependence of the Fermi level position E_F as well as the state density $\nu(E_F)$, accompanied with the structure parameters α and a upon the change of the admixture concentration c [8, 9]. It is easily to suppose that the admixture's atoms volume has the spherical form and a radius R_c is linked with the concentration as:

$$(qR_c)^{-3} = (qr_s)^{-3}c, \tag{12}$$

where r_s —the electron gas parameter.

Further one could write the standard Poisson equation [8] for the screened potential $V(r)$ near an admixture:

$$\Delta V(r) = q^2\{V(r) - \Delta E_F\} \tag{13}$$

which has the with the elementary solution of the following type:

$$V(r, R_c) - \Delta E_F = [-Z_v e/r]\{qR_c ch[q(R_c - r)] + sh[q(R_c - r)]\}/[qR_c ch(qR_c) + sh(qR_c)] \tag{14}$$

Here Z_ν is a difference of the components valences. Finally, the value $\nu(E_F)$ for the binary metallic alloy can be expressed as follows [8]:

$$\nu(E_F) = \nu(E_F) + \Delta\nu(E_F). \tag{15}$$

The Fermi level position and respectively the metallic alloy catalytic activity is quantitatively dependent upon the components concentration.

In Refs. [8, 25] there are presented the concrete examples of the modeling the catalytic activity of metallic alloy in relation of the H ionization reaction for different metallic alloys. For example, the corresponding estimates show that the alloy *Ni-Cu* with a small concentration of *Cu* (till 16%) is a good catalyst for the hydrogen ionization reaction, however situation is changed in the opposite direction with a growth of ζ. This is in good agreement with experiment [1, 6, 7].

The similar approach can be generalized to find a link between the alloy structure parameters and their catalytic activity for oxygen electro-restoring reaction. One should again to solve the Schrödinger equation for system "O_2-electron" in the potential Φ (r). If the negative ion system has the bound state for given values of (6)–(9), then the studied material is a good catalyst for the reaction. In order to solve the Schrödinger equation in a case of diatomic molecule it is necessary to use the two-centre approximation [27–32]. In the standard elliptic coordinates:$\mu = (r_a + r_b)/R_{ab}$, $\eta = (r_a - r_b)/R_{ab}$ (R_{ab} is a distance between nuclei) the variables' in the corresponding Schrödinger equation are separated if one use the formula: $1/2R_{ab}(\mu + \eta) \approx 1/2\ R_{ab}\mu$. Then the potential (10) in the molecular approximation can be written as follows:

$$\Phi(\mu, \eta) = -2\mu a\, exp[-\alpha q_F R_{ab}\mu] cos[\alpha q_F R_{ab}\mu]/\left[R_{ab}\left(\mu^2 - \eta^2\right)\right] = g(\mu)/(\mu^2 - \eta). \tag{16}$$

Further one can write the corresponding Schrödinger equation which is reduced to the differential equations system [8, 28]:

$$\begin{gathered}\{d/d\mu(\mu^2 - 1)d/d\mu - [\lambda_{ml} + m^2/(\mu^2 - 1) + \mu^2 c^2 + R_{ab} g(\mu)/2]\}T_{nlm} = 0, \\ \{d/d\eta(-\eta^2 + 1)d/d\eta + [\lambda_{ml} + \eta^2 c^2 m^2(1 - \eta^2)]\}S_{lm} = 0, \\ \{d^2/d\varphi^2 + m^2\}\Sigma_m = 0.\end{gathered} \tag{17}$$

where $\Psi_{nlm} = \mathrm{T}_{nlm}(\mu)\, S_{lm}(\eta)\Sigma_m(\varphi)$ is a wave function; $E = -2c^2/R_{ab}^2$ is one-electron energy, λ_{ml} is a coupling constant. In order to fine the numerical solution of the Eqs. (17) one could use the standard approaches such as the Numerov method or matrix technique of the Newton-Rafson etc. (see [27–32]).

In Refs. [8, 25] the authors has presented the results of application of the described model to estimating the catalytic properties for the alloy *Ni-Cu* in relation to the oxygen electro-restoring reaction. In a case of the small content of *Cu* [$c(Cu) = 20\%$; $c(Ni) = 80\%$; $E(1\pi_g) = 1{,}3$ eV] the compound *Ni-Cu* is an effective catalyst. With changing c, the situation is changing into opposite side. Catalytic activity of material decreases if $c(Cu) = 50\%$, $c(Ni) = 50\%$ [$E(1\pi_g) = 1{,}1$ eV]. These facts correlate well to experimental data [6, 7]. If $c(Cu) = 90\%$, $c(Ni) = 10\%$ [$E(1\pi_g) = 1{,}5$ eV], the catalytic activity is quite high again.

In order to generalize the above presented approach on a case of description of the catalytic processes on semiconductors and determine a link between their electron structure parameters and the catalytic activity in the relation to model H ionization reaction one could use the known Resta model in the Thomas-Fermi theory [33, 34]. The master Poisson equation has the standard form:

$$V(r) = q\{V(r) - A\}, \tag{18}$$

where $q = 4k_F/\pi a_B$ and A is a constant. Beyond the radius R the point charge Z_e potential is equal to:

$$V(R) = -Ze^2/[\varepsilon(0)r], r > R, \tag{19}$$

where $\varepsilon(0)$ is a static dielectric permeability. The general solution with taking into account for the corresponding continuity and boundary conditions looks as:

$$V(r) = -\{Ze^2/r\}\{sh[q(R-r)]/sh[qR] - Ze/\varepsilon(0)R, r < R \tag{20}$$

The link between the screening parameter and $\varepsilon(0)$ is determined by the relationship: $\varepsilon(0) = sh[qR]/qR$. Let us note that if $\varepsilon(0) > 1$, R is equal to finite value comparable with distance to the nearest atoms (for example, for *NiO*, *CuO*, *ZnO*, *ZnS*, *ZnTe* semiconductors this value is 4.8–6.1 Å) [7, 35]. The next step is the numerical solution of the Schrödinger equation with the potential (20), which allows to determine the corresponding energy spectrum in dependence upon $\varepsilon(0)$, k_F (E_F) and then to find a link between the semiconductors structure parameters and their catalytic activity likely above described approach. In ref. [8, 24] there are presented the concrete estimates of the catalytic activity for the *CuO*, *ZnO* semiconductors in the hydrogen ionization (oxidation) reaction. The general conclusion is that the *CuO* and *ZnO* semiconductors are good catalysts for the hydrogen ionization reaction. This fact is in a good agreement with the known experimental data [6, 7].

In conclusion of this section we present a brief generalization of the model to the case of the influence of an external electric field. To determine the corresponding spectrum of the quasi-bound states one needs to carry out the diagonalization of the energy matrice, calculated between states with the same main quantum number (e.g.

[36–39]). Let us note that the corresponding Hamiltonian can be written in the cylindrical coordinates (ρ, z), namely:

$$H\psi(\rho, z) = E\psi(\rho, z) \tag{21}$$

$$H = -1/2\left(\partial^2/\partial\rho^2 + 1/\rho\partial/\partial\rho + \partial^2/\partial z^2 - 1/\left[(z+R/2)^2+\rho^2\right]^{1/2}\right.$$
$$\left. - 1/\left[(z-R/2)^2+\rho^2\right]^{1/2} - Fz + V(\rho, z)\right.$$

Here F denotes an electric field strength; V is an effective potential (of the type (16), (20)). To solve the Eq. (21) one should use the finite differences method (look [40]). The eigen-values of the Hamiltonian can be computed by means of the inverse iterations method. The corresponding system of inhomogeneous equations can be solved by the Thomas method. However, to date any of the similar theoretical model application are missing.

3 Conclusions

We have presented an advanced quantum-mechanical and electrodynamical approaches in the electron theory of catalysis to determine quantitatively the catalytic activity for metals, binary metallic alloys and semiconductor materials. The solutions of some model tasks associated with stabilization of ionic states of atomic hydrogen and molecular oxygen in effective electron gas are given. Within the approach the quantitative correlation between the electron structure parameters of the materials and their catalytic activity is found on example of simple model reactions. Surely, the catalytic activity properties of different materials are directly connected with the corresponding electronic processes inside and on the surface of the materials. Naturally, to carry out more advanced approach one should take into account a whole number of the additional important factors (for example the electrolyte influence, surface effects, the electrodes potential, electron concentration in the surface layer etc. [1–10]). A great interest attracts the establishment of correlation between the presented approach and recent theories of the catalysis on the metals and semiconductors. In conclusion let us underline that the presented approach ideas can be quite easily generalized in order to take into consideration a whole number of such additional physical and chemical factors such as a direct electric, magnetic or laser field effect on the catalytic processes and governing by these processes, search and approbation of new classes of nanocluster catalysts and so on [38, 39, 41, 42].

Acknowledgements Authors are thankful to Prof. E. Brändas, J. Maruani, A. Tadjer, R. Pavlov for the invitation to present an invited report at the workshop QSCP-XX (Bulgaria, 2015). The help in editing the manuscript by Mr. Heike Rossel and Ms. Heena Naveen (Springer) is very much appreciated.

References

1. Waite RE, Bockris J, Conway BE (1993) Modern aspects of electrochemistry, vol 21. Plenum Press, N-Y.; Paunovic M (ed) (2006) Fundamentals of electrochemical deposition, the electrochemical society series. John Wiley & Sons, Inc., N.-Y.
2. Thomas JM, Thomas WJ (1996) Principles and practice of heterogeneous catalysis. Wiley-VCH, N.-Y.; Richards R (2006) Surface and nanomolecular catalysis. CRC Press, N.-Y.
3. Fornasiero P, Montini T, Graziani M, Zilio S, Succi M (2008) Catal Today 137:475–482; Ma Z, Yin H, Overbury SH, Dai S (2008) Catal Lett 126:20–30
4. Albonetti S, Bonelli R, Mengou JE, Femoni C, Tiozzo C, Zacchini S, Trifirò F (2008) Catal Today 137:483–488; Baddeley CJ, Kyriakou G (eds) (2016) Surface science for heterogeneous catalysis, a special issue in Honour of Richard Lambert, Surf Sci 646:1–360 (Elsevier)
5. Austin S (1996) Material solid state batteries. World Scientific, Singapore; Kiselev VF, Krylov OV (eds) (1987) Electronic phenomena in adsorption and catalysis on semiconductors and dielectrics. Springer series in surface science, vol 7. Springer, Berlin
6. Royter VA (1968) Catalytic properties of substances. Moscow, Nauka Publisher
7. Wolkenstein FF (1991) Electron Processes on surface of semiconductors under chemosorbtion. Nauka Publisher, Mosco; Meylikhov EZ, Lazarev SD (1987) Electrophysical properties of semiconductors. Nauka Publisher, Moscow
8. Glushkov AV (1991) Electrochemistry 27:131–135; ibid. (1993) 29:369–376; Glushkov AV, Kondratenko PA, Lepikh YI, Fedchuk AP, Svinarenko AA, Lovett L (2009) Int J Quantum Chem 109:3473–3481
9. Lundqvist S, March N (1983) Theory of Inhomogeneous electron gas. Plenum Press, N-Y., Ch. 5
10. Schwab G-M (1969) Surface Sci 13:198–200; Aubry S (2007) J Phys Condens Matter 19:255204
11. Wilson S (2007) Handbook on molecular physics and quantum chemistry. Wiley, Chichester, 650 p; Glushkov AV (2008) Relativistic quantum theory. quantum mechanics of atomic systems. Astroprint, Odessa, 900 p
12. Slater JC (1974) The self-consistent field method for molecules and solids: quantum theory of molecules and solids. McGraw-Hill, N.-Y.
13. Lidorenko IS, Izmestiev AA, Medvedev IG, Muchnik GF (1975) Dokl Akad Nauk USSR 223:639–642
14. Hohenberg P, Kohn W (1964) Phys Rev B 136:864–878; Kohn W, Sham LJ (1965) Phys Rev A 140:1131–1145
15. Maruani J, Lefebvre R, Brändas E (eds) (2004) Advanced topics in theoretical chemical physics, progress in theoretical chemistry and physics, vol 12. Springer, Berlin
16. Gidopoulos NI, Wilson S (eds) (2004) The fundamentals of electron density, density matrix and density functional theory in atoms, molecules and the solid state, series: progress in theoretical chemistry and physics, vol 14. Springer, Berlin
17. Bardeen J (1964) Surf Sci 2:381–394; Lang ND, Kohn W (1973) Phys Rev B 7:3541–3552
18. Theophilou AK (1972) J Phys F 2:1124–1132; Vannimenus J, Budd HF (1975) Solid State Commun 10:1157–1168
19. Lundqvist BI, Norskov JK, Hjelmberg H (1979) Surf Sci 80:441–448; Gunnarsson O, Norskov JK, Hjelmberg H (1980) Phys Scr 22:165–178
20. Piccinin S, Stampfl C, Scheffler M (2008) Phys Rev B 77:075426; Li W-X, Stampfl C, Scheffler M (2003) Phys Rev Lett 90:256102; Stampfl C, Kreuzer JJ, Payne SH, Scheffler M (1999) Appl Phys A 69:471–480
21. Gracia L, González-Navarrete P, Calatayud M, Andrés J (2008) Catal Today 139:214–220; Schwab G-M (1969) Surf Sci 13:198–200

22. Föhlisch A, Nyberg M, Hasselström J, Karis O, Pettersson LGM, Nilsson A (2002) Phys Rev Lett 89:276102; Wallinder D, Hultquist G, Tvenen B, Hörlund E (2001) Corr Sci 43: 1267–1282; Wallinder D, Hörlund E, Hultquist G (2002) J Electrochem Soc 149:B393
23. Glushkov AV, Fedchuk AP, Khetselius OY (1998) In: Karwowski J (ed) Proceedings of the European science foundation REHE school and workshop on "Spin-Orbit Coupling in Chemical Reactions". Torun University Publisher, Torun, Poland, pp 11–12
24. Glushkov AV (2000) In: Proceedings of the 220th national meeting of american chemical society: catalysis and plasma technologies. ACS Publisher, Washington, USA, pp 221–224
25. Glushkov AV (2000) In: Proceedings of the 12th international congress on catalysis. Granada, Spain, Th31
26. Kraeft W-D, Kremp D, Ebeling W, Ropke G (1986) Quantum statistics of charged particle systems. Akad.-Verlag, Berlin
27. Bonch-Bruevich VL, Glasko VB (1959) Sov Phys Dokl 4:147–150; Kar S, Ho Y-K (2006) Int J Quantum Chem 106:814–820; Ghoshal A, Ho Y-K (2009) J Phys B At Mol Opt Phys 42:075002
28. Komarov IV, Ponomarev LI, Slavyanov SY (1982) Spheroidal and coulomb spheroidal functions. North Holland, Amsterdam
29. Glushkov AV Russ J Phys Chem 64:3100–3106 (1990); ibid. (1991) 65:2970–2978; ibid. (1992) 66:589–596; ibid. (1992) 66:1516–1524
30. Glushkov AV (1991) Opt Spectr 71:395–398; Ibid 72:55–61 (1992); ibid. (1994) 77:5–10; ibid. (1996) 80:60–67
31. Glushkov AV (1990) Russ J Struct Chem 31:3–8; ibid. (1993) 34:3–12; ibid. (1993) 34:13–24
32. Miller KJ, Green AE (1974) J Chem Phys 60:2617–2624
33. Resta R (1977) Phys Rev B 16:2717–2726
34. Mott NF (1974) Metal-insulator transitions. Taylor & Francis, London
35. Glushkov AV, Malinovskaya SV (2003) In: Fazio G, Hanappe F (eds) New projects and new lines of research in nuclear physics. World Scientific, Singapore, pp 242–280; Glushkov AV, Rusov VD, Ambrosov SV, Loboda AV ibid, pp 146–160
36. Glushkov AV, Khetselius OY, Malinovskaya SV (2008) Eur Phys J ST 160:195–204; Glushkov AV, Khetselius OY, Malinovskaya SV (2008) Mol Phys (UK) 106:1257–1260
37. Glushkov AV, Khetselius OY, Loboda AV, Svinarenko AA (2008) Frontiers in quantum systems in chemistry and physics, progress in theoretical chemistry and physics. In: Wilson S, Grout PJ, Maruani J, Delgado-Barrio G, Piecuch P(eds), vol 18. Springer, Berlin, pp 523–540
38. Glushkov AV, Ivanov LN (1992) Phys Lett A 170:33–37; (1993) J Phys B At Mol Opt Phys 26:L379–389
39. Glushkov AV, Ambrosov SV, Ignatenko AV, Korchevsky DA (2004) Int J Quantum Chem 99:936–939
40. Glushkov AV (2005) Atom in an electromagnetic field. TEC, Odessa
41. Glushkov AV (2005) Low Energy Antiproton Phys, AIP Serie, 796:206–210; Glushkov AV, Loboda AV, Gurnitskaya EP, Svinarenko AA (2009) Phys Scr T135:014022; Khetselius OY (2009) Phys Scr T135:014023
42. Glushkov AV, Malinovskaya SV, Svinarenko AA, Chernyakova YG (2004) Int J Quantum Chem 99:879–896; Glushkov AV, Ambrosov SV, Loboda AV, Gurnitskaya EP, Prepelitsa GP (2005) Int J Quantum Chem 104:562–569

Non-Linear Chaotic Dynamics of Quantum Systems: Molecules in an Electromagnetic Field and Laser Systems

A.V. Glushkov, V.V. Buyadzhi, A.S. Kvasikova, A.V. Ignatenko, A.A. Kuznetsova, G.P. Prepelitsa and V.B. Ternovsky

Abstract We present a general, uniform chaos-geometric formalism to analysis, modelling and forecasting a non-linear chaotic dynamics of quantum systems (such as diatomic molecules in an electromagnetic infrared field, laser and quantum generators system etc.). The approach is based on using quantum-mechanical and kinetical models for quantum and laser systems combined with the advanced techniques such as a multi-fractal method, mutual information approach, correlation integral and false nearest neighbour algorithms, the Lyapunov exponent's (LE) analysis, and surrogate data method, prediction models etc. The results of the numerical studying a chaotic dynamics of diatomic molecules (on example of the GeO molecule in an infrared field) and some laser systems are listed. There are presented data on the topological and dynamical invariants, in particular, the correlation, embedding, Kaplan-Yorke dimensions, LE, Kolmogorov's entropy etc.

Keywords Quantum and laser systems · Molecules in electromagnetic field · Non-linear analysis technique · Low- and high-dimensional chaos

1 Introduction

The past decade has witnessed a great number of studies employing the ideas gained from the science of chaos to characterize, model, and forecast a dynamics of various systems (see [1–36]). The outcomes of such studies are very perspective as they not only revealed that the dynamics of the apparently irregular phenomena could be understood from a chaotic deterministic viewpoint but also reported physically reasonable predictions using such an approach for different systems.

A.V. Glushkov (✉) · V.V. Buyadzhi · A.S. Kvasikova · A.V. Ignatenko · A.A. Kuznetsova · G.P. Prepelitsa · V.B. Ternovsky
Odessa State Environmental University, L'vovskaya str. 15, Odessa-9 65016, Ukraine
e-mail: glushkovav@gmail.com; dirac13@mail.ru

A. Tadjer et al. (eds.), *Quantum Systems in Physics, Chemistry, and Biology*, Progress in Theoretical Chemistry and Physics 30,
DOI 10.1007/978-3-319-50255-7_10

In a case of quantum systems, using of chaos constructions may seem self-contradictory in many respects (see e.g. [27–35]). In any case quantum chaos now mainly refers to the study of the consequences, for a quantum system, of the more or less chaotic nature of the dynamics of its classical analogue. It has followed two main avenues. The first one is based on semiclassical techniques - specifically the use of semiclassical Green's functions in the spirit of Gutzwiller's trace formulae, which provides a link between a quantum system and its $h \rightarrow 0$ limit, the second is associated with the Bohigas-Giannoni-Schmit conjecture or related approaches [37], which states that the spectral fluctuations of classically chaotic systems can be described using the proper ensembles of random matrices. Some of the beauty of quantum chaos is that it has developed a set of tools which have found applications in a large variety of different physical contexts, ranging from atomic, molecular and nuclear physics, optical or microwave resonators and mesoscopic physics and others (see [1–35]). It should be mentioned interpretation of the chaotic phenomena in quantum systems through the mechanism of strong nonlinear interaction and overlapping of the resonances (overlapping and merging resonances, "snapshots" of the gas of resonances, stochastization of vibrational motion in molecules etc.). In this chapter we present an uniform chaos-geometric formalism to analysis, modelling and forecasting the non-linear chaotic dynamics of quantum systems (such as diatomic molecules in an electromagnetic infrared field, laser and quantum generators systems). The formalism is based on using quantum-mechanical models for molecular and laser systems combined with the advanced techniques such as the wavelet and multi-fractal analysis, mutual information and correlation integral, false nearest neighbour algorithms, the LE analysis, and surrogate data method, prediction models etc.

2 Chaos–Geometric Approach to Modelling and Prediction Dynamics of the Complex Systems: General Formalism

The key topic of quantitative studying chaotic dynamics of the complex systems is an numerical analysis of the characteristic time series. We formally consider scalar measurements $s(n) = s(t_0 + n\Delta t) = s(n)$, where t_0 is the start time, Δt is the time step, and is n the number of the measurements. Since processes resulting in the chaotic behaviour are fundamentally multivariate, it is necessary to reconstruct phase space using as well as possible information contained in the $s(n)$. Such a reconstruction results in a certain set of d-dimensional vectors $\mathbf{y}(n)$ replacing the scalar measurements. Packard et al. [16] proposed the method of using time-delay coordinates to reconstruct the phase space of an real dynamical system. The direct use of the lagged variables $s(n + \tau)$, where τ is some integer to be determined, leads to a coordinate system in which the structure of orbits in phase space can be captured. Using a collection of time lags to create a vector in d dimensions,

$$\mathbf{y}(n) = [s(n), s(n+\tau), s(n+2\tau), \ldots, s(n+(d-1)\tau)], \tag{1}$$

the required coordinates can be provided. The dimension d is called the embedding dimension, d_E. One should note that the choice of proper time lag τ is very important for the further subsequent reconstruction of phase space [12, 16, 17]. There are two widespread methods to compute it. First approach is based on determination the linear autocorrelation function

$$C_L(\delta) = \frac{\frac{1}{N}\sum\limits_{m=1}^{N}[s(m+\delta)-\bar{s}][s(m)-\bar{s}]}{\frac{1}{N}\sum\limits_{m=1}^{N}[s(m)-\bar{s}]^2}, \tag{2}$$

$$\bar{s} = \frac{1}{N}\sum_{m=1}^{N} s(m) \tag{3}$$

and further to find that time lag where $C_L(\delta)$ first passes through zero. Another approach is based on the average mutual information [10]. If we consider two systems, A and B, with measurements a_i and b_k, the amount one learns in bits about a measurement of a_i from a measurement of b_k is given by the arguments of information theory as

$$I_{AB}(a_i, b_k) = \log_2\left(\frac{P_{AB}(a_i, b_k)}{P_A(a_i)P_B(b_k)}\right), \tag{4}$$

Here we introduce the probabilities of observing a out of the set of all A $[P_A(a_i)]$, and finding b in a measurement B $[P_B(b_i)]$, and joint probability of the measurement of a and b $[P_{AB}(a_i, b_k)]$. Further let us suppose of the sets of measurements $s(n)$ as the A and of the measurements a time lag τ later, $s(n + \tau)$, as B set. Then the average mutual information between observations at n and $n + \tau$ is as follows

$$I_{AB}(\tau) = \sum_{a_i, b_k} P_{AB}(a_i, b_k) I_{AB}(a_i, b_k) \tag{5}$$

At last one should choose that τ where the first minimum of $I(\tau)$ occurs.

The further step is determine the embedding dimension and reconstruct an Euclidean space R^d large enough so that the number of points d_A can be unfolded without ambiguity [4, 11]. There are a few methods to reconstruct the attractor dimension (e.g., [11, 12]). One of the widespread methods to study the signatures of chaos in a time series is given by a correlation integral approach. To calculate the correlation integral, one can use the known algorithm by Grassberger and Procaccia [20]. The correlation integral is as follows:

$$C(r) = \lim_{N \to \infty} \frac{2}{N(n-1)} \sum_{\substack{i,j \\ (1 \le i < j \le N)}} H(r - \|\mathbf{y}_i - \mathbf{y}_j\|) \tag{6}$$

where H is the Heaviside step function with $H(u) = 1$ for $u > 0$ and $H(u) = 0$ for $u \le 0$, r is the radius of sphere centered on $\mathbf{y}_i$ or $\mathbf{y}_j$, and N is the number of data measurements. If the time series is characterized by an attractor, then $C(r)$ is connected with a radius r as follows:

$$d = \lim_{\substack{r \to 0 \\ N \to \infty}} \frac{\log C(r)}{\log r}, \tag{7}$$

where d is correlation exponent. To check the data obtained by the correlation integral analysis, one should use another method which is known as a surrogate data method [21, 22].

The next step of studying the nonlinear system is computing the LE, which are the dynamical invariants [11–14, 22–25]. In a chaos theory, the LE spectrum is treated as a measure of the effect of perturbing the initial conditions of a system. Other important invariants are the Kolmogorov entropy K and attractor's dimension. The value K measures an average rate at which information about the state is lost with time and is the sum of the positive LE. The inverse of the Kolmogorov entropy is equal to the average limit of predictability $\mathrm{Pr}_{\max}$. The Kaplan and Yorke dimension is:

$$d_L = j + \sum_{\alpha=1}^{j} \lambda_\alpha / |\lambda_{j+1}|, \tag{8}$$

Table 1 A chaos-geometric approach to nonlinear analysis and prediction of chaotic dynamics of the complex systems

I.	Preliminary study and assessment of the presence of chaos:
1.	Test by Hottvod-Melben: $K \rightarrow 1$ – chaos;
2.	Fourier decompositions, irregular nature of change – chaos;
3.	Spectral analysis, Energy spectra statistics, the Wigner distribution, the spectrum of power, "Spectral rigidity";
II.	The geometry of the phase space. Fractal Geometry:
4.	Computing time delay τ using autocorrelation function or mutual information;
5.	Determining embedding dimension d_E by the method of correlation dimension or algorithm of false nearest neighbouring points;
6.	Computing multi-fractal spectra. Wavelet analysis;
III.	Prediction:
7.	Computing global Lyapynov dimensions LE: λ_α; Kaplan-Yorke dimension d_L, KE, average predictability measure $\mathrm{Pr}_{\max}$;
8.	Determining the number of nearest neighbour points NN for the best prediction results;
9.	Methods of nonlinear prediction; neural network algorithm…;

where j is such that $\sum_{\alpha=1}^{j} \lambda_\alpha > 0$ and $\sum_{\alpha=1}^{j+1} \lambda_\alpha > 0$, and the LE λ_α are taken in descending order. To compute the LE one can use the method with the linear fitted map by Sano and Sawada or neural networks algorithm [11, 12, 24].

Table 1 reflects the main blocks of an universal approach to numerical studying non-linear dynamics of complex systems [11, 12].

3 Non-Linear Dynamics of Molecular Systems in an Electromagnetic Field and Laser Systems with Elements of a Chaos

3.1 Molecular Systems in an Electromagnetic Field

Here we present the results of analysis of the chaotic dynamics for diatomic molecules in an electromagnetic (infrared) field. The cited analysis is based on the numerical solution of the time-dependent Schrödinger equation and realistic Simons-Parr-Finlan model for the potential of diatomic molecule $U(x)$ (the quantum unit). Secondly, it is based on using an universal approach to analysis of nonlinear chaotic dynamics (chaos-geometric unit). The Simons-Parr-Finlan formulae [31] is:

$$U(r) = B_0[(r - r_e)/r]^2 \left\{ 1 + \sum_n b_n [(r - r_e)/r]^n \right\} \quad (9)$$

or introducing $x = r - r_0$:

$$U(r) = B_0[x/(x + r_0]^2 \left\{ 1 + \sum_n b_n [x/(x + r_0]^n \right\} \quad (10)$$

where the coefficients b_i are linked with corresponding molecular constants.

The problem of dynamics of diatomic molecules in an infrared field is reduced to solving the Schrödinger equation:

$$i\partial\Psi/\partial t = [H_0 + U(x) - d(x)E_M\varepsilon(t)\cos(\omega_L t)]\Psi, \quad (11)$$

where E_M—the maximum field strength, $\varepsilon(t) = E_0 cos(\nu t)$ corresponds the pulse envelope (chosen equal to one at the maximum value of electric field). A molecule in field gets an induced polarization and its high-frequency component is as

$$P_x(t) = p_c^{(x)}(t) cos\omega t + p_s^{(x)}(t) sin\omega t, \quad (12)$$

$$P_y(t) = p_c^{(y)}(t) cos\omega t + p_s^{(y)}(t) sin\omega t, \quad (13)$$

$$p_c^{(x,y)}(t) = \left(\frac{1}{T}\right) \oint \langle \psi(t) | \hat{d}_{x,y} | \psi(t) \rangle \cos \omega t dt, \tag{14}$$

where *T*—period of the external field, *d*—dipole moment. The power spectrum can be further determined as usually $S(\omega) = |F[p(t)]|^2$. To avoid the numerical noise during the Fourier transformation, the attenuation technique used, i.e. at $t > t_p$, p (t) is replaced by

$$p(t) \cos^2\{\pi(t - t_p)/[2(T - t_p)]\}, \quad (t_p < t < T) \text{ with } T = 1.6 t_p. \tag{15}$$

In the regular case of molecular dynamics, a spectrum will consist of small number of the well resolved lines. In the case of chaotic dynamics situation changes essentially. The corresponding energy of interaction with the field is much higher than anharmonicity constant $W > xh\Omega$. It is obvious that a spectrum in this case become more complicated [28]. We have carried out the numerical computing dynamics of the diatomic molecule GeO in the electromagnetic field (the molecule and field parameters are as follows: $\hbar\Omega = 985.8$ cm^{-1}, $y\hbar\Omega = 4.2$ cm^{-1}, $B = 0.48$ cm^{-1}, $d_O = 3.28$ D, M = 13.1 a.e.м.; the field intensity is 2.5–25 GW/cm^2, respectively: W = 3.39–10.72 cm^{-1}). The Chirikov parameter in this case is as: $\delta n = 2(Ed/B)^{\frac{1}{2}} \gg 1$. According to classical treating [28], these parameters correspond to chaotic regime. The principle of quantum mechanics enter, of course, into the mixed interpretation in terms of classical trajectories [29]. The final answers are at least understandable intuitively, from other one they are result of numerical analysis of complex molecular dynamics, which involve a superposition of high-order energy transitions, intensive interaction of non-linear resonances and chaotic motion of a molecule [28, 29]. In Fig. 1 we list the computed theoretical time dependence of polarization for GeO molecule in a field in a chaotic regime.

The further step is an analysis of the corresponding time series (with the time step $\Delta t = 4 \times 10^{-14}$ s). In Table 2 we list the computed values of the correlation dimension d_2, embedding dimension d_N, which are computed on the basis of the of false nearest neighbouring points algorithm with noting (%) of false points for different values of the lag time τ. In Table 3 we list the computed values of the Kaplan-Yorke attractor dimension (d_L), LE (λ_i, i = 1–3) and the Kolmogorov entropy (K_{entr}).

Fig. 1 The computed characteristic time dependence of polarization of the GeO molecule in a field in a chaotic regime (see parameters in text)

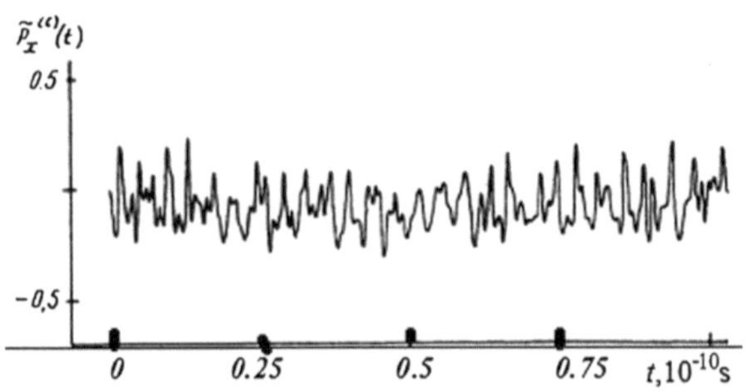

Table 2 Correlation dimension d_2, embedding dimension d_N, computed on the basis of false nearest neighbouring points algorithm with noting (%) of false points for different values of τ

τ	d_2	(d_N)
42	3.04	5 (4.2)
4	2.73	3 (1.1)
6	2.73	3 (1.1)

Table 3 The Kaplan-Yorke attractor dimension (d_L), LE (λ_i, $i = 1{-}3$) and the Kolmogorov entropy (K_{entr})

λ_1	λ_2	λ_3	d_L	K_{entr}
0.146	0.0179	−0.321	2.51	0.16

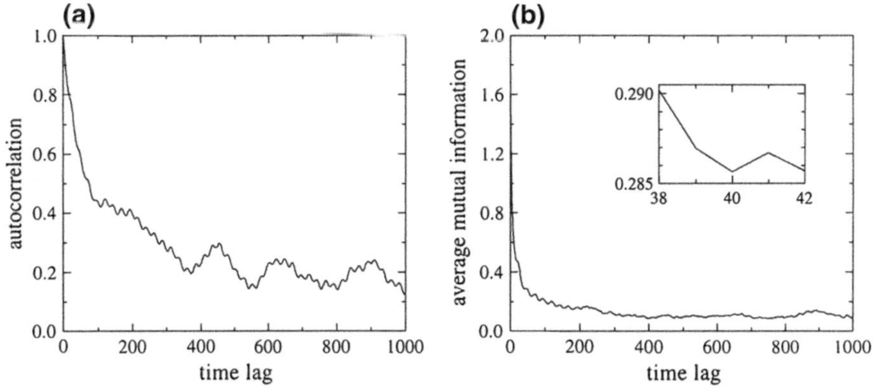

Fig. 2 **a** Autocorrelation function and **b** average mutual information depending on time delay

Analysis of the LE, correlation, Kaplan Yorke dimensions, Kolmogorov entropy data shows that the dynamics of the GeO molecule in an electric field has the elements of a deterministic chaos (low-D strange attractor) and this conclusion is entirely agreed with the results of the classical-dynamical treating [28].

3.2 *Non-Linear Analysis of Chaotic Oscillations in a Grid of Quantum Generators*

Here we present results of non-linear analysis of the chaotic oscillations in a grid o two quantum generators. Dynamics of this systems has intensively studied from the viewpoint of the corresponding differential equations solutions [3, 10]). In Refs. [8, 26, 27] the series for the characteristic vibration amplitude are presented in a case of two lasers connected through general resonator. We have analysed the parameter series in a chaotic regime (4096 imput points). In Fig. 2 we present the variations of the autocorrelation coefficient.

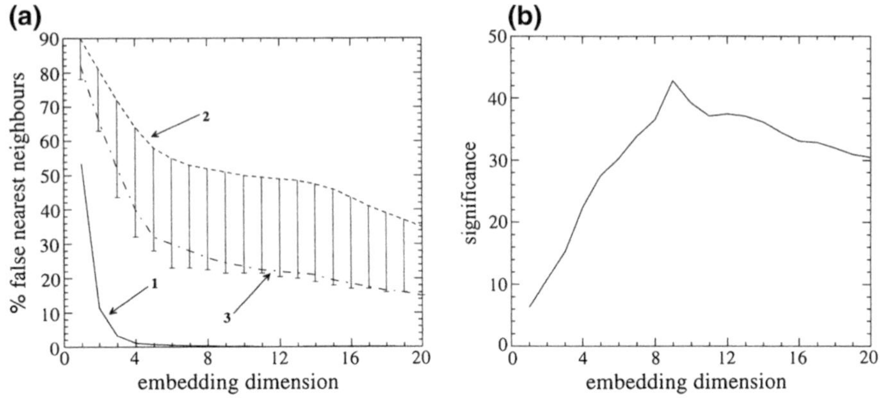

Fig. 3 **a** Embedding dimension estimation by false nearest neighbour method for original t series (line 1), mean values of surrogate data sets (2), one surrogate realization (3). *Error bars* show min % of false nearest neighbour among all realizations of surrogate data; b Relationship between significance values of correlation dimension and embedding dimension"

Table 4 The correlation dimension d_2, embedding dimension d_N, which are computed on the basis of the false nearest neighbouring points algorithm with noting (%) of false points for different values of the lag time τ

τ	d_2	(d_N)
64	7.9	10 (12)
10	7.1	8 (1.2)
12	7.1	8 (1.2)

Figure 3 displays the percentage of false nearest neighbours that was computed for the series, for phase-spaces reconstructed with embedding dimensions from 1 to 20. In Table 4 we list the computed values of the correlation dimension d_2, embedding dimension d_N, which are computed on the basis of the false nearest neighbouring points algorithm with noting (%) of false points for different values of the lag time τ. Accordingly in Table 5 we list the computed values of the Kaplan-York attractor dimension (d_L), LE (λ_I, i = 1–3) and the Kolmogorov entropy (K_{entr}).

3.3 A Chaos Generation Analysis in the Semiconductor GaAs/GaAlAs Laser System with Retarded Feedback

Here we present the results the numerical studying the low- and high dimensional dynamics of a chaos generation in the semiconductor GaAs/GaAlAs laser with the retarded feedback. Fischer et al. [36] have carried out the experimental studying

Table 5 The Kaplan-Yorke attractor dimension (d_L), LE (λ_I, i = 1–3) and the Kolmogorov entropy (K_{entr}) for the system of two semiconductors lasers connected through general resonator

λ_1	λ_2	λ_3	d_L	K_{entr}
0.515	0.198	−0.146	6.9	0.745

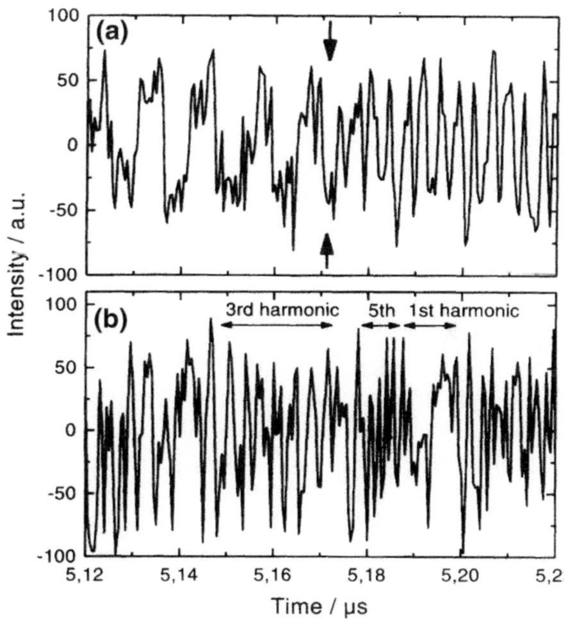

Fig. 4 The time series of intensity in the GaAs/GaAlAs Hitachi HLP1400 laser (the measured data, from Ref. [36])

dynamics of a chaos generation in the semiconductor GaAs/GaAlAs Hitachi HLP1400 laser; an instability is generated by means of the retarded feedback during changing the control parameter such as the feedback strength μ (or in fact an injection current).

Of course, depending on the system μ there is appeared a multi-stability of different states with the modulation period: $T_n = 2\tau/(2n + 1)$, n = 0, 1, 2, ... The state of n = 0 is called as a ground one. With respect to the frequency modulation, other states are called as the third harmonic, fifth harmonic and so on. In the Fig. 4 we list the measured data on the time-dependent intensities for a semiconductor laser device with feedback: (a) the time series, which illustrates a chaotic wandering between the ground state and the state of the third harmonic; (b) the time series for a system in a state of the global chaotic attractor. In the Table 6 we present our original data on the correlation dimension d_2, the embedding dimension, computed on the basis of the false nearest neighboring points algorithm (d_N) with percentage of false neighbors (%) which are calculated for different lag times τ. The data are presented for two regimes: I. chaos and II. hyperchaos.

Table 6 The dimension d_2, embedding dimension, computed by the false nearest neighboring points algorithm (d_N) with percentage of false neighbors (%), calculated for different lag times τ

Chaos regime (I)			Hyperchaos regime (II)		
τ	d_2	(d_N)	τ	d_2	(d_N)
58	3.4	5 (8.1)	67	8.4	11 (15)
6	2.2	4 (1.05)	10	7.4	8 (3.4)
8	2.2	4 (1.05)	12	7.4	8 (3.4)

Table 7 The LE: λ_1–λ_3, the Kaplan-Yorke attractor dimension d_L and the entropy K_{entr}

Regime	λ_1	λ_2	λ_3	d_L	K_{entr}
Chaos (I)	0.151	0.00001	−0.188	1.8	0.15
Hyperchaos (II)	0.517	0.192	−0.139	7.1	0.71

In Table 7 we present our original data on the LE, Kaplan-Yorke attractor dimensions, the Kolmogorov entropy K_{entr}. One can see that there are the LE positive and negative values. The resulting Kaplan-Yorke dimensions in both cases are very similar to the correlation dimension, which is computed using the Grassberger-Procaccia algorithm [20]. The Kaplan-Yorke dimension is less than the embedding dimension that confirms the correct choice of the latter. A scenario of chaos generation is in converting initially periodic states into individual chaotic states with increasing the parameter μ through a sequence of the period doubling bifurcations. Further there is appeared a global chaotic attractor after merging an individual chaotic attractors according a few complicated scenario (see details in Refs. [11, 12, 36]).

4 Conclusions

Here we presented a general chaos-geometric formalism to analysis, modelling and prediction of the non-linear dynamics of complex systems and the results of analysis of the chaotic dynamics for some quantum (diatomic molecule in a field) and laser (grid of two semiconductor lasers) systems. One should note that the mutual information approach provided a time lag which is needed to reconstruct phase space. Such an approach allowed concluding the possible nonlinear nature of process resulting in the amplitude level variations. The correlation dimension method provided a low (or high-) fractal-dimensional attractor thus suggesting a possibility of the existence of chaotic behaviour. It has been shown that the systems (diatomic molecule in an field and laser systems) dynamics in chaotic regime exhibit a nonlinear behaviour with elements of a low-and a high dimensional chaos. The LE analysis does support this conclusion.

Acknowledgements Authors are thankful to Prof. E. Brändas, J. Maruani, A. Tadjer, R. Pavlov for the invitation to present an invited report at the workshop QSCP-XX (Bulgaria, 2015). The help in editing the manuscript by Mr. Heike Rossel and Ms. Heena Naveen (Springer) is very much appreciated.

References

1. Prigogine I (1980) From being to becoming. Freeman, New York; Resonances, Springer Lecture notes in physics, Brändas E, Elander N (eds) (1989) vol 325. Springer, Berlin
2. Mandelbrot B (1983) The fractal geometry of nature. W.Freeman & Co, San Francisco
3. Arnold V (1978) Mathematical methods of classical mechanics. Academic Press, New York
4. Brändas E (1997) Advances in chemical physics: resonances, instability, and irreversibility. In: Prigogine I, Rice SA, vol 99. Wiley, pp 211–244
5. Engelmann AR, Natiello MA, Höghede M, Engdahl E, Brändas E (1986) Int J Quantum Chem 30:84
6. Blümel R, Reinhardt WP (2005) Chaos in atomic physics. Cambridge University Press, pp 1–344; Chirikov BV (1979) Phys Rep 52:263–379
7. Abarbanel H, Brown R, Sidorowich J, Tsimring L (1993) Rev Mod Phys 65:1331–1392
8. Vedenov A, Ezhov A, Levchenko E (1987) Non-linear waves. Structure and bi-furcations. In: Gaponov-Grekhov A, Rabinovich M, Moscow, pp 53–69
9. Gutzwiller M (1990) Chaos in classical and quantum mechanics. Springer, New York; Ott E (2002) Chaos in dynamical systems. Cambridge University Press, Cambridge
10. Gallager R (1986) Information theory and reliable communication. Wiley, New York; Glushkov AV (2008) Modern theory of a chaos. OSENU, Odessa
11. Glushkov A, Khetselius O, Brusentseva S, Zaichko P, Ternovsky V (2014) Advances in neural networks. Fuzzy systems and artificial intelligence. In: Balicki J (ed) Recent Adv. in computer engineering, vol 21. WSEAS, Gdansk, pp 69–75
12. Glushkov A, Svinarenko A, Buyadzhi V, Zaichko P, Ternovsky V (2014) Advances in neural networks. Fuzzy systems and artificial intelligence: Recent Advancement. In: Balicki J (ed) Computer engineering, vol 21. WSEAS, Gdansk, pp 143–150
13. Glushkov AV, Khokhlov VN, Tsenenko IA (2004) Atmospheric teleconnection patterns: wavelet analysis. Nonlinear proceedings in geophysics, vol 11, pp 285–296
14. Glushkov AV, Prepelitsa G, Svinarenko A, Zaichko P (2013) Dynamical Systems Theory, vol T1. In: Awrejcewicz J, Kazmierczak M, Olejnik P, Mrozowski J (eds) Lodz, Polland, pp 467–487
15. Lorenz EN (1963) J Atm Sci 20:130–141
16. Packard N, Crutchfield J, Farmer J, Shaw R (1988) Phys Rev Lett 45:712–716
17. Kennel M, Brown R, Abarbanel H (1992) Phys Rev A 45:3403–3412
18. Takens F (1981) Dynamical systems and turbulence. In: Rand D, Young L. Springer, Berlin, pp 366–381
19. Mañé R (1981) Dynamical systems and turbulence. In: Rand D, Young L (eds) Springer, Berlin, pp 230–242
20. Grassberger P, Procaccia I (1983) Physica D 9:189–208
21. Theiler J, Eubank S, Longtin A, Galdrikian B, Farmer J (1992) Phys. D. 58:77–94
22. Fraser A, Swinney H (1986) Phys Rev A 33:1134–1140
23. Havstad J, Ehlers C (1989) Phys Rev A 39:845–853
24. Sano M, Sawada Y (1995) Phys Rev Lett 55:1082–1086
25. Schreiber T (1999) Phys Rep 308:1–64
26. Glushkov A, Bunyakova Y, Fedchuk A, Serbov N, Svinarenko A, Tsenenko I (2007) Sensor Electr Microsyst Technol 3(1):14–22

27. Prepelitsa GP, Glushkov AV, Lepikh YI, Buyadzhi VV, Ternovsky VB, Zaichko PA (2016) Sensor Electr Microsyst Technol (Copernicus) 11(4):43–58
28. Berman GP, Kolovskii A (1989) JETP 85:1552–1562
29. Zhang C, Katsouleas T, Joshi C (1993) Proceedings of short wavelength physics with intense laser pulses, San-Diego, USA, pp 21–28
30. Sakai M, Miyazaki N (1995) Appl Phys B 61:493–498
31. Simons G, Parr RG, Finlan JM (1973) J Chem Phys 59:3229
32. Connerade JP (1997) J Phys B: Atom Mol Opt Phys 30:L31–L36
33. Krug A, Buchleitner A (2002) Phys Rev A 66:053416
34. Cheng T, Liu J, Chen S, Guo H (2000) Phys Lett A 265:384390
35. Benvenuto F, Casati G, Shepelyansky D (1994) Z Phys B 94:481–486
36. Fischer I, Hess O, Elsaber W, Gobel E (1994) Phys Rev Lett 73:2188–2191
37. Ullmo D (2008) Rep Prog Phys 71: 026001

Landau Quantisation of Electron Motion in the Crust of Highly Magnetised Neutron Stars

N. Chamel, Y.D. Mutafchieva, Zh.K. Stoyanov, L.M. Mihailov and R.L. Pavlov

Abstract With observed surface magnetic fields up to $\sim 10^{15}$ G, neutron stars—the stellar remnants of gravitational core-collapse supernovae explosions—are unique cosmic laboratories for probing the properties of matter under extreme conditions. The outermost layer of a neutron star is thought to consist of a solid crust, whose atoms are fully ionised by the huge pressure. Electrons are expected to be highly degenerate and form an essentially ideal relativistic Fermi gas. With increasing depth, nuclei become progressively more neutron rich by capturing electrons until at some point, neutrons start to drip out of nuclei. At about half the density found in heavy atomic nuclei, the crust dissolves into a neutron liquid with a small admixture of protons and electrons. The composition and the equation of state of highly magnetised neutron-star crusts are shown to be significantly affected by Landau quantisation of electron motion.

1 Introduction

Neutron stars are the remnants of massive stars (with a mass $M > 8M_{\odot}$, $M_{\odot}$ being the mass of the Sun) at the end point of their evolution (see, e.g., Ref. [1]). Neutron

Electronic supplementary material The online version of this chapter (doi:10.1007/978-3-319-50255-7_11) contains supplementary material, which is available to authorized users.

N. Chamel (✉)
Institute of Astronomy and Astrophysics, Université Libre de Bruxelles, CP 226, Boulevard du Triomphe, 1050 Brussels, Belgium
e-mail: nchamel@ulb.ac.be

Y.D. Mutafchieva · Zh.K. Stoyanov · R.L. Pavlov
Institute for Nuclear Research and Nuclear Energy, Bulgarian Academy of Sciences, 72 Tsarigradsko Chaussee, 1784 Sofia, Bulgaria

L.M. Mihailov
Institute of Solid State Physics, Bulgarian Academy of Sciences, 72 Tsarigradsko Chaussee, 1784 Sofia, Bulgaria

A. Tadjer et al. (eds.), *Quantum Systems in Physics, Chemistry, and Biology*, Progress in Theoretical Chemistry and Physics 30,
DOI 10.1007/978-3-319-50255-7_11

stars have a mass between one and two times that of the Sun, but packed into a space only 20 km across (100,000 times smaller than the Sun's diameter). The average density of a neutron star can thus be higher than that found inside the heaviest atomic nuclei. Neutron stars are not only the most compact observed stars in the Universe, but they are also endowed with the strongest magnetic fields known. Neutron-star observations thus offer the unique opportunity to explore the properties of matter under extreme conditions, which cannot be reproduced in terrestrial laboratories.

Most neutron stars discovered so far have been detected as radio pulsating stars—*pulsars* for short—with surface magnetic fields reaching $10^{13} - 10^{14}$ G [2] (as compared to $\sim 10^{-1}$ G for the Earth's magnetic field). A distinct class of neutron stars with even higher magnetic fields was proposed by Duncan and Thomson in 1992 [3]. It is nowadays widely accepted that soft-gamma ray repeaters (SGRs) and anomalous x-ray pulsars (AXPs) represent different facets of these so called *magnetars* (see e.g., Ref. [4] for a review). At the time of this writing, 11 SGRs and 12 AXPs have been already identified [5]. The magnetic fields measured at the surface of these stars attain about $10^{14} - 10^{15}$ G [5–7], and various observations indicate the presence of much higher fields in the stellar interiors [8–14]. According to numerical simulations, neutron stars may potentially possess internal magnetic fields as high as 10^{18} G (see, e.g., Refs. [15, 16] and references therein).

The interior of a neutron star is thought to be stratified into distinct layers, which can be classified as follows with increasing depth (see, e.g., Ref. [1]). A very thin atmosphere of light elements (mainly hydrogen and helium though heavier elements like carbon may also be present [17]) possibly surrounds a Coulomb liquid of electrons and ions. The matter beneath consists of a solid crust (see, e.g., Ref. [18] for a review). The pressure is so high that atoms are ionised, and free electrons form a highly degenerate quantum gas. With further compression, nuclei become progressively more neutron rich by capturing electrons (see, e.g., Ref. [19]). Above some threshold density $\sim 10^{11}$ g cm^{-3}, neutrons start to drip out of nuclei thus delimiting the outer and inner regions of the crust (see, e.g., Refs. [20, 21]). The crust extends up to a density $\sim 10^{14}$ g cm^{-3}, above which it dissolves into an homogeneous liquid mixture of nucleons and electrons. The innermost part of the core may contain other particles such as hyperons or even deconfined quarks.

The properties of the outermost regions of a neutron star can be drastically modified by a sufficiently high magnetic field [22–24]. In this paper, we study the composition and the equation of state of the outer crust of a highly magnetised neutron star using the most recent experimental atomic mass measurements from the 2012 Atomic Mass Evaluation (AME) [25], complemented by the microscopic atomic mass model HFB-24 [26]. After briefly reviewing Landau quantisation in Sect. 2, our model of neutron-star crusts is described in Sect. 3. Our results are presented and discussed in Sect. 4.

2 Landau Quantisation of Electron Motion

The importance of the magnetic field for the structure of atoms in the neutron-star atmosphere can be assessed by introducing the "atomic" magnetic field, defined by

$$B_{\rm at} = \frac{m_e^2 e^3 c}{\hbar^3} \simeq 2.35 \times 10^9 \text{ G}, \tag{1}$$

where m_e is the electron mass, e the elementary electric charge, c the speed of light and $\hbar$ the Planck-Dirac constant. This is the magnetic field strength B for which the electron cyclotron energy $\hbar\omega_c = eB/(m_e c)$ is equal to the typical atomic Coulomb energy e^2/a_0, where $a_0 = \hbar^2/(m_e e^2)$ is the Bohr radius. Most neutron stars are endowed with much higher magnetic fields $B \gg B_{\rm at}$. Therefore, the magnetic field is expected to change significantly the properties of atoms in the atmosphere of a neutron star. The electron motion perpendicular to the magnetic field lines is quantised into Landau orbitals with a characteristic magnetic length scale (see, e.g., Ref. [1])

$$a_m = a_0 \sqrt{\frac{B_{\rm at}}{B}}\,. \tag{2}$$

The atomic magnetic field $B_{\rm at}$ is such that $a_m = a_0$. For $B \gg B_{\rm at}$, we thus have $a_m \ll a_0$. As a consequence, atoms are expected to adopt a very elongated shape along the magnetic field lines and to form linear chains. The attractive interaction between these chains could lead to a transition into a magnetically condensed phase.

In the region of the outer crust on which we focus, atoms are fully ionised, and free electrons behave essentially as an ideal relativistic Fermi gas. The main correction arises from electrostatic interactions of point like ions in a uniform charge compensating electron background (see, e.g., Ref. [27] for a discussion of higher-order corrections). We assume that the temperature T is lower than the crystallization temperature T_m and that ions are arranged in a regular crystal lattice. Considering structures made of ions with proton number Z and mass number A, the crystallization temperature is given by

$$T_m = \frac{e^2}{a_e k_{\rm B} \Gamma_m} Z^{5/3} \simeq 1.36 \times 10^6 a_0 n_e^{1/3} \left(\frac{Z}{40}\right)^{5/3} \frac{175}{\Gamma_m} \text{ K}, \tag{3}$$

where $a_e = (3/(4\pi n_e))^{1/3}$ is the electron-sphere radius, n_e is the electron number density, $k_{\rm B}$ is Boltzmann's constant, and Γ_m is the Coulomb coupling parameter at melting (see, e.g., Ref. [1]). Since T_m is typically much lower than the electron Fermi temperature defined by

$$T_{\rm Fe} = \frac{\mu_e - m_e c^2}{k_{\rm B}} \simeq 5.93 \times 10^9 \left(\frac{\mu_e}{m_e c^2} - 1\right) \text{ K}, \tag{4}$$

where μ_e is the electron Fermi energy, electrons are highly degenerate except in the outermost envelope of the star. With increasing magnetic field strength, the Landau orbitals shrink. As the magnetic field strength reaches the value $B_{\rm rel}$ given by

$$B_{\rm rel} = \left(\frac{m_e c^2}{\alpha \lambda_e^3}\right)^{1/2} \simeq 4.41 \times 10^{13}\,{\rm G}\,, \tag{5}$$

where $\alpha = e^2/(\hbar c)$ is the fine structure constant, the characteristic magnetic length scale a_m becomes comparable to the electron Compton wavelength $\lambda_e = \hbar/(m_e c) = \alpha a_0$ (or equivalently the electron cyclotron energy becomes comparable to the electron rest-mass energy). As a consequence, the Landau quantisation of electron motion is modified by relativistic effects. Surface magnetic fields $B > B_{\rm rel}$ have been inferred in SGRs, AXPs, as well as in a few radio pulsars.

The quantising effects of the magnetic field on the properties of the outer crust of a neutron star are most important when only the first Landau level is occupied. The magnetic field will be referred to as strongly quantising. This situation arises when $n_e < n_{eB}$ and $T < T_B$ with (see, e.g., Ref. [1])

$$n_{eB} = \frac{B_\star^{3/2}}{\sqrt{2}\pi^2 \lambda_e^3}\,, \tag{6}$$

$$T_B = \frac{m_e c^2}{k_{\rm B}} B_\star \simeq 5.93 \times 10^9\, B_\star\ {\rm K}\,, \tag{7}$$

where $B_\star \equiv B/B_{\rm rel}$. Since matter is electrically charge neutral, the number density n_N of nuclei is simply given by $n_N = n_e/Z = n/A$, where n is the average nucleon number density. To the electron density (6) thus corresponds the average nucleon number density

$$n_B = \frac{A}{Z} n_{eB} \simeq 1.24 \times 10^{-9}\, \frac{A}{Z} B_\star^{3/2}\ {\rm fm}^{-3}\,, \tag{8}$$

and the mass density

$$\rho_B = \frac{A}{Z} m n_{eB} \simeq 2.06 \times 10^6\, \frac{A}{Z} B_\star^{3/2}\ {\rm g\ cm}^{-3}\,, \tag{9}$$

where m is the average mass per nucleon, approximated here by the unified atomic mass unit. Conversely, for a given mass density ρ the strongly quantising regime corresponds to magnetic field strengths

$$B_\star > \left(\frac{\rho Z \lambda_e^3 \sqrt{2}\pi^2}{A m}\right)^{2/3} \approx 180 \left(\frac{2Z}{A}\rho_{10}\right)^{2/3}\,, \tag{10}$$

where $\rho_{10} = \rho/(10^{10}\ {\rm g\ cm}^{-3})$.

3 Microscopic Model of Highly Magnetised Neutron-Star Crusts

Formed in the aftermath of gravitational core-collapse supernova explosions, neutron stars are initially very hot (with a typical temperature $T \sim 10^{10}$ K) and fully fluid. Newly born neutron stars cool rapidly by emitting neutrinos copiously produced in their interior and after a few months their outer layer solidifies. Following Refs. [28, 29], it is assumed that all possible reaction rates remain sufficiently high to keep matter close to the full thermodynamic equilibrium until the star eventually becomes cold and fully "catalysed". Accretion of matter from a companion star may change substantially the internal constitution of the neutron-star crust. We shall not consider this possibility here since observed magnetars are generally isolated.

In the magnetar theory, neutron stars are born with very high magnetic fields of order $B \sim 10^{16} - 10^{17}$ G, which decay on a typical time scale of order 10^3 years [30]. The magnetic fields are therefore sustained for a very long time as compared to the cooling time of the star, and can therefore alter the formation of neutron-star crusts. For simplicity, we shall consider that the magnetic field is spatially uniform at the microscopic scale of interest here.

As discussed in the appendix of Ref. [19], the equilibrium composition of each layer of the outer crust at a given pressure P is found by minimizing the Gibbs free energy per nucleon defined by

$$g = \frac{\mathscr{E} + P}{n}, \tag{11}$$

where $\mathscr{E}$ is the average energy density of matter, given by the sum of the energy density $\mathscr{E}_N$ of nuclei, the energy density $\mathscr{E}_e$ of a relativistic electron Fermi gas, and the electron-electron and electron-ion interaction energy density $\mathscr{E}_L$ (usually referred to as the lattice energy density). Ignoring the effects of the magnetic field on nuclei, we have

$$\mathscr{E}_N = n_N M'(A, Z)c^2 , \tag{12}$$

where $n_N = n/A$ is the number density of nuclei, and $M'(A, Z)$ their mass (including the rest mass of Z protons, and $A - Z$ neutrons). The nuclear mass $M'(A, Z)$ can be obtained from the *atomic* mass $M(A, Z)$ after subtracting out the rest mass energy and the binding energy of the atomic electrons (see Eq.(A4) of Ref. [31]). In this paper, we made use of the most recent experimental atomic mass measurements from the 2012 AME [25]. For the masses that have not yet been measured, we employed the microscopic atomic mass model HFB-24 [26]. This model, based on the self-consistent Hartree-Fock-Bogoliubov method using a generalized Skyrme effective nucleon-nucleon interaction (see, e.g., Ref. [32] for a short review), fits the 2353 measured masses of nuclei with N and $Z \geq 8$ appearing in the 2012 AME with a root-mean square deviation of 549 keV. It is also compatible with various other experimental and astrophysical constraints [33]. According to the Bohr-van Leeuwen theorem [34], the lattice energy density is independent of the magnetic field and is given by (see, e.g., Ref. [1])

$$\mathscr{E}_L = C\alpha\hbar c n_e^{4/3} Z^{2/3} , \tag{13}$$

where $C < 0$ is the dimensionless crystal structure constant. In this expression, the finite-size of the ions is neglected, as well as the small contribution due to quantum zero-point motion of ions off their equilibrium position [35]. The lattice contribution to the pressure can be readily obtained from Eq. (13) and is given by $P_L = \mathscr{E}_L/3$. Expressions for the electron energy density $\mathscr{E}_e$ and pressure P_e can be found in Ref. [1]. The total matter pressure is $P = P_e + P_L$. Note that ions do not contribute to the pressure apart from the lattice term. In the following, we shall consider that ions are arranged on a body-centred cubic lattice. In this case, the crystal structure constant is given by $C = -1.44423$ [36].

4 Equilibrium Composition and Equation of State

Assuming that the magnetic field is strongly quantising at the surface of the star, the average nucleon density is approximately given by [22]

$$n_s \approx \frac{A_s}{\lambda_e^3} \left[\frac{|C|\alpha B_\star^2}{4\pi^4 Z_s} \right]^{3/5} , \tag{14}$$

with Z_s and A_s the corresponding proton number and the mass number of the equilibrium nuclide. Considering that the surface of a neutron star is made of iron ($Z_s = 26$ and $A_s = 56$) leads to

$$n_s \simeq 2.50 \times 10^{-10} B_\star^{6/5}\, \text{fm}^{-3} , \tag{15}$$

and for the mass density

$$\rho_s \simeq 3.87 \times 10^{8} B_\star^{6/5}\ \text{g cm}^{-3} . \tag{16}$$

It can be easily seen that $n_s < n_B$ therefore our assumption of strongly quantising magnetic field was justified. Equation (15) shows that the nucleon density (hence also the mass density) does not vanish at the surface of a highly magnetised neutron star, and the higher the magnetic field is, the higher is the surface density. It should be stressed however that Eq. (15) provides only an approximate estimate of the surface density because of our neglect of the nonuniformity of the electron gas [37]. Moreover, this expression is only valid for nonrelativistic electrons, or equivalently

$$B_\star \ll \frac{2\pi^2}{Z_s^2 |C|^3 \alpha^3} \simeq 2.5 \times 10^4 . \tag{17}$$

In deeper regions of the outer crust, the density n in a layer at pressure P can be approximately expressed as [23]

$$n \approx n_s \left(1 + \sqrt{\frac{P}{P_0}}\right), \tag{18}$$

where

$$P_0 = m_e c^2 \frac{n_s^2 \pi^2 \lambda_e^3}{B_\star} \left(\frac{Z}{A}\right)^2 \simeq 1.82 \times 10^{-11} B_\star^{7/5} \left(\frac{Z}{A}\right)^2 \text{ MeV fm}^{-3}. \tag{19}$$

At densities $n \gg n_B$, many Landau levels are populated so that the quantisation effects disappear and the equation of state matches smoothly with that obtained in the absence of magnetic fields [23].

The composition of the outer crust changes with density and with the magnetic field strength (see Tables 1–4 in the Online Resource). For the "weak" magnetic fields prevailing in most pulsars $B_\star < 1$, the sequence of equilibrium nuclides in their outer crust is the same as that obtained in the absence of magnetic fields. However the highest density at which each nuclide can be found is increased, especially in the shallow region of the crust where the effects of Landau quantisation are the most important. For the high fields expected to exist in magnetars $B_\star \gg 1$, the sequence of equilibrium nuclides may change. Table 1 indicates the magnetic field strength above which a nuclide appears or disappears. For sufficiently high magnetic fields, almost the entire outer crust is made of only one element [24]. With our crust model, this element is found to be ^{90}Zr. Transitions from one nuclear species to another occur at some pressure $P_{1\to2}$ such that $g(A_1, Z_1, P_{1\to2}) = g(A_2, Z_2, P_{1\to2})$. These transitions are accompanied by a discontinuous change δn_e of the electron density given to first order in α by

Table 1 Magnetic field strength $B_\star = B/B_{\rm rel}$ for the appearance (+) or the disappearance (-) of a nuclide in the outer crust of a cold nonaccreting neutron star. See text for details

Nuclide	$B_\star$
^{58}Fe(-)	9
^{66}Ni(-)	67
^{88}Sr(+)	859
^{126}Ru(+)	1031
^{80}Ni(-)	1075
^{128}Pd(+)	1445
^{78}Ni(-)	1610
^{79}Cu(-)	1617
^{64}Ni(-)	1668
^{130}Cd(+)	1697
^{132}Sn(+)	1989

$$\delta n_e = \left[P_L(n_e, Z_1) - P_L(n_e, Z_2)\right] \left(\frac{\partial P_e}{\partial n_e}\right)^{-1} . \tag{20}$$

Expanding $g(A_2, Z_2, P_{1\to2})$ to first order in α, the threshold condition for the onset of the transition $1 \to 2$ is approximately given by

$$\mu_e + C\alpha\hbar c n_e^{1/3} \left(\frac{Z_2^{5/3}}{A_2} + \frac{1}{3}\frac{Z_2 Z_1^{2/3}}{A_2} - \frac{4}{3}\frac{Z_1^{5/3}}{A_1}\right) \left(\frac{Z_2}{A_2} - \frac{Z_1}{A_1}\right)^{-1} = \mu_e^{1\to2} , \tag{21}$$

where we have introduced the threshold electron Fermi energy

$$\mu_e^{1\to2} \equiv \left(\frac{M'(A_1, Z_1)c^2}{A_1} - \frac{M'(A_2, Z_2)c^2}{A_2}\right) \left(\frac{Z_2}{A_2} - \frac{Z_1}{A_1}\right)^{-1} + m_e c^2 . \tag{22}$$

With increasing pressure, nuclei become progressively more neutron rich until g equals $m_n c^2$ (m_n being the neutron mass) for some pressure $P_{\rm drip}$. Ignoring neutron-band structure effects [38, 39], further compression leads to the appearance of free neutrons [21]. In this particular case, the threshold condition (21) must be replaced by [21]

$$\mu_e + \frac{4}{3} C e^2 n_e^{1/3} Z_d^{2/3} = \mu_e^{\rm drip} , \tag{23}$$

where

$$\mu_e^{\rm drip} \equiv \frac{-M'(A_d, Z_d)c^2 + A_d m_n c^2}{Z_d} + m_e c^2 , \tag{24}$$

Z_d and A_d denoting the proton and mass numbers of the equilibrium nuclide at the neutron-drip transition. In the strongly quantising regime assuming that electrons are ultrarelativistic, the threshold density and pressure for the transition $1 \to 2$ are approximately given to zeroth order in α by

$$n_{1\to2} \approx \frac{A_1}{Z_1} \frac{\mu_e^{1\to2}}{m_e c^2} \frac{B_\star}{2\pi^2 \lambda_e^3} , \tag{25}$$

$$P_{1\to2} \approx \frac{B_\star (\mu_e^{1\to2})^2}{4\pi^2 \lambda_e^3 m_e c^2} . \tag{26}$$

The neutron-drip density and pressure can be calculated similarly by substituting $\mu_e^{1\to2}$ with $\mu_e^{\rm drip}$. The density and the pressure at the interface between different layers are thus expected to increase almost linearly with the magnetic field strength. Accounting for electron-ion and electron-electron interactions introduce small non-linearities (see, e.g., Ref. [21]). These considerations are confirmed by the numerical results shown in Fig. 1. Equations (25) and (26) are only valid if the magnetic field strength exceeds some value, which we shall denote by $B_\star^{1\to2}$ for transitions $1 \to 2$

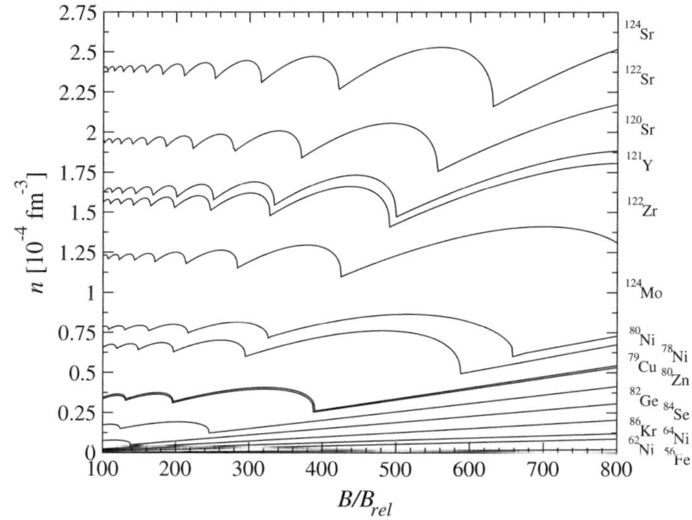

Fig. 1 Maximum average nucleon density at which a given nucleus (indicated on the *right-hand side*) is present in the outer crust of a cold nonaccreting neutron star, as a function of the magnetic field strength

and $B_\star^{\rm drip}$ for the neutron-drip transition. Using Eq. (6), we find

$$B_\star^{1\to 2} \approx \frac{1}{2}\left(\frac{\mu_e^{1\to 2}}{m_e c^2}\right)^2 , \quad B_\star^{\rm drip} \approx \frac{1}{2}\left(\frac{\mu_e^{\rm drip}}{m_e c^2}\right)^2 . \tag{27}$$

Since the electron Fermi energy increases with density, $B_\star^{1\to 2}$ is the lowest for the transition ^{56}Fe$\to$^{62}Ni in the shallow region of the neutron-star crust ($B_\star^{^{56}{\rm Fe}\to{}^{62}{\rm Ni}} \approx$ 1.8), and the highest for the transition ^{122}Sr$\to$^{124}Sr in the deepest region ($B_\star^{^{122}{\rm Sr}\to{}^{124}{\rm Sr}} \approx 1140$). For the same reason, $B_\star^{\rm drip} > B_\star^{1\to 2}$ for any transition $1 \to 2$. In the weakly quantising regime for which many Landau levels are populated, Eqs. (21) and (23) must be solved numerically. As shown in Fig. 1, Landau quantisation of electron motion leads in this case to magnetic oscillations of the densities at the interface between two different layers. Because the amplitude of these oscillations increases with $B_\star$ (see Fig. 1), Landau quantisation effects on the structure of neutron-star crusts are negligible if $B_\star \ll B_\star^{^{56}{\rm Fe}\to{}^{62}{\rm Ni}}$. Note that $B_\star^{^{56}{\rm Fe}\to{}^{62}{\rm Ni}}$ is of the same order as $B_{\rm rel}$. On the contrary, the magnetic field is strongly quantising in any layer of the outer crust of a neutron star if $B_\star > B_\star^{\rm drip}$. Using Eqs. (24) and (27) for ^{124}Sr, we find $B_\star^{\rm drip} \approx 1179$. Including electron-electron and electron-ion interactions increases $B_\star^{\rm drip}$ by about 9% [21].

In order to assess the range of validity of our zero-temperature treatment, let us estimate the values of the characteristic temperatures for strongly quantising magnetic fields. At the surface of the neutron star (electrons are nonrelativistic), we find

$$T_{\rm Fe} \approx 5.2 \times 10^7 B_\star^{2/5}\ {\rm K}\,, \qquad T_m \approx 1.7 \times 10^7 B_\star^{2/5}\ {\rm K}\,, \tag{28}$$

where we have adopted the value $\Gamma_m \approx 175$ for an unmagnetised one-component plasma [1]. At the neutron-drip transition, we obtain

$$T_{\rm Fe} \approx 2.8 \times 10^{11}\ {\rm K}\,, \qquad T_m \approx 1.2 \times 10^8 B_\star^{1/3}\ {\rm K}\,. \tag{29}$$

The lowest temperature is thus found to be T_m. However, these values of T_m are probably underestimated since highly-magnetised Coulomb crystals tend to melt at a higher temperature [40].

5 Conclusion

We calculated the composition and the equation of state of the outer crust of cold nonaccreting neutron stars endowed with very high magnetic fields $B \gg m_e^2 c^3/(e\hbar) \simeq 4.4 \times 10^{13}$ G, as measured in soft-gamma ray repeaters, anomalous x-ray pulsars and even in a few radio pulsars [2, 5]. For this purpose, we made use of the most recent experimental atomic mass data [25] complemented with the Hartree-Fock-Bogoliubov atomic mass model HFB-24 [26]. The Landau quantisation of electron motion caused by the magnetic field changes the crustal composition and leads to characteristic magnetic oscillations of the threshold density at the interface between two different layers. Moreover, the equation of state is considerably stiffened at low pressures due to magnetic condensation. If $B > 5.7 \times 10^{16}$ G, electrons are confined to the lowest Landau level in any region of the outer crust for the model considered here. In this case, the threshold densities and pressures increase almost linearly with B, and are approximately given by Eqs. (25) and (26). For sufficiently high magnetic fields, the outer crust is predicted to be almost entirely composed of a pure body-centred cubic crystal of ^{90}Zr, with much heavier elements in the deepest layers. However, for such fields the crustal properties may be further modified by Landau quantisation of nucleon motion [24], an effect not included in this work but which warrants further studies.

By changing the crustal composition, the presence of a high magnetic field also modifies crustal properties like elastic coefficients. The present results may thus have implications for the interpretation of the quasiperiodic oscillations observed in soft gamma-ray repeaters. Likewise, the change of thermal and electrical conductivities may impact the thermal and magnetic field evolution of magnetars

Acknowledgements This work was mainly supported by Fonds de la Recherche Scientifique—FNRS (Belgium), and the Bulgarian National Science Fund under contract No. DFNI-T02/1. This work was also partially supported by the European Cooperation in Science and Technology (COST) Action MP1304 *NewCompStar*.

References

1. Haensel P, Potekhin AY, Yakovlev DG (2007) Neutron stars 1: equation of state and structure. Springer
2. Ng CY, Kaspi VM (2011) AIP Conf Proc 1379:60
3. Thompson C, Duncan RC (1992) Astrophys J 392:L9
4. Woods PM, Thompson C (2006) In: Lewin WHG, van der Klis M (eds) Compact stellar x-ray sources. Cambridge University Press, Cambridge, pp 547
5. Olausen SA, Kaspi VM (2014) Astrophys J Suppl Ser 212:6
6. Tiengo A et al (2013) Nature 500:312
7. An Hongjun et al (2014) Astrophys J 790:60
8. Stella L, DallOsso S, Israel GL (2005) Astrophys J 634:L165
9. Kaminker AD, Yakovlev DG, Potekhin AY, Shibazaki N, Shternin PS, Gnedin OY (2007) Astrophys Sp Sci 308:423
10. Watts AL, Strohmayer TE (2007) Adv Sp Res 40:1446
11. Samuelsson L, Andersson N (2007) Mon Not Roy Astr Soc 374:256
12. Vietri M, Stella L, Israel GL (2007) Astrophys J 661:1089
13. Rea N, Esposito P, Turolla R, Israel GL, Zane S, Stella L, Mereghetti S, Tiengo A, Götz, Göğüş E, Kouveliotou C (2010) Science 330:944
14. Makishima K, Enoto T, Hiraga JS, Nakano T, Nakazawa K, Sakurai S, Sasano M, Murakami H (2014) Phys Rev Lett 112:171102
15. Pili AG, Bucciantini N, Del Zanna L (2014) Mon Not R Astron Soc 439:3541
16. Chatterjee D, Elghozi T, Novak J, Oertel M (2015) Mon Rot R Astron Soc 447:3785
17. Ho WCG, Heinke CO (2009) Nature 462:71
18. Chamel N, Haensel P (2008) Physics of neutron star crusts. Living Rev Relativ 11:10. http://www.livingreviews.org/lrr-2008-10
19. Chamel N, Fantina AF (2015) Phys Rev D 92:023008
20. Chamel N, Fantina AF, Zdunik JL, Haensel P (2015) Phys Rev C 91:055803
21. Chamel N, Stoyanov ZhK, Mihailov LM, Mutafchieva YD, Pavlov RL, Velchev ChJ (2015) Phys Rev C 91:065801
22. Lai D, Shapiro SL (1991) Astrophys J 383:745
23. Chamel N, Pavlov RL, Mihailov LM, Velchev ChJ, Stoyanov ZhK, Mutafchieva YD, Ivanovich MD, Pearson JM, Goriely S (2012) Phys Rev C 86:055804
24. Basilico D, Peña Arteaga D, Roca-Maza X, Coló G (2015) Phys Rev C 92:035802
25. Audi G, Wang M, Wapstra AH, Kondev FG, MacCormick M, Xu X, Pfeiffer B (2012) Chin Phys C 36:002
26. Goriely S, Chamel N, Pearson JM (2013) Phys Rev C 88:024308
27. Chamel N, Fantina AF (2016) Phys Rev D 93:063001
28. Harrison BK, Wheeler JA (1958) Onzième Conseil de Physique Solvay. Stoops, Bruxelles, Belgium
29. Harrison BK, Thorne KS, Wakano M, Wheeler JA (1965) Gravitation theory and gravitational collapse. The University of Chicago Press
30. Dall'Osso S, Shore SN, Stella L (2009) Mon Not Roy Astr Soc 398:1869
31. Lunney D, Pearson JM, Thibault C (2003) Rev Mod Phys 75:1021
32. Chamel N, Pearson JM, Fantina AF, Ducoin C, Goriely S, Pastore A (2015) Acta Physica Polonica B 46:349
33. Pearson JM, Chamel N, Fantina AF, Goriely S (2014) Eur Phys J A 50:43
34. Van Vleck JH (1932) The theory of electric and magnetic susceptibilities. Oxford University Press, London
35. Baiko DA (2009) Phys Rev E 80:046405
36. Coldwell-Horsfall RA, Maradubin AA (1960) J Math Phys 1:395
37. Lai D (2001) Rev Mod Phys 73:629
38. Chamel N (2006) Nucl Phys A 773:263
39. Chamel N, Naimi S, Khan E, Margueron J (2007) Phys Rev C 75:055806
40. Potekhin AY, Chabrier G (2013) Astron Astrophys 550:A43

Part III
Atomic and Molecular Properties

Electronic and Nuclear Dynamics for a Non-Equilibrium Electronic State: The Ultrafast Pumping of N_2

J. Šmydke, J. Ajay, F. Remacle and R.D. Levine

Abstract Ultrashort laser pulses that span only a small number of optical cycles have a broad energy spectrum. Interaction of such a broad energy ultraviolet pulse with a molecule excites a coherent superposition of several electronic states that are within the pulse energy window. Such a coherent electronic wavepacket necessarily evolves on more than one Born-Oppenheimer potential energy curve at a time. We study the short time electron nuclei dynamics induced by an ultrashort XUV pulse in the three lowest Σ and three lowest Π excited states of the nitrogen molecule that are accessible from the electronic ground state.

Keywords Non-stationary electronic states · Ultrafast molecular dynamics · Coherent electronic state · Diabatic electronic states · Dipole moment of a homonuclear molecule

---

Electronic supplementary material The online version of this chapter (doi:10.1007/978-3-319-50255-7_12) contains supplementary material, which is available to authorized users.

J. Šmydke (✉) · J. Ajay · F. Remacle · R.D. Levine
The Fritz Haber Center for Molecular Dynamics and Institute of Chemistry, The Hebrew University of Jerusalem, Jerusalem 9190401, Israel
e-mail: jan.smydke@gmail.com

J. Šmydke
Department of Radiation and Chemical Physics, Institute of Physics, Academy of Sciences of the Czech Republic, 18221 Praha 8, Czech Republic

F. Remacle
Department of Chemistry, B6c, University of Liege, B4000 Liege, Belgium

R.D. Levine
Crump Institute for Molecular Imaging and Department of Molecular and Medical Pharmacology, David Geffen School of Medicine and Department of Chemistry and Biochemistry, University of California, Los Angeles, CA 90095, USA

A. Tadjer et al. (eds.), *Quantum Systems in Physics, Chemistry, and Biology*, Progress in Theoretical Chemistry and Physics 30,
DOI 10.1007/978-3-319-50255-7_12

1 Introduction

In traditional photochemistry it is assumed that a molecule in its ground state is excited by a pulse of a well defined energy to a single particular excited electronic state. On the new electronic state the nuclei become non-stationary as they suddenly experience a different electronic force field. Consequently the nuclei start moving on the excited potential energy surface. The time scale of the nonstationary vibronic motion is typically tens of femtoseconds and such processes were subject of many pump and probe experiments for the past 20 years [1, 2]. Recently ultrashort sub-femtosecond pulses are becoming more widely available [3–10]. The essential point is that an ultrafast pulse, where ultrafast means an energy span of the order of the spacings of excited electronic states, can pump a coherent wavepacket of electronic states [11]. Such a state can exhibit electronic dynamics not possible when dealing with stationary electronic states. For a complete dynamical description one needs to include both the electronic degrees of freedom that are pumped out of equilibrium and the responding nuclear motion. This is currently possible for diatomic molecules [12–19]. For larger molecules one is nowadays typically confined to a description for frozen nuclei [11, 20–26].

In this paper, for a model system of the nitrogen N_2 molecule, we explain how ultrashort pulses affect both the electronic and the nuclear motion. Specifically we include the pump pulse as part of the Hamiltonian so that the creation of the excited non equilibrium state is treated exactly. We use a grid method [27, 28] to describe the nuclear dynamics. The paper is organized as follows: In Sect. 2 we introduce the nonstationary electronic states and explain their consequence for the nuclear dynamics. Section 3 describes the model system that we used—the nitrogen molecule and the ultrashort pulse. Finally, the results of the simulations are discussed in Sect. 4 by means of the motion of the nuclear density, population of states and the dipole moment behavior. In future work we will discuss how to probe [9, 29, 30] the nonequilibrium dipole moment of the homonuclear N_2 molecule [31, 32].

2 Non-Stationary Electronic States and the Nuclear Dynamics

If the pulse duration is short, the pulse energy spectrum will broaden from a sharper value (for a long pulse duration) to a distribution. For femtosecond and sub-femtosecond ultrashort ultraviolet pulses, the pulse energy width can encompass several excited electronic states. When a molecular system interacts with such an ultrashort pulse, the result is a coherent superposition of the several optically active electronically excited states that lie within the pulse energy width

$$\Psi^{el}(r;R,t) = \sum_k c_k \Phi_k^{el}(r;R)\exp(-(i/\hbar)E_k(R)t) \tag{1}$$

where $\Phi_k^{el}(r;R)$ are the stationary solutions of the time-independent Schroedinger equation for the electronic Hamiltonian

$$\hat{H}^{el}(R)\Phi_k^{el}(r;R) = E_k(R)\Phi_k^{el}(r;R) \tag{2}$$

where r denotes the electronic coordinates and R the nuclear coordinates that is a parameter so that $\Phi_k^{el}(r;R)$ are the Born-Oppenheimer electronic states. The $\exp(-i/\hbar)E_k(R)t)$ is the phase factor that is required to make $\Psi^{el}(t)$ a (non-stationary) eigenfunction of the time dependent Schroedinger equation for the electronic Hamiltonian

$$i\hbar\frac{\partial}{\partial t}\Psi^{el}(r;R,t) = \hat{H}^{el}(R)\Psi^{el}(r;R,t) \tag{3}$$

A wavepacket that consists of at least two stationary states with different phases $E_k/\hbar$ is nonstationary as the expectation values of various operators are varying in time with beating frequencies between the individual component phases

$$\begin{aligned}\langle\Psi^{el}(r;R,t)|\hat{O}|\Psi^{el}(r;R,t)\rangle &= \sum_k |c_k|^2\langle\Phi_k^{el}(r;R)|\hat{O}|\Phi_k^{el}(r;R)\rangle \\ &+ \sum_{k\neq l} c_k^* c_l \exp(-(i/\hbar)(E_l(R) - E_k(R))t)\langle\Phi_k^{el}(r;R)|\hat{O}|\Phi_l^{el}(r;R)\rangle\end{aligned} \tag{4}$$

For such a wave packet physical observables like charge distribution or dipole moment can be varying in time.

When the molecular electronic state is nonstationary as in Eq. (1), the nuclei see the field of several electronic states and the nuclear part of the total wavefunction thus evolves on more than one potential energy surface at a time. The involved electronic states can also be coupled, either by non adiabatic coupling in geometries where the Born-Oppenheimer approximation fails, a coupling that is quite important in N_2 [33] or other coupling like spin-orbit interaction or when the pump or probe laser fields are on. Below we use the diabatic electronic states that do not diagonalize the electronic Hamiltonian. When the electronic states are adiabatically or nonadiabatically coupled the dynamics of the molecular system becomes even more interesting since not only is the electronic wave function that is pumped by the pulse nonstationary and the nuclei move on several potential energy surfaces, but also the population of the individual electronic components is varying in time.

The total wave function that includes both the electronic and the nuclear part can be written in the following form

$$\Psi(r,R,t) = \sum_{j,k} C_{jk}(t)\Phi_j^{el}(r;R)\chi_{jk}(R) \tag{5}$$

where r denotes the electronic and R the nuclear coordinates, the electronic state index j spans the ground as well as the excited electronic states and $C_{jk}(t)$ is the time dependent expansion coefficient of the electronic and the nuclear basis functions. Note that unlike the electronic wave function of Eq. (1) here the wave function depends on both r and R as dynamical variables. Instead of using vibronic eigenfunctions of the electronic states as the basis for the nuclear motion, it is advantageous here to expand the nuclear wavefunction component on the grid of the nuclear coordinate R_i so that

$$\chi_{jk}(R_i) = \delta_{ki} \tag{6}$$

with the result that these basis states are orthogonal

$$\langle \chi_{jk}(R) | \chi_{lm}(R) \rangle = \delta_{km} \tag{7}$$

A direct measure of a superposition of electronic states is an oscillating electronic dipole moment

$$\boldsymbol{\mu}(t) = \langle \Psi(r,R;t) | \hat{\boldsymbol{\mu}} | \Psi(r,R;t) \rangle = \sum_{jl} \sum_{km} C^*_{jk}(t) C_{lm}(t) \langle \chi_{jk}(R) | \boldsymbol{\mu}_{jl}(R) | \chi_{lm}(R) \rangle \tag{8}$$

where $\hat{\boldsymbol{\mu}} = -\hat{r}$ is the electric dipole moment operator and $\boldsymbol{\mu}_{jl}$ denotes the transition electronic dipole moments connecting states j and l. For the nitrogen molecule, which has the $D_{\infty h}$ symmetry, and by using the grid expansion of the nuclear wave function component [6], the formula simplifies to

$$\mu(t) = \sum_{j} \sum_{m} C^*_{0m}(t) C_{jm}(t) \mu_{0j}(R_m) + \text{c.c.} \tag{9}$$

where 0 denotes the ground electronic state and the dipole can thus be decomposed into the contributions of each electronically excited state j. Since the ground state nuclear wave function is confined to a narrow Franck-Condon region, the transition dipole elements contribute only when the excited state nuclear wavepacket is nonzero in that region. Hence the dipole is spatially sensitive to the location of the nuclear motion. Note also that the superposition of electronic states leads to an oscillating dipole moment even if the molecule does not have a permanent dipole, provided there is a nonzero transition dipole moment between the ground and excited states.

3 The Model System

We study a neutral N_2 molecule in the excitation energy range 12–15 eV. The ground electronic state $X^1\Sigma_g^+$ is described by a Morse potential [34]. The excited states used are b', c' and e' states of $^1\Sigma_u^+$ symmetry and b, c, and o states of $^1\Pi_u$ symmetry.

We are using the quantum chemical results of Spelsberg and Meyer, Ref. [35], that report the potential energy curves in a diabatic representation with fine adjustments to match experimental spectroscopic data.

A direct output of quantum chemistry are potential energy curves in the adiabatic representation with possible regions of avoided crossings between states, see [35, 36]. In these regions, there can be non-negligible nonadiabatic couplings between the mutually approaching states. These nonadiabatic couplings come from the nuclear kinetic energy operator $\hat{T}_N(R)$ of the total Hamiltonian

$$\hat{H}(R) = \hat{T}_N(R) + \hat{H}^{el}(R) = -\frac{\hbar^2}{2\mu}\frac{\partial^2}{\partial R^2} + \hat{H}^{el}(R) \tag{10}$$

acting on the electronic component of the wave function

$$\hat{T}_N(R)\left\{\Phi_j^{el}(R)\chi_{jk}(R)\right\} = -\frac{\hbar^2}{2\mu}\left\{\frac{\partial^2}{\partial R^2}\Phi_j^{el}(R) + 2\frac{\partial}{\partial R}\Phi_j^{el}(R)\frac{\partial}{\partial R} + \Phi_j^{el}(R)\frac{\partial^2}{\partial R^2}\right\}\chi_{jk}(R) \tag{11}$$

where μ denotes the reduced mass of the diatomic molecule. The first two terms of Eq. (11) can couple different electronic states. Hence, the nuclear kinetic energy term is not diagonal in the adiabatic representation. In the Born-Oppenheimer approximation the two terms are neglected. To fully treat these couplings, which is necessary for the nitrogen molecule [33], while avoiding the inconvenience of a coupling by the nuclear momentum one can use the diabatic approach [27, 37–44]. This is an alternative representation, in which the nuclear kinetic energy operator becomes diagonal and the nonadiabatic couplings transform into off-diagonal diabatic potential curves $V_{ij}(R)$ that couple electronic states i and j. Linear combinations of different adiabatic states in the avoided crossing regions are formed so as to result in smooth diagonal diabatic potentials which can cross each other. The diabatic representation thus moves the coupling from the nuclear kinetic energy term to electronic potential energy that is off diagonal.

For a short pulse with a 12–15 eV energy window, six low lying diabatic excited states, namely b', c', e' $^1\Sigma_u^+$ and b, c, o $^1\Pi_u$ are directly optically coupled to the ground electronic state. Their potential energy curves are shown in Fig. 1. Due to symmetry, the $^1\Sigma_u^+$ and $^1\Pi_u$ states do not interact either via nonadiabatic couplings or through the transition dipole moment. The b' and b states have valence character with a broad potential well, while the remaining c', e', c and o states are Rydberg excited with more tightly bound narrow potential wells. For convenience, the ground state potential curve dressed by the mean pulse energy 13.61 eV is also shown in Fig. 1.

Our intention is to simulate a system in which both the $^1\Sigma_u^+$ and $^1\Pi_u$ states are optically pumped and the electronic Hamiltonian fully includes the significant diabatic interaction between the valence and the Rydberg excited states in each of the two symmetries. The mean pulse energy 13.61 eV has been chosen, which

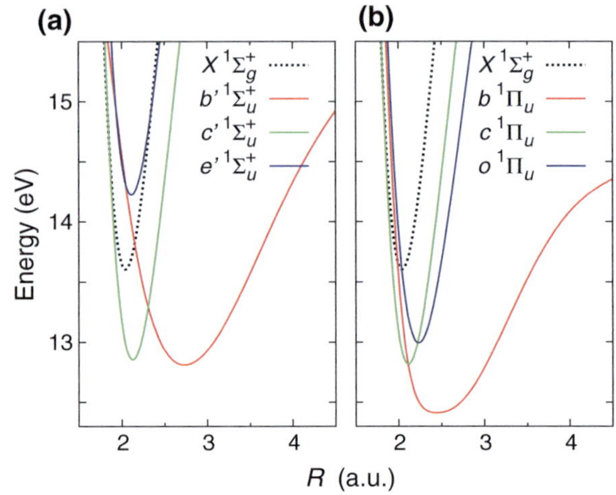

Fig. 1 The diabatic potential energy curves of the N_2 molecule as calculated in Ref. [35]. Panel **a** shows the $^1\Sigma_u^+$ states, panel **b** the $^1\Pi_u$ states. Both panels include the ground state $X^1\Sigma_g^+$ Morse potential [34] dressed by the mean pulse energy 13.61 eV

yielded more or less balanced $^1\Sigma_u^+$ and $^1\Pi_u$ excitation and led to relatively strong diabatic interaction between the involved electronic states (see the Supporting Information for details).

The laser pulse is described as an oscillating electric field confined to a Gaussian envelope

$$E(t) = E_0 e \exp(-(t-t_0)^2/2\sigma^2)\cos(\omega_0 t + \phi_0) \tag{12}$$

where E_0 is the electric field intensity, $\boldsymbol{e}$ is a unit vector that defines the pulse polarization, t_0 is the center of the pulse, σ^2 is the variance of the Gaussian envelope, ω_0 is the pulse carrier frequency and ϕ_0 is the initial phase. In our simulations we used the field intensity 10^{11} W cm^{-2}, pulse centered at 12 fs, FWHM of the Gaussian envelope 3 fs, the carrier frequency corresponding to 13.61 eV (0.5 a.u.) and the zero initial phase.

The Hamiltonian of the system contains the dipole interaction between the pulse and the molecule

$$\hat{W} = -E(t)\hat{\boldsymbol{\mu}} \tag{13}$$

From the dipole selection rules the $^1\Sigma_u^+$ states are obtained if the laser pulse is polarized in parallel with the molecular bond axis while the $^1\Pi_u$ states are excited by a pulse of perpendicular polarization. For a randomly oriented molecule, however, both the $^1\Sigma_u^+$ and $^1\Pi_u$ states are excited.

The Hamiltonian matrix elements between the electronic states i and j thus read

$$\hat{H}_{ij}(R,t) = \frac{-\hbar^2}{2\mu}\frac{\partial^2}{\partial R^2} + V_{ij}(R) - E(t)\boldsymbol{\mu}_{ij}(R) \tag{14}$$

where μ is the reduced mass of the molecule, $V_{ij}(R)$ are the diabatic potentials, $E(t)$ is the electric field intensity of the pulse and

$$\boldsymbol{\mu}_{ij}(R) = \langle \Phi_i^{el}(R) | \hat{\boldsymbol{\mu}} | \Phi_j^{el}(R) \rangle \tag{15}$$

is the transition dipole moment which was also obtained from Ref. [35].

The nuclear coordinate was expanded on the grid with 1024 points between 1.5 and 8 a.u. and the time-dependent Schroedinger equation was integrated by the Gear's predictor-corrector algorithm with a short time step 0.05 a.u. so as to accurately describe the response to the strong laser field.

4 Results and Discussion

In our simulations, the laser pulse acted on an initial wave function, which was the zeroth vibronic level of the ground $X^1\Sigma_g^+$ electronic state. During the pulse, all the involved excited states became populated and for each such excited wavepacket component the nuclear motion started to evolve on its own potential energy curve.

The evolution of each nuclear wavepacket component is shown in Fig. 2 by heatmaps of the nuclear probability density as a function of the internuclear distance and time. On Panels **a** and **d** we can see that on the valence b' and b states the bond is elongated until reaching the outer turning point on the potential well at about 3.5 −4.0 a.u. and then moving back to the Franck-Condon region with a period of about 50 fs. We can also observe secondary chunks that are delayed from the main component and emerge as a result of the population flows from the Rydberg states via the diabatic coupling as we explain below. The Rydberg wavepacket components (Panels **b**, **c**, **e**, **f**), by contrast, oscillate in their tight potential wells. The e' state is the least populated of all (as its energy is almost 1 eV above the mean pulse energy) and since its potential closely resembles the ground state, its motion is confined to the narrow range of the Franck-Condon region.

To better understand the nuclear motion in the Fig. 2 we can also plot Fig. 3, which shows how the populations evolve as a result of the diabatic couplings. From the calculations of [35] we know that the couplings are strongest for the internuclear distance between 1.8 and 2.6 a.u., which is somewhat wider than the Franck-Condon region and corresponds with the range of the Rydberg vibronic motion. Therefore, we also plot the population of the valence b' and b states only within that region and denote them by b_{Ry}' and b_{Ry}' respectively.

Consider first the $^1\Sigma_u^+$ states. We notice that the b' state is dominant, being about twice as much populated as the sum of both the Rydberg states, and that the population of the e' state is almost negligible. When the pulse starts, the b' and c' states get immediately populated. The c' state reaches its peak population and then suddenly depopulates—obviously in favour of the b' and also the e' state. The e' state is thus primarily populated via the diabatic coupling to the c' state rather than

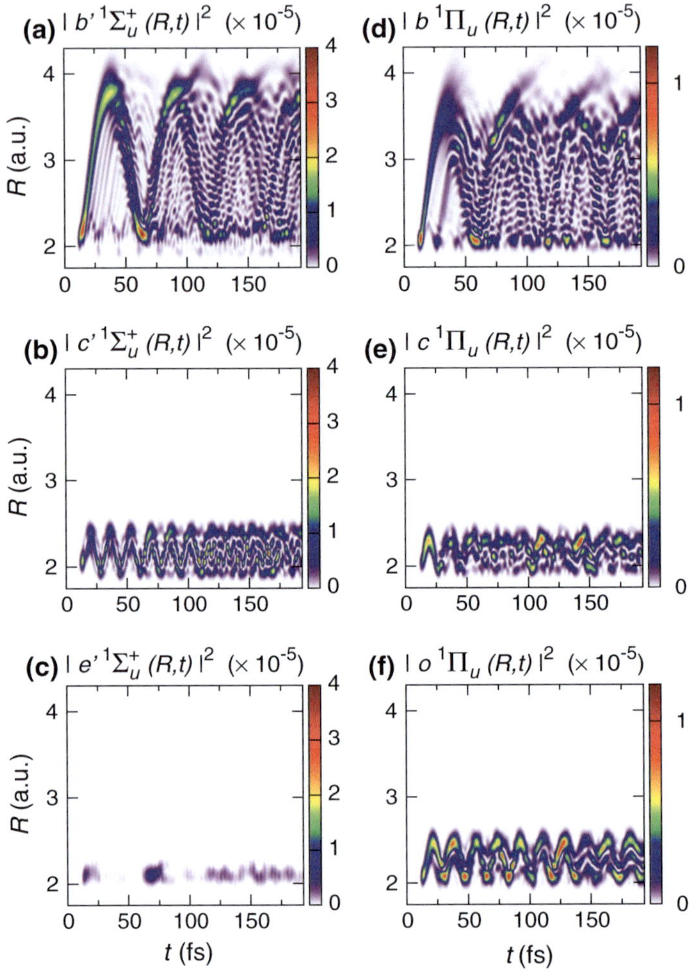

Fig. 2 Nuclear motion of the excited N_2 molecule as the evolution of the nuclear density components in time. Panels **a–c** show the $^1\Sigma_u^+$, panels **d–f** the $^1\Pi_u$ symmetry

directly by the pulse, because the pulse mean energy is too low and the transition dipole moment of the e' state is the weakest [35]. Nevertheless, the c' state gets populated again from the b' and e' states while the b' state starts to move out of the Franck-Condon region, which is observed as the rapid drop of the local b_{Ry}' population (the dotted line).

Between 20 and 50 fs we can see several tiny peaks on the b_{Ry}' curve with corresponding fluctuations in the total b' population. These suggest the secondary chunks of the b' wavepacket leaving the Franck-Condon region with a delay after the main component as is observed in Fig. 2.

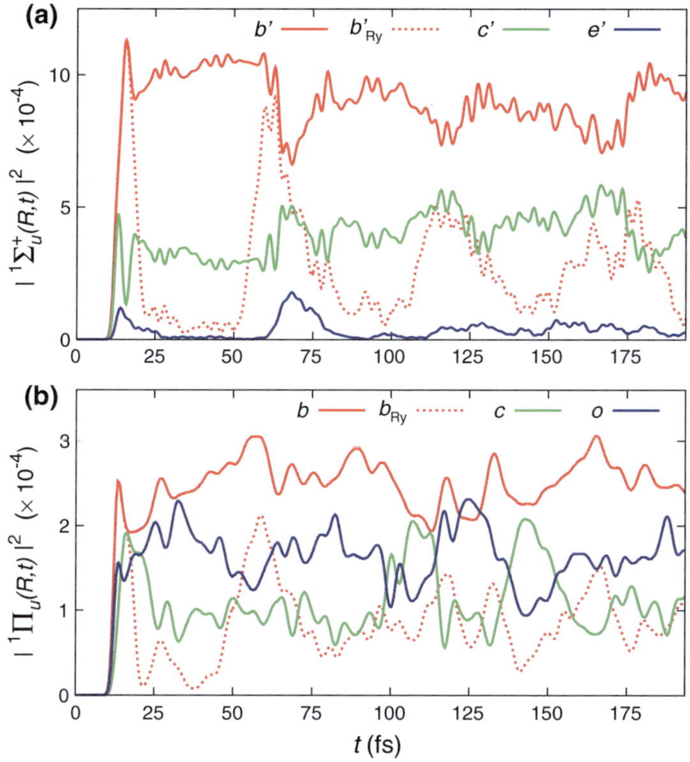

Fig. 3 Populations of the excited wavepacket components as functions of time for $^1\Sigma_u^+$ (Panel **a**) and $^1\Pi_u$ states (Panel **b**). The dotted lines show the valence state populations in the region occupied by the Rydberg states, which has strong diabatic couplings. This region spans the internuclear distance between 1.8 and 2.6 a.u.

At 50 fs the main chunk of the b' wavepacket arrives back to the Franck-Condon region which increases the population of the c' and e' states. The sudden rise of the e' population from zero level before 75 fs is remarkable in the corresponding heatmap in the Fig. 2. The Rydberg states are then again depopulated in favour of the valence b' state which is for the second time leaving the Franck-Condon region. We can again notice two little double peaks on the dotted b_{Ry}' curve at 75 and 90 fs with corresponding shapes on the b' curve and also counter peaks on the c' curve, which suggests a population exchange between the b' and c' states.

After 100 and 150 fs we still can observe a clear rise of b_{Ry}' population as the b' state approaches the Franck-Condon region, however, the b' population in that region then becomes permanent which leads to nonzero e' population as observed in Fig. 2.

We can do an analogous analysis for the $^1\Pi_u$ states. The populations are, however, more equal, though the valence b state is still dominant. Because both the Rydberg states are similarly populated, the population flows are more complicated than in the $^1\Sigma_u^+$ states. Nevertheless, we still can observe some clear patterns.

After all the states are populated by the pulse, the valence b state leaves the Franck-Condon region. At 25 fs the c state rapidly depopulates in favour of the b state in the Franck-Condon area, which again leaves the region as can be seen in the Fig. 3. The b state is further populated by the decrease of o state until the b state returns back and feeds both the c and o states at about 60 fs. At about 100 fs, when the b state returns back for the second time, it first populates the c state, then the c population steeply decreases in favour of both the b and o states. Between 130 and 140 fs the c population increases again, borrowing from the o and b states.

One can also notice in the Fig. 2 two sudden rises of intensity in c state after 100 and before 150 fs as well as a big peak in the o state at 125 fs. We can even recognize some counter peaks between a pair of states, but the population flows in the $^1\Pi_u$ symmetry mostly involve all three excited states, one feeding the other two for a while and then quickly changing their roles. The population transfers are thus complex and their interpretation not straightforward.

In Fig. 4, we show separately the time evolution of the dipole for the $^1\Sigma_u^+$ and the $^1\Pi_u$ states. We plot the absolute value of the total dipole which shows fast oscillations modulated by envelope peaks (Panels **a** and **d**). The fast oscillations are present for the entire time scale ($\approx$200 fs) investigated. They correspond to the beating frequencies ($\approx$1 fs) between the GS and the excited states and reflect the electronic coherences. The nuclear motion on the different potentials and the diabatic couplings between them are responsible for the modulation of the fast oscillations. To distinguish between the individual components we plot the envelopes of the dipole signal corresponding to each excited state contribution (Panels **b** and **e**). For easier assignment of the peaks, the dipole components for a system in which the diabatic couplings are set to zero and no population flow between the diabatic states is allowed are plotted in panels **c** and **f**.

In the uncoupled $^1\Sigma_u^+$ system (Panel **c**) we can notice how the b' wavepacket reaches the Franck-Condon region while oscillating in its broad potential energy well with the period of about 50 fs. The c' Rydberg state potential is close to the Franck-Condon area and we clearly see the effect of the wavepacket oscillations in the narrow space having a permanent population in the Franck-Condon region. The potential of the e' Rydberg state is closely matching with the shape of the ground state potential curve so there are no oscillations of the dipole signal. Due to the very low population, the uncoupled e' dipole signal is however negligible.

By comparing to the diabatically coupled system (Panel **b**), we can see b' peaks emerging at 25 and before 50 fs. These correspond to the population transfer in the Franck-Condon region due to the diabatic coupling as we described in Fig. 3 and leads to secondary oscillations of the envelope of the dipole moment. Similarly for the signal at 75 fs and before 100 fs. The coupled c' dipole is mostly resembling the uncoupled system with irregularities like the one at 65 fs which arise from the

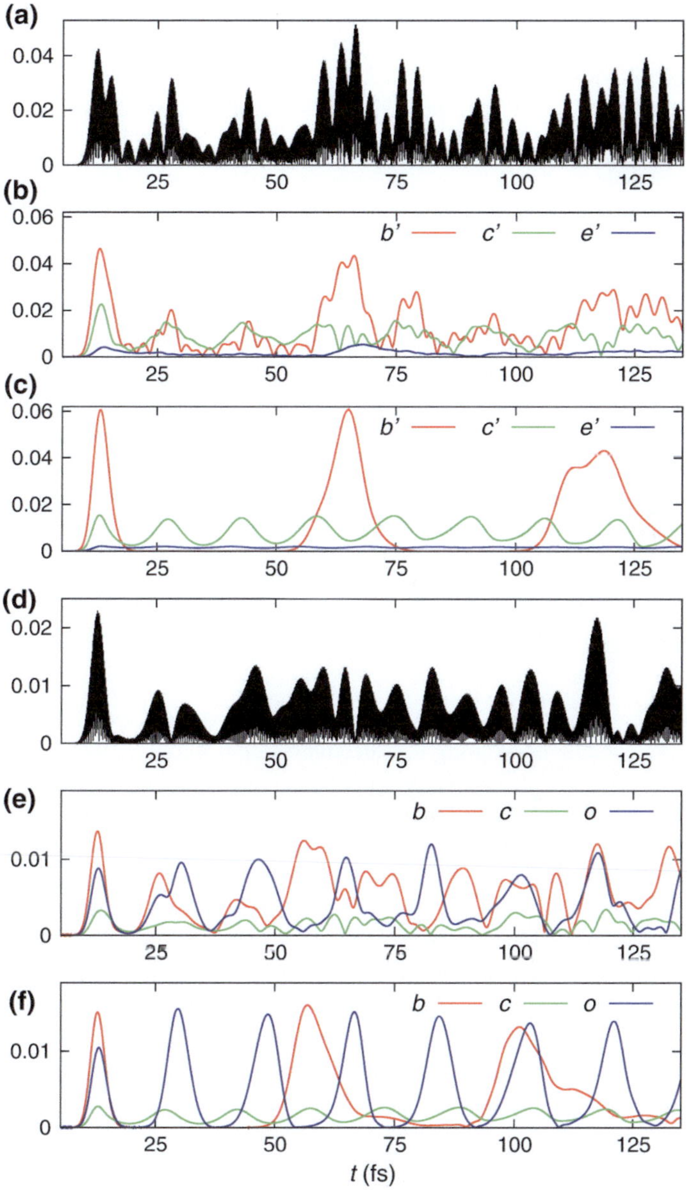

Fig. 4 $^1\Sigma_u^+$ (Panels **a–c**) and $^1\Pi_u$ (Panels **d–f**) dipole moment evolution of the excited N_2 wavepacket. Panels **a** and **d** show the absolute value of the total dipole for the given symmetry. Panels **b** and **e** show envelopes of the dipole component signals. Panels **c** and **f** show envelopes of the dipole components for a model system, in which all the diabatic couplings have been set to zero

interaction with the b' wavepacket coming back to the Franck-Condon area. We can also notice a remarkable peak for c' state at 65 fs as an evidence of the steep population rise shown in Figs. 2 and 4.

The shape of the total $^1\Pi_u$ dipole signal (Panel **d**) is noticeably simpler than for the $^1\Sigma_u^+$ states. The uncoupled $^1\Pi_u$ system resembles the $^1\Sigma_u^+$ in b (b') and c (c') states. The o state is, however, more populated than the e' state and the two states also quite differ in their potentials. The o state is deeper and broader, which is why we can see a strong regular signal of the uncoupled o state matching with the wavepacket moving in and out of the Franck-Condon area.

The coupled $^1\Pi_u$ dipole also shows the presence of the secondary valence b wavepacket population transfer in the Franck-Condon region at 25 and before 50 fs as is observed in Figs. 2 and 3. Similarly the b peaks at 75 and before 100 fs seem to correspond to the periodic motion of those secondary population (cmp. Fig. 2). In the Panel **e** of the Fig. 2 we see that the c wavepacket does not move very regularly after 50 fs and the same is observed in the dipole behavior. The o dipole component more or less keeps the pattern of the uncoupled system with tiny irregularities coming from the population flows like the one at 25 fs.

5 Conclusion

On the model system of the nitrogen molecule N_2 we have seen that an ultrashort laser pulse is capable of exciting a molecular system into a coherent superposition of several electronic states. Such an excited state drives the nuclear motion simultaneously on more than a single potential energy curve. Due to the coupling of the diabatic states the nuclear dynamics is not a simple superposition of independent wavepacket evolutions on different potentials, but is strongly influenced by population exchanges between the diabatic states. These population flows are specific to certain regions in space given by the couplings. An electric dipole moment of the molecule induced by the ultrafast pulse excitation is highly space sensitive to the nuclear wavepacket motion. A direct correlation between the dipole behaviour and the nuclear dynamics can be observed. Lastly, despite the complex nuclear motion the electronic coherence persists throughout the time scale of our computation which covers several periods of vibration.

Acknowledgements This work is supported by the U.S. Department of Energy (DOE), Office of Science, Basic Energy Sciences (BES) under Award # DE-SC0012628 and in part by the Einstein Foundation of Berlin. We benefited from our participation in the COST actions MOLIM CM1405 and CM1204 XLIC.

References

1. Mukamel S (1990) Femtosecond optical spectroscopy—a direct look at elementary chemical events. Annu Rev Phys Chem 41:647–681
2. Zewail AH (1994) Femtochemistry: ultrafast dynamics of the chemical bond, World Scientific
3. Hentschel M et al (2001) Attosecond metrology. Nature 414:509
4. Aseyev SA, Ni Y, Frasinski LJ, Muller HG, Vrakking MJJ (2003) attosecond angle-resolved photoelectron spectroscopy. Phys Rev Lett 91(22)
5. Kienberger R et al (2004) Atomic transient recorder. Nature 427(6977):817
6. Sansone G et al (2006) Isolated single-cycle attosecond pulses. Science 314(5798):443
7. Goulielmakis E et al (2007) Attosecond control and measurement: lightwave electronics. Science 317(5839):769
8. Goulielmakis E, et al (2010) Real-time observation of valence electron motion. Nature 466 (7307):739-U737
9. Neidel C, et al (2013) Probing time-dependent molecular dipoles on the attosecond time scale. Phys Rev Lett 111(3):033001–033001
10. Lepine F, Sansone G, Vrakking MJJ (2013) Molecular applications of attosecond laser pulses. Chem Phys Lett 578:1–14
11. Remacle F, Levine RD (2006) An electronic time scale in chemistry. Proc Natl Acad Sci USA 103(18):6793–6798
12. Bandrauk AD, Chelkowski S, Kawai S, Lu HZ (2008) Effect of nuclear motion on molecular high-order harmonics and on generation of attosecond pulses in intense laser pulses. Phys Rev Lett 101(15)
13. Bandrauk AD, Chelkowski S, Lu HZ (2009) Signatures of nuclear motion in molecular high-order harmonics and in the generation of attosecond pulse trains by ultrashort intense laser pulses. J Phys B-At Mol Opt Phys 42(7)
14. Muskatel BH, Remacle F, Levine RD (2012) Ultrafast Predissociation Mechanism of the (1)Pi (u) States of N-14(2) and Its Isotopomers upon Attosecond Excitation from the Ground State. J Phys Chem A 116(46):11311–11318
15. Muskatel BH, Remacle F, Levine RD (2014) AttoPhotoChemistry. Probing ultrafast electron dynamics by the induced nuclear motion: The prompt and delayed predissociation of N-2. Chem Phys Lett 601:45–48
16. Braun H et al (2014) Coupled electron-nuclear wavepacket dynamics in potassium dimers. J Phys B 47:124015
17. Medišauskas L, Morales F, Palacios A, González-Castrillo A, Plimak Lev, Smirnova, O, Martín F, Ivanov MY (2015) Signatures of attosecond electronic-nuclear dynamics in the one-photon ionization of molecular hydrogen: analytical model versus ab initio calculations. New J Phys 17:053011
18. Ranitovic P et al (2014) Attosecond vacuum UV coherent control of molecular dynamics. Proc Natl Acad Sci USA 111(3):912–917
19. Chang BY, Shin S, Palacios A, Martín F, Sola IR (2015) Oscillating molecular dipoles require strongly correlated electronic and nuclear motion. J Phys B 48:43001
20. Hennig H, Breidbach J, Cederbaum LS (2005) Electron correlation as the driving force for charge transfer: Charge migration following ionization in N-methyl acetamide. J Phys Chem A 109(3):409–414
21. Remacle F, Levine RD (2006) The time scale for electronic reorganization upon sudden ionization of the water and water-methanol hydrogen bonded dimers and of the weakly bound NO dimer. J Chem Phys. 125(13):133321–133327
22. Nest M, Remacle F, Levine RD (2008) pump and probe ultrafast electron dynamics in LiH: a computational study. N J Phys 10:025019
23. Kuleff AI & Cederbaum LS (2014) Ultrafast correlation-driven electron dynamics. J Phys B-At Mol Opt Phys 47(12)

24. Mignolet B, Levine RD, Remacle F (2012) Localized electron dynamics in attosecond-pulse-excited molecular systems: probing the time-dependent electron density by sudden photoionization. Phys Rev A 86(5)
25. Kus T, Mignolet B, Levine RD, Remacle F (2013) Pump and probe of ultrafast charge reorganization in small peptides: a computational study through sudden ionizations. J Phys Chem A 117(40):10513–10525
26. Mignolet B, Levine RD, Remacle F (2014) Charge migration in the bifunctional PENNA cation induced and probed by ultrafast ionization: a dynamical study. J Phys B-At Mol Opt Phys 47(12)
27. Tannor DJ (2007) Introduction to quantum mechanics. A time-dependent perspective. University Science Book, Sausalito
28. Kosloff D, Kosloff R (1983) A fourier method solution for the time dependent schroedinger equation as a tool in molecular dynamics. J Comput Phys 52:35
29. Beck AR, et al (2014) Attosecond transient absorption probing of electronic superpositions of bound states in neon: detection of quantum beats. New J Phys 16
30. Beck AR, Neumark DM, Leone SR (2015) Probing ultrafast dynamics with attosecond transient absorption. Chem Phys Lett 624:119–130
31. Medišauskas L, et al (2015) Initial electronic coherence in molecular dissociation induced by an attosecond pulse. Phys Rev A—At Mol Opt Phys 92(5)
32. Okino T, et al (2015) Direct observation of an attosecond electron wave packet in a nitrogen molecule. Sci Adv 1(8)
33. Stahel D, Leoni M, Dressler K (1983) Non-adiabatic representations of the 1-Sigma-U+ and 1-Pi-U states of the N-2 molecule. J Chem Phys 79(6):2541–2558
34. Konowalow DD, Hirschfelder JO (1961) Morse potential parameters for OO, NN, and NO interactions. Phys Fluid 4:637-
35. Spelsberg D, Meyer W (2001) Dipole-allowed excited states of N2: Potential energy curves, vibrational analysis, and absorption intensities. J Chem Phys 115(14):6438–6449
36. Hochlaf M, Ndome H, Hammoutene D, Vervloet M (2010) Valence-Rydberg electronic states of N−2: spectroscopy and spin-orbit couplings. J Phys B-At Mol Opt Phys 43(24)
37. Smith FT (1969) diabatic and adiabatic representations for atomic collision problems. Phys Rev179(1):111
38. Nikitin EE (1974) Theory of elementary atomic and molecular processes in gases. Clarendon, Oxford
39. Faist MB, Levine RD (1976) Collisional ionization and elastic-scattering in alkali-halogen atom collisions. J Chem Phys 64(7):2953–2970
40. Desouter-Lecomte M et al (1985) Nonadiabatic unimolecular reactions of polyatomic molecules. J Phys Chem 89:214
41. Desouter-Lecomte M, Dehareng D, Lorquet JC (1987) Constructing approximately diabatic states from LCAO-SCF-CI calculations. J Chem Phys 86:1429
42. Lorquet JC, Desouter M (1972) Excited states of gaseous ions. transitions to and predissociation of the C 2Sigma U+ state of N2+. Chem Phys Lett 16:136
43. Mead CA, Truhlar DG (1982) Conditions for the definition of a strictly diabatic electronic basis for molecular systems. J Chem Phys 77(12):6090–6098
44. Baer M (2006) Beyond born-oppenheimer: electronic nonadiabatic coupling terms and conical intersections. Wiley

On the Inter-Ring Torsion Potential of 2,2′-Bithiophene: A Review of Open Problems and Current Proposals

Olga A. Guskova

Abstract In this review, we describe the results of the most important theoretical studies of rotation around the bond connecting two thiophene rings in 2,2′-bithiophene. The review first summarizes the earlier studies (since late 1960s), in which the most energetically favourable conformations of the molecule have been characterized for the first time. It then examines the one-dimensional potentials of internal rotation calculated using semiempirical, Hartree-Fock (HF), post-Hartree-Fock methods, and Kohn-Sham density functional theory (throughout the 1990s to the present), as well the torsion potential functions $V(\theta)$. Three directions in recent studies are highlighted: (i) the development and testing of force fields supplemented with new parameters of torsion interactions in thiophene-containing materials, (ii) the application of new hybrid, exchange-correlation and long-range corrected functionals for describing the inter-ring rotation and through-space (non-valent) intramolecular interactions which stabilize either *cis* (*syn*)- or *trans* (*anti*)-rotamer, and (iii) the torsion-dependent properties of bithiophene-containing systems. A concluding part gives a brief outlook on further studies in the field and offers a road map for novel research directions that are required to realize new breakthroughs in thiophene-based device performance in the future.

Keywords Conjugated polymer · Conformation · *cis*- and *trans*-Rotamers · 2,2′-Bithiophene · Internal rotation barriers · Intramolecular interactions · Theoretical study

---

Electronic supplementary material The online version of this chapter (doi:10.1007/978-3-319-50255-7_13) contains supplementary material, which is available to authorized users.

O.A. Guskova (✉)
Institute Theory of Polymers, Leibniz Institute of Polymer Research Dresden, Hohe Str. 6, 01069 Dresden, Germany
e-mail: guskova@ipfdd.de

O.A. Guskova
Dresden Center for Computational Materials Science (DCMS), Technische Universität Dresden, 01062 Dresden, Germany

A. Tadjer et al. (eds.), *Quantum Systems in Physics, Chemistry, and Biology*, Progress in Theoretical Chemistry and Physics 30,
DOI 10.1007/978-3-319-50255-7_13

1 Introduction

The planarity of 2D aromatic structures is the most important feature of π-conjugated low-molecular substances. This property relies upon maximal overlap of *p*-atomic orbitals. For polymers consisting of such structural units, i.e. macromolecules with extended π-delocalization along the backbone, there is one specific characteristic, so called effective conjugation length ℓ_c, which characterizes a number of coplanar rings in the polymer chain (Fig. 1). The rotation of the rings, twisting or bending of the chain reduces the effective conjugation length, changing the electrical and optical materials properties, for instance, it increases the optical gap and decreases the polymer hole mobility.

The deviation from coplanarity can be treated as structural defect, since the conformational disorder leads to the localization/trapping of the charge carriers, which diminishes both their non-linear optical response and charge carrier mobility. Such defects can be permanent, resulting from crosslinking or mislinkages during synthesis [1] or temporary, which are induced by changes in external conditions (temperature, pressure, confinement, etc.). The packing of the chains in a crystalline solid-state, i.e. the presence of ordered adjacent molecules held together via intermolecular interactions, e.g. π-π stacking, generates more coplanar structures and enlarges the ℓ_c value. However, the conformations of the chain in solution or in amorphous phase are defined by increased entropy, which disorders the chains and leads to moderate conjugation as compared to the fully extended chains in crystalline phases.

The question about conformational properties of the thiophene-based polymers and copolymers is of particular current interest because these materials are being basic elements of organic electronic devices. These polymers have even been called

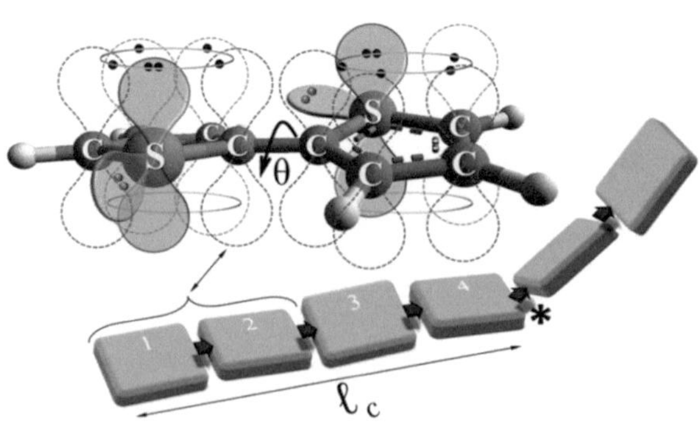

Fig. 1 Effective conjugation length ℓ_c of a conjugated polymer and the aromaticity of 2,2′-bithiophene (BT). Each sulphur atom has two lone electron pairs, but only one pair on unhybridized *p*-orbital overlaps with dumbbell-shaped p_z-orbitals of carbon atoms in the five-membered ring. The resulting π-bond is shown as ellipsoid above/below the rings. The torsion angle θ is the angle of the internal rotation. Star codes the conjugation break. All orbitals are shown as 2D-objects for simplicity

"the workhorse of plastic electronics" [2]. Experiments have found, that the conjugation length ℓ_c of thiophene derivatives varies from several [3] up to 96 [4] sequentially connected monomeric units (the simplified scheme of Fig. 1 shows only four monomeric units in conjugation). This range of values may be associated with regioregularity of the polymer chain, properties of the substituents in thiophene rings, their sterics, packing, etc. [5, 6].

In order to define the impact of all aforementioned factors on the conformational behaviour of thiophene-based chains, it is necessary to understand properties of the simplest reference system of 2,2′-bithiophene (2,2′-bithienyl), which comprises only two unsubstituted fife-membered rings (Fig. 1). The torsion angle θ here is a key variable of the internal rotation potential $V(\theta)$ for this "model system".

Figure 2 schematically depicts a modern representation of the torsion potential curve for 2,2′-bithiophene and provides the nomenclature, which will be used throughout the paper. There are five specific points on the energy profile. If the torsional angle $\theta = 0°$, which means that both sulphur atoms are on the same side from the bond connecting two rings, it is cisoid (*cis–*) conformation; in some papers a term "*synperiplanar*" (*syn–*) conformation has been used. The local minimum corresponds to *syn–gauche* conformation. A maximum on the potential energy curve at $\theta = 90°$ coincides with *gauche–gauche–* conformation, sometimes called a propeller-like state [7]. The global minimum belongs to *anti–gauche–* conformation, and finally, if $\theta = 180°$ the conformation is transoid (*trans–*) or "*antiperiplanar*" (*anti–*) one.

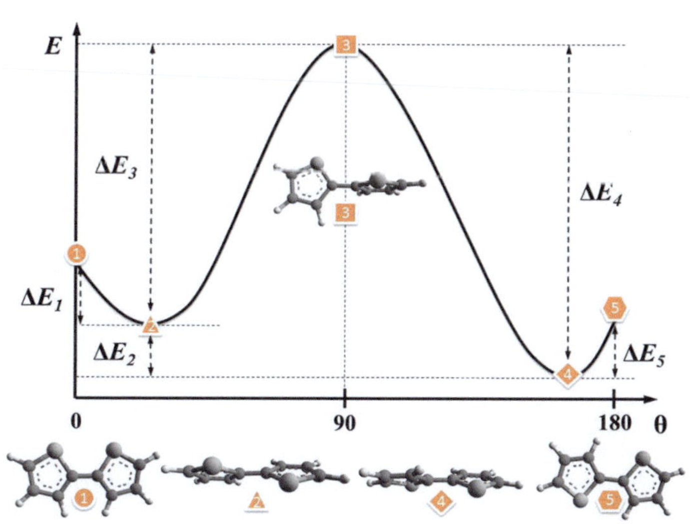

Fig. 2 Conformers of 2,2′-bithiophene following the Klyne-Prelog terminology: *1—cis* (*syn* or *synperiplanar*), *2—synclinal*, *gauche* or *skew*, *3—gauche-gauche*, *4—anticlinal*, *gauche* or *skew*, *5—trans* (*anti*, or *antiperiplanar*), and their location on the potential energy surface; the energies ΔE_1–ΔE_5 are the barriers of transitions for different conformational states

Various aspects of the problem of internal rotation in 2,2′-bithiophene have been discussed in experimental studies (Table 1, see Supplementary Information for more details) [8–22]: the population of *cis*- and *trans*- conformers in gas phase or the appearance of coplanar conformations in the solid state, the energy required for the *trans*-*cis*-interconversion, etc. For instance, it has been known since 1958 that bithiophene molecule in the gas phase has nonplanar geometry with twisting angle 146° [8]. In the solid state however both rings of the molecule appear to be anti-coplanar due to intermolecular forces [9, 10]. The conformations of partially oriented 2,2′-bithiophenes have been studied in NMR experiments [13, 14, 16, 18, 22]. The experiments suggested that molecule exists as a mixture of two conformers, which differ in energy [17] and that the interring rotation is sterically hindered [16–21].

Despite important achievements, it is still challenging to obtain the complete potential of internal rotation from purely experimental data: on the one side, only the most stable conformers can be experimentally detected, on the other side experiments bring additional variables—temperature, solvent, packing or confinement. Their influence on the conformational equilibrium can be significant, if not decisive. There are two major advantages to performing computer calculations of

Table 1 Experimental studies of conformational properties of 2,2′-bithiophene

The torsion angle θ for *trans*- or *anti-gauche* conformer	The torsion angle θ for *cis*- or *syn-gauche* conformer	Experimental method
Molecule is nonplanar, the rotational angle θ is 146°		Gas electron diffraction [8]
180° Molecule is planar, *trans*-conformer predominates	–	X-ray crystallography [9, 10]
148(3)°, 56(4)% The energy difference between two stable conformers is 0.2 kcal/mol	36(5)°, 44(4)%	Gas electron diffraction [11]
159° The torsional barrier at anti-planar conformer is 0.072 kcal/mol	21°	Fluorescence spectroscopy [12]
146°, 35–80%	34°	PMR, nematic solvent [13]
180°, 61 ± 3%	24 ± 4°	PMR, nematic solvent [14]
157 ± 10°	–	Electric birefringence (molecular Kerr constants) [15]
64%	36%	NMR, nematic solvent [16]
The barrier to the internal rotation is 5.1 ± 2 kcal/mol		
The energy difference of two stable conformers is 1.16 ± 0.13 kcal/mol		Fluorescence spectroscopy [17]
156.3–158.4°	22.6–29.5°	NMR, nematic solvent [18]
54%	46%	Terahertz spectroscopy [19]

the interring rotation in bithiophene. First of all, the analysis of the full potential energy surface including energetically unfavourable conformations and interconversion barriers can be performed (Fig. 2). Additionally, the factors influencing the stabilization of particular conformational state can be treated separately. Secondly, the impact of environment can be excluded or taken into account depending on the research question.

Knowledge of torsion potential is very important for different kinds of purposes, first of all for estimating the intrinsic properties of the system, because the structure of a molecule often has a great effect on its reactivity. This structural information is needed for the calculation of spectra, excited states, and the corresponding transition energies [23–28]. Additionally, such "model" molecules as 2,2′-bithiophene are regarded as a reference system for the description of more complex objects, e.g. homologues or substituted derivatives. On the other side, the study of "model systems" is an intermediate step toward atomistic molecular dynamics (MD) simulation of conjugated polymers and copolymers, because bithiophene fragment is a principal building block to extend the conjugated framework of different functional units. Classical microscopic modelling in turn provides an opportunity to investigate the samples on different time and length scales ranging from isolated molecules in vacuum to molecular ensembles. Accordingly, force field (FF) parameters for this simulation technique should be improved and adjusted to the appropriate condensed phase properties. For thiophene-containing chains this parameter set includes the equilibrium values of bonds (bond-stretching term) and angles (angle-bending term) and the force constants of respective potentials, the force constants for the potential of internal rotation $V(\theta)$, partial charges, parameters for improper dihedrals and out-of-plane motion of atoms, etc.

Nowadays, the parameters for these terms are obtained either from ab initio studies of small "model" compounds or from geometry characteristics calculated from the vibrational spectra of these "model" compounds in the gas phase (FTIR and Raman spectroscopy), complemented if necessary with ab initio quantum calculations (QM). The force-field based MD improved by additional parameters is extensively used for the prediction of persistence and conjugation length of polythiophenes, the conformation of the chains in solution, at the surfaces [29] or interfaces [30], for characterization of the self-assembly or crystallization [31] processes in thiophene-containing materials for plastic electronics.

2 Methods for the Calculation of Torsion Potential

For BT single rotors composed of two aromatic moieties linked by a single bond, two approximations to the rotation can be applied: rigid frame model and relaxed rotor. The first approach assumes that at each specified torsion angle θ (SCCS, Fig. 1), the geometry of the molecule is held constant without the freedom to vary any internal coordinate and each state, i.e. each point on the potential energy curve (Fig. 2) is computed as a single-point energy. In contrast, one-dimensional potential energy curve is obtained in relaxed rotor approximation by computing partially optimized energies at each constrained value of the central torsional angle θ, i.e. in this model all internal degrees of freedom are allowed to adjust during rotation except four atoms defining the torsional angle θ which remain fixed. Differences between two model potentials and their predicted torsional frequencies are sought to quantify the relaxation effects. The values of θ are varied usually from 0° to 180° or 360° by a 5°–30° step, i.e. by scaling the torsional angle along the potential energy surface.

The torsion potential in modern consistent force fields developed for organic molecules [32] is represented by a Fourier cosine series. One functional form is:

$$E_\theta = \sum_n V_n[1 + \cos(n\theta + \delta)], \tag{1}$$

where E_θ is the energy term from the deformation of the dihedral angle θ, V_n gives a qualitative information of the relative barrier to rotation, n is multiplicity associated with the torsional angle θ, δ—the phase factor; $n\theta$ shows the changes in rotational angle. The number of terms needed in the Fourier series depends on the complexity of the torsional potential and the desired accuracy. For modelling organic compounds three terms n are generally used in the series:

$$E_\theta = V_1(1 - \cos\theta) + V_2(1 - \cos 2\theta) + V_3(1 - \cos 3\theta). \tag{2}$$

The V_1 term is attributed to residual dipole-dipole interactions, van der Waals interactions, or any other direct interactions between atoms not otherwise accounted for. The V_2 term is attributed to conjugation or hyperconjugation, which is geometrically related to p orbitals. The V_3 term is attributed to steric or bonding/antibonding interactions.

The coefficients V_n mentioned above are fitted to calculated points of a potential function (Fig. 2). The Fourier expansion can include up to six terms, which corresponds to six V_n coefficients. If V_n does not exceed 0.1 kJ/mol (0.024 kcal/mol), the corresponding term of the expansion can be neglected [33]. The obtained coefficients for the torsion quartets are included into the force fields, and finally, the refined FFs are validated using a set of experimental data [34, 35].

Next chapters highlight the theoretical efforts aimed at understanding the internal rotation of bithiophene. To the best of our knowledge, the first publications have been published in late 1960s [36, 37], and today modern scientists report their results on conformations of 2,2′-bithiophene in high-impact scientific journals (Fig. 3). Thus, this paper describes a 50-year history of ideas and suggestions that have been put forward over the years. In these publications one can trace the development and evolution of electronic structure calculation methods of quantum chemistry as well.

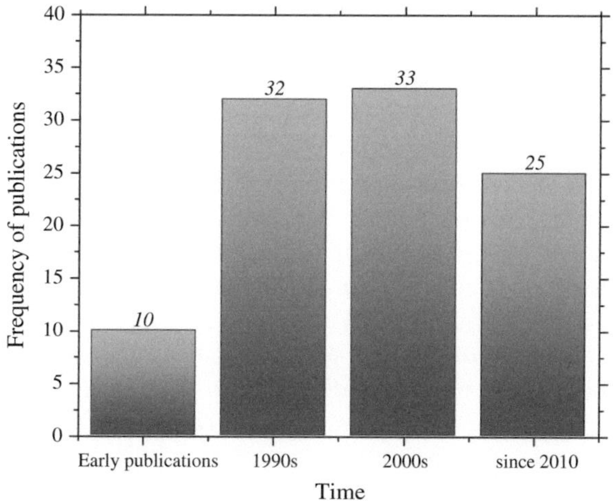

Fig. 3 Number of theoretical studies of torsion interactions and potentials of bithiophene over time (1964–2015)

3 A Review of Early Publications

A theoretical characterization of the bicyclic substances was initially motivated by the fact, that these molecules enter the group of important naturally-occurring compounds [36]. On the other hand, experimental structural investigations (electron diffraction, X-ray analysis and electric dipole measurements) [8, 9, 13, 15, 16] reported significant information, providing a reference point in order to test the existing methods of computational chemistry and to compare the experimental and calculated structural properties. Additionally, the relationship of stereochemistry of BTs and their electronic properties has aroused a great theoretical interest.

The computational elucidation of bithiophene conformers began in 1964 by Wachters and Davies [36]. Using the method of self-consistent molecular orbitals (SCMO), the authors have calculated the UV spectra of isomeric BTs assuming their planar structure. The angle of twist for 2,2′-bithiophene obtained from the analysis of charge distribution, dipole moment and oscillator strengths is found from 79° to 97°. The electronic spectra of isomeric biheterocycles, including 2,2′-bithiophene in *synperiplanar*, *antiperiplanar* conformations have been predicted by Trinajstić et al. [37, 38] applying the same semiempirical SCMO (Pariser-Parr approximation) method. The authors concluded that the spectral data are not sufficient enough to predict the most preferred conformation of the molecules.

The barrier to internal rotation of 2,2′-bithiophene was, so far as we aware, first demonstrated by Skancke in self-consistent theory (SCF) study [39]. Three contributions to the total energy of the molecule have been taken into account: the π-electron energy, the core energy and the van der Waals interactions for the series of angles from 0° to 180° with 15° step. From these calculations, the molecule is characterized as having nearly free rotation around the central bond, and the both planar forms are predicted as being less favourable as compared to twisted one. The minimal energy corresponds to $\theta = 60°$ with the energy barrier of 1 kcal/mol. Analysing the theoretical issues, the author emphasized that this result does not correspond to the experimental picture [8].

In attempt to rationalise the relative importance of two opposing factors influencing the conformational preferences, namely π-electron conjugation and nonbonded repulsions from both H· · ·H short contacts and lone pair on sulphur atoms, Galasso and Trinajstić applied the Extended-Hückel MO approach to isomeric bithiophenes [40]. In this paper, the dependence of extended Hückel energy versus angle of twist θ reveals the barrier height between *anti-gauche* and *syn-gauche* forms of 3.3 kcal/mol. The stabilisation energy of *anti-* over *synclinal* conformer of ca. 0.3 kcal/mol indicates that the *anticlinal* conformer is more favourable, unless the difference, as the authors pointed out, is negligible.

In papers by Abu-Eittah et al. [41, 42], the electronic absorption spectra of three isomers of bithiophene are studied within SCF localized orbital method on the planar and nonplanar geometries. The energy profile depicts that the *trans-* and *cis-*conformers are energetically indistinguishable and there is no barrier at $\theta = 0°$ and 180°. Both planar geometries are much more stable than the twisted one at $\theta = 90°$. The height of the potential barrier for the conversion from *cis-* to *trans-*conformer (7.14 kcal/mol) suggests that the 2,2′-bithiophene assumes a rigid or quasi-rigid all-planar geometry, which is explained by π-electron conjugation.

Ab initio Hartree-Fock and valence effective Hamiltonian calculations have been used by Brédas et al. [43]. The motivation for the calculation of the one-dimensional energy rotational profile (rigid rotor approximation), ionization potential, band gap and other electronic properties has been changed from fundamental aspect to the applied one. For the first time authors discussed material properties as a function of torsion angle between adjacent rings in bithiophene. Moreover, in this paper bithiophene was considered as a “model” system of polythiophene. It was found that the absolute minimum on the surface of potential energy corresponds to *antiperiplanar*-conformation, which is stabilized by 1.3 kcal/mol as compared to the *synperiplanar* one. The perpendicular orientation of two rings corresponds to TS on the potential energy surface with the barrier height of 4.2 kcal/mol.

In 1986 Barone et al. [33] analysed the conformational behaviour of bithiophenes by means of ab initio STO-3G and STO-3G* computations within relaxed rotor model, i.e. optimizing the most significant geometrical parameters for each value of the dihedral angle. It is found that energy minima occur at planar conformations and the *trans*-conformer is more stable by 0.88 kcal/mol, which corresponds to 14% *cis*-population at 300 K. The geometry optimization of the molecules, which was performed on planar conformers, shows only minor effects on the relative stabilities of both states, but generally reduces the transition barrier (Table 2); this value is in a good agreement with NMR data of Bucci et al. 5.1 ± 2 kcal/mol [16]. Furthermore, the potential function $V(\theta)$ was expanded in a Fourier series with six terms. Analysing the results, authors concluded that three terms are sufficient for an accurate description of the rotational potential. Another important conclusion is that the conformational behaviour of systems composed of

Table 2 Energetic barriers to the transition state (TS) at $\theta = 90°$ and stabilisation energies of *gauche*-conformers of 2,2′-bithiophene (for the definitions see Fig. 2)

The energy difference between the TS at $\theta = 90°$ and *anticlinal* minimum ΔE_4, kcal/mol	Stabilisation energy ΔE_2 *anticlinal* versus *synclinal*, kcal/mol	Details of the calculations (functionals and basis sets) and references
3.46	0.58	Extended Hückel MO approach [40]
7.14	0	SCF localised orbitals method [41]
4.2	1.3	HF/STO-3G [43]
5 ± 2	0.88	HF/STO-3G, STO-3G* [33]
4.68	0.33	Extended-Hückel MO [44]

two five-membered rings is always dominated by the conjugative contribution (V_2) to the torsional potential $V(\theta)$.

The conformational properties of bithiophene for the prediction of helical conformations of polythiophene in neutral and doped state in solution have been studied by Cui and Kertesz [44] using semiempirical modified neglect of diatomic overlap (MNDO) method and Extended-Hückel MO approach. It was observed that the first method fails in predicting the planar structures, and therefore cannot be used for characterization of the inter-ring rotation, whereas the second one gives the rotational barrier of 4.68 kcal/mol, which is in line with experimental findings [16].

Later on the relatively large magnitude of the rotational barrier obtained experimentally has been argued and associated with a possible deformation of the bithiophene geometry in a liquid crystal medium. It was also debated that the methods mentioned in Table 2 overestimate heights and positions of the peaks on the potential energy surface. Nevertheless taken together, these early papers provided the necessary groundwork, giving the first basic blueprint for further theoretical characterization of the internal rotation in bithiophene.

4 The 1990s Publications

Several innovative strategies in macromolecular synthesis and experimental characterization were developed by that time, which were responsible for the chemical revolution of the semiconducting polymer field. For instance, it was recognized that supramolecular packing of the conjugated polymeric chains induces changes in very important physical parameters. That is why the researchers went into the rational design of molecular structures via a proper choice of the monomers, of the side chain lengths and their sterics and composition. These and many others trends in experimental chemistry and physics reflected on the theoretical approaches, giving rise to computational design of small conjugated species and to further development of QM methods. It is interesting to note that at this period the joint experimental-theoretical publications appeared for the first time [11, 17], and that the theoretical studies themselves became more systematic.

Several research groups have made substantial contribution to this field: Leclerc and Durocher et al. [45–51], Brédas et al. [52–54], Hernández and López Navarrete

Table 3 Energetic barriers to the transition state (TS) at $\theta = 90°$ and stabilisation energies of *gauche*-conformers of 2,2′-bithiophene (for the definitions see Fig. 2)

The energy difference between the TS at $\theta = 90°$ and *anticlinal* minimum ΔE_4, kcal/mol	Stabilisation energy ΔE_2 *anticlinal* versus *synclinal*, kcal/mol	Details of the calculations (functionals and basis sets) and references
2.6	2	HF/3-21G [52]
4.26	1.2 (two planar conformers)	HF/STO-3G [63]
1.1	0.3	MNDO [55]
0.63	–	AM1 [64]
1.9	0.7	SCF, MIDI-4*, ACPF [59]
1.7	0.8	HF/3-21G and HF/DZP [53]
1.7 (*a*)	0.7 (*a*), 0.2 (*b*)	*a* – HF/6-31G*, *b* – experiment [11]
1.58	0.64	HF/3-21G* [66]
4.6 (*a*), 5.7 (*b*)	2.5 (*a*), 0.46 (*b*)	*a* – AM1/AMPAC, *b* – INDO/S [45]
1.4 (*a*), 1.7 (*b*)	0.6 (*a*), 0.7 (*b*)	*a* – HF/3-21G*, *b* – HF/6-31G** [57]
1.46 (*a*), 1.51 (*b*)	0.55 (a), 0.48 (*b*)	*a* – HF/6-31G*, *b* – MP2/6-31G* [70]
1.4	0.6	HF/3-21G(d) [67]
1.7 (*a*), 1.4 (*b*)	0.9 (*a*), 0.8 (*b*)	*a* – HF/6-31G*, *b* – HF/6-311G** [60]
1.336	0.717	HF/6-31G** [58]
0.5	0.2	AM1 [71]
0.5–1	1 (*a*, *b*), 0.6–0.7 (*c*)	*a* – MP2/6-31G(d), *b* – MP2/6-311++G(d,p), *c* – MP2/6-311++G(2d,2p) [61]
0.8–1.3	1.3–1.6	HF SCF/6-31G(d), 6-311++G (d,p) or 6-311++G(2d,2p) [61]
1.8–2.6	0.5–0.7	B3LYP/6-31G(d), 6-311++G(d, p) or 6-311++G(2d,2p) [61]
1.51 (*a*), 2.5 (*b*)	0.48 (*a*), 0.69 (*b*)	*a* – MP2/6-31G*, *b* – B3-PW91/6-31G* [54]
0.43	0.22	AM1; PM3 and INDO/S [47, 49]
1.5	0.55	HF/3-21G* [46]
1.69 (*a*), 4.3 (*b*)	0.7 (*a*), 1.2 (*b*)	*a* – HF/6-31G*, *b* – HF/STO-3G [49, 50]
1.7 (*a*), 1.3 (*b*), 2.7 (*c*)	0.8 (*a*), 0.5 (*b*), 0.6 (*c*)	*a* – HF/6-31G(d), *b* – MP2/6-31G(d), *c* – B3LYP/6-31G(d) [62]

et al. [55–58], Alemán and Karpfen et al. [59–62] and others [63–73]. Table 3 summarizes the data from some of these papers in chronological order of publication.

Leclerc et al. [45–51] have conducted the most extensive research on this topic in 1990s. They began with semiempirical treatment of conformational analysis of bithiophene molecule within the rigid-rotor approximation [45] taking into account the effect of dielectric continuum on the absorption spectra of the samples. In these calculation the authors predicted the coexistence of two conformers with $\theta = 140°$ and 40° in a vapour phase and more twisted conformations in non-polar media. A theoretical design of polythiophenes with specific thermochromic properties has been performed in 1997 and 1998 [46, 48]. The correlation of the rotational energy barrier and the blue shift observed at high temperatures for a series of bithiophenes decorated with some functional groups ($–C_2H_5$, $–OCH_3$ [51] or $–SCH_3$ [50]) was rationalized.

Not only molecules in their ground state, but also the conformations of bithiophene and its homologues in the first excited single and triplet states have been studied [47]. Performed semiempirical AM1 calculations reproduced both minima on the potential energy surface (Fig. 2), whereas PM3 method failed to predict the conformational behaviour of bithiophene in the ground state. In both S_1 and T_1 states, bithiophene molecule relaxed to planar conformation and the barriers to the internal rotation in these excited states were predicted to be higher, as compared to the previously observed value [45] for the molecule in the ground state.

The comparative study on the predictive power of semiempirical AM1, PM3 and ab initio STO-3G, HF/3-21G* and HF/6-31G* methods is described in the Ref. [49] by Leclerc et al. Authors have found that HF/3-21G* and HF/6-31G* give very similar potential energy surfaces with the global minimum corresponding to *anti*-conformer at $\theta = 146°$. By contrast, STO-3G method predicts the *anti*-conformer to be the most stable one, but in its planar geometry. In other words, this minimal basis set cannot elaborate the conformational issues precisely; semiempirical AM1 method can reproduce the picture only qualitatively.

In order to conclude about the persistent length and the length of conjugation for a set of conjugated polymers, Brédas and Heeger [52] performed the HF/3-21G ab initio study of dimeric molecules including bithiophene. Three conformers were considered $\theta = 180°$, 90° and 0°, for which the full optimization of geometry has been performed, and three conformers with $\theta = 90°$, 150° and 120° were calculated as single point energies. The *anti*-conformer was found to be energetically more stable by 2 kcal/mol with the barrier height of 2.6 kcal/mol at $\theta = 90°$ (Table 3). Restricted HF calculations with a double-zeta polarized basis set [53] showed the energy barrier for rotation to be smaller than those previously determined [52]. The effect of electron correlations taken into account in this study by MP2 (Møller–Plesset) method demonstrated lower stabilization energies for *syn*- and *anti*- conformers, suggesting the coexistence of both forms in solution or gas phase. A number of density functional theory (DFT) and MP2 calculations were carried out by Viruela, Brédas et al. [54] in order to assess the optimal conformations of bithiophene. Besides this issue, the methodological questions about the validity and limitations of different DFT functionals, namely S, VWN, LYP, B3LYP, B3-P86, and B3-PW91 were discussed. DFT-predicted optimized geometries were found to be very close to the MP2 geometries. In many cases a four-fold potential to the internal rotation was observed: the energy difference between conformers does not exceed 0.69 kcal/mol with the barrier height of 2.5 kcal/mol (e.g. for B3-PW91/6-31G*). These DFT functionals predicted nearly identical conformational properties of bithiophene. However, some functionals failed in precise predicting the energetics of the inter-ring rotation, for instance they increased the barrier through the propeller-like conformation and the conjugative planarization was sometimes overestimated, as compared to MP2 results.

López Navarrete et al. [55] reported on the decisive role of intra- and intermolecular interactions in biheterocyclic systems. The intramolecular forces have a great importance for the molecules in vacuum. It has been discussed that the rotational potential in this case has two terms: steric conflicts and a contribution from delocalization of p_z electrons. The first one favours the twisted conformation, the second one—the planarization of the backbone. The balance of these competing forces yields a structure with $0° < \theta < 90°$. If the molecules are packed forming an ordered lattice of the crystal, the intermolecular contribution predominates and the molecules are characterized by coplanar structure.

The problems of transferability of semiempirical quadratic force fields for polythiophenes have been addressed in Ref. [56] by Hernández, López Navarrete et al. The role of sterics in methyl-substituted dimers of bithiophene has been investigated in Ref. [57]. It was found that the rotational potentials for substituted bithiophene derivatives have absolute minimum for orthogonal conformers, i.e. the steric hindrance causes significant changes on rotation. The effects of surrounding medium (solvents: ethyl acetate and acetonitrile) on the electronic properties and internal rotation were resolved in Ref. [58]. As a result, all conformational equilibria were found to be rather insensitive to the dielectric permittivity of the solvent. Nevertheless, the *syn*-conformer with a larger dipole moment is preferentially stabilized in polar environment (acetonitrile).

A small energetic effect of geometry relaxation on the torsion potential was demonstrated by Karpfen et al. [59]. This work was continued in 1997 [61] in a systematic study of flexible rotor torsion potentials for different ab initio and DFT functionals (Table 3). The qualitative picture of inter-ring potentials was found to be correctly described at all levels of approximation, the energies of stabilization of *anti*-conformer over *syn*-form have been reproduced as well. However, the minima at *anti-gauche* and *syn-gauche* conformations were underestimated in DFT calculations, whereas the rotational barriers at propeller-like state were substantially larger. The authors obtained the dependence of both the position of minima and the well depths on the basis set applied, and concluded that additional studies of this problem are needed. The fitting of calculated potentials has been performed as well, and a four-parameter representation was concluded to be sufficient for the detailed description of the curves. Similar simulation setup has been applied by Alemán et al. [60, 62] in conformational studies of bithiophene and its substituted derivatives. In these publications the energy contributions V_n to the internal rotation have been analysed by barriers decomposition into a truncated Fourier expansion by varying the basis sets in computations. It was found that the well heights are strongly dependent on the basis set, indicating that the HF/6-31G* is the required minimum for such research tasks. Two years later [62] these authors expanded the list of the objects of the study to terthiophene and considered the 2D energy maps for the simultaneous rotation of flanked rings around the central thiophene unit.

A joint experimental-theoretical approach has been used by Distefano et al. [63, 66] in order to characterize the electronic and geometrical structure of thiophene oligomers and to extrapolate the results to polymeric chains. Dos Santos et al. [64, 65] identified the optimized conformations and the internal rotation of bithiophene using semiempirical calculations taking into consideration the solid-state contribution to the total energy. In papers [68, 69] the authors have described the electrical and spectroscopic or scanning probe microscopy studies on oligo-and polythiophene thin films supplemented by theoretical calculations of bithiophene conformations in vacuum or in partially hydrated state. Ortí et al. [70] described each V_n term in a Fourier decomposition of the torsional curve obtained at MP2/6-31G* level. Six terms were considered, but only V_1–V_4 terms were found to significantly contribute to the potential: V_1 causes mainly the *cis-trans*-energy difference, V_2 encompasses the planarization of the backbone, V_3 term contributes to destabilization of the planar *cis*-conformation and V_4 promotes the twisted form of the molecule.

Thus, systematic evaluation of the conformational preferences of bithiophene during this period consisted primarily in applying semiempirical [45, 64, 71], ab initio [11, 49, 50, 57, 72] and DFT [54, 61, 73] methods for theoretical design of intramolecular interactions via "chemical" decor of bithiophene backbone with short side chains. The general qualitative picture of the inter-ring rotation in the system under study has already been created by the end of the 90s.

5 Publications Since 2000

At the millennium, when A.J. Heeger, A.G. MacDiarmid, and H. Shirakawa received the Nobel Prize in Chemistry for discovery and development of conducting polymers, the computational prediction of structure and properties of organic semiconducting molecules has become a rapidly-expanding area of chemistry and physics. On the other side, a rapid increase in computational power has provided a great boost for further progress in theoretical modelling of oligo- and polythiophenes. Indeed, along with development of more sophisticated computational methods and schemes in quantum chemistry and the methods of computer simulation of polymers (force-field based MD), it became possible to investigate molecular systems containing thousands of atoms overcoming the problem of different time and length scales involved in the physics of charge transport in conjugated materials and solving the morphological issues. The relationship of these two aspects is of paramount importance for the modern materials science [2].

Table 4 Energetic barriers to the transition state (TS) at $\theta = 90°$ and stabilisation energies of *gauche*-conformers of 2,2′-bithiophene (for the definitions see Fig. 2)

The energy difference between the TS at $\theta = 90°$ and *anticlinal* minimum ΔE_4, kcal/mol	Stabilisation energy ΔE_2 *anticlinal* versus *synclinal*, kcal/mol	Details of the calculations (functionals and basis sets) and references
–	0.68 (*a*), 0.49 (*b*), 0.69 (*c*), 0.68 (*d*), 0.74 (*e*)	*a* – HF/6-31G*, *b* – MP2/6-31G*, *c* – B3LYP/6-31G*, *d* – BLYP/6-31G*, *e* – SVWN [74]
1.7 (*a*), 1.3 (*b*), 1.9 (*c*), 1.3 (*d*)	0.6 (*a*), 0.48 (*b*), 0.45 (*c*,*d*)	*a* – HF/6-31G*, *b* – MP2/6-311G**, *c* – B3LYP/6-311G**, *d* – CCSD/6-311G** [75]
1.7	0.5	MP2/cc-pVDZ [76]
1.88	0.63	MP2/cc-pVDZ [78]
1.72	0.68	HF, MP2/6-31G** [79]
5.37	0.96	DMol3, LDA/DND [81]
2.47 (*a*), 2.75 (*b*)	0.53 (*a*), 0.63 (*b*)	*a* – B3LYP/aug-cc-pVTZ, *b* – B3LYP/6-31G** [82]
1.94 (*a*), 1.64 (*b*)	0.36 (*a*), 0.49 (*b*)	*a* – MP2/aug-cc-pVTZ, *b* – MP2/6-31G** [18, 34, 82–84]
1.77 (*a*), 1.49 (*b*)	0.49 (*a*), 0.48 (*b*)	*a* – CCSD(T)/cc-pVDZ, *b* – CCSD/6-31G** [82]
2.8	0.25	CFF91 [84]
1.8 (*a*), 1.8 (*b*), 2.7 (*c*)	0.4 (*a*), 0.47 (*b*), 0.55 (*c*)	*a* – OB95/cc-pVTZ, *b* – O1B95/cc-pVTZ, *c* – OHandHB95 [85]
2.3 (*a*), 34 (*b*), 23 (*c*), 27 (*d*)	0.8 (*a*), 18 (*b*), ~ 0 (*c*, *d*)	*a* – B3LYP/6-311++G**, *b* – PCFF; *c* – CVFF, *d* – UFF [86]
0.2	0.198	PM5 [87]
1.42 (*a*), 0.84 (*b*), 2.3 (*c*), 2.5 (*d*), 1.8 (*e*), 1.9 (*f*)	0.62 (*a*), 0.29 (*b*), 0.47 (*c*), 0.5 (*d*), 0.26 (*e*), 0.49 (*f*)	*a* – HF/DIDZ, *b* – MP2/DIDZ, *c* – MPW1 k/MG3, *d* – BB1 k/MG3, *e* – MC3MPW, *f* – MC3BB [88]
1.7	0.4	MP2/cc-pVDZ [89]
2.1 (*a*), 3.2 (*b*), 1.4 (*c*)	0.4 (*a*), 0.6 (*b*), 0.4 (*c*)	*a* – CC2/cc-pVTZ [from Ref. 82], *b* – S/cc-pVTZ, *c* – LHF/cc-pVTZ [90]
3	0.7	B3LYP/SVP [91]
0.88–3.26	0.3–1.3	HF, B3LYP and MP2 [92]
2.2	1.1	HF/6-31G* [94]
1.9	0.9	[95] and CFF 91 [60]
1.84	0.69	B3LYP/6-311G(d,p) [96]
2.3	0.5	BHHLYP/cc-pVTZ [97]
2.2 (*a*), 2.5 (*b*), 1.55 (*c*,*d*)	0.48 (*a*), 0.72 (*b*), 0.31 (*c*), 0.24 (*d*)	*a* – B3LYP/6-311++G(2d,2p), *b* – B3LYP/aug-cc-pVDZ, *c* – MP2/6-311++G(2d,2p), *d* – MP2/aug-cc-pVDZ [19]

(continued)

Table 4 (continued)

The energy difference between the TS at $\theta = 90°$ and *anticlinal* minimum ΔE_4, kcal/mol	Stabilisation energy ΔE_2 *anticlinal* versus *synclinal*, kcal/mol	Details of the calculations (functionals and basis sets) and references
2.3	0.73	B3LYP-D2/6-31+G(d,p) [104]
2.7	0.6	B97-D/TZV(2d,2p) [105]
1.6	0.2	LMP2/cc-pVTZ(-f) [106]
1.8	0.8	B3LYP/6-31G* [107]
0.8	0	ONIOM/B3LYP/6-31++G(d,p) [108]
2.18	0.44	Focal point analysis/MP2/cc-pVQZ [113]
2.2 (*a*), 2.7 (*b*), 2.6 (*c*), 2.4 (*d*), 1.6 (*e*)	0.45 (*a*), 0.46 (*b*), 0.46 (*c*), 0.4 (*d*), 0.4 (*e*)	*a* – B3LYP, M062X/6-311+G(2df,p), *b* – CCSD(T)/cc-pVTZ, *c* – wB97x/6-311+G(2df,p), *d* – MP2/6-311+G(2df,p), *e* – wB97xd/6-311+G(2df,p) [114]
2.76	0.46	B3LYP/6-31G* [115]

Since 2000, the article's titles often contained such words as "bithiophene revisited", "re-examination of rotation", "improved force field" and "reinvestigated inter-ring rotation" emphasizing that the researchers were looking for a new solution to the old problem using new methods and approaches [74–116]. Table 4 summarizes the observables from some of these papers in chronological order of publication.

In particular, De Almeida et al. [74, 75] have applied a set of correlated MP2, MP4 and CCSD (coupled cluster with single and double excitations) ab initio and DFT methods for a detailed characterisation of potential energy surface of bithiophene (Table 4). Interestingly, these authors found out a good quantitative agreement between the potential energy surfaces predicted by means of MP4/cc-pVDZ and B3LYP/6-311G**; the latter can be referred to as the most computationally efficient procedure for studying the bithiophene.

The MP2/cc-pVDZ calculations for rigid rotation by Siebbeles et al. [76, 89] and within relaxed rotor approximation by Biskupič et al. [78] predicted the *cis-trans* interconversion behaviour of bithiophene (Table 4) with the stabilisation energy and the barrier height at $\theta = 90°$ similar to the findings by De Almeida et al. [75] and Pinto et al. [79, 80]. A slightly different picture was observed by Kearley et al. [81] in local density approximation (LDA) calculations with double numerical DND basis set. These authors noticed a quite systematic trend toward greater delocalization and a higher torsional barrier at $\theta = 90°$ for longer thiophene oligomers, whereas the impact of non-valent interactions remained unchanged.

A comprehensive study of atomistic force field parameters for thiophene-based conjugated polymers has been performed by Raos et al., which resulted in a series of papers [18, 34, 82–84, 100, 111]. These authors developed and validated several force fields for crystalline oligo-thiophenes and for the related polymeric chains, e.g. poly(3-n-butylthiophene) and poly(3-(S)-2-methylbutylthiophene). The first simulation work [82], which was devoted to the systematic description of the torsional energies of bithiophene, suggested that high-level ab initio methods are necessary for reliable results. The torsion potential calculated at MP2/aug-cc-pVTZ level was used subsequently in FF-based all-atom MD simulations [34, 84]. Later on, this torsion term for bithiophene was criticised for its lower rotational barrier at $\theta = 90°$ and a higher tendency to distort away from planarity toward an *anti-gauche* state [111]. In this new study, thiophene tetramer has been used as a "model" system for evaluation of inter-ring torsion potential, which has a more extensive conjugation than dithienyl [81, 96, 104, 111].

Many researchers have pointed out that force fields e.g. universal force field (UFF), consistent valence force field (CVFF) and polymer consistent force field (PCFF) (Table 4) lose the local minimum on the potential energy surface for *anti-gauche* conformation (dihedral angle is ca. 150°) and do not reproduce the magnitude of the potential barriers and the positions of the extrema. In some papers it was found, that extended electronic conjugation makes the inter-ring rotation correlated up to the second-nearest neighbour rings [108]. Thereafter, these force fields have been supplemented with a new set of interring torsion parameters (s. Sect. 2) for all-atom and coarse-grained simulations of various thiophene-containing systems [31, 35, 86, 91, 93–96, 98, 102, 104, 106–109]. For example, amorphous phases of oligothiophenes were modelled using DREIDING [31], PCFF [86, 93], CFF91 [95]. In these papers the torsional rotation/chain distortions and the oligomer length were identified as the main contributors to structural disorder in the samples. The behaviour of a single short poly(alkylthiophene) chain in aqueous and organic implicit solvents has been described by Kurnikova et al. [94] using improved and augmented version of AMBER force field. Oligothiophene liquid-crystalline phases [102] and crystals [30, 31, 35, 98, 107, 109] of poly (alkylthiophenes) have been modelled using original QM-based versions of torsional potential or adapted one from Ref. [34], aiming at microscopic description of mesophases, definition of transition temperatures (melting temperature T_m, glass transition temperature T_g) and conformational disorder in pure systems or at the organic/organic, organic/inorganic interfaces [29, 30].

Both rapid increase in computational resources and the application of efficient parallelization schemes and algorithms have advanced further progress in the development of new hybrid, exchange-correlation/long-range corrected DFT functionals as well as modern wavefunction-based post-HF methods, e.g. many-body perturbation theory, coupled-cluster theory, configurational interaction and multi-reference approaches for the system under study [85, 88, 90, 113, 114, 117–119]. Many authors stressed the significance of dispersion interactions, delocalization error and strong static correlation for the characterisation of small conjugated molecules using DFT, which might have particularly dramatic consequences for the properties predictions. Specifically, the absence of dispersion term leads to incorrect π-π thiophene-thiophene stacking or thiophene-substituent interactions, suffering from delocalization error gives overestimations of torsional barriers and results in an overstabilized planar structure.

For instance, the OPTX exchange-correlation functional and its hybrid versions have been tested for the description of torsion potentials in π-conjugated molecules by Sancho-García and Cornil [85]. The general conclusion is that the hybrid OB95 approach reproduces well both barriers for *anti-* to *anti-gauche* and for *anti-gauche* to TS at $\theta = 90°$, as compared to experimental findings for bithiophene [12]. Another example studied by Sancho-García [88] describes the advantages of multicoefficient MC3MPW and MC3BB methods over MPW1k and BB1k hybrid functionals, e.g. the latter methods overestimate the main barrier height and give larger locally calculated errors for the torsion potential of dithienyl. Localized HF potential in combination with LYP exchange term was used by Fabiano and Della Sala [90]. The performance of twelve conventional and modern DFT functionals with six basis sets have been tested in focal point analysis studies by Bloom and Wheeler [113]. The authors concluded that the most accurate predictions of conformational behaviour and the best performance give B2PLYP, 06-HF ωB97X, ωB97X-D and M06-HF regardless of the basis set employed.

Among wavefunction-based post-HF methods, coupled-cluster theory with single, double and perturbative triple excitations (CCSD(T)) has been used by many researchers as “gold standard” in calculations of nonbonded interactions and rotational barriers of 2,2′-bithiophene [74, 75, 82, 114] and in predictions of conductive and optical properties of thiophene oligomers [26b, 100, 105, 113, 114, 117–119].

The third aspect—focusing on torsion-dependent properties of materials—has received renewed attention [26a, 97, 99, 101, 103, 105, 110–112, 114, 115, 120–126]. As many early publications, recent papers shed light on optical, electronic and photovoltaic properties [26a, 97, 99, 103, 110, 115, 120, 122, 123, 125, 126], non-valent [101, 111, 112, 114, 124] or directional interactions in the solid state [105, 121] of *trans*-bithiophene.

Altogether, these studies described the correlation between the molecular conformations and different experimentally observed intriguing phenomena [e.g. 103, 115, 120], state-of-the-art methods of calculations helped to discover competing intra- and intermolecular CH-π interactions [111, 112] in substituted bithiophenes and P3HT and predict deactivation pathways in oligothiophene-based materials for photophysical applications [119]. A better understanding of key factors influencing the coplanarity and torsion interactions [114] also helped to systematize and control the effects of substituents, their sterics and positions in the ring, the backbone conjugation and the weak non-valent interactions. This knowledge may also facilitate interpretation of basic mechanisms (for instance the torsion dependence of binding energy [126]) behind the processes in oligo- and polythiophene thin films for flexible electronic devices.

6 Conclusions and Outlook

The theoretical studies of inter-ring rotation in 2,2′-bithiophene since the first reports in late 1960s have been reviewed. Thanks to great progress in chemical synthesis and characterisation techniques, and application of high-level quantum chemistry, this field of research has seen the tremendous advances over the past fifteen years. The current interest has soared in the structural properties of pristine molecule, in rational design of new derivatives and development of new methods and models. We have considered how the lessons learned from modelling of isolated bithiophene molecule are being used to adjust modern force fields for classical simulations of longer molecules and polymers containing bithiophene fragments. The problem of parameters transferability, e.g. dependence of the torsional barrier height on chain length [81, 96, 104, 111] has been described as well, although this topic is still being debated today.

Concluding the review, we highlight the actual scientific challenges and directions that are closely related to bithiophene inter-ring rotation. First of all, further ab initio calculations in theoretical and computational nanoscience are needed as guidance to identify and characterize newer bithiophene-based objects, e.g. functional torsional switches [127] and quinoidal bithiophenes [128]. Other alternatives that could be explored include the study of stability of global and local minima on the potential energy surface using pseudopotential Car-Parrinello molecular dynamics or theoretical search of the "perfect twist" in oligothiophene chains [129]. The large-scale MD simulation of thin film morphology and polymers with π-extended bithiophene donor units (for instance, diketopyrrolopyrrole- or benzothiadiazole-based chains) at various interfaces is now impossible without taking into account the twisting of the backbone and torsion correlations along the chain. Intermolecular interactions in bithiophenes solid state have recently attracted extensive attention, but many questions remain unresolved and many answers are still incomplete. Breakthroughs are expected to come in the field of biological organic electronics which utilizes bioinspired substances that contain short conjugated thiophene and bio-active segments, for instance for nanowire-based cell stimulation or for creation of innovative biocompatible materials [29a–c, 130].

The collected data will serve as a good introduction for both experimentalists and theoreticians and can provide some guidance to those focused on molecular properties of bithiophene-containing materials for advanced flexible electronics.

Acknowledgements The author is very grateful to Prof. Dr. Jens-Uwe Sommer and PD Dr. Marina Grenzer (Institute Theory of Polymers, Leibniz Institute of Polymer Research Dresden) for many useful discussions. The funding by ESF Young Investigators Group "*CoSiMa – Computer Simulations for Materials Design*" at DCMS, Technische Universität Dresden (http://dcms.tu-dresden.de/de/), Project "Boosting functional design of ambipolar organic semiconductors for advanced flexible electronics" is highly appreciated.

References

1. Bäuerle P, Pfau F, Schlupp H, Würtner F, Gaudl K-U, Balparda Caro M, Fischer P (1993) J Chem Soc Perkin Trans II 2:489–494
2. (a) Poelking C, Daoulas K, Troisi A, Andrienko D (2014) In: Ludwigs S (ed) P3HT revisited: from molecular scale to solar cell devices. Adv Polym Sci 265. Springer, Heidelberg, New York, Dordrecht, London, p 139. (b) Andrienko D (2015) In: Koch N (ed) Supramolecular materials for opto-electronics. RSC Smart Mater 12. The Royal Society of Chemistry, Cambridge, p 309
3. Ma J, Li S, Jiang Y (2002) Macromolecules 35(3):1109–1115
4. Kishino S, Ueno Y, Ochiai K, Rikukawa M, Sanui K, Kobayashi T, Kunugita H, Ema K (1998) Phys Rev B 58(20):R13430(1–4)
5. Marseglia EA, Grepioni F, Tedesco E, Braga D (2000) Mol Cryst Liq Cryst 348:137–151
6. Lukevics E, Barbarella G, Arsenyan P, Belyakov S, Pudova O (2000) Chem Heterocycl Comp 36(6):630–662
7. Gronowitz S (ed) (2009) Thiophene and its derivatives. John Wiley & Sons, New York Chichester Brisbane Toronto Singapore
8. Almenningen A, Bastiansen O, Svendsås P (1958) Acta Chem Scand 12(6):1671–1674
9. Chaloner PA, Gunatunga SR, Hitchcock PB (1994) Acta Cryst Sec C Cryst Struct Commun 50(12):1941–1942
10. Pelletier M, Brisse F (1994) Acta Cryst Sec C Cryst Struct Commun 50(12):1942–1945
11. Samdal S, Samuelsen EJ, Volden HV (1993) Synth Met 59:259–265
12. Takayanagi M, Gejo T, Hanazaki I (1994) J Phys Chem 98(49):12893–12898
13. Khetrapal CL, Kunwar AC (1974) Mol Phys 28(2):441–446
14. Ter Beek LC, Zimmerman DS, Burnell EE (1991) Mol Phys 74(5):1027–1035
15. Aroney MJ, Lee HK, Le Fèvre RJW (1972) Aust J Chem 25:1561–1564
16. (a) Veracini CA, Macciantelli D, Lunazzi L (1973) J Chem Soc Perkin Trans II 6:751–754. (b) Bucci P, Longeri M, Veracini CA, Lunazzi L (1974) J Am Chem Soc 96(5):1305–1309
17. Chadwick JE, Kohler BE (1994) J Phys Chem 98(14):3631–3637
18. Concistré M, De Lorenyo L, De Luca G, Longeri M, Pileio G, Raos G (2005) J Phys Chem A 109(44):9953–9963
19. Fedor AM, Allis DG, Korter TM (2009) Vib Spectrosc 49:124–132
20. Kimura Y, Katano Y, Tanaka S, Yoshinari T, Itoh H, Kuriyama Y, Nagasaka S (2010) Synth Met 160:1131–1135
21. Berardi R, Spinozzi F, Zannoni C (1994) Liq Cryst 16(3):381–397
22. Cinacchi G (2010) J Phys Chem A 114(31):8114–8118
23. Rubio M, Merchán M, Pou-Amérigo R, Ortí E (2003) ChemPhysChem 4:1308–1315
24. Zhou J, Yu W, Bragg AE (2015) J Phys Chem Lett 6(17):3496–3502
25. Andrzejak M, Witek HA (2011) Theor Chem Acc 129:161–172
26. (a) Prlj A, Curchod BFE, Corminboeuf C (2015) Phys Chem Chem Phys 17:14719–14730. (b) Prlj A, Curchod BFE, Fabrizio A, Floryan L, Corminboeuf C (2015) Phys Chem Lett 6 (1):13–21
27. Alparone A (2013) Adv Phys Chem 2013:1–8
28. Breza M, Lukeš V, Vrábel I (2001) J Mol Struct (Theochem) 572:151–160
29. (a) Gus'kova OA, Khalatur PG, Khokhlov AR (2009) Macromol Theory Simul 18 (4–5):219–246. (b) Gus'kova OA, Mena-Osteritz E, Schillinger E, Khalatur PG, Bäuerle P, Khokhlov AR (2007) J Phys Chem C 111(19):7165–7174. (c) Gus'kova OA, Khalatur PG, Bäuerle P, Khokhlov AR (2008) Chem Phys Lett 461(1):64–70. (d) Guskova O, Schünemann C, Eichhorn K-J, Walzer K, Levichkova M, Grundmann S, Sommer J-U (2013) J Phys Chem C 117(33):17285–17293. (e) Gus'kova OA, Schillinger E, Khalatur PG, Bäuerle P, Khokhlov AR (2009) Polym Sci Ser A 51(4):430–445
30. Liu T, Cheung DL, Troisi A (2011) Phys Chem Chem Phys 13:21461–21470
31. Alexiadis O, Mavrantzas VG (2013) Macromolecules 46(6):2450–2467

32. Maple JR, Hwang M-J, Stockfisch TP, Dinur U, Waldman M, Ewig CS, Hagler AT (1994) J Comput Chem 15(2):162–182
33. Barone V, Lelj F, Russo N, Toscano M (1986) J Chem Soc Perkin Trans II 6:907–910
34. Marcon V, Raos G (2004) J Phys Chem B 108(46):18053–18064
35. To TT, Adams S (2012) Nanosci Nanotechnol Lett 4(7):703–711
36. Wachters AJH, Davies DW (1964) Tetrahedron 20:2841–2849
37. Trinajstić N, Hinchliffe A (1968) Croat Chem Acta 40(3):163–169
38. Milun M, Trinajstić N (1973) Spectrosc Lett 6(6):329–346
39. Skancke A (1970) Acta Chem Scand 24:1389–1397
40. Galasso V (1972) Tetrahedron 28(16):4419–4429
41. Abu-Eittah R, Al-Sageir F (1978) Int J Quantum Chem 13(5):565–577
42. Abu-Eittah RH, Al-Sugeir FA (1985) Bull Chem Soc Jpn 58(7):2126–2132
43. Brédas JL, Street GB, Thémans B, André JM (1985) J Chem Phys 83(3):1323–1329
44. Cui CX, Kertesz M (1989) Phys Rev B 40(14):9661–9670
45. Belletête M, Leclerc M, Durocher G (1994) J Phys Chem 98(88):9450–9456
46. Di Césare N, Belletête M, Durocher G, Leclerc M (1997) Chem Phys Lett 275:533–539
47. Belletête M, Di Césare N, Leclerc M, Durocher G (1997) J Mol Struct (Theochem) 391 (1–2):85–99
48. Raymond F, Di Césare N, Belletête M, Durocher G, Leclerc M (1998) Adv Mater 10 (8):599–602
49. Di Césare N, Belletête M, Leclerc M, Durocher G (1998) A conformational study of ethyl-substituted bithiophenes. Synth Met 94:291–298
50. Di Césare N, Belletête M, Raymond F, Leclerc M, Durocher G (1998) J Phys Chem A 102 (16):2700–2707
51. Di Césare N, Belletête M, Leclerc M, Durocher G (1999) J Mol Struct (Theochem) 467:259–273
52. Brédas JL, Heeger AJ (1990) Macromolecules 23(4):1150–1156
53. Quattrocchi C, Lazzaroni R, Brédas JL (1993) Chem Phys Lett 208(1–2):120–124
54. Viruela PM, Viruela R, Ortí E, Brédas J-L (1997) J Am Chem Soc 119(6):1360–1369
55. López Navarrete JT, Tian B, Zerbi G (1990) Synth Met 38:299–312
56. Ramírez FJ, Hernández V, López Navarrete JT (1994) J Comput Chem 15(4):405–423
57. Hernandez V, López Navarrete JT (1994) J Chem Phys 101(2):1369–1377
58. Hernández V, López Navarrete JT (1996) Synth Met 76:221–224
59. Kofranek M, Kovář T, Lischka H, Karpfen A (1992) J Mol Struct (Theochem) 259:181–198
60. Alemán C, Julia L (1996) J Phys Chem 100(5):1524–1529
61. Karpfen A, Choi CH, Kertesz M (1997) J Phys Chem 101(40):7426–7433
62. Alemán C, Domingo VM, Fajarí L, Jiliá L, Karpfen A (1998) J Org Chem 63(4):1041–1048
63. Jones D, Guerra M, Favaretto L, Modelli A, Fabrizio M, Distefano G (1990) J Phys Chem 94 (15):5761–5766
64. dos Santos DA, Galvão DS, Laks B, dos Santos MC (1991) Chem Phys Lett 184 (5–6):579–583
65. dos Santos DA, Galvão DS, Laks B, dos Santos MC (1992) Synth Met 51:203–209
66. Distefano G, Dal Colle M, Jones D, Zambianchi M, Favaretto L, Modelli A (1993) J Phys Chem 97(14):3504–3509
67. Padilla-Campos L, Toro-Labbé A (1995) J Mol Struct (Theochem) 330:223–229
68. Benincori T, Brenna E, Sannicolo F, Trimarco L, Moro G, Pitea D, Pilati T, Zerbi G, Zotti G (1995) J Chem Soc Chem Commun 8:881–882
69. Porter TL, Minore D, Zhang D (1995) J Phys Chem 99(35):13213–13216
70. Ortí E, Viruela PM, Sánchez-Marín J, Tomás F (1995) J Phys Chem 99(14):4955–4963
71. Kiliç GB, Toppare L, Yurtsever E (1996) Synt Met 78:19–25
72. Forni A, Sironi M, Raimondi M, Cooper DL, Gerratt J (1997) J Phys Chem A 101 (24):4437–4443
73. Bongini A, Bottoni A (1999) J Phys Chem A 103(34):6800–6804

74. De Oliveira MA, Duarte HA, Pernaut J-M, De Almeida WB (2000) J Phys Chem A 104 (35):8256–8262
75. Duarte HA, dos Santos HF, Rocha WR, De Almeida WB (2000) J Chem Phys 113 (10):4206–4215
76. Grozema FC, van Duijnen PTh, Berlin YA, Ratner MA, Siebbeles LDA (2002) J Phys Chem B 106(32):7791–7795
77. Zhao J (2002) Synth Met 128:261–266
78. Lukeš V, Breza M, Biskupič S (2002) J Mol Struct (Theochem) 618:93–100
79. Diaz-Quijada GA, Weinberg N, Holdcroft S, Pinto BM (2002) J Phys Chem A 106 (7):1266–1276
80. Diaz-Quijada GA, Weinberg N, Holdcroft S, Pinto BM (2002) J Phys Chem A 106 (7):1277–1285
81. van Eijck L, Johnson MR, Kearley GJ (2003) J Phys Chem A 107(42):8980–8984
82. Raos G, Famulari A, Marcon V (2003) Chem Phys Lett 379:364–372
83. Raos G, Famulari A, Meille SV, Gallazzi MC, Allegra G (2004) J Phys Chem A 108 (4):691–698
84. Marcon V, Raos G, Allegra G (2004) Macromol Theory Simul 13:497–505
85. Sancho-García JC, Cornil J (2004) J Chem Phys 121(7):3096–3101
86. Zhang G, Pei Y, Ma J, Yin K, Chen C-L (2004) J Phys Chem B 108(22):6988–6995
87. Gombojav B, Yoshinari T, Itoh H, Nagasaka S-i, Kuriyama Y, Koyama K (2004) J Phys Soc Jpn 73(11):3166–3170
88. Sancho-García JC (2005) J Phys Chem A 109(15):3470–3475
89. Siebbeles LDA, Grozema FC, de Haas MP (2005) Warman JM 72:85–91
90. Fabiano E, Della Sala F (2006) Chem Phys Lett 418:496–501
91. Westenhoff S, Beenken WJD, Yartsev A, Greenham NC (2006) J Chem Phys 125:154903 (1–7)
92. Zhao J, Li P, Li Y, Huang Z (2007) J Mol Struct (Theochem) 808:125–134
93. Zhang G, Ma J, Wen J (2007) J Phys Chem B 111(40):11670–11679
94. Widge A, Matsuoka Y, Kurnikova M (2008) J Mol Graph Model 27(1):34–44
95. Vukmirović N, Wang L-W (2009) J Phys Chem B 113(2):409–415
96. Darling SB, Sternberg M (2009) J Phys Chem B 113(18):6215–6218
97. Macchi G, Medina BM, Zambianchi M, Tubino R, Cornil J, Barbarella G, Gierschner J, Meinardi F (2009) Phys Chem Chem Phys 11:984–990
98. Cheung DL, McMahon DP, Troisi A (2009) J Phys Chem B 113(28):9393–9401
99. Bouzzinea SM, Hamidi M, Bouachrine M (2009) J Appl Chem Res 11:40–46
100. (a) Arosio P, Moreno M, Famulari A, Raos G, Catellani M, Meille SV (2009) Chem Mater 21(1):78–87. (b) Moreno M, Casalegno M, Raos G, Meille SV, Po R (2010) J Phys Chem B 114(4): 1591–1602
101. Sánchez-Sanz G, Alkorta I, Elguero (2011) J Comput Theor Chem 974:37–42
102. Pizzirusso A, Savini M, Muccioli L, Zannoni C (2011) J Mater Chem 21:125–133
103. Tang S, Zhang J (2011) Int J Quantum Chem 111(9):2089–2098
104. Bhatta RS, Yimer YY, Tsige M, Perry DS (2012) Comput Theor Chem 995:36–42
105. Vujanovich EC, Bloom JWG, Wheeler SE (2012) J Phys Chem A 116:2997–3003
106. DuBay KH, Hall ML, Hughes TF, Wu C, Reichman DR, Friesner RA (2012) J Chem Theory Comput 8(11):4556–4569
107. Łużny W, Piwowarczyk K (2013) Synth Met 179:1–9
108. Bhatta RS, Perry DS (2013) Comput Theor Chem 1008:90–95
109. Bhatta RS, Yimer YY, Perry DS, Tsige M (2013) J Phys Chem B 117(34):10035–10045
110. Bjorgaard JA, Köse ME (2013) J Phys Chem A 117(18):3869–3876
111. Baggioli A, Meille SV, Raos G, Po R, Brinkmann M, Famulari A (2013) Int J Quantum Chem 113(8):2154–2162
112. Baggioli A, Famulari A (2014) Phys Chem Chem Phys 16(9):3983–3994
113. Bloom JW, Wheeler SE (2014) J Chem Theory Comput 10(9):3647–3655
114. Lin T-J, Lin S-T (2015) Phys Chem Chem Phys 17(6):4127–4136

115. Böckmann M, Schemme T, de Jong DH, Denz C, Heuer A, Doltsinis NL (2015) Phys Chem Chem Phys 17(43):28616–28625
116. Vikramaditya T, Saisudhakar M, Sumithra K (2015) J Mol Struct 1081:114–123
117. Sitkiewicz SP, Mikołajczyk MM, Toman P, Zaleśny R, Bartkowiak W (2013) Chem Phys Lett 566:67–70
118. Verma P, Perera A, Morales JA (2016) Mol Phys 114(3–4):547–561
119. Kölle P, Schnappinger T, de Vivie-Riedle R (2016) Phys Chem Chem Phys 18:7903–7915
120. Donohoo-Vallett PJ, Bragg AE (2015) J Phys Chem B 119(8):3583–3594
121. Kim J, Kwon O-P, Jazbinsek M, Park YC, Lee YS (2015) J Phys Chem C 119 (22):1258–12607
122. Andrzejak M, Orzeł Ł (2014) Phys Chem Chem Phys 16(12):5605–5612
123. Improta R, Ferrer FJ, Stendardo E, Santoro F (2014) ChemPhysChem 15(15):3320–3333
124. Einkauf JD, Mathivathanan L, de Lill DT (2016) J Mol Struct 1104:33–39
125. Oftadeh M, Moshfegh M, Abdallah HH (2016) Phys Chem Res 4(1):35–46
126. Bhatta RS, Tsige M (2014) Polymer 55(11):2667–2672
127. Claveau A (2015) Synthesis and investigation of photochromic torsional switches [Master's thesis]. [EPFL: Switzerland] École Polytechnique Fédérale de Lausanne
128. Casado J, Ortiz RP, López Navarrete JT (2012) Chem Soc Rev 41(17):5672–5686
129. Doval DA, Dal Molin M, Ward S, Fin A, Sakai N, Matile S (2014) Chem Sci 5 (7):2819–2825
130. Besar K, Ardona HA, Tovar JD, Katz HE (2015) ACS Nano 9(12):12401–12409

Theoretical Exploration of the Vibrational Structure and IVR of S_0 Thiophosgene at High Excitation Energies

Svetoslav Rashev and David C. Moule

Abstract In this work, we review our recent theoretical work on the vibrational level structure and vibrational mixing characteristics of very highly vibrationally excited S_0 thiophosgene (in the range of dissociation at ~20000 cm^{-1}), using our recently developed vibrational variational calculation method and a recently derived refined and modified quartic potential energy surface. We also present and discuss some new results, especially concerning the statistical treatment of the computationally obtained data that complement and extend the previously obtained picture. We compare the results from our calculations to the available experimentally measured dataset (derived from SEP and LIF spectra) and draw conclusions about the character and extent of vibrational mixing in this highly interesting range of vibrational excitation energies.

Keywords Vibrational spectra · Intramolecular vibrational energy redistribution (IVR) · Thiophosgene · Potential energy surface (PES) · Variational vibrational calculations

1 Introduction

Understanding the nature and characteristics of very highly excited vibrational states in polyatomic molecules, in the range of dissociation or even higher, is of practical interest for chemistry, but also of theoretical interest for statistical math-

---

Electronic supplementary material The online version of this chapter (doi:10.1007/978-3-319-50255-7_14) contains supplementary material, which is available to authorized users.

S. Rashev (✉)
Institute of Solid State Physics, Bulgarian Academy of Sciences, Tsarigradsko chaussee 72, 1784 Sofia, Bulgaria
e-mail: rashev@issp.bas.bg; svrashev@gmail.com

D.C. Moule
Department of Chemistry, Brock University, St. Catharines, ON L2S3A1, Canada

A. Tadjer et al. (eds.), *Quantum Systems in Physics, Chemistry, and Biology*, Progress in Theoretical Chemistry and Physics 30,
DOI 10.1007/978-3-319-50255-7_14

ematics (theories of ergodicity and chaotic behavior in a system with many degrees of freedom) and physics (spectroscopy, dynamical behavior of a highly excited quantum system). This problem, known by the acronym IVR (intramolecular vibrational energy redistribution), has been very extensively explored both experimentally as well as theoretically during the last half century [1–5]. In the theoretical studies both quantum and classical approaches have been widely employed. From a quantum perspective, the problem has a stationary and a dynamical aspect, both aspects being equivalent mathematically and quantum mechanically. Indeed, knowing the quantum stationary states of the system in the range of interest, its dynamics can in principle be extracted and it will depend exclusively on the preparation of the initial state. Thus, if we take a single quantum stationary state, it will stay constant and not evolve in time [2]. If we synchronously excite a few eigenstates, a quantum beating pattern will ensue. And if we excite simultaneously a large number of closely spaced states (a dense bunch) then we could observe relaxation dynamical behavior.

A most important feature of very highly excited molecular vibrational states is that they are very densely spaced, which is the major reason for the extreme difficulties in their theoretical exploration. The vibrational stationary quantum states of the system are the eigenstates of a full molecular vibrational Hamiltonian. A theoretical study of the vibrational quantum states consists in trying to derive these states as linear combinations in terms of a conveniently defined basis set of vibrational states, by diagonalizing the full molecular vibrational Hamiltonian in the chosen basis set. The density of basis states in the energy range of interest should practically coincide with the density of stationary states. It is very important to decide which basis set is to be used for representation of the eigenstates, because there are many possibilities which are determined by the coordinates used for description of intramolecular vibrational motion. (Using a basis set usually amounts to considering a system of coupled anharmonic oscillators.) In particular, one may use a basis set of localized modes (LM) wavefunctions, which is in fact our approach. These are defined e.g. to correspond to localized single bond stretches or to interbond angle bends. Alternatively, one may use normal mode (NM) basis set (using NM vibrational coordinates), as was done by many other authors. As a rule a NM basis wave function is usually more strongly delocalized over the molecule than a LM basis wavefunction. Therefore, e.g. by finding out that a particular eigenstate contains a large contribution from a NM state, this does not directly imply that a certain bond stretch is overexcited and hence the molecule would be likely to dissociate along this bond, as would be the case for the LM basis set.

Which are the possible situations that may occur for highly excited vibrational states of a polyatomic molecule? It is possible that all stationary states (eigenstates) could be expressed as linear combinations of practically all underlying vibrational basis states (whatever the chosen basis set). This would mean that even accessing a single eigenstate (if this were possible) this state would contain comparable contributions from all molecular vibrational types. This would be the case of complete vibrational mixing and IVR. However it might happen that there exists some sparse distribution of eigenstates (feature states), superimposed on the dense submanifold of uniformly vibrationally mixed states, which are expressible as linear combinations of only small part of the basis states and therefore contain predominant excitation in a certain vibrational mode that might turn out to be localized on a particular molecular bond. Such a situation should be of interest for the purposes of bond selective chemistry, if it were possible to physically access such localized states.

The older theories of chemical reactions, were based on the assumption that a large highly vibrationally excited polyatomic molecule, should generally have statistical distribution of the vibrational energy among all molecular vibrational modes. However recently Chowdary and Gruebele predicted [6], that there should exist many well isolated "feature" states close to dissociation limit whose density should grow with molecular size. This view is supported by the many spectroscopically observed sharp spectral lines in S_0 thiophosgene (TP), close to and above dissociation (at ~20000 cm^{-1}) [7, 8].

It is quite difficult to study theoretically the character of highly excited vibrational states (in the range of dissociation energies) in a molecule with more than three atoms. As mentioned above, the main difficulty comes from the extremely high density of vibrational states. It is impossible to diagonalize the relevant extremely large Hamiltonian matrices, in order to obtain the vibrational eigenstates. In order to overcome this difficulty we have designed a specific vibrational variational calculation method that is based on a specific search/selection procedure employed to select a subset of relevant symmetrically adapted vibrational basis states from the existing huge manifold of basis states of all symmetries. Using our method we have been able to perform converged vibrational calculations at very high vibrational excitation energies for the four atomics formaldehyde and thiophosgene [9–11].

During the last decades, thiophosgene was recognized as a suitable model molecular system for studying characteristic spectral and photophysical features of polyatomic molecules [12–18]. E.g., for studying intramolecular vibrational energy redistribution (IVR) in the ground electronic state S_0 [7, 16, 19, 20], as well as for exploring basic concepts of electronic radiationless transitions [12–15]. The vibrational spectroscopy of thiophosgene has been studied in detail [21–26]. Ultra-high resolution spectroscopy (using synchrotron radiation) was used to study the lower energy excitation range [25, 26]. However, despite these spectroscopic studies, the fundamental mode ν_3 could not be observed and measured directly and its frequency is unknown. In Table 1 are shown the 6 fundamentals of S_0 thiophosgene—frequencies and symmetries. Many sharp spectral features have been observed and even assigned (using LIF and SEP [7, 8, 20, 27]), ranging up to very high vibrational excitation energies of ~23000 cm^{-1} [7, 8]. These spectral features

Table 1 Fundamental vibrational frequencies (in cm^{-1}) for S_0 $^{35}Cl_2CS$ with symmetries

Mode (symmetry)	Shape	Experimentally observed
ν_3 (A_1)	Symmetric Cl–C–Cl bend	292.80
ν_6 (B_2)	Asymmetric Cl–C–Cl bend	305
ν_4 (B_1)	Out-of-plane bend	471.04267
ν_2 (A_1)	Symmetric C–Cl stretch	503.80679
ν_5 (B_2)	Asymmetric C–Cl stretch	819.614
ν_1 (A_1)	C–S stretch	1139.70

were shown to form regular progressions and could be assigned in terms of normal mode (NM) quantum numbers [7, 8, 19].

In our work [10] we obtained a refined quartic PES, that described successfully both the experimentally observed fundamental as well as some higher excited frequencies in S_0 TP [15, 16, 25, 26]. Using this field, in our subsequent work [11], we explored theoretically the vibrational structure of TP in the entire range of vibrational excitation energies up to dissociation. For this purpose our quartic PES has been adapted to display the correct asymptotic behavior when approaching the C–Cl bond dissociation energies, which was not true for the original quartic field [10].

In this work we review our results from our large scale calculations on the highly excited vibrational level structure and character in the range of dissociation at ~20000 cm^{-1}. We complement the previous results with new findings and draw some new and more sophisticated conclusions from them.

2 Basic Features of our Vibrational Calculation Method

Our Hamiltonian formalism for large scale vibrational calculations, has been already described in due detail [9, 10], therefore we give here only a rather brief account of its essential features. In our Hamiltonian description we use internal curvilinear vibrational coordinates. We use the exact kinetic energy expression T of Handy in such coordinates [28], while the potential energy surface (PES) V is presented as a quartic Taylor expansion of the coordinate shifts from equilibrium to fourth order, the three bond stretch coordinates being presented in Simons-Parr-Finlan (SPF) form [9, 10]. Based on this curvilinear coordinate representation, we define a vibrational (6D) basis set in separable product (symmetrized) form that represents the primitive state space (PS). Thiophosgene has C_{2v} symmetry. Our vibrational treatment of the molecule is fully symmetrized. This means that the employed 6D vibrational basis set functions will always belong to one of the C_{2v} point group symmetry species and are obtained as products of symmetrized components, expressed as suitable linear combinations of the 1D functions corresponding to the separate vibrational coordinates. The set of 1D basis functions along each curvilinear coordinate are defined as harmonic oscillators for the three stretches and the out-of-plane bend and as Legendre polynomials for the two in-plane S–C–Cl bends. These basis functions are subsequently prediagonalized to match as closely as possible the molecular vibrational levels. This prediagonalization serves to diminish the magnitude of the nondiagonal matrix elements in the full Hamiltonian matrix representation and thus reduce the dimensionality of the required Hamiltonian matrix itself. Since we take a large number of vibrational 1D basis functions along each coordinate (30 or more), our entire PS of 6D basis states is extremely large.

To build the Hamiltonian matrix of a particular vibrational calculation problem, we employ our specific search/selection procedure (described in due detail previously [9, 10]) that serves to select an optimally small however representative active space (AS) of basis states, from the huge available PS. A search/selection procedure is started from a particular basis state $|0\rangle$, chosen to be the best zeroth-order representation of the molecular vibrational state that we want to explore. A very important characteristic of our procedure is that it is symmetrically adapted, i.e. all selected basis states will necessarily have the same symmetry species as the chosen initial state of the search $|0\rangle$. Starting from $|0\rangle$, the search/selection algorithm proceeds to probe systematically (at steadily increasing distance in state space from $|0\rangle$), the available basis states from the primitive space, by calculating the relevant coupling matrix elements, and to select only those states, that have substantial (direct or indirect) coupling strength to $|0\rangle$. For this purpose we need a very fast method to calculate matrix elements. This we have achieved by pre-calculating specific arrays of 1D matrix elements corresponding to each 1D basis set and storing them in computer core memory, so that each actual calculation of a matrix element is reduced to a series of simple arithmetic operations and no numerical integrations. The search/selection procedure is controlled by a number of parameters whose values have to be fixed at the outset. Among them are the two major parameters, determining the scope and the quality of the search, C and R, that have been defined and discussed in detail in our previous work [9, 10]. R (a filtering parameter) is a window around the zeroth energy of the initial basis state $|0\rangle$ which acts to promote the selection of basis states that are located within the energetic R vicinity of $|0\rangle$ and to prevent the selection of too far displaced basis states. In our calculations of the lower excited vibrational states, where the accuracy must be higher, R has been usually chosen as ~1500 cm^{-1} [10], but it had to be reduced to $R = 100\ cm^{-1}$ when exploring the higher excited vibrational states, to make these calculations feasible. The second parameter C is the limiting value for selection of basis states, characterized by an evaluation function value f_k (ascribed to each selected basis state $|k\rangle$), accounting for its (both direct and indirect) coupling strength to $|0\rangle$ (a probed basis state for which the calculated $f_k > C$ is selected while if $f_k < C$, the state will not be selected) [9, 10]. According to the values chosen for the parameters C and R, the search/selection procedure will select a varying number of basis states, which results in varying precision of the calculation, according to the requirements. In the course of the implementation of the search/selection procedure, the Hamiltonian matrix H is built and stored in memory, which contains all diagonal and nondiagonal Hamiltonian matrix elements of the selected basis states from the AS. Our search/selection algorithm is symmetrized, therefore all selected basis states in the AS are of the same symmetry species as the initial state $|0\rangle$, which also contributes to the reduced size of the Hamiltonian matrix.

Next we tridiagonalize the obtained Hamiltonian matrix H employing a conventional Lanczos iteration without reorthogonalization [29, 30], starting with the vector $|0\rangle$. For a Hamiltonian matrix H with dimensionality N (where N is the dimensionality of the selected AS), we usually perform $3N$ Lanczos iterations to be sure to include all true eigenvalues of the H. Then we diagonalize the obtained tridiagonal Lanczos matrix using the routine *tqli*() from Numerical recipes [31], in slightly modified form. From this calculation we obtain the eigenvalues E_i and first components C_i (corresponding to basis state $|0\rangle$) of the eigenvectors $|i\rangle$ of H ($|C_i|^2$ is the spectral distribution in the explored range of vibrational excitations).

3 Potential Energy Surface

As mentioned above, the vibrational structure of S_0 thiophosgene has been studied repeatedly spectroscopically in considerable detail. The obtained ample experimental data have enabled the construction of several potential fields for the ground electronic state [19, 32, 33]. In our recent work [10], we designed a quartic PES that reproduced correctly the lower excited experimentally measured vibrational frequencies. First, using the *nwchem* suite of ab initio programs [34], we performed ab initio computations to determine a full set of quartic force constants. Next, using our vibrational variational method, we performed calculations to adjust the obtained ab initio force constants so that the calculated vibrational levels should most closely reproduce the experimentally measured vibrational frequencies. As a result of this adjustment we were able to achieve a satisfactory fit of the calculated with the experimentally measured vibrational frequencies, in the range up to ~3000 cm^{-1} [10]. The results from our vibrational calculations in this lower excited range and a comparison to the experimentally measured frequencies, are presented in Table 1 from our recent work [10], while the values of all refined quartic force constants (in units aJ and radian) are also given in the Supplement to this work as part of the c++ code for calculation of the PES.

Looking at the results from our calculations shown in Table 1 of Ref. [10] it is seen, that our refined quartic field [10] can correctly describe the vibrational level structure in S_0 thiophosgene up to ~5000 cm^{-1} of vibrational excitation energy, but this is not true for the higher vibrationally excited domain and in particular at energies close to dissociation (breaking of the C–Cl bond at ~20000 cm^{-1}). In Fig. 1 (broken line) is displayed the 1D cross section of our quartic PES along the C–Cl stretch coordinate, given by the following expression (in the quartic PES):

$$U_{C-Cl}(x) = Z\left\{\frac{(F_{11}+F_{55})}{2}\left(\frac{x}{x+x_0}\right)^2 + \frac{(F_{111}+F_{155})}{2\sqrt{2}}\left(\frac{x}{x+x_0}\right)^3 + \frac{(F_{1111}+F_{1155}+F_{5555})}{2\sqrt{2}}\left(\frac{x}{x+x_0}\right)^4\right\} \quad (1)$$

where F_{11}, F_{55}, F_{111}, F_{155}, F_{1111}, F_{1155}, F_{5555} are the relevant force constants from our refined quartic force field [10] (given in units aJ); Z = 1/(*hc*)—a factor to convert to wavenumbers; *x* is the shift of the C–Cl coordinate from the equilibrium value of x_0 = 1.73 A. This expression (1) dissociates at 114656 cm^{-1}, being much higher than the measured dissociation limit of the C–Cl bond, at ~20000 cm^{-1}. We need to change this expression in order to achieve a better representation of the true molecular dissociation limit. For this purpose, in our PES we have replaced expression 1 with another one, $U'_{C-Cl}(x)$ (2), that dissociates at the correct limit of *Dis* = 20000 cm^{-1}, however having the same values for the second, third and fourth derivatives in *x* at *x* = 0 as Eq. 1, in order to preserve the same calculated fundamental frequencies as the original quartic PES. Such an expression can take a Morse-like form, but having two additional parameters, to make it more flexible:

$$U'_{C-Cl}(x) = Dis\left[1 - \exp\left(-ax - bx^2 - cx^3\right)\right]^2 \quad (2)$$

The values of the parameters *Dis*, *a*, *b* and *c* in this expression have been determined as follows. The parameter *Dis* was set to 20000 cm^{-1}. We calculated analytically the 2nd, 3rd and 4th derivatives of expressions 1 and 2 with respect to *x*, at *x* = 0 (1st derivative is zero), and set the respective derivatives equal to each other. From the three equations thus obtained we determined the values of the parameters *a*, *b* and *c*, in the expression for $U'_{C-Cl}(x)$ (2), that correspond to the values of the force constants F_{11}, F_{55}, F_{111}, F_{155}, F_{1111}, F_{1155}, F_{5555} in expression 1:

$$\begin{aligned}
&a^2 = \mathrm{A}/(x_0^2 Dis);\ b = a^2 - \mathrm{A}/(a x_0^3 Dis) + \mathrm{B}/(2a x_0^3);\\
&c = (3\mathrm{A} + \mathrm{C})/(2a x_0^4 Dis) - b^2/(2a) + 1.5ab - 7a^3/24;\\
&\text{where } \mathrm{A} = \mathrm{Z}(F_{11}+F_{55})/2,\ \mathrm{B} = \mathrm{Z}(F_{111}+F_{155})/(2\sqrt{2}),\ \mathrm{C} = \mathrm{Z}(F_{1111}+F_{1155}+F_{5555})/4.
\end{aligned} \quad (3)$$

Next, we replace the expression 1 in our original quartic PES with the Eq. 2, thus ensuring correct behavior both at dissociation and at equilibrium. This expression 2 is displayed in Fig. 1, solid curve. This improved expression (2) was incorporated in our vibrational code (replacing expression 1) and used in all our subsequent calculations on the higher vibrationally excited states of S_0 thiophosgene (Fig. 2).

Of course, a reasonable question might arise, why aren't we using a more sophisticated PES expression for S_0 TP in our calculations, e.g. an n-mode representation, especially when trying to explore the very highly excited vibrational domain? Furthermore, the employment of quartic and up to sextic Taylor

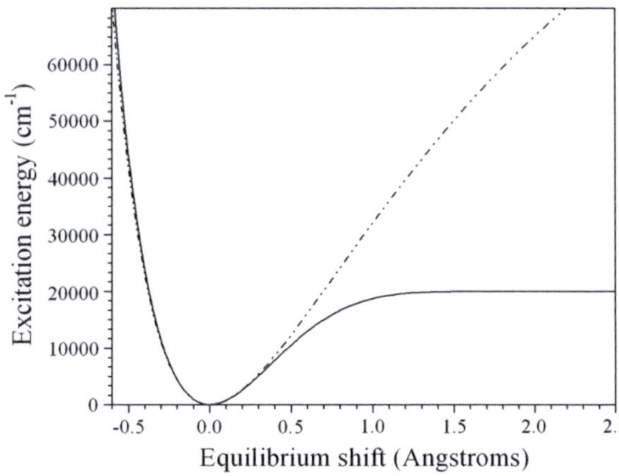

Fig. 1 1D cuts through the refined PES for thiophosgene [10], along the C–Cl stretch coordinate, for the original quartic expression (Eq. 1, *broken line*) and for the improved expression (Eq. 2, *solid line*)

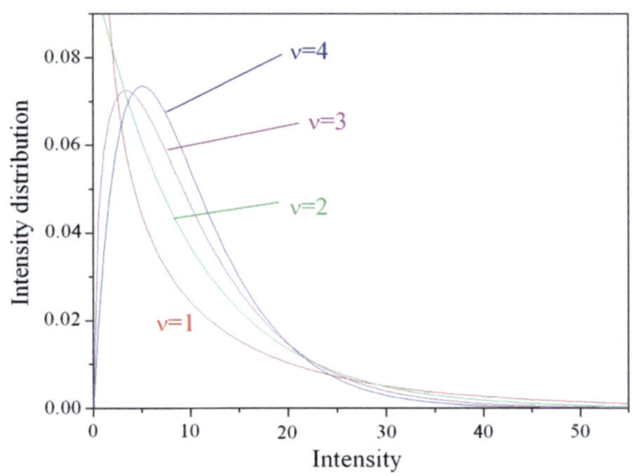

Fig. 2 Spectral distributions according to the Brody expression (Eq. 6) for several different values of the parameter ν

expansions at the equilibrium molecular configuration has been shown to lead to certain difficulties in the variational calculations, when going to the higher excited domain, where these expansions might have unphysical behavior [35, 36]. We can provide following arguments in support of our choice of a quartic PES. First of all the quartic expansion is very simple and convenient to work with (and the Morse-like update considered above makes it even more convenient), especially suitable for our code that requires very fast calculation of the matrix elements involved, vide supra. Such fields have been very widely employed in both perturbational as well as variational vibrational calculations by many authors and their utility discussed repeatedly, e.g. Refs. [33, 35–37], to name just a few. Next, the employed refined quartic PES in this work was shown to reproduce quite satisfactorily the experimentally measured frequencies of S_0 TP up to comparatively high vibrational excitation energies [10]. As a further test of the quality of our PES, we have calculated 1D cuts along all curvilinear coordinates (both stretching as well as bending) and found that they have regular behavior at the large displacements from equilibrium and do not display any unphysical characteristics that might lead to unwanted distortions of the results from the variational calculations. This is partially due to the employment of SPF coordinates for the stretches, as pointed out in [35, 36]. Here we do not claim to have built a global PES, which is a very demanding task, and for that reason of course we do not claim to be able to reproduce the exact vibrational level structure at excitation energies comparable to dissociation. This would be an unrealistic task. We just believe to have designed a suitable model PES that allows us to obtain a crude however realistic picture of the vibrational level structure and to judge about the basic features of effective level density, vibrational level mixing and dynamics at dissociation energies. As a result of all this we believe, that our updated quartic field is a good choice for the aims of the present work.

4 Vibrational Calculations at the Higher Vibrationally Excited States in S_0 Thiophosgene and the Relevant Vibrational Level Densities

Now we proceed to analyze the results from our large scale calculations performed in the extremely interesting highly excited energy range at ~20000 cm^{-1}, where dissociation of one or two C–Cl bonds can occur. We have performed calculations on five different feature states in this energy range: $1_{14}4_{12}$ at 20016 cm^{-1}, $1_{12}3_{1}4_{16}$ (20041 cm^{-1}), $1_{15}2_{1}3_{1}4_{8}$ (20059 cm^{-1}), $1_{16}3_{2}4_{6}$ (20077 cm^{-1}) and $1_{11}2_{1}4_{18}$ (20084 cm^{-1}). Careful convergence tests on the results from such calculations have been performed in our previous work [11] where it was shown that our calculations can reproduce qualitatively correctly the experimentally measured vibrational level structure and the obtained picture does not depend critically on some limitations to the values of our parameters introduced in the search to make the calculations feasible (in particular the reduced parameter value of R = 100 cm^{-1}).

In Figs. 3 and 4 are displayed the overall shapes $|C_i|^2$ of all 5 calculated spectra. For the calculations on each of these states (at parameter values $C = 10^{-12}$, R = 100 cm^{-1}), our search/selection algorithm selected an AS of N = 149512, 278922, 170256, 170194, 236201 basis states of A_1 symmetry, respectively. All spectral shapes are seen to be quite broad, which is a result of the choice of our initial state as a LM state instead of a NM state. These overall shapes are however of no particular interest and only the detailed level structure in a narrow spectral

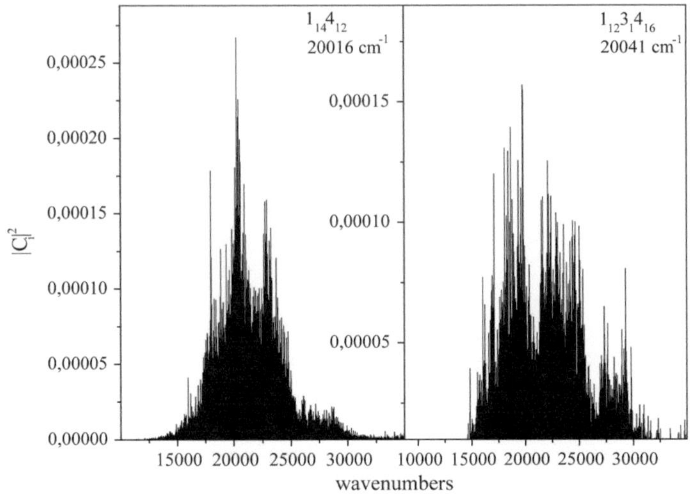

Fig. 3 Overall view of the calculated spectral distribution $|C_i|^2$ for the two states: $1_{14}4_{12}$ at 20016 cm^{-1} and $1_{12}3_14_{16}$ at 20041 cm^{-1}

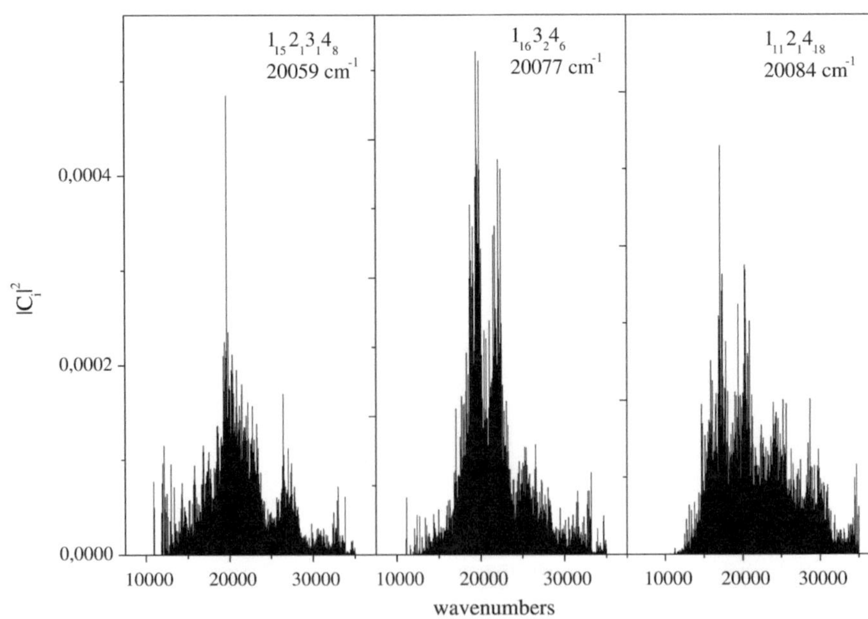

Fig. 4 Overall view of the calculated spectral distribution $|C_i|^2$ for the three states: $1_{15}2_13_14_8$ (20059 cm^{-1}), $1_{16}3_24_6$ (20077 cm^{-1}) and $1_{11}2_14_{18}$ (20084 cm^{-1})

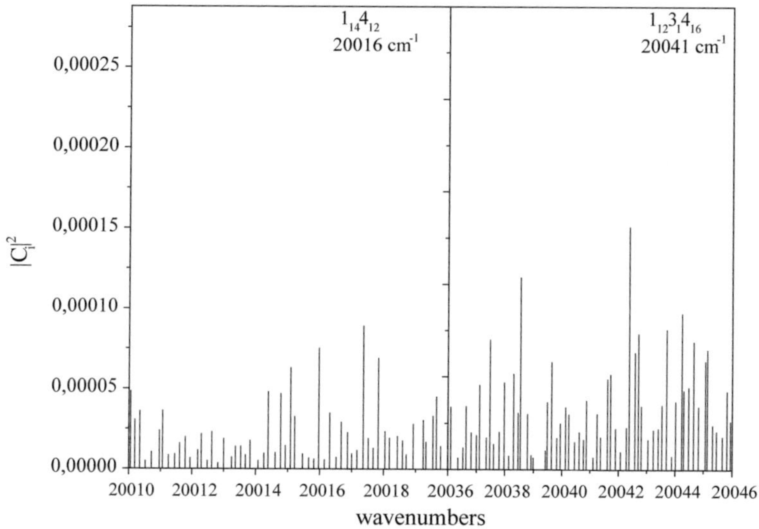

Fig. 5 Detailed view of a narrow interval of the calculated spectral distribution $|C_i|^2$ for the two states: $1_{14}4_{12}$ at 20016 cm^{-1} and $1_{12}3_14_{16}$ (20041 cm^{-1})

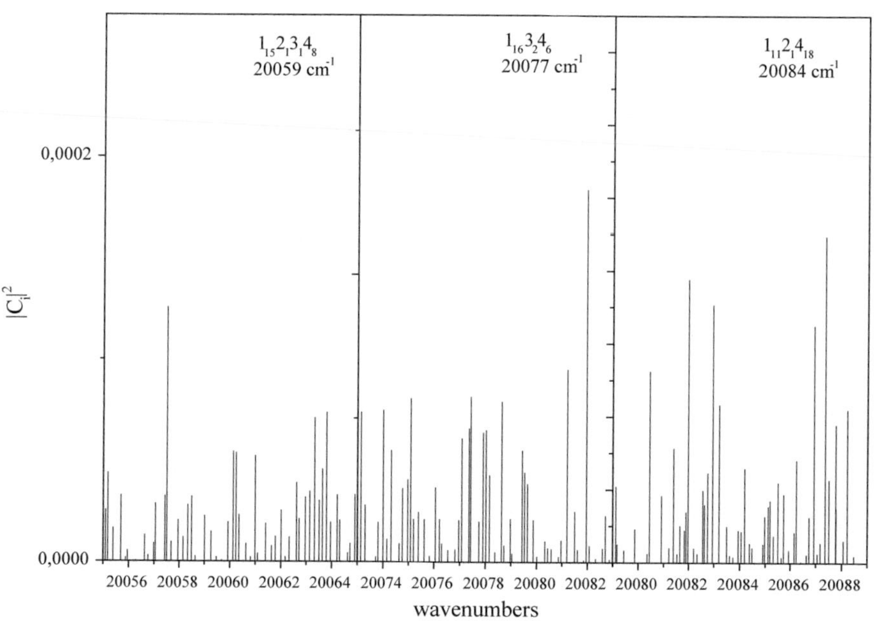

Fig. 6 Detailed view of a narrow interval of the calculated spectral distribution $|C_i|^2$ for the three states: $1_{15}2_13_14_8$ (20059 cm^{-1}), $1_{16}3_24_6$ (20077 cm^{-1}) and $1_{11}2_14_{18}$ (20084 cm^{-1})

range of relevance could yield valuable information about the mechanisms and extent of vibrational level mixing and IVR.

For this purpose, in Figs. 5 and 6 are displayed narrow sections of the same five calculated spectra to visualize their detailed eigenvalue structure in the range of interest. From these images, the effective vibrational level density involved in the vibrational mixing at the states of interest $|0\rangle = 1_{14}4_{12}$ (20016 cm^{-1}), $1_{12}3_{1}4_{16}$ (20041 cm^{-1}), $1_{15}2_{1}3_{1}4_{8}$ (20059 cm^{-1}), $1_{16}3_{2}4_{6}$ (20077 cm^{-1}), $1_{11}2_{1}4_{18}$ (20084 cm^{-1}), can be estimated by direct count. The effective vibrational level density is a very important measure for the extent of vibrational level mixing and IVR. It shows how much of the full available vibrational level density is effectively coupled to the initial basis state $|0\rangle$ through anharmonic and kinetic interactions in the molecular Hamiltonian. From Figs. 5 and 6 it is readily seen, that the selected effective level density for all cases is in the range of ~5–10 l/cm^{-1}. This observed density is certainly much lower than the entire available A_1 density in thiophosgene at $E_v \sim 20000$ cm^{-1}. Our estimation of the full available level density at $E_v \sim 20000$ cm^{-1}, calculated by an anharmonic level count, was $\rho \sim 1800$ l/cm^{-1}, and the A_1 level density was $\rho_{A1} \sim 450$ l/cm^{-1}. We point out that the effective level density $\rho_{eff} \sim 6$ l/cm^{-1}, of protected states found in this work, amounts to ~0.013 of the $\rho_{A1} \sim 450$ l/cm^{-1} value obtained by our anharmonic count.

The major result from our calculations as well as from the estimations by the authors [8] is that at high vibrational excitation energies in the range of dissociation there exists a sparse submanifold of vibrational "feature" states that are very well isolated from the remaining very dense background manifold (not involved in extensive vibrational mixing and IVR). These feature states are characterized by predominant excitation in the ν_1 (C–S stretch) and ν_4 (out-of-plane) modes. On the other hand the dense background level manifold consists of basis states that contain high excitation in the ν_2, ν_5 (C–Cl stretch) and ν_3, ν_6 (S–C–Cl bend) modes. To check whether this latter manifold is strongly vibrationally intermixed we have tried to perform a calculation, starting with the $|0\rangle = 5_{10}6_{10}$ state, located at ~10000 cm^{-1}, i.e. halfway to dissociation. This calculation could not be finalized at the usual parameter values of the search because the algorithm proceeded to select too many basis states, which is evidence for strong vibrational level mixing. In our recent work [11] it has been demonstrated that the existence and density of an isolated sparse manifold of feature states is not affected and limited by the scope and parameters of a particular calculation. The available background vibrational density is not involved in vibrational mixing with the feature states and it should remain completely "dark", assuming that the latter states are "bright". This may be regarded as a theoretical justification of both the theoretical predictions of Chowdary and Gruebele [6] about the existence of a considerable (and growing with molecular size) number of protected feature states at very high vibrational excitation energies as well as the experimental observation of numerous assignable feature states in S_0 thiophosgene.

5 Statistical Methods and Calculations

At the higher vibrational excitation energies, where the spectrum is too complex to be understood and assigned in terms of mode quantum numbers, it can better be analyzed in terms of statistical distributions [38, 39]. The major purpose of statistical treatments is to establish how strong and extensive is the mixing among the vibrational basis states (eigenfunctions of a separable Hamiltonian) and how close is the considered system to quantum ergodic character.

A sensitive measure for the extent of level mixing comes from the analysis of the level spacing distribution of nearest neighbors in the eigenstate spectrum. Indeed, it has been demonstrated that the eigenvalues of a separable Hamiltonian yield a Poisson distribution of nearest neighbor spacing [40]:

$$P_{\mathrm{P}}(x) = \frac{1}{\langle x \rangle} exp\left(- \frac{x}{\langle x \rangle}\right), \tag{4}$$

where $\langle x \rangle$ is the mean spacing between adjacent levels. This distribution has maximum at zero and falls off exponentially with increasing spacing x. This is the distribution, which should be obtained for the spacing of basis states energies, irrespective of the basis set used, because the basis states are eigenstates of a separable Hamiltonian. On the other hand, the eigenvalues of a strongly mixed Hamiltonian system possess the same level spacing distribution as the Gaussian orthogonal ensemble (GOE) of random matrices, which is characterized by "level repulsion" and has been approximated by Wigner with the expression [41]:

$$P_{\mathrm{W}}(x) = \frac{\pi x}{2\langle x \rangle^2} exp\left(- \frac{\pi x^2}{4\langle x \rangle^2}\right), \tag{5}$$

There is a useful expression, the Brody distribution [42] that covers the range between these two extremes of level coupling situations:

$$P_{\mathrm{B}}(x) = ax^q exp\left(- bx^{q+1}\right), \tag{6}$$

where $a = b(q + 1)$ and $b = \left\{ \frac{\Gamma\left[\frac{q+2}{q+1}\right]}{\langle x \rangle} \right\}^{q+1}$; for $q = 0$, the Poisson distribution is obtained while for $q = 1$, the Wigner distribution. By comparing our calculated distribution with the Brody distribution, we can determine the value of q which is suitable for describing our distribution and hence the extent of vibrational mixing and stochasticity which is greater, the closer is q to 1.

The spectral distributions $x = |C_i|^2$ of the calculated spectrum provide another good measure of stochasticity among highly excited vibrational levels. This is achieved by fitting the available spectral distribution to a chi-squared distribution with ν degrees of freedom [39]:

$$P_{\nu}(x) = \frac{\nu}{2\bar{I}\Gamma(q/2)}\left(\frac{\nu x}{2\bar{I}}\right)^{\frac{\nu}{2}-1} exp\left(-\frac{\nu x}{2\bar{I}}\right), \tag{7}$$

where $\bar{I}$ is the mean $|C_i|^2$ and Γ is the gamma function. A quantum system can be considered ergodic, if the observed spectral distribution resembles a chi-squared distribution with one degree of freedom $\nu = 1$, i.e. a Porter-Thomas distribution [38, 39]. In those cases, when the approximating chi-squared distribution has more than one degree of freedom $\nu > 1$, this should be considered as evidence for incomplete vibrational mixing and nonergodic IVR behavior. The larger the ν value, the further from ergodicity is the system although, as discussed by Polik et al. [39], ν cannot be regarded simply as the number of independent channels for decay of the initial state. In Fig. 2 is displayed for illustration the shape of the spectral distribution (Eq. 7) for several different values of the parameter ν.

We also note that in a recent paper Keshavamurthy [43] has examined the ergodic character of highly excited eigenstates in thiophosgene (at dissociation) using very sensitive measures of quantum ergodicity. He found the existence of localized eigenstates, suggesting nonstatistical energy flow dynamics. In particular, even feature states with considerable excitation in the ν_2 and ν_3 modes have shown deviations from ergodicity that were not among the spectroscopically observed features by Berrios et al. [8].

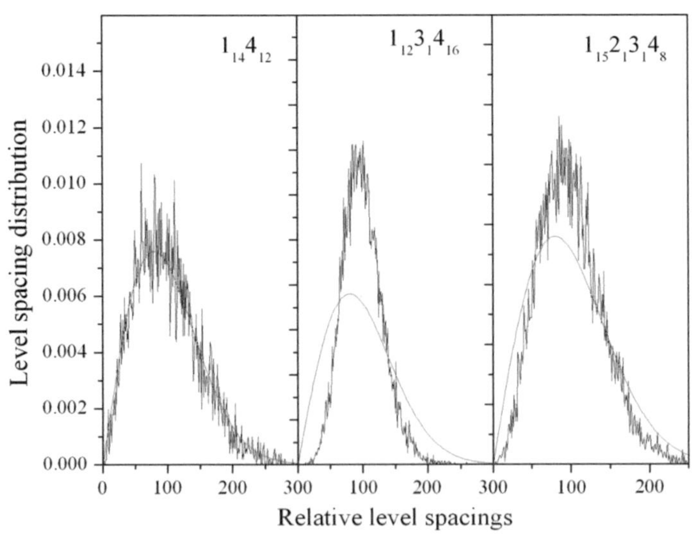

Fig. 7 Nearest neighbor level spacing distributions in the calculated spectra for the three states $1_{14}4_{12}$ at 20016 cm^{-1}, $1_{12}3_14_{16}$ at 20041 cm^{-1} and $1_{15}2_13_14_8$ (20059 cm^{-1}); to obtain this distribution, we have included in the analysis all calculated levels in the range 19000–21000 cm^{-1}; the mean level spacing for the three spectra is 0.173, 0.164 and 0.192 cm^{-1} respectively, which appear as 100 in the displayed *curves*; the Wigner distribution is plotted in *red* on each image for comparison

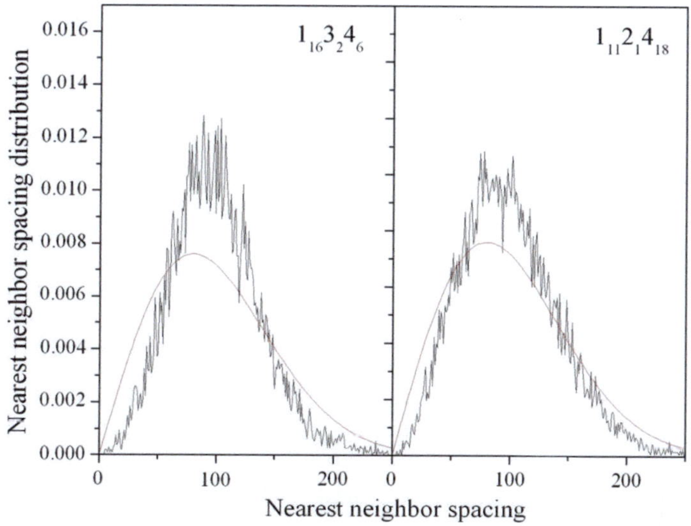

Fig. 8 Nearest neighbor level spacing distributions in the calculated spectra for the two states $1_{16}3_{2}4_{6}$ (20077 cm^{-1}) and $1_{11}2_{1}4_{18}$ (20084 cm^{-1}); the mean level spacing for the two spectra is 0.186 and 0.151 cm^{-1} respectively, which appear as 100 in the displayed *curves*; the Wigner distribution is plotted in *red* on each image for comparison

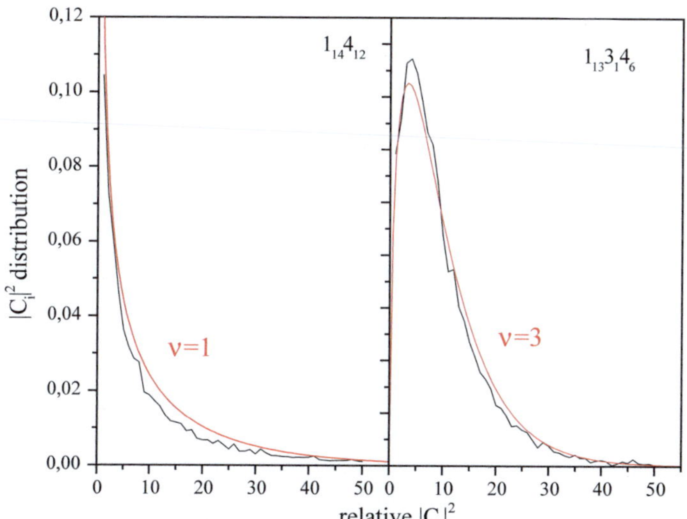

Fig. 9 Distribution of *line* $|C_i|^2$ values in the calculated spectra of the $1_{14}4_{12}$ (20016 cm^{-1}) and $1_{12}3_{1}4_{16}$ (20041 cm^{-1}) states, in units $I/\bar{I}$ (black); $\bar{I}$ is the mean $|C_i|^2$ value ($\bar{I} = 7.5 \times 10^{-6}$ and 1.4×10^{-5}, respectively), which appears on the graphs at $I = 10$. To calculate these distributions, we have included in the analysis all calculated levels in the range 19000–21000 cm^{-1} (20291 and 12208 states, respectively). Plots of model Brody distributions (Eq. 6) with $\nu = 1$ and $\nu = 3$ degrees of freedom respectively are also displayed in *red*

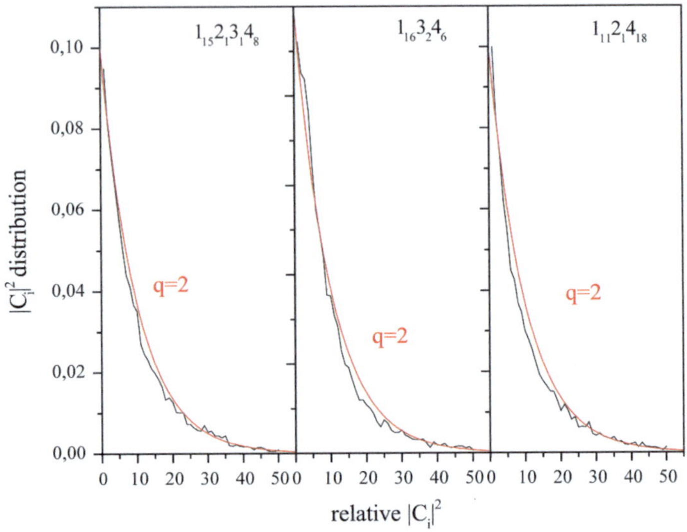

Fig. 10 Distribution of *line* $|C_i|^2$ values in the calculated spectra of the $1_{15}2_13_14_8$ (20059 cm^{-1}), $1_{16}3_24_6$ (20077 cm^{-1}) and $1_{11}2_14_{18}$ (20084 cm^{-1}) states, in units $I/\bar{I}$ (black); $\bar{I}$ is the mean $|C_i|^2$ value ($\bar{I} = 2.3 \times 10^{-5}$, 2.5×10^{-5} and 1.3×10^{-5} respectively), which appears on the graphs at $I = 10$. To calculate these distributions, we have included in the analysis all calculated levels in the range 19000–21000 cm^{-1}. Plots of model Brody distributions (Eq. 6) with $\nu = 2$ degrees of freedom respectively are also displayed in *red* on each graph

In Figs. 7 and 8 are presented the calculated nearest neighbor spacing distributions for all five calculated spectra of the states $1_{14}4_{12}$ (20016 cm^{-1}), $1_{12}3_14_{16}$ (20041 cm^{-1}), $1_{15}2_13_14_8$ (20059 cm^{-1}), $1_{16}3_24_6$ (20077 cm^{-1}), $1_{11}2_14_{18}$ (20084 cm^{-1}). To calculate these distributions, in each case we have included in the analysis all calculated eigenstates in the range 19000–21000 cm^{-1} (20291, 12208, 10420, 10754, 13274 states, respectively). The mean spacing for each distribution has been calculated as 0.173, 0.164, 0.192, 0.186 and 0.151 cm^{-1} respectively, which appears as 100 in the displayed curves. The Wigner distribution is plotted in red on each image for comparison. It is seen that this distribution resembles to a (more or less) satisfactory extent the calculated distributions which could be regarded as evidence for quite good and extensive vibrational level mixing and IVR among the levels belonging to the effective vibrational level manifold.

Next, in Figs. 9 and 10 are presented the line distributions $|C_i|^2$ in our five calculated spectra of the states $1_{14}4_{12}$ (20016 cm^{-1}), $1_{12}3_14_{16}$ (20041 cm^{-1}), $1_{15}2_13_14_8$ (20059 cm^{-1}), $1_{16}3_24_6$ (20077 cm^{-1}), $1_{11}2_14_{18}$ (20084 cm^{-1}), in units x/$\bar{I}$. $\bar{I}$ is the mean $|C_i|^2$ ($\bar{I} = 7.5 \times 10^{-6}$, 1.4×10^{-5}, 2.3×10^{-5}, 2.5×10^{-5} and 1.3×10^{-5} respectively), which appears on the graphs at x = 10. To calculate these distributions, we have included in the analysis all calculated eigenstates in the range 19000–21000 cm^{-1} (vide supra). As seen from Figs. 9 and 10, these spectral distribution could be best fit by the plot (red curve) of expression (Eq. 7) with $\nu = 1$

degrees of freedom for the $1_{14}4_{12}$ state, $\nu = 3$ degrees of freedom for the $1_{12}3_14_{16}$ state and $\nu = 2$ for the $1_{15}2_13_14_8$, $1_{16}3_24_6$ and $1_{16}3_24_6$ states. Thus only $1_{14}4_{12}$ shows convincing evidence for extensive vibrational mixing and ergodicity while the remaining four states appear to be intermixed to a lesser extent.

6 Conclusions

In this work, using our recently developed vibrational variational calculation method, we have carried out large scale calculations on five selected experimentally observed and assigned [8] feature (basis) states, in the higher range of vibrational excess energies in S_0 thiophosgene, where dissociation occurs: $1_{14}4_{12}$ (20016 cm^{-1}), $1_{12}3_14_{16}$ (20041 cm^{-1}), $1_{15}2_13_14_8$ (20059 cm^{-1}), $1_{16}3_24_6$ (20077 cm^{-1}) and $1_{11}2_14_{18}$ (20084 cm^{-1}). The employed PES for the calculations was originally based on a quartic Taylor expansion in terms of internal curvilinear vibrational coordinates, whose force constants have been determined in our previous work [10], to yield correct values for the lower excited experimentally measured frequencies. This PES expression has been partially updated, in order to describe correctly the C–Cl bond dissociation limit at ~20000 cm^{-1}. This improvement was required for a more realistic description of the highly excited vibrational states and their density.

Our general conclusion from the calculated spectral distributions at 20000 cm^{-1} and their detailed analysis is that there exists a sparse vibrational level submanifold of "feature" states (whose density is $\rho_{\text{eff}} \sim$ 5–10 l/cm^{-1}) that is superimposed on much denser background vibrational level manifold. The feature states contain predominant excitation in the ν_1 and ν_4 modes, while the dense background levels contain high excitations in the ν_2, ν_3, ν_5 and ν_6 modes. The "feature" vibrational levels from the sparse submanifold are rather strongly vibrationally intermixed among themselves but comparatively well isolated from the levels belonging to the dense background vibrational manifold. These latter levels also appear to be strongly vibrationally mixed among themselves. In close relation to this analysis we have examined and discussed in some detail the methods usually employed for estimation of vibrational level densities in molecules at high excitation energies and the difficulties that may be encountered. We have also performed statistical analysis on the calculated spectral distributions using both nearest neighbor spacing distributions as well as spectral distributions. From the calculated nearest neighbor spacing distributions and their comparison with theoretically derived expressions it was found, that our distributions were more or less satisfactorily described by the Wigner expression (Eq. 5), which was considered as evidence for extensive vibrational mixing and ergodic behavior. On the other hand, the calculated spectral distributions were less straightforward to analyze. In most considered cases they showed some deviations from completely stochastic behavior. Further calculations and analyses are required to gain more detailed and unambiguous information on the vibrational level mixing and IVR of highly vibrationally excited S_0 thiophosgene.

Acknowledgements This research was supported by the National Research and Engineering Council of Canada and by the Project INERA, under contract REGPOT 316309.

References

1. Felker PM, Zewail AH (1988) Adv Chem Phys 70:265
2. Uzer T, Miller WH (1991) Phys Rep 199:73
3. Gruebele M, Bigwood R (1998) Int Rev Phys Chem 17:91
4. Gruebele M, Bigwood R (2000) Adv Chem Phys 114:193
5. Gruebele M, Wolynes PG (2004) Acc Chem Res 37:261
6. Chowdary PD, Gruebele M (2008) Phys Rev Lett 101:250603
7. Strickler B, Gruebele M (2004) Chem Phys Phys Chem 6:3786
8. Berrios E, Pratt S, Tripathi P, Gruebele M (2013) J Phys Chem A 117:12082
9. Rashev S, Moule DC (2012) Spectrochim Acta A 87:286
10. Rashev S, Moule DC (2015) Spectrochim Acta A 140:305
11. Rashev S, Moule DC (2016) J Theor Chem Comput 15:1650005
12. Maciejewski A, Steer RP (1993) Chem Rev 93:67
13. Moule DC, Burling IR, Liu H, Lim EC (1999) J Chem Phys 111:5027
14. Moule DC, Lim EC (1999) J Chem Phys 110:9341
15. Fujiwara T, Lim EC, Judge RH, Moule DC (2006) J Chem Phys 124:124301
16. Sibert EL, Gruebele M (2006) J Chem Phys 124:024317
17. Jung C, Taylor HS, Sibert EL (2006) J Phys Chem A 110:5317
18. Keshavamurthy S (2012) J Chem Sci 124:291
19. Bigwood R, Milam B, Gruebele M (1998) Chem Phys Lett 287:333
20. Lee S, Engel M, Gruebele M (2006) Chem Phys Lett 420:151
21. Lowell RJ, Jones EA (1960) J Mol Spectrosc 4:173–189
22. Downs J (1963) Spectrochim Acta 19:1165
23. Brand JCD, Callomon JH, Moule DC, Tyrrell J, Goodwin TH (1965) Trans Faraday Soc 61:2365
24. Frenzel CA, Blick KE, Bennett CR, Niedenzu K (1970) J Chem Phys 53:198
25. McKellar ARW, Billinghurst BE (2010) J Mol Spectrosc 260:66
26. McKellar ARW, Billinghurst BE (2015) J Mol Spectrosc 315:24
27. Chowdary PD, Strickler B, Lee S, Gruebele M (2007) Chem Phys Lett 434:182
28. Handy NC (1987) Mol Phys 61:207
29. Cullum JK, Willowghby RA (1985) Lanczos algorithms for large symmetric eigenvalue computations, vols I, II. Birkhauser, Boston
30. Wyatt RE (1989) Adv Chem Phys 73:231
31. Press WH, Flannery BP, Teukolsky SA, Vetterling WR (1988) Numerical recipes in C. Cambridge University Press, Cambridge
32. Rashev S, Bivas I, Moule DC (2007) Chem Phys Lett 438:153
33. Davisson JL, Brinkmann NR, Polik WF (2012) Mol Phys 110:2587
34. Valiev M, Bylaska EJ, Govind N, Kowalski K, Straatsma TP, van Dam HJJ, Wang D, Nieplocha J, Apra E, Windus TL, de Jong WA (2010) Comput Phys Commun 181:1477
35. Csaszar AG, Mills IM (1997) Spectrochim Acta A 53:1101
36. Czako G, Furtenbacher T, Csaszar AG, Szalay V (2004) Mol Phys 102:2411
37. Fortenberry RC, Huang X, Yachmenev A, Thiel W, Lee TJ (2013) Chem Phys Lett 574:1
38. Porter CE, Thomas RG (1956) Phys Rev 104:483

39. Polik WF, Guyer DR, Miller WH, Moore CB (1990) J Chem Phys 92:3471
40. Berry MV, Tabor M (1977) Proc Roy Soc London Ser A 356:375
41. Wigner EP (1932) Phys Rev 40:749
42. Brody TA (1972) Nuovo Cimento Lett 7:482
43. Keshavamurthy S (2013) J Phys Chem A 117:8729

Why is the "Donor-Acceptor" Stretching a *sine qua non* in Understanding the Vibrational Signatures of Ionic Hydrogen Bonds?

Jake A. Tan and Jer-Lai Kuo

Abstract Vibrational coupling between O–O stretch and ionic hydrogen bond (IHB) stretch were carefully examined for the following proton-bound Zundel dimers: $H_5O_2^+$, $(MeOH)_2H^+$, and $(Me_2O)_2H^+$. Two-dimensional vibrational calculations were performed at MP2/aug-cc-pVDZ level using the method of discrete variable representation (DVR). It was found that in bare Zundel dimers, the first overtone of O–O stretch does not couple with IHB fundamental, ruling out the possibility of Fermi resonance. Meanwhile, the O–O stretch combination tone with IHB stretch strongly couples with the IHB fundamental, resulting in an observable combination band and a red shifted intense band in comparison with the zero-order picture. Breaking the inherent symmetry of these Zundel dimers by Ar tagging relaxes the selection rule against Fermi resonance which is dramatically exhibited by $ArH_5O_2^+$. These results demonstrate the critical role of the hydrogen bond "donor-acceptor" coordinate in modulating the IHB stretch frequency and intensity distribution. Such role was found to be sensitive to the presence of the messenger atom.

Keywords Ionic hydrogen bond • Infrared spectroscopy • Zundel

---

Electronic supplementary material The online version of this chapter (doi:10.1007/978-3-319-50255-7_15) contains supplementary material, which is available to authorized users.

J.A. Tan
Department of Chemistry, National Tsing-Hua University,
No. 101 Kuang-Fu Road, Section 2, Hsinchu 30013,
Taiwan, Republic of China

J.A. Tan • J.-L. Kuo (✉)
Molecular Science and Technology Program,
Taiwan International Graduate Program, Academia Sinica,
No. 128 Academia Road, Section 2, Nangang District,
Taipei 11529, Taiwan, Republic of China
e-mail: jlkuo@pub.iams.sinica.edu.tw

J.A. Tan • J.-L. Kuo
Institute of Atomic and Molecular Sciences, Academia Sinica,
No. 1 Roosevelt Road, Section 4, Da-an District, Taipei 10617,
Taiwan, Republic of China

A. Tadjer et al. (eds.), *Quantum Systems in Physics, Chemistry, and Biology*,
Progress in Theoretical Chemistry and Physics 30,
DOI 10.1007/978-3-319-50255-7_15

1 Introduction

The smallest and ubiquitous cation that any chemist could think of is the proton (H^+). Compared with hydrogen (H), which is the smallest atom, a proton is more than a thousand times smaller than the H's radius [1, 2]. Such combination of small radius and charge enables proton to possess a unique high charge density, causing electron pairs of surrounding molecules to be distorted. In an environment comprised of polar molecules, a proton never exists in isolation and readily associates with the neighboring molecule's electronegative atoms [2]. For example, in aqueous media, H^+ associates with water molecules to form $H^+(H_2O)_n$ clusters. The first two smallest clusters are H_3O^+ and $H_5O_2{}^+$. The former is termed as Eigen ion, while the latter is called Zundel ion and is likely to be a structure that serves as intermediate in proton transfer between two water molecules [3–5]. Furthermore, Zundel ion is an example of a strong ionic hydrogen bond (IHB).

Prior to understanding the kinetics of proton transfer, a knowledge on the vibrational regime is essential. It is possible to imagine that one of the vibrational modes has a significant O–H^+–O stretching, strong oscillator strength, and corresponds to proton transfer. Experimental measurements via direct absorption for such ionic clusters are not well resolved due to low ion density in the ion trap chamber [6]. To circumvent this limitation, Lee and co-workers [7, 8] pioneered a method which relies on dissociation as the consequence of photon absorption. These methods are known as infrared multiple photon dissociation (IRMPD) and "messenger technique" action spectroscopy.

In IRMPD, a high frequency mode is excited followed by a transfer of the excitation energy to the background states via intramolecular vibrational relaxation (IVR). Succeeding absorption of photon raises the energy of the cluster until it reaches a cut off value where dissociation can occur [6]. A powerful laser is required for this experiment to be successful. Meanwhile in "messenger technique" action spectroscopy, noble gas atoms like Ar and Ne are typically used as tagging agents [6, 8–11]. The dissociation of interest is along the van der Waals coordinate between the cluster and messenger. Compared with IRMPD, the energy requirement for dissociation along the van der Waals coordinate is smaller. As a result, messenger atoms can probe lower frequencies. Between the two techniques, utilization of messenger atom usually produces sharper peaks than that of IRMPD.

The challenge in understanding the vibrational signatures of Zundel type IHB clusters can be attributed to the following: (1) Strong anharmonic character of IHB due to the proton's large amplitude vibrations [12–18]. (2) Possible intermode coupling between overtones and combination bands of low frequency modes [14, 19–23], and (3) Possible intermode coupling between fundamental bands [14–16, 24, 25].

As an illustration, considering Zundel $H_5O_2{}^+$, there has been a general consensus that the IHB stretch vibrational signature is located around 1000 cm^{-1} [26–28]. Examination of $H_5O_2{}^+$ shows that there is only one possible way for proton to stretch, which intuitively leads to a single strong band around 1000 cm^{-1}. However, measurements on $NeH_5O_2{}^+$ reveal an intense doublet instead of a singlet centered at the said frequency [9, 29]. The assigning of the doublet was found to be so difficult that it made Bowman and co-workers [13] to construct a full-dimensional potential at the CCSD(T)/aug-cc-pVTZ level of theory and basis, which was eventually used by Meyer and co-workers [20–24] for their Multidimensional Time Dependent Hartree (MCTDH) calculation to unmask the doublet's identity. The nature of the doublet is an example of the second complication above. A dark vibrational background state which has characters of O–O stretch and water wagging happened to be near the IHB fundamental [19, 21, 30]. Along with a small energy mismatch and satisfying symmetry constraints, the dark background state couples with the IHB band leading to an intensity distribution, which accounts for the observed doublet.

In our previous work [25] on $(MeOH)_2H^+$, we found that the triplet signature [14, 31] centered around 1000 cm^{-1} is caused by strong intermode coupling between IHB stretch and flanking group motions namely: out-of-phase C–O stretch and out-of-phase in-plane Me- rock. Unlike $H_5O_2^+$, where coupling occurs between IHB stretch and combination bands of low frequency modes, a different form of coupling exists in $(MeOH)_2H^+$. In particular, $(MeOH)_2H^+$ exhibits the third complication above, where IHB couples with other fundamental modes. Given these facts, the question arises on how many degrees of freedom are needed to fully describe the nature of IHB stretch?

Among the three challenges mentioned above, the first two are always encountered in Zundel IHBs, while the third one is system dependent. One of the low frequency modes universal among Zundel ions is the stretching along the hydrogen bond "donor-acceptor" coordinate. In one of the reviews of Marx [32], the one-dimensional double minimum potential, which is often used to describe hydrogen bond mediated proton transfer, is scrutinized. The shape of the proton transfer coordinate is greatly affected by the distance of the "donor-acceptor" coordinate. For short distances, the potential curve is symmetrical and single well, implying a very strong hydrogen bond. As the "donor-acceptor" coordinate elongates at intermediate distances, the potential curve morphs to a low-barrier double well. Further elongation leads to a high barrier double well. Such findings suggest that at least two degrees of freedom are needed to qualitatively describe IHB stretch.

The variation in the potential experienced by the shared proton relative to the "donor-acceptor" coordinate is consistent with our findings with Zundel type amine dimers [33]. In these cases, the "donor-acceptor" coordinate is the N–N axis. By varying the degrees of methylation from $(NH_3)_2H^+$ to $(Me_3N)_2H^+$, the equilibrium N–N distance increases as a consequence of steric effects. A rigid scan along the proton transfer coordinate reveals that the barrier increases with the degree of methylation. In terms of the IHB vibrational signature, a counter-intuitive red shift was observed from $(NH_3)_2H^+$ to $(Me_3N)_2H^+$. Furthermore, a two-dimensional potential (IHB and N–N stretches) shows a transition from single well to double well for the IHB coordinate as the N–N distance is varied. Overall the 2D potential is a double well. In addition to this, N–N stretch has a unique ability to induce a red shift on the IHB stretch [33].

In this chapter, we report our investigations on another homologous series, namely: $H_5O_2^+$, $(MeOH)_2H^+$, and $(Me_2O)_2H^+$. This time we will examine the origin of O–O stretch's capability in inducing a red shift on IHB stretch; the reason why the combination band of IHB stretch with O–O stretch is intense; and the possible selection rules that allow overtone Fermi resonance and combination band as a consequence of symmetry breaking by Ar tagging. Though our calculations in this work are limited to two-dimensional cases, we hope that our insights will be interesting for the reader. For higher level and higher dimensional treatments, the reader is advised to check the references cited in this work.

2 Calculation Methods

2.1 Geometry Optimization and Harmonic Frequency Calculations

Optimization for the geometry of the singlet ground electronic state of $H_5O_2^+$, $(MeOH)_2H^+$, and $(Me_2O)_2H^+$ were performed at MP2/aug-cc-pVDZ level using Gaussian 09 [34, 35]. From the optimized structure, a relaxed scan along the dihedral angle was performed. It was found out that there is only one stable conformer for $H_5O_2^+$ and $(Me_2O)_2H^+$, while $(MeOH)_2H^+$ has two conformers related by a twist along the ∠HOOH dihedral angle. By virtue of Boltzmann distribution, the ratio of the more stable conformer to the less stable at 68 K is 96:4. As a result, we only considered the more stable conformer of $(MeOH)_2H^+$ in this study. A schematic of the optimized structures can be found in Fig. 1a–c. In all cases, the proton lies at the midpoint of the O–O axis—a strong feature of Zundel IHBs [32].

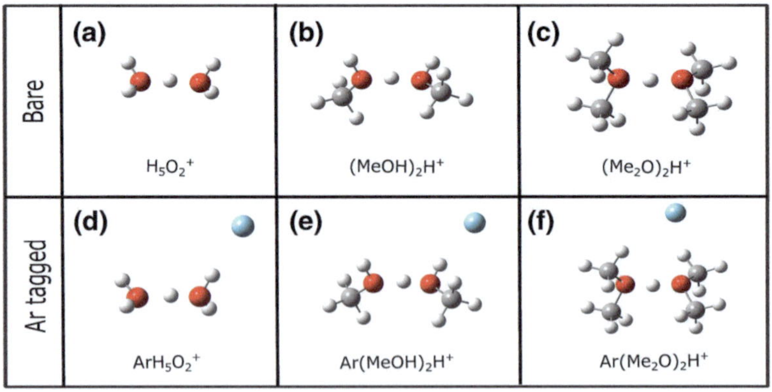

Fig. 1 Schematic representations for the minimum structures of $H_5O_2^+$, $(MeOH)_2H^+$, $(Me_2O)_2H^+$, and their Ar tagged counterparts. The point groups of all bare and Ar tagged structures are C_2 and C_1 respectively

From the minimum structure, normal mode coordinates were obtained by diagonalization of the Hessian matrix. This was done by running a frequency job in Gaussian 09 [34], which yields harmonic frequencies. Normal modes corresponding to O–O and IHB stretches were selected upon visualization of the frequency calculation results. Lastly, similar calculations were performed for the Ar tagged species: $ArH_5O_2^+$, $Ar(MeOH)_2H^+$, and $Ar(Me_2O)_2H^+$ to investigate the possible role of symmetry breaking. Their minimum structures are also shown in Fig. 1d–f.

2.2 Anharmonic Vibrational Calculations

The harmonic frequency calculations described in Sect. 2.1 lack two important aspects. First, it did not account the anharmonic nature of O–O stretch and IHB stretch. Second, it did not include any coupling terms. To account for the anharmonicity, we performed a one-dimensional (1D) anharmonic vibrational calculation for each of these modes. We chose to use normal mode (NM) coordinates to describe their vibration and will be called as NM1D. The 1D Hamiltonian operator for O–O stretch and IHB stretch can be written as follows:

$$\begin{aligned} H_{\mathrm{O-O}}^{\mathrm{NM1D}} &= \frac{-\hbar^2}{2\mu_{\mathrm{O-O}}}\frac{d^2}{dQ_{\mathrm{O-O}}^2} + \sum_i \alpha_i Q_{\mathrm{O-O}}^i \\ H_{\mathrm{IHB}}^{\mathrm{NM1D}} &= \frac{-\hbar^2}{2\mu_{\mathrm{IHB}}}\frac{d^2}{dQ_{\mathrm{IHB}}^2} + \sum_j \beta_j Q_{\mathrm{IHB}}^j \end{aligned} \qquad (1)$$

Where $\hbar$ is the reduced Planck's constant; Q_{i} and μ_{i} are the displacement vector and effective mass for the *i*th mode respectively. The summation terms in Eq. (1) contains harmonic and anharmonic contributions for the potential energy.

To account for the intermode coupling between O–O stretch and IHB stretch, we performed a two-dimensional (2D) anharmonic vibrational calculation. Similar with NM1D, we cast the 2D Hamiltonian in NM coordinates and will be called as NM2D. In the normal mode picture, the kinetic energy components are separable, and the coupling terms are confined in the potential energy part. In this coordinate, the 2D Hamiltonian operator can be easily constructed as follows:

$$H^{\mathrm{NM2D}} = \frac{-\hbar^2}{2}\left[\frac{1}{\mu_{\mathrm{O-O}}}\frac{\partial^2}{\partial Q_{\mathrm{O-O}}^2} + \frac{1}{\mu_{\mathrm{IHB}}}\frac{\partial^2}{\partial Q_{\mathrm{IHB}}^2}\right] + V(Q_{\mathrm{O-O}}, Q_{\mathrm{IHB}}) \qquad (2)$$

where $V(Q_{\mathrm{O–O}}, Q_{\mathrm{IHB}})$ refers to the 2D potential, which contains anharmonic contributions for each mode as well as cross-terms in the potential.

The potential energy surface is constructed by scanning along the two NM coordinates considered in this study, these are: O–O stretch ($Q_{\mathrm{O–O}}$) and IHB stretch (Q_{IHB}). The corresponding vibrational Schrödinger equations were solved using the method of discrete variable representation (DVR), and harmonic oscillators were used as basis functions [36–39]. The implementation of our DVR code is described elsewhere [40–43]. The DVR Hamiltonian was diagonalized via Davidson's algorithm [44]. Readers who are interested in the numerical aspects and further details of DVR are advised to refer to the earlier works of Light as well as Colbert and Miller [36–39]. For each degree of freedom, we used 7 DVR grids.

From the eigenvectors and eigenvalues, the transition intensities relative to the ground vibrational state were calculated based on Fermi-Golden rule [45].

$$\mathrm{A}_{fo} = \left(\frac{\pi N_A}{3c\varepsilon_o\hbar}\right)\nu_{fo}\left|\langle\Psi_f\middle|\mu(\vec{R})\middle|\Psi_o\rangle\right|^2 \tag{3}$$

Where A_{fo} is the integrated absorption coefficient (S.I. km mol^{-1}), N_A is the Avogadro's constant; c is the speed of light; ε_o is the permittivity of free space; $\hbar$ is the reduced Planck's constant; Ψ_o and Ψ_f are the initial and final vibrational states; $\mu(\vec{R})$ is the dipole moment surface in Debye; ν_{fo} is the $\Psi_o \rightarrow \Psi_f$ excitation energy in cm^{-1}. Substitution of the corresponding numerical values for the physical constants results in the following prefactor, with A_{fo} is in units of km mol^{-1}.

$$\mathrm{A}_{fo} = 2.50664\nu_{fo}\left|\langle\Psi_f\middle|\mu(\overrightarrow{R})\middle|\Psi_o\rangle\right|^2 \tag{4}$$

The dipole moment surface was constructed in a similar manner to that of the potential energy surface. Both surfaces were built by scanning along the NM coordinates at MP2/aug-cc-pVDZ.

One might wonder about the indeterminacy of the dipole moment surface, $\mu(\overrightarrow{R})$. In particular, the dipole moment of a charge species is dependent on the choice of the coordinate's origin. However, despite of such indeterminacy, the corresponding transition dipole moment $\langle\Psi_f\middle|\mu(\overrightarrow{R})\middle|\Psi_o\rangle$ is unique for charge species, a proof is available in the supporting information.

2.3 Calculation of Hamiltonian Matrix Elements

The choice for a basis of representation is crucial in highlighting the extent of interaction between vibrational states. A good representation should be able to express the physics and chemistry of the Hamiltonian with the fewest basis functions as possible. Since the goal is to understand the coupling between vibrational states arising from O–O and IHB stretches, we decided to pick the direct products of their NM1D vibrational states and use it as a representation basis for the NM2D Hamiltonian.

For clarity, we will use $H^{\mathrm{NM1D}}_{\mathrm{O-O}}$, $H^{\mathrm{NM1D}}_{\mathrm{IHB}}$ and H^{NM2D} as operator notations. Their corresponding matrix representation in DVR basis will be $\mathbf{H}^{\mathrm{NM1D}}_{\mathrm{O-O}}$, $\mathbf{H}^{\mathrm{NM1D}}_{\mathrm{IHB}}$ and

$\mathbf{H}^{\mathrm{NM2D}}$. Their respective DVR basis will be denoted as $\{|\phi^{\mathrm{O-O}}\rangle\}$, $\{|\phi^{\mathrm{IHB}}\rangle\}$ and $\{|\phi^{2D}\rangle\}$. The latter DVR basis is related from the first two by a direct product $|\phi^{\mathrm{O-O}}\rangle\otimes|\phi^{\mathrm{IHB}}\rangle$.

Furthermore, we define $|n_{\mathrm{O\text{–}O}}>$ and $|n_{\mathrm{IHB}}>$ to be the NM1D eigenvectors for $Q_{\mathrm{O\text{–}O}}$ and Q_{IHB} respectively. These eigenvectors are solutions of Eq. (1) obtained via DVR method, that is,

$$\begin{aligned}\mathbf{H}^{\mathrm{NM1D}}_{\mathrm{O-O}}|n_{\mathrm{O-O}}\rangle &= \varepsilon_{n_{\mathrm{O-O}}}|n_{\mathrm{O-O}}\rangle\\ \mathbf{H}^{\mathrm{NM1D}}_{\mathrm{IHB}}|n_{\mathrm{IHB}}\rangle &= \varepsilon_{n_{\mathrm{IHB}}}|n_{\mathrm{IHB}}\rangle\end{aligned} \tag{5}$$

The direct product of $|n_{\mathrm{O\text{–}O}}>$ and $|n_{\mathrm{IHB}}>$ is a 2D eigenvector, which corresponds to a case of two anharmonically corrected 1D eigenvectors that do not couple with each other. With such attribute, we will term these direct products as "pure states" and is defined in Eq. (6). A set containing all possible direct products of $|n_{\mathrm{O\text{–}O}}>$ and $|n_{\mathrm{IHB}}>$ is equal to the set of "pure state" basis, Eq. (7).

$$|n_{\mathrm{O-O}}, n_{\mathrm{IHB}}\rangle = |n_{\mathrm{O-O}}\rangle\otimes|n_{\mathrm{IHB}}\rangle \tag{6}$$

$$\{|\chi^{\mathrm{PureState}}\rangle\} = \{|n_{\mathrm{O-O}}, n_{\mathrm{IHB}}\rangle = |n_{\mathrm{O-O}}\rangle\otimes|n_{\mathrm{IHB}}\rangle\} \tag{7}$$

Since $|n_{\mathrm{O\text{–}O}}>$ and $|n_{\mathrm{IHB}}>$ are represented in $\{|\phi^{\mathrm{O-O}}\rangle\}$ and $\{|\phi^{\mathrm{IHB}}\rangle\}$, the corresponding "pure states" is also represented in the same DVR basis for $\mathbf{H}^{\mathrm{NM2D}}$, $\{|\phi^{2D}\rangle\}$. Moreover, the components of $|\chi_p^{\mathrm{PureState}}\rangle$ corresponds to the projections of $\langle\phi_k^{2D}|$ in $|\chi_p^{\mathrm{PureState}}\rangle$ as shown in Eq. (8), relating the DVR basis to the "pure state" basis.

$$|\chi_p^{\mathrm{PureState}}\rangle = \sum_k |\phi_k^{\mathrm{2D}}\rangle\langle\phi_k^{\mathrm{2D}}|\chi_p^{\mathrm{PureState}}\rangle = \sum_k |\phi_k^{\mathrm{2D}}\rangle\Omega_{kp} \tag{8}$$

Where we defined $\Omega_{kp} = \langle\phi_k^{\mathrm{2D}}|\chi_p^{\mathrm{PureState}}\rangle$. Its corresponding matrix $\mathbf{\Omega}$ is a unitary matrix and will be briefly proven below:

$$\langle\chi_p^{\mathrm{PureState}}|\phi_k^{\mathrm{2D}}\rangle = \langle\phi_k^{\mathrm{2D}}|\chi_p^{\mathrm{PureState}}\rangle^* = \Omega_{kp}^* = \Omega_{pk}^\dagger \tag{9}$$

$$\delta_{ij} = \langle\phi_i^{\mathrm{2D}}|\phi_j^{\mathrm{2D}}\rangle = \sum_p \langle\phi_i^{\mathrm{2D}}|\chi_p^{\mathrm{PureState}}\rangle\langle\chi_p^{\mathrm{PureState}}|\phi_j^{\mathrm{2D}}\rangle = \sum_p \Omega_{ip}\Omega_{pj}^\dagger \tag{10}$$

Hence,

$$\mathbf{1} = \mathbf{\Omega}\mathbf{\Omega}^\dagger \tag{11}$$

Where $\mathbf{1}$ is a unit matrix.

From here, we can perform a change of basis from DVR basis to "pure state" basis. Let $\mathbf{H}^{\mathrm{NM2D}}$ and $\mathbf{H}^{\mathrm{PureState}}$ be the 2D Hamiltonian matrix in DVR and pure state basis respectively. The relationship between $\mathbf{H}^{\mathrm{NM2D}}$ and $\mathbf{H}^{\mathrm{PureState}}$ can be derived as follows:

$$\begin{aligned} \mathbf{H}_{pq}^{\mathrm{PureState}} &= \langle \chi_p^{\mathrm{PureState}} | \mathbf{H}^{\mathrm{NM2D}} | \chi_q^{\mathrm{PureState}} \rangle \\ &= \sum_{ij} \langle \chi_p^{\mathrm{PureState}} \left| \phi_i^{2D} \rangle \langle \phi_i^{2D} | \mathbf{H}^{\mathrm{NM2D}} | \phi_j^{2D} \rangle \langle \phi_j^{2D} \right| \chi_q^{\mathrm{PureState}} \rangle \sum_{ij} \Omega_{pi}^{\dagger} \mathbf{H}_{ij}^{\mathrm{NM2D}} \Omega_{jq} \end{aligned} \tag{12}$$

Hence, $\mathbf{H}^{\mathrm{NM2D}}$ and $\mathbf{H}^{\mathrm{PureState}}$ are related by the following unitary transformation.

$$\mathbf{H}^{\mathrm{PureState}} = \mathbf{\Omega}^{\dagger} \mathbf{H}^{\mathrm{NM2D}} \mathbf{\Omega} \tag{13}$$

3 Results and Discussion

The rest of the chapter is organized as follows: In Sect. 3.1, a short discussion on the structural aspects characterizing IHB at equilibrium is presented. Meanwhile, the origin of O–O stretch's capability in modulating the IHB stretch frequency is explained in Sect. 3.2. Lastly, the consequences of symmetry breaking which relaxes the coupling schemes explored in Sect. 3.2 is examined in Sect. 3.3.

3.1 Structural Aspects from ab initio Calculations

In cases where the hydrogen bond is very strong, the proton is likely to be located at the midpoint of the "donor-acceptor" coordinate [32]. Comparing with the neutral counterparts, IHBs could be at least six times stronger in the gas phase [46]. With such information, we predict that the shared proton will be at the midpoint of the O–O axis and this was confirmed by our optimizations at MP2/aug-cc-pVDZ. A schematic of the minimum structures is found in Fig. 1a–c.

Furthermore, we also probe the effect of Ar tagging on the IHB's equilibrium structure. In the earlier work of Hammer and co-workers [9], an investigation on the binding site for Ar and Ne was reported for $H_5O_2^+$ at MP2/aug-cc-pVDZ. The Ar atom favorably binds on the free OH moieties over the shared proton. This behavior is consistent with our optimization for $ArH_5O_2^+$ and $Ar(MeOH)_2H^+$, as shown in Fig. 1d–f. Meanwhile, for $(Me_2O)_2H^+$ there are no free OH. The Ar atom was found to attach on top of an oxygen atom, Fig. 1f. Table 1 shows the effect of Ar tagging on the O–H^+ distances. Across all the species, it is evident that messenger tagging induces symmetry breaking not only on the point group level but also on

Table 1 Structural parameter of the O–H$^+$–H moiety for $H_5O_2^+$, $(MeOH)_2H^+$, $(Me_2O)_2H^+$, and their Ar-tagged counterparts. These parameters are obtained from MP2/aug-cc-pVDZ level

	O_a–H (Å)	O_b–H (Å)	O–O (Å)	∠O_a–H$^+$–O_b (°)
$H_5O_2^+$	1.202	1.202	2.399	173.22
$ArH_5O_2^+$	1.121	1.299	2.416	173.43
$(MeOH)_2H^+$	1.201	1.201	2.394	170.95
$Ar(MeOH)_2H^+$	1.129	1.286	2.407	170.94
$(Me_2O)_2H^+$	1.199	1.199	2.397	175.15
$Ar(Me_2O)_2H^+$	1.202	1.197	2.396	175.19

Note O_a refers to oxygen of the molecule where Ar messenger is attached

the O–H$^+$–O structure. In all bare Zundel structures, the point group is C_2, and the corresponding single Ar-tagged is C_1.

3.2 *Which State Couples with IHB Stretch O–O Overtone or Combination Bands?*

In order to assess the effects of anharmonicity and intermode coupling, we calculate the vibrational frequencies and intensities at harmonic, normal mode one-dimensional (NM1D), and normal mode two-dimensional (NM2D) treatments. The results are shown in Table 2. Harmonic vibrational frequencies and intensities were obtained from the output file of a Gaussian 09 frequency job.

Meanwhile, vibrational frequencies and intensities for NM1D and NM2D were obtained by solving Eqs. (1) and (2) by DVR approach and evaluation of integrated absorption coefficients by Fermi's Golden rule Eq. (4). Comparison between harmonic and NM1D for O–O fundamental stretch reveal its nearly harmonic behavior. It was found to be consistent across all bare cases. Meanwhile, a significant blue shift (>250 cm^{-1}) is observed for IHB fundamental stretch from harmonic to NM1D, revealing the importance of higher even order terms in the potential. Lastly, overtones and combination bands under harmonic treatments have zero intensity by virtue of the selection rules [45].

Focusing our attention on NM2D, it is evident that the IHB fundamental is red shifted relative to NM1D, Table 2. Across the homologous series, the extent of red shift decreases from $H_5O_2^+$ to $(Me_2O)_2H^+$. The 2D potential energy surfaces and wave functions for O–O and IHB fundamental stretch is shown in Fig. 2. These potential surfaces were constructed by scanning along normal mode displacements in real space as described in reference [33].

In addition, the first overtone of O–O stretch is a dark state while its combination band with IHB stretch is a bright state. Since the combination band's intensity must have come from the IHB fundamental, then it follows that IHB stretch couples with the combination band, while the overtone does not.

Table 2 Selected vibrational frequencies (cm^{-1}) and intensities ($km\ mol^{-1}$) calculated via harmonic approximation, normal mode one-dimensional (NM1D), and normal mode two-dimensional (NM2D) for bare Zundels

$H_5O_2^+$						
	Harmonic approx.		NM1D		NM2D	
	Frequency	Intensity	Frequency	Intensity	Frequency	Intensity
$\nu_{O–O}$	607	0.3	605	0.3	581	0.3
$2\nu_{O–O}$	1214	0.0	1208	0.0	1155	0.0
ν_{IHB}	808	2981.5	1251	2700.8	1064	2329.3
$\nu_{O–O} + \nu_{IHB}$	1415	0.0	–	–	1601	323.7
$(MeOH)_2H^+$						
	Harmonic approx.		NM1D		NM2D	
	Frequency	Intensity	Frequency	Intensity	Frequency	Intensity
$\nu_{O–O}$	577	2.4	578	2.4	565	2.4
$2\nu_{O–O}$	1154	0.0	1156	0.0	1127	0.0
ν_{IHB}	702	2502.8	1048	2351.5	930	2137.0
$\nu_{O–O} + \nu_{IHB}$	1279	0.0	–	–	1473	190.0
$(Me_2O)_2H^+$						
	Harmonic approx.		NM1D		NM2D	
	Frequency	Intensity	Frequency	Intensity	Frequency	Intensity
$\nu_{O–O}$	544	1.5	545	1.5	536	1.5
$2\nu_{O–O}$	1088	0.0	1089	0.0	1071	0.0
ν_{IHB}	737	3011.1	1064	2823.9	979	2610.1
$\nu_{O–O} + \nu_{IHB}$	1281	0.0	–	–	1500	187.7

To further understand this behavior, we performed a change of basis from DVR to pure state on the vibrational Hamiltonian. The corresponding matrix elements are shown in Table 3 for $H_5O_2^+$ and Tables S1–S2 of the supporting information for $(MeOH)_2H^+$ and $(Me_2O)_2H^+$ respectively. Across all three Zundel ions, it is consistent that pure states corresponding to O–O stretch with IHB stretch in its ground state, $|n_{O–O}, 0\rangle$ couples with themselves as shown in the upper leftmost block diagonal in Tables 3, S1 and S2. In other words, $\{|n_{O–O}, 0\rangle\}$ forms a single block. Meanwhile, states corresponding to IHB fundamental and its combination with O–O stretch (i.e. $\{|n_{O–O}, 1\rangle\}$) do form a separate block from $\{|n_{O–O}, 0\rangle\}$ and hence do not couple with each other. The $\langle 0, 1|H|1, 1\rangle$ is nonzero so $|0, 1\rangle$ and $|1, 1\rangle$ do couple, while $\langle 0, 1|H|2, 0\rangle$ is zero so $|0, 1\rangle$ and $|2, 0\rangle$ do not couple. Hence, Fermi resonance between $|0, 1\rangle$ and $|2, 0\rangle$ is forbidden.

A simplified vibrational state interaction diagram involving: $|0, 1\rangle$, $|1, 1\rangle$, and $|2, 1\rangle$ is shown in Fig. 3. The origin of the red shift corresponding to the IHB stretch from NM1D to NM2D together with the intensities of the combination band is explained using a two-state system, Fig. 3a. Both IHB stretch fundamental, $|0, 1\rangle$ and combination band, $|1, 1\rangle$ couple to form two new states: $|+\rangle$ and $|-\rangle$. Each of these new states has substantial $|0, 1\rangle$ character causing transitions from the

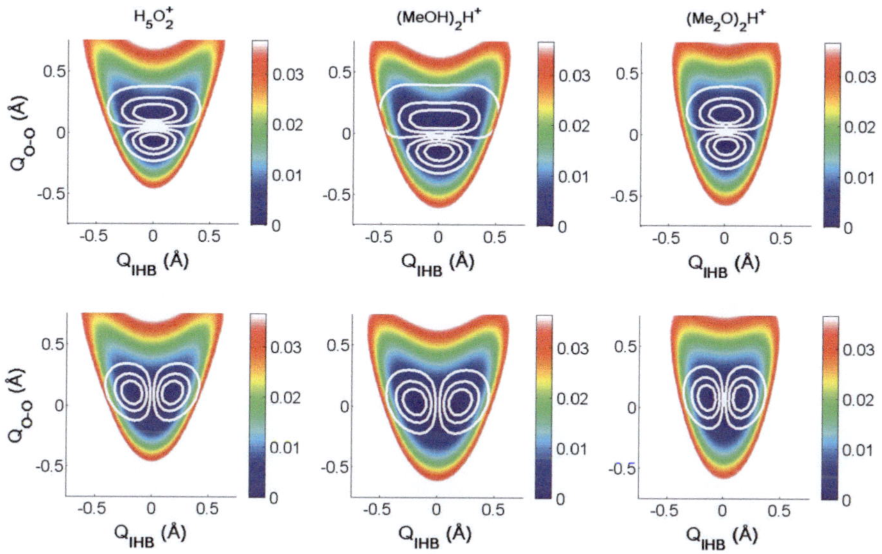

Fig. 2 Two-dimensional (2D) potential energy surface and wave functions for the O–O (*top row*) and IHB (*bottom row*) stretch fundamentals for the bare Zundels. The colored contours refer to the 2D potential, with the *color bars* expressed in Hartree units. The white contours refer to the vibrational wave functions. The potential was constructed by scanning along normal mode displacements in real space as described in reference [33]

Table 3 Hamiltonian matrix elements (cm^{-1}) in pure state basis $|n_{O-O,}n_{IHB}>$ for $H_5O_2^+$

H	|0, 0>	|1, 0>	|2, 0>	|0, 1>	|1, 1>	|2, 1>	|0, 2>	|1, 2>	|2, 2>
<0, 0|	0	130	23	0	0	0	3	178	31
<1, 0|	130	610	182	0	0	0	178	10	250
<2, 0|	23	182	1218	0	0	0	31	250	16
<0, 1|	0	0	0	1254	340	59	0	0	0
<1, 1|	0	0	0	340	1872	476	0	0	0
<2, 1|	0	0	0	59	476	2488	0	0	0
<0, 2|	3	178	31	0	0	0	2912	427	73
<1, 2|	178	10	250	0	0	0	427	3531	599
<2, 2|	31	250	16	0	0	0	73	599	4149

ground vibrational state to |+> and |–> states to manifest intensity. Between the two coupled states, |+> has more |0, 1> character while |–> is dominated by |1, 1> . This manner of mixing allows us to assign |+> state as the "IHB stretch" and the |–> state as the "combination band". Since |+> is red shifted relative to |0, 1> , the two-state model is able to capture the red shifting of IHB stretch from NM1D to NM2D in Table 2.

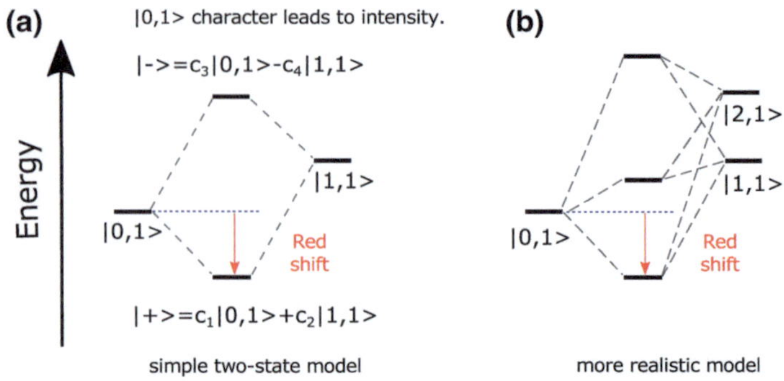

Fig. 3 A vibrational state interaction diagram which accounts for the observed red shift of IHB stretch from NM1D to NM2D via a) oversimplified two-state system and b) a more realistic model which accounts the interaction between |1, 1> and |2, 1>. The interaction between |0, 1> and |1, 1> leads to two coupled states |+> and |−>, which both contains substantial |0, 1> character. This accounts for the observed intensity of the combination band |1, 1>

By looking at the matrix elements in Tables 3, S1 and S2, the |1, 1> and |2, 1> combination couples strongly. This interaction is not captured by the simple two-state model mentioned above. Instead a more realistic picture can be found in Fig. 3b where the |2, 1> state mixes with |1, 1> .

One might wonder about the origin of such block matrices in Tables 3, S1, and S2. In order to further understand the results, the 2D vibrational Hamiltonian in Eq. (2) can be partitioned into three terms: (1) Anharmonically corrected 1D Hamiltonian for O–O stretch, $H_{\mathrm{O-O}}^{\mathrm{NM1D}}$; (2) Anharmonically corrected 1D Hamiltonian for IHB stretch, $H_{\mathrm{IHB}}^{\mathrm{NM1D}}$ and (3) Cross-terms in the potential part, ΔV.

$$H^{\mathrm{NM2D}} = H_{\mathrm{O-O}}^{\mathrm{NM1D}} + H_{\mathrm{IHB}}^{\mathrm{NM1D}} + \Delta V \tag{14}$$

$$\Delta V = \sum_{i \neq 0, j \neq 0} a_{i,j} Q_{\mathrm{O-O}}^{i} \, Q_{\mathrm{IHB}}^{j} \tag{15}$$

It is easy to show that for two pure states $|n'_{\mathrm{O-O}}, n'_{\mathrm{IHB}}\rangle$ and $|n_{\mathrm{O-O}}, n_{\mathrm{IHB}}\rangle$, the corresponding contribution of the first two Hamiltonian terms in Eq. (14) can be expressed as follows:

$$\langle n'_{\mathrm{O-O}}, n'_{\mathrm{IHB}} | H_{\mathrm{O-O}}^{\mathrm{NM1D}} | n_{\mathrm{O-O}}, n_{\mathrm{IHB}} \rangle = \varepsilon_{\mathrm{O-O}}^{n_{\mathrm{O-O}}} \, \delta_{n'_{\mathrm{O-O}}, n_{\mathrm{O-O}}} \delta_{n'_{\mathrm{IHB}}, n_{\mathrm{IHB}}} \tag{16}$$

$$\langle n'_{\mathrm{O-O}}, n'_{\mathrm{IHB}} | H_{\mathrm{IHB}}^{\mathrm{NM1D}} | n_{\mathrm{O-O}}, n_{\mathrm{IHB}} \rangle = \varepsilon_{\mathrm{IHB}}^{n_{\mathrm{IHB}}} \, \delta_{n'_{\mathrm{O-O}}, n_{\mathrm{O-O}}}, \delta_{n'_{\mathrm{IHB}}, n_{\mathrm{IHB}}} \tag{17}$$

These two equations imply that both $H_{\mathrm{O-O}}^{\mathrm{NM1D}}$ and $H_{\mathrm{IHB}}^{\mathrm{NM1D}}$ do not contribute to the off-diagonal terms of the Hamiltonian matrix in pure state basis. It is only along the main diagonal that the above terms will have a nonzero contribution. On the other hand, the contributions of ΔV, $\langle n'_{\mathrm{O-O}}, n'_{\mathrm{IHB}} | \Delta V | n_{\mathrm{O-O}}, n_{\mathrm{IHB}} \rangle$ can be simplified if we expanded ΔV as a group of odd and even terms relative to $Q_{\mathrm{O-O}}$ and Q_{IHB} we can write:

$$\Delta V = \sum_{i}^{odd}\sum_{j}^{odd} a_{i,j} Q_{\mathrm{O-O}}^{i} Q_{\mathrm{IHB}}^{j} + \sum_{i}^{even}\sum_{j}^{odd} a_{i,j} Q_{\mathrm{O-O}}^{i} Q_{\mathrm{IHB}}^{j} + \sum_{i}^{odd}\sum_{j}^{even} a_{i,j} Q_{\mathrm{O-O}}^{i} Q_{\mathrm{IHB}}^{j} + \sum_{i}^{even}\sum_{j}^{even} a_{i,j} Q_{\mathrm{O-O}}^{i} Q_{\mathrm{IHB}}^{j} \quad (18)$$

Since the 2D potential is even with respect to Q_{IHB}, that is $\Delta V(Q_{\mathrm{O-O}}, Q_{\mathrm{IHB}}) = V(Q_{\mathrm{O-O}}, -Q_{\mathrm{IHB}})$ (see Fig. 2) then it follows that all coefficients involving odd powers of Q_{IHB} will vanish. Therefore, the first two terms will be zero. With these simplifications, the cross-terms contribution to the matrix elements can be written as:

$$\langle n'_{\mathrm{O-O}}, n'_{\mathrm{IHB}} | \Delta V | n_{\mathrm{O-O}}, n_{\mathrm{IHB}} \rangle = \sum_{i}^{odd}\sum_{j}^{even} a_{i,j} \langle n'_{\mathrm{O-O}} | \mathbf{Q}_{\mathrm{O-O}}^{i} | n_{\mathrm{O-O}} \rangle \langle n'_{\mathrm{IHB}} | \mathbf{Q}_{\mathrm{IHB}}^{j} | n_{\mathrm{IHB}} \rangle + \sum_{i}^{even}\sum_{j}^{even} a_{i,j} \langle n'_{\mathrm{O-O}} | \mathbf{Q}_{\mathrm{O-O}}^{i} | n_{\mathrm{O-O}} \rangle \langle n'_{\mathrm{IHB}} | \mathbf{Q}_{\mathrm{IHB}}^{j} | n_{\mathrm{IHB}} \rangle \quad (19)$$

At this point, mere evaluation of the integrals relative to $dQ_{\mathrm{O-O}}$ and dQ_{IHB} is not trivial to be deciphered unless we naïvely think in terms of group theory. We assume that the anharmonically corrected eigenvectors have the same symmetry as their harmonic counterparts. Then, it is possible to tentatively assign the following Mulliken labels.

$$\Gamma_{\mathbf{Q}_{\mathrm{O-O}}^{i}} = A \qquad \forall i \in \mathbb{Z}^{+} \cup \{0\} \qquad \Gamma_{\mathbf{Q}_{\mathrm{IHB}}^{j}} = \begin{cases} A & \text{if } j \text{ is even} \\ B & \text{if } j \text{ is odd} \end{cases} \quad (20)$$

$$\Gamma_{|n_{\mathrm{O-O}}\rangle} = A \qquad \forall n_{\mathrm{O-O}} \in \mathbb{Z}^{+} \cup \{0\} \qquad \Gamma_{|n_{\mathrm{IHB}}\rangle} = \begin{cases} A & \text{if } n_{\mathrm{IHB}} \text{ is even} \\ B & \text{if } n_{\mathrm{IHB}} \text{ is odd} \end{cases} \quad (21)$$

By means of the method of vanishing integrals [45] it follows that $< n'_{\mathrm{O-O}}|\mathbf{Q}^{i}_{\mathrm{O-O}}|n_{\mathrm{O-O}}>$ will not vanish as the corresponding direct product will always correspond to a totally symmetric representation A. Meanwhile, the integral $< n'_{\mathrm{IHB}}|\mathbf{Q}^{j}_{\mathrm{IHB}}| n_{\mathrm{IHB}}>$ will vanish when $|n_{\mathrm{IHB}}>$ and $|n'_{\mathrm{IHB}}>$ belong to different irreducible representations as $\mathbf{Q}^{j}_{\mathrm{IHB}}$ will always belong to a totally symmetric representation A for even values of j. This simple line of thinking enables us to understand why the set $\{|n_{\mathrm{O-O}}, n_{\mathrm{IHB}}>\}$ couples with each other provided that they have the same quantum number along IHB. Two sets of block matrices: $\{|n_{\mathrm{O-O}}, n_{\mathrm{IHB}}>\}$ and $\{|n_{\mathrm{O-O}}, n_{\mathrm{IHB}\pm}1>\}$ do not couple with each other while $\{|n_{\mathrm{O-O}}, n_{\mathrm{IHB}}>\}$ and $\{|n_{\mathrm{O-O}}, n_{\mathrm{IHB}\pm}2>\}$ do couple with each other.

Table 4 Selected terms for $\langle n'_{O\text{–}O}|\mathbf{Q}^i_{O\text{–}O}|n_{O\text{–}O}\rangle$ and $\langle n'_{IHB}|\mathbf{Q}^j_{IHB}|n_{IHB}\rangle$ taken form $H_5O_2^+$. The numbers are expressed in atomic units

$\langle n'_{O\text{–}O}|\mathbf{Q}^i_{O\text{–}O}|n_{O\text{–}O}\rangle$	$\mathbf{Q}^i_{O\text{–}O}$				
	$Q_{O\text{–}O}$	$Q^2_{O\text{–}O}$	$Q^3_{O\text{–}O}$	$Q^4_{O\text{–}O}$	$Q^5_{O\text{–}O}$
$\langle 0|\mathbf{Q}^i_{O\text{–}O}|0\rangle$	0.012	0.022	0.001	0.001	0.000
$\langle 1|\mathbf{Q}^i_{O\text{–}O}|0\rangle$	−0.147	−0.006	−0.010	−0.001	−0.001
$\langle 2|\mathbf{Q}^i_{O\text{–}O}|0\rangle$	−0.006	0.031	0.002	0.004	0.001
$\langle 3|\mathbf{Q}^i_{O\text{–}O}|0\rangle$	0.000	0.003	−0.008	−0.001	−0.002
$\langle 4|\mathbf{Q}^i_{O\text{–}O}|0\rangle$	0.000	0.000	−0.001	0.002	0.000
$\langle 5|\mathbf{Q}^i_{O\text{–}O}|0\rangle$	0.000	0.000	0.000	0.000	0.001
$\langle n'_{IHB}|\mathbf{Q}^j_{IHB}|n_{IHB}\rangle$	$\mathbf{Q}^j_{IHB}$				
	$\mathbf{Q}^2_{IHB}$	$\mathbf{Q}^4_{IHB}$	$\mathbf{Q}^6_{IHB}$	$\mathbf{Q}^8_{IHB}$	$\mathbf{Q}^{10}_{IHB}$
$\langle 0|\mathbf{Q}^j_{IHB}|0\rangle$	0.044	0.005	0.001	0.000	0.000
$\langle 1|\mathbf{Q}^j_{IHB}|0\rangle$	0.000	0.000	0.000	0.000	0.000
$\langle 2|\mathbf{Q}^j_{IHB}|0\rangle$	0.056	0.011	0.003	0.001	0.000
$\langle 3|\mathbf{Q}^j_{IHB}|0\rangle$	0.000	0.000	0.000	0.000	0.000
$\langle 2|\mathbf{Q}^j_{IHB}|1\rangle$	0.000	0.000	0.000	0.000	0.000

Alternatively, we can evaluate a few integrals and inspect the trend. We decided to look at a few contributing integrals for $H_5O_2^+$, which are compiled in Table 4. Similar behavior was also found for $(MeOH)_2^+$ and $(Me_2O)_2H^+$. As can be seen in the entries of Table 4, the blocks of zeroes in Tables 3, S1, and S2 are from $\langle 1|\mathbf{Q}^j_{IHB}|0\rangle$, and $\langle 2|\mathbf{Q}^j_{IHB}|1\rangle$ which is zero across $j = 2$ to 10. Such behavior is consistent with the simple group theoretical picture discussed above.

3.3 *Consequences of Symmetry Breaking: Fermi Resonance*

In most experimental conditions, the vibrational spectra of proton bound Zundel dimers are measured with messenger tagging. In Sect. 3.2 we examined the causality of the red shift in IHB stretch upon inclusion of the O–O stretch. An explanation was also offered for the IHB intensities being shared on the combination band and not on the O–O stretch overtone. These two results are consequences of the bare Zundel's symmetry on the vibrational wave functions and normal modes. In this section, we will break the above C_2 symmetry by tagging Ar and will examine the effects of such symmetry breaking on the predicted spectral features.

As mentioned earlier, tagging Ar atom changes the point group from C_2 to C_1. Such descending on the molecular symmetry results to changing all *B* labelled vibrational normal modes to *A* and would certainly have an effect on the potential energy curve's topology. In order to have a clear picture on how the tagging distorts the potential surface, we performed similar NM2D vibrational calculations that are shown in Fig. 4. Comparison with the corresponding 2D potential of their bare

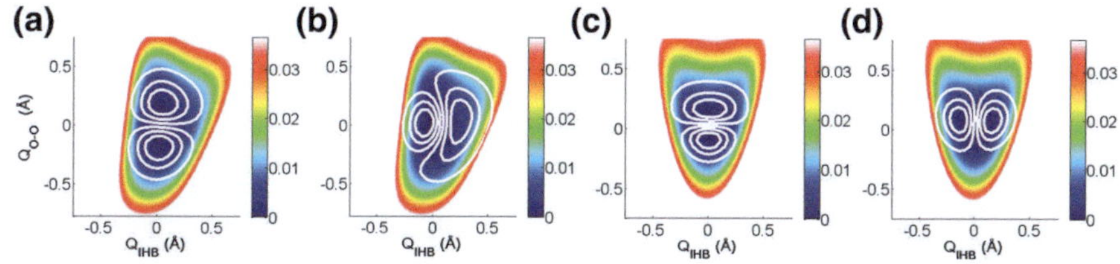

Fig. 4 Two-dimensional (2D) potential energy surface and wave functions for the O–O and IHB stretch fundamentals **a–b** for $Ar(H_5O_2)^+$ and **c–d** for $Ar(Me_2O)_2H^+$. The colored contours refer to the 2D potential, with the *color bars* expressed in Hartree units. The white contours refer to the vibrational wave functions. The potential was constructed by scanning along normal mode displacements in real space as described in reference [33]

counterparts in Fig. 2 shows that both $ArH_5O_2^+$ experienced a more dramatic change in the potential in comparison with $Ar(Me_2O)_2H^+$. Such observation can be rationalized on the binding sites of Argon, which is at the O–H moieties for $H_5O_2^+$; while for $(Me_2O)_2H^+$, Ar binds on top of the oxygen atom.

Meanwhile, in terms of the spectroscopic aspects, the main effect of perturbing the potential's topology via Ar tagging is to relax the symmetry constraints from the previous section. The results of our NM2D calculations are shown in Table 5. One obvious effect of symmetry breaking in Zundel dimer is relaxing the constraints against Fermi resonance between O–O stretch overtone and IHB stretch. The most obvious case is that of $ArH_5O_2^+$ where O–O stretch overtone is more intense than the fundamental. The 2D wave functions for the IHB O–O stretch overtone, IHB stretch, and their combination bands are presented in Fig. 5.

The Fermi resonance for $Ar(Me_2O)_2H^+$ is weaker than $ArH_5O_2^+$. The reason for such observation is that, Ar tagging in $(Me_2O)_2H^+$ does not significantly perturb the topology of the potential surface as can be seen in Fig. 5 (right column). Furthermore, comparison between $Ar(Me_2O)_2H^+$ and $(Me_2O)_2H^+$ reveals minor differences ($\lesssim 2$ cm^{-1}) at the peak positions. Although such perturbations are too small, nevertheless, it lifted the symmetry constraints for Fermi resonance by allowing $Ar(Me_2O)_2H^+$ to exhibit a very weak overtone.

Table 5 Selected vibrational frequencies (cm^{-1}) and intensities (km mol^{-1}) calculated two-dimensional normal mode (NM2D) for $ArH_5O_2^+$, $Ar(MeOH)_2H^+$, and $Ar(Me_2O)_2H^+$

NM2D						
	$ArH_5O_2^+$		$Ar(MeOH)_2H^+$		$Ar(Me_2O)_2H^+$	
	Frequency	Intensity	Frequency	Intensity	Frequency	Intensity
ν_{O-O}	547	105.4	551	177.1	536	1.7
$2\nu_{O-O}$	1131	149.1	1167	1.2	1071	0.4
ν_{IHB}	1254	1797.5	933.5	681.1	977	2487.2
$\nu_{O-O} + \nu_{IHB}$	1850	37.5	1502	1.0	1499	177.1

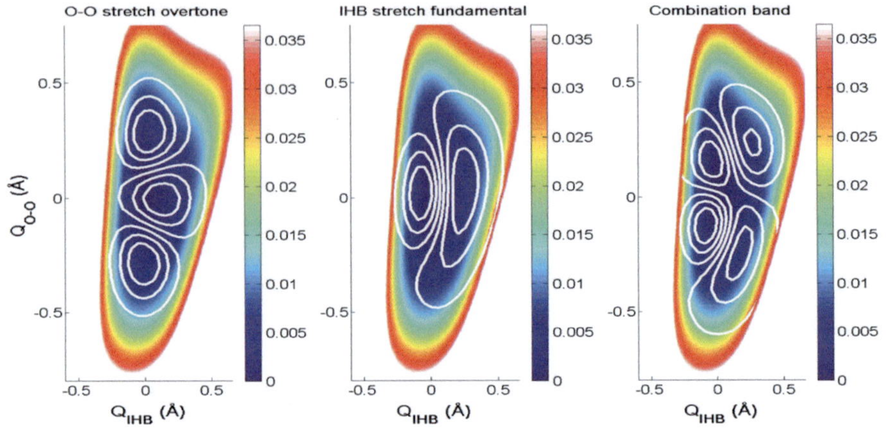

Fig. 5 Two-dimensional potential energy surface and wave functions for the O–O stretch overtone (*left*), IHB stretch fundamental (*middle*), and combination band (*right*) of $ArH_5O_2^+$. The colored contours refer to the 2D potential, with the *color bars* expressed in Hartree units. The white contours refer to the vibrational wave functions. The potential was constructed by scanning along normal mode displacements in real space as described in reference [33]

Table 6 Hamiltonian matrix elements (cm^{-1}) in pure state basis $|n_{O-O}, n_{IHB}>$ for $ArH_5O_2^+$

H	|0, 0>	|1, 0>	|2, 0>	|0, 1>	|1, 1>	|2, 1>	|0, 2>	|1, 2>	|2, 2>
<0, 0|	0	7	39	−40	80	−58	−27	29	−38
<1, 0|	7	568	−3	80	−113	126	29	−77	47
<2, 0|	39	−3	1187	−58	126	−183	−38	47	−121
<0, 1|	−40	80	−58	1277	−83	125	81	−262	121
<1, 1|	80	−113	126	−83	1955	−147	−262	231	−384
<2, 1|	−58	126	−183	125	−147	2676	121	−384	371
<0, 2|	−27	29	−38	81	−262	121	2740	−248	202
<1, 2|	29	−77	47	−262	231	−384	−248	3512	−384
<2, 2|	−38	47	−121	121	−384	371	202	−384	4319

In order to see the effect of symmetry breaking on the coupling, the Hamiltonian matrix represented in pure state basis for Ar tagged *Zundel* dimers are tabulated in Table 6 and S3 for $ArH_5O_2^+$ and $Ar(Me_2O)_2H^+$. Compared with the Hamiltonian matrix of their bare counterparts in Table 3, S1 and S2, it is obvious that the symmetry breaking allows the coupling between $\{|n_{O-O}, n_{IHB}>\}$ and $\{|n_{O-O}, n_{IHB} + 1>\}$.

4 Conclusions

In this chapter, the rules of quantum theory governing the coupling between IHB stretch and O–O stretch were explored. For bare Zundel cases, the overtone of O–O stretch will not couple with the IHB stretch, ruling out the possibility of Fermi resonance. However, the combination band of O–O stretch with IHB stretch efficiently borrows intensity from the IHB fundamental stretch. The above selection rules are consequences of the symmetry of the vibrational wave functions and normal mode displacements. By means of Ar tagging, the molecular symmetry of the bare Zundel descends from C_2 to C_1, thereby relaxing the selection rules for Fermi resonance. Going back to the question why the hydrogen bond "donor-acceptor" coordinate is a *sine qua non* in understanding the vibrational signatures of ionic hydrogen bond? We offer the following answers: (1) The "donor acceptor" coordinate has the capability of inducing a red shift on the IHB stretch. This effect can be critical in predicting accurate peak positions. (2) IHB stretch couples strongly with its combination band with O–O stretch. Such interaction causes the combination band to possess appreciable intensity, making it observable in experimental measurements. Lastly, these interactions are found to be sensitive with the environment of the cluster as in the case of messenger tagging.

Acknowledgements This work was financially supported by the Academia Sinica and the Taiwan Ministry of Science and Technology (MOST) under grant MOST101-2113-M-001-023-MY3 and MOST104-2113-M-001-017. JAT would also thank MOST for conference travel support under MOST-103-2922-I-007-213. Computational resources are supported in part by the National Center for High Performance Computing. We wish to thank contributions from Jheng-Wei Li and Dr. Masato Morita for the DVR scanner and solver respectively. Fruitful discussions and helpful advice from Dr. Kaito Takahashi are also acknowledged.

References

1. Bell RP (1973) The proton in chemistry. Chapman and Hall, London
2. Cotton F, Wilkinson G, Gaus PL (1994) Basic inorganic chemistry, 3rd edn. Wiley, Canada
3. Jeffrey GA (1997) An introduction to hydrogen bonding. Oxford University Press, New York
4. Huggins ML (1936) Hydrogen bridges in ice and liquid water. J Phys Chem 40:723–731. doi:10.1021/j150375a004
5. Bethell DE, Sheppard N (1953) The infrared spectrum of the H_3O^+ ion in acid hydrates. J Chem Phys 21:1421–1422. doi:10.1063/1.1699260
6. Polfer NC (2011) Infrared multiple photon dissociation spectroscopy of trapped ion. Chem Soc Rev 40:2211–2221. doi:10.1002/mas.20217
7. Okumura M, Yeh LI, Lee YT (1988) Infrared spectroscopy of the cluster ions $H_{3+}(H_2)$n. J Chem Phys 88:79. doi:10.1063/1.454488
8. Yeh LI, Okumura M, Myers JD, Price JM, Lee YT (1989) Vibrational spectroscopy of the hydrated hydronium cluster ions H3O+. (H2O)n (n = 1, 2, 3). J Chem Phys 91:7319–7330. doi:10.1063/1.457305
9. Hammer NI, Diken EG, Roscioli JR, Johnson MA, Myshakin EM, Jordan KD, McCoy AB, Huang X, Bowman JM, Carter S (2005) The vibrational predissociation spectra of the H_5O_{2+}

RGn(RG = Ar, Ne) clusters: correlation of the solvent perturbations in the free OH and shared proton transitions of the Zundel ion. J Chem Phys 122:244301. doi:10.1063/1.1927522
10. Diken EG, Headrick JM, Roscioli JR, Bopp JC, Johnson MA, McCoy AB (2005) Fundamental excitations of the shared proton in the H_3O_{2-} and H_5O_{2+} complexes. J Phys Chem A 109:1487–1490. doi:10.1021/jp044155v
11. Headrick JM, Bopp JC, Johnson MA (2004) Predissociation spectroscopy of the argon-solvated H_5O_{2+} "zundel" cation in the 1000–1900 cm^{-1} region. J Chem Phys 121:11523–11526. doi:10.1063/1.1834566
12. Moore DT, Oomens J, van der Meer L, von Helden G, Meijer G, Valle J, Marshall AG, Eyler JR (2004) Probing the vibrations of shared, OH^+ O-bound protons in the gas phase. Chem Phys Chem 5:740–743. doi:10.1002/cphc.200400062
13. Huang X, Braams BJ, Bowman JM (2005) Ab initio potential energy and dipole moment surfaces for H_5O_{2+}. J Chem Phys. doi 10(1063/1):1834500
14. Roscioli JR, McCunn LR, Johnson MA (2007) Quantum structure of the intermolecular proton bond. Science 316:249–254. doi:10.1126/science.1138962
15. Fridgen TD, McMahon TB, MacAleese L, Lemaire J, Maitre P (2004) Infrared spectrum of the protonated water dimer in the gas phase. J Phys Chem A 108:9008–9010. doi:10.1021/jp040486w
16. Fridgen TD, MacAleese L, Maitre P, McMahon TB, Boissel P, Lemaire J (2005) Infrared spectra of homogeneous and heterogeneous proton-bound dimers in the gas phase. Phys Chem Chem Phys 7:2747–2755
17. Stoyanov ES, Reed CA (2006) IR Spectrum of the $H_5O_2{}^+$ cation in the context of proton disolvates L–H+–L, 12992–13002
18. Asmis KR, Pivonka NL, Santambrogio G, Brümmer M, Kaposta C, Neumark DM, Wöste L (2003) Gas-phase infrared spectrum of the protonated water dimer. Science 299:1375–1377. doi:10.1126/science.1081634
19. McCunn LR, Roscioli JR, Johnson MA, McCoy AB (2008) An H/D isotopic substitution study of the $H_5O_2{}^+$. Ar vibrational predissociation spectra: exploring the putative role of Fermi resonances in the bridging proton fundamentals. J Phys Chem B 112:321–327. doi:10.1021/jp075289m
20. Vendrell O, Gatti F, Lauvergnat D, Meyer H-D (2007) Full-dimensional (15-dimensional) quantum-dynamical simulation of the protonated water dimer. I. Hamiltonian setup and analysis of the ground vibrational state. J Chem Phys 127:184302. doi:10.1063/1.2787588
21. Vendrell O, Gatti F, Meyer H-D (2007) Full dimensional (15-dimensional) quantum-dynamical simulation of the protonated water dimer. II. Infrared spectrum and vibrational dynamics. J Chem Phys 127:184303. doi:10.1063/1.2787596
22. Vendrell O, Brill M, Gatti F, Lauvergnat D, Meyer H-D (2009) Full dimensional (15-dimensional) quantum-dynamical simulation of the protonated water-dimer III: mixed Jacobi-valence parametrization and benchmark results for the zero point energy, vibrationally excited states, and infrared spectrum. J Chem Phys 130:234305. doi:10.1063/1.3152488
23. Vendrell O, Gatti F, Meyer H-D (2009) Full dimensional (15 dimensional) quantum-dynamical simulation of the protonated water-dimer IV: isotope effects in the infrared spectra of $D(D_2O)_2^+$, $H(D_2O)_2^+$, and $D(H_2O)_2^+$ isotopologues. J Chem Phys 131:034308. doi:10.1063/1.3183166
24. Li X, Moore DT, Iyengar SS (2008) Insights from first principles molecular dynamics studies toward infrared multiple-photon and single-photon action spectroscopy: Case study of the proton-bound dimethyl ether dimer. J Chem Phys. doi:10.1063/1.2903446
25. Tan JA, Kuo J-L (2015) Strong quantum coupling in the vibrational signatures of a symmetric ionic hydrogen Bond: The Case of $(CH_3OH)_2H^+$. J Phys Chem A 119:11320–11328. doi:10.1021/acs.jpca.5b10554
26. McCoy AB, Huang X, Carter S, Landeweer MY, Bowman JM (2005) Full-dimensional vibrational calculations for $H_5O_2{}^+$ using an ab initio potential energy surface. J Chem Phys 122:061101. doi:10.1063/1.1857472
27. Valeev EF, Schaefer HF (1998) The protonated water dimer: Brueckner methods remove the spurious C_1 symmetry minimum. J Chem Phys 108:7197. doi:10.1063/1.476137

28. Vener MV, Kühn O, Sauer J (2001) The infrared spectrum of the O•••H•••O fragment of $H_5O_2^+$: Ab initio classical molecular dynamics and quantum 4D model calculations. J Chem Phys 114:240. doi:10.1063/1.1330748
29. Olesen SG, Guasco TL, Roscioli JR, Johnson MA (2011) Tuning the intermolecular proton bond in the $H_5O_2^+$ "Zundel ion" scaffold. Chem Phys Lett 509:89–95. doi:10.1016/j.cplett.2011.04.060
30. Vendrell O, Gatti F, Meyer HD (2007) Dynamics and infrared spectroscopy of the protonated water dimer. Angew Chemie Int Ed 46:6918–6921. doi:10.1002/anie.200702201
31. Fridgen TD, Macaleese L, McMahon TB, Lemaire J, Maitre P (2006) Gas phase infrared multiple-photon dissociation spectra of methanol, ethanol and propanol proton-bound dimers, protonated propanol and the propanol/water proton-bound dimer. Phys Chem Chem Phys 8:955–966. doi:10.1039/b516661f
32. Marx D (2006) Proton transfer 200 years after von Grotthuss: insights from ab initio simulations. Chem Phys Chem 7:1848–1870. doi:10.1002/cphc.200600128
33. Tan JA, Li J-W, Kuo J-L (2015) Proton quantum confinement on symmetric dimers of ammonia and lower amine homologs. In: Nascimento MA, Maruani J, Brandas EJ, Delgado-Barrio G (eds) Frontiers in quantum methods and applications in chemistry and physics. Selected proceedings of QSCP-XVIII (Paraty, Brazil, December, 2013). Springer, pp 77–89
34. Frisch MJ, Trucks GW, Schlegel HB, Scuseria GE, Robb MA, Cheeseman JR, Scalmani G, Barone V, Mennucci B, Petersson GA et al (2009) Gaussian 09, Revision D.01
35. Møller C, Plesset MS (1934) Note on an approximation treatment for many-electron systems. Phys Rev 46:618–622. doi:10.1103/PhysRev.46.618
36. Colbert DT, Miller WH (1992) A novel discrete variable representation for quantum mechanical reactive scattering via the S-matrix Kohn method. J Chem Phys 96:1982. doi:10.1063/1.462100
37. Light JC, Hamilton IP, Lill JV (1985) Generalized discrete variable approximation in quantum mechanicsa. J Chem Phys 82:1400. doi:10.1063/1.448462
38. Bacic Z, Light JC (1989) Theoretical methods for rovibrational states of floppy molecules. Annu Rev Phys Chem 40:469–498. doi:10.1146/annurev.physchem.40.1.469
39. Light JC, Carrington T (2003) Discrete variable representations and their utilization. Adv Chem Phys
40. Morita M, Takahashi K (2013) Multidimensional local mode calculations for the vibrational spectra of $OH^-(H_2O)_2$ and $OH^-(H_2O)_2 \cdot Ar$. Phys Chem Chem Phys 15:14973. doi:10.1039/c3cp51903a
41. Morita M, Takahashi K (2013) Multidimensional local mode calculations for $OH^-(H_2O)_3$—importance of intermode anharmonicity. Phys Chem Chem Phys 15:14973. doi:10.1039/c3cp51903a
42. Ogata Y, Kawashima Y, Takahashi K, Tachikawa M (2015) Effect of quantum distribution and vibrational. Phys Chem Chem Phys 17:25505–25515. doi:10.1039/C5CP03632A
43. Li J-W, Morita M, Takahashi K, Kuo J-L (2015) Features in vibrational spectra induced by Ar-tagging for $H_3O^+Ar_m$, $m = 0$–3. J Phys Chem A, 151022145830001. doi:10.1021/acs.jpca.5b08898
44. Stathopoulos A, Fischer CF (1994) A Davidson program for finding a few selected extreme eigenpairs of a large, sparse, real, symmetric matrix. Comput Phys Commun 79:268–290. doi:10.1016/0010-4655(94)90073-6
45. Atkins PW, Friedman RS (2005) Molecular quantum mechanics, 4th edn. Oxford University Press, New York
46. Meot-Ner M (2005) The ionic hydrogen bond. Chem Rev 105:213–284. doi:10.1021/cr9411785

Relativistic Many-Body Perturbation Theory Calculations of the Hyperfine Structure and Oscillator Strength Parameters for Some Heavy Element Atoms and Ions

O.Yu. Khetselius, P.A. Zaichko, A.V. Smirnov, V.V. Buyadzhi, V.B. Ternovsky, T.A. Florko and V.F. Mansarliysky

Abstract The formalism of the relativistic many-body perturbation theory with an optimized zeroth approximation is applied to computing the energies and hyperfine structure constants for some heavy Li-like multicharged ions and alkali (caesium) atoms. The exchange-correlation, nuclear and radiative corrections are correctly and effectively taken into account. The modified Uehling-Serber approximation is used to take into account for the Lamb shift polarization part. In order to take into account the contribution of the Lamb shift self-energy part we have used the generalized non-perturbative procedure, developed by Ivanov-Ivanova et al. The energies and oscillator strengths of radiation transition in spectra of some Li-like ions (Z = 20 – 70) and Cs are computed on the basis of the combined relativistic energy approach (S-matrix formalism) and relativistic many-body perturbation theory. The data on oscillator strengths of radiative transitions from the ground state to the low-excited and Rydberg states $2s_{1/2} - np_{1/2,3/2}$, $np_{1/2,3/2}$-$nd_{3/2,5/2}$ (n = 2 – 12) in the Li-like ions are presented. Some results are obtained at first. It is performed an analysis of the computed oscillator strength values with available theoretical and experimental results.

Keywords Hyperfine structure · Oscillator strengths · Relativistic perturbation theory · Energy approach · Correlation, nuclear, radiative corrections

---

O.Yu. Khetselius (✉) · P.A. Zaichko · A.V. Smirnov · V.V. Buyadzhi
V.B. Ternovsky · T.A. Florko · V.F. Mansarliysky
Odessa State Environmental University (OSENU), L'vovskaya str. 15,
Odessa-9 65016, Ukraine
e-mail: okhetsel@gmail.com

A. Tadjer et al. (eds.), *Quantum Systems in Physics, Chemistry, and Biology*,
Progress in Theoretical Chemistry and Physics 30,
DOI 10.1007/978-3-319-50255-7_16

1 Introduction

It is well known that studying the spectroscopic and radiation characteristics, including the spectral lines hyperfine structure, for heavy elements and multicharged ions is of a great interest for the further development as atomic and nuclear theories and spectroscopy of multicharged ions (see, for example, Refs. [1–18]). Besides, the corresponding data on the spectroscopic and structural properties of the heavy neutral and highly ionized atoms are of a great importance for many fields of a modern atomic physics, plasma physics and chemistry, astrophysics, laser physics, quantum electronics and so on. The precise determination of the hyperfine structure constants allows to refine available data on nuclear magnetic moments of different elements (isotopes). At last, precise computing these constants for the neutral and highly ionized atoms is very important from the viewpoint of check as consistency of theoretical description of the nuclear effects and other as an accuracy of the various atomic computational codes.

The traditional and widespread methods to computing spectral and radiation characteristics of multi-electron systems with a large nuclear charge are provided by the multi-configuration relativistic Hartree-Fock (RHF) and Dirac-Fock (DF) approaches (see, for example, Refs. [3–5, 8–18]). At present time there are actively used three very general and important computer systems for relativistic and QED calculations of atomic and molecular properties developed in the Oxford and German-Russian groups etc. ("GRASP", "Dirac"; "BERTHA", "QED", "Dirac"). From the QED viewpoint the useful overview of the relativistic electronic structure theory is presented in Refs. [1–12].

In the present chapter we briefly present the key aspects of the relativistic many-body perturbation theory (PT) formalism with an optimized zeroth approximation and results of its application to computing the energies and hyperfine structure constants for some heavy Li-like ions and atoms. The main exchange-correlation, nuclear and radiative corrections are correctly taken into account [1, 12–14]. The contribution of the magnetic inter-electron interaction is considered in the lowest order on α^2 (α is the fine structure constant) parameter. The modified Uehling-Serber approximation is used to take into account for the Lamb shift polarization part. In order to take into account the contribution of the Lamb shift self-energy part we have used the generalized non-perturbative scheme by Ivanov-Ivanova et al. The oscillator strengths of radiation transition in spectra of some Li-like ions ($Z = 20 - 70$) and Cs are computed on the basis of the combined relativistic energy approach [1, 10] and relativistic many-body PT with the zeroth optimized one-particle approximation. The data on oscillator strengths of radiative transitions from the ground state to the low-excited and Rydberg states $2s_{1/2} - np_{1/2,3/2}$, $np_{1/2,3/2}$-$nd_{3/2,5/2}$ ($n = 2 - 12$) in the Li-like ions are presented.

2 Relativistic Method to Computing Hyperfine Structure Parameters of Atoms and Multicharged Ions

Here we present a brief description of the key moments of our approach (more details can be found in Refs. [1, 11–14]). The relativistic electron wave functions are determined from solution of the relativistic Dirac equation with a general potential. The latter includes ab initio mean-field potential, electric, polarization potentials of a nucleus.

Let us consider in details more simple case of the Li-like ion. The charge distribution in the Li-like ion nucleus $\rho(r)$ can be described by the Gaussian function:

$$\rho(r|R) = \left(4\gamma^{3/2}/\sqrt{\pi}\right)\exp\left(-\gamma r^2\right) \tag{1a}$$

(here $\gamma = 4/\pi R^2$ and R is the effective nucleus radius) or by the Fermi function:

$$\rho(r) = \rho_0/\{1 + \exp[(r - c)/a)]\}, \tag{1b}$$

where the parameter $a = 0.523$ fm, the parameter c is chosen by such a way that it is true the following condition for average-squared radius:

$$< r^2 >^{1/2} = (0.836 \cdot \mathrm{A}^{1/3} + 0.5700)\mathrm{fm}. \tag{1c}$$

Further one should use the formulas for the finite size nuclear potential and its derivatives on the nuclear radius. The Coulomb potential for the spherically symmetric density $\rho(r|R)$ is:

$$V_{nucl}(r|R) = -((1/r)\int_0^r dr' r'^2 \rho\left(r'|R\right) + \int_r^\infty dr' r' \rho\left(r'|R\right) \tag{2}$$

Within the differential equations method by Ivanov-Ivanova et al. it is determined by the system of differential equations with the corresponding boundary conditions, which are in details presented in Refs. [1, 12].

For definiteness let us further to consider the DF type equations for a three-electron system 1 s^2nlj, which formally fall into the one-electron relativistic Dirac equations for the orbitals 1s and nlj with the potential:

$$V(r) = 2V(r|1s) + V(r|nlj) + V_{ex}(r) + V(r|R) \tag{3}$$

It is important to note that the potential (3) includes the electrical and the polarization potentials of a nucleus and the standard Hartree potential and the potential.

V_{ex} of the exchange inter-electron interaction too. The main exchange effect is accounted for if in the equation for the valent electron orbital we assume

$$V(r) = V(r|core) + V(r|nlj) \tag{4}$$

and in the equation for the *nlj* orbital:

$$V(r) = 2V(b, r|core), \tag{5}$$

where b is the optimization parameter (see below). The rest of the exchange and correlation effects are considered in the first two orders of the PT by the total inter-electron interaction [9–12].

To take into account the important (for heavy atomic systems) radiative QED corrections one should use the procedure, which is in details described in the Refs. [1, 11–14]. An effect of the vacuum polarization effect is usually taken into account in the first PT theory order by means of the Uehling-Serber approximation. In particular, the Uehling potential can be written as follows:

$$U(r) = -\frac{2\alpha}{3\pi r}\int_1^{\infty} dt\, \exp(-2rt/\alpha Z)\left(1 + 1/2t^2\right)\frac{\sqrt{t^2-1}}{t^2} \equiv -\frac{2\alpha}{3\pi r}C(g), \tag{6}$$

where $g = r/(\alpha Z)$. More correct and consistent approach is presented in Ref. [1]. The Uehling potential, determined as a quadrature (6), may be approximated with high precision by a simple analytical function. The use of new approximation of the Uehling potential permits one to decrease the calculation errors for this term down to 0.5–1%. A method for calculation of the self-energy part of the Lamb shift is based on an idea by Ivanov-Ivanova (see Ref. [9]). In an atomic system the radiative shift and the relativistic part of energy are, in principle, defined by one and the same physical field. One could suppose that there exists some universal function that connects the self-energy correction and the relativistic energy. The self-energy correction for the states of a hydrogen-like ion was presented by Mohr [7]:

$$E_{SE}(H|Z, nlj) = 0.027148\frac{Z^4}{n^3}F(H|Z, nlj) \tag{7}$$

These results are modified here for the states $1s^2\,nlj$ of Li-like ions. It is supposed that for any ion with *nlj* electron over the core of closed shells the sought value may be presented in the form:

$$E_{SE}(Z, nlj) = 0.027148\frac{\xi^4}{n^3}f(\xi, nlj)\ \left(cm^{-1}\right) \tag{8}$$

where parameter $\xi = (E_R)^{1/4}$, E_R is the relativistic part of the bounding energy of the outer electron; the universal function $f(\xi, nlj)$ does not depend on the composition of the closed shells and the actual potential of the nucleus. The general computational scheme in a case of Li-like ions with a finite nucleus consists of a few steps: (i) computing the values E_R and ξ for the states *nlj* of H-like ions with the point nucleus; (ii) construction of an approximating function $f(\xi, nlj)$ by the found reference Z and the appropriate $F(H|Z, nlj)$; (iii) computing E_R and ξ for the states *nlj* of *Li*-like ions with a finite nucleus; (iv) computing value of E_{SE} for the sought states by the formula (8). The energies of the states of Li-like ions were calculated twice: with a conventional constant of the fine structure $\alpha = 1/137.03597$ and with $\tilde{\alpha} = \alpha/1000$. A detailed evaluation of the computational accuracy may be made only after a complete calculation of $E^n_{SE}(Li\,Z, nlj)$. It may be stated that the above extrapolation method is more justified than using the widely spread expansions by the parameter αZ.

In order to describe the hyperfine structure of atomic system let us remind that the energies of electric quadruple and magnetic dipole interactions are defined by a standard way with the hyperfine structure constants, usually expressed through the standard radial integrals:

$$A = \{[(4,32587)10^{-4}Z^2\chi g_I]/(4\chi^2 - 1)\}(RA)_{-2}, \tag{9a}$$

$$B = \{7.2878\,10^{-7}Z^3Q/[(4\chi^2 - 1)I(I - 1)\}(RA)_{-3}, \tag{9b}$$

Here g_I is the Lande factor, Q is a quadruple momentum of nucleus (in Barn); $(RA)_{-2}$, $(RA)_{-3}$ are the radial integrals usually defined as follows:

$$(RA)_{-2} = \int_0^\infty drr^2F(r)G(r)U(1/r^2, R), \tag{10a}$$

$$(RA)_{-3} = \int_0^\infty drr^2[F^2(r) + G^2(r)]U(1/r^3, R). \tag{10b}$$

The radial parts F and G of the Dirac two components function for electron, which moves in the potential $V(r, R) + U(r, R)$, are determined by solution of the Dirac equations. To define the hyperfine interaction potentials $U(1/r^n, R)$, we use the method by Ivanov et al. [9]. Other details can be found in [1, 11–14].

3 Relativistic Energy Approach to Computing Oscillator Strengths for Multicharged Ions

Let us remind that an initial general energy formalism combined with an empirical model potential method in a theory of atoms and multicharged ions has been developed by Ivanov-Ivanova et al. [9], further more general ab initio gauge-invariant relativistic approach has been presented by Glushkov-Ivanov [10]. We use the optimized version of this formalism which is presented in [1, 11, 18]. In the energy approach [9, 10] the imaginary part of electron energy shift of an atom is connected with the radiation decay possibility (transition probability). The total energy shift of the state is usually presented in the form:

$$\Delta E = \mathrm{Re}\Delta E + i\Gamma/2 \tag{11}$$

where Γ is interpreted as the level width. For the α-n radiation transition ImΔE in the lowest order of the PT is determined as [9]:

$$\mathrm{Im}\Delta E = -\frac{1}{4\pi} \sum_{\substack{\alpha > n > f \\ [\alpha < n \leq f]}} V_{\alpha n \alpha n}^{|\omega_{\alpha n}|}, \tag{12}$$

where $\omega_{\alpha n}$ is a frequency of the α-n radiation, (α > n > f) for electron and (α < n < f) for vacancy. The matrix element *V* is determined as follows:

$$V_{ijkl}^{|\omega|} = \iint dr_1 dr_2 \Psi_i^*(r_1)\Psi_j^*(r_2)\frac{\sin|\omega|r_{12}}{r_{12}}(1-\alpha_1\alpha_2)\Psi_k^*(r_2)\Psi_l^*(r_1) \tag{13}$$

where α_1, α_2 are the Dirac matrices. The separated terms of the sum in (13) represent the contributions of different channels and a probability of the dipole transition is:

$$\Gamma_{\alpha n} = \frac{1}{4\pi} \cdot V_{\alpha n \alpha n}^{|\omega_{\alpha n}|}$$

$$V_{1234}^{\omega} = [(j_1)(j_2)(j_3)(j_4)]^{1/2} \sum_{\lambda\mu}(-1)^{\mu}\begin{pmatrix} j_1 j_3 & \lambda \\ m_1 - m_3 & \mu \end{pmatrix} \times \mathrm{Im}[Q_\lambda^{\mathrm{Cul}}(1243) + Q_\lambda^{\mathrm{Br}}(1243)] \tag{14}$$

where j_i is the total single electron momentums, m_i—the projections; Q^{Cul} is the Coulomb part of interaction, Q^{Br}—the Breit part. For example, the imaginary part Q_λ^{Cul} contains the radial R_λ and angular S_λ integrals as follows:

$$\begin{aligned} \mathrm{Im}\, Q_\lambda^{Cul}(12;43) = Z^{-1}\mathrm{Im}\{&R_\lambda(12;43)S_\lambda(12;43) + R_\lambda(\tilde{1}2;43)S_\lambda(\tilde{1}2;43) \\ &+ R_\lambda(12;\tilde{4}3)S_\lambda(12;\tilde{4}3) + R_\lambda(\tilde{1}2;\tilde{4}3)S_\lambda(\tilde{1}2;\tilde{4}3)\}. \end{aligned} \tag{15}$$

The detailed expressions for the Coulomb and Breit parts can be found in Refs. [9, 19], in particular, the detailed expressions for the radial R_λ integrals and angular S_λ parts. The corresponding oscillator strength: $gf = \lambda^2\Gamma_{\alpha n}/6.67 \times 10^{15}$, where g is the degeneracy degree, λ is a wavelength in angstrems (Å). Their detailed definitions are presented in Refs. [9, 10]. The relativistic wave functions are calculated by solution of the Dirac equation with the potential, which includes the “outer electron-ionic core” potential and polarization potential [1, 11]. In order to describe interaction of the outer electron with the He-like core the modified model potential (4) has been used. The calibration of the single model potential parameter b has been performed on the basis of the special ab initio procedure within relativistic energy approach [10] (see also [1, 11, 18]). In Ref. [10] the lowest order multi-electron effects, in particular, the gauge dependent radiative contribution Im $\delta E_{\rm ninv}$ for the certain class of the photon propagator calibration is treated. This value is considered to be representative for the correlation effects, whose minimization is a reasonable criterion in the searching for the optimal one-electron basis of the many-body PT. The minimization of the density functional Im $\delta E_{\rm ninv}$ leads to the integral-differential equation that can be solved using one of the standard codes. It provides the construction of the optimized one-particle representation. All calculations are performed with using the numeral code Superatom-ISAN (version 93).

4 Results and Conclusions

Below we present the results of computing the oscillator strengths for the radiation transitions (particularly, $2s_{1/2} - np_{1/2,3/2}$, $np_{1/2,3/2} - nd_{3/2,5/2}$; n = 2 – 12) in spectra of some Li-like multicharged ions (Z = 20 – 70). In order to perform a test the obtained results we compare our computed data on the oscillator strengths for some Li-like ions with the known theoretical and tabulated results [17, 18]. In Table 1 we list the computed oscillator strength values for the $2s_{1/2}$–$2p_{1/2,3/2}$ transitions in Li-like ions Ca^{17+}, Kr^{33+}, Zr^{37+}, Mo^{39+}, Sn^{47+}, Cs^{52+}, Ba^{53+}, Tm^{66+}, Yb^{67+}. The corresponding DF data by Zilitis [17] and the "best" compillated (experimental) data [17] for the low-Z Li-like ions are listed in Table 1 too.

One should note that hitherto the experimental data on the oscillator strengths for many (especially, high-Z) Li-like ions are absent. The analysis shows that there is a physically reasonable agreement between the listed data. Our theoretical data are in a good agreement with available experimental results [17, 18] that is provided by

Table 1 Oscillator strengths of the $2s_{1/2}$–$2p_{1/2,3/2}$ transitions in Li-like ions

Method	DF [17]	DF [17]	[17]	[17]	Our data	Our data
Ion	$2s_{1/2}$–$2p_{1/2}$	$2s_{1/2}$–$2p_{3/2}$	$2s_{1/2}$–$2p_{1/2}$	$2s_{1/2}$–$2p_{3/2}$	$2s_{1/2}$–$2p_{1/2}$	$2s_{1/2}$–$2p_{3/2}$
Ca^{17+}	0.0234	0.0542	0.024	0.054	0.0236	0.0541
Kr^{33+}	–	–	0.0127	0.0491	0.0131	0.0514
Zr^{37+}	0.0114	0.0543	–	–	0.0118	0.0540
Mo^{39+}	–	–	0.011	0.056	0.0107	0.0556
Sn^{47+}	0.0092	0.0686	–	–	0.0095	0.0684
Cs^{52+}	–	–	–	–	0.0085	0.0761
Ba^{53+}	–	–	–	–	0.0084	0.0788
Tm^{66+}	–	–	–	–	0.0071	0.1140
Yb^{67+}	0.0067	0.1170	–	–	0.0069	0.1167

Table 2 The reduced dipole matrix elements (a.u.) of some transitions in the Cs (see text)

Transition	SD [4]	Scaled [4]	DF [4]	RHF [4]	RHF [4]	QDA [13]	RMPT Our data	Exp.
$6p_{1/2}$-6s	4.482	4.535	4.510	4.494	–	4.282	4.486	4.4890(7)
$6p_{3/2}$-6s	6.304	6.382	6.347	6.325	–	5.936	6.320	6.3238(7)
$7p_{1/2}$-6s	0.297	0.279	0.280	0.275	0.2825	0.272	0.283	0.284(2)
$7p_{3/2}$-6s	0.601	0.576	0.576	0.583	0.582	0.557	0.582	0.583(9)
$8p_{1/2}$-6s	0.091	0.081	0.078	–	–	0.077	0.087	–
$8p_{1/2}$-6s	0.232	0.218	0.214	–	–	0.212	0.225	–
$6p_{1/2}$-7s	4.196	4.243	4.236	4.253	4.237	4.062	4.231	4.233(22)
$6p_{3/2}$-7s	6.425	6.479	6.470	6.507	6.472	6.219	6.478	6.479(31)
$7p_{1/2}$-7s	10.254	10.310	10.289	10.288	10.285	9.906	10.308	10.308(15)
$7p_{3/2}$-7s	14.238	14.323	14.293	14.295	14.286	13.675	14.322	14.320(20)

Table 3 The calculation results of the nuclear finite size correction into energy (cm $^{-1}$) of the low transitions for Li-like ions and values of the effective radius of nucleus (10 $^{-13}$ cm)

Z	$2s_{1/2}$-$2p_{1/2}$	$2s_{1/2}$–$2p_{3/2}$	R
20	–15.1	–15.5	3.26
69	–20 690.0	–21 712.0	4.93
79	–62 315.0	–66 931.0	5.15
92	–267 325.0	–288 312.0	5.42

using the optimized one-quasi-particle representation and accounting for the main exchange-correlation (polarization) effects. The estimate of the gauge-non-invariant contributions into the radiation width (speech is about the difference between the oscillator strengths values computed with using the transition operator in the form of "length" and "velocity") is about 0.3%, i.e., the results obtained with the photon propagator gauges (Coulomb, Babushkin, Landau) are very close.

In Table 2 we present our results (RMPT) of computing the reduced matrix elements (in atomic units) of the radiative transitions in the ^{133}Cs spectrum. The experimental (Exp) and other theoretical (SD- the results of computing within the relativistic DF single-double approximation [4]; DF, RHFc—the Dirac-Fock and relativistic Hartree—Fock method data with accounting for the second order correlation corrections; QDA—the data by the PT with the quantum defect approximation) [4, 13, 18] data are listed too. In Table 3 we present the results calculation of the nuclear finite size correction to energy (cm^{-1}) of some transitions for Li-like ions and values of the effective radius of nucleus (10^{-13} cm). Our calculation showed also that a variation of the nuclear radius on a few % could lead to changing transition energies on dozens of thousands 10^3 cm^{-1}.

In Table 4 we present the calculated data of the hyperfine structure constants for some Li-like ions (the parameters: $A = Z^3 g_I \bar{A}$, $B = Z^3 Q\bar{B}/[I(2I-1)]$; in cm^{-1}).

Table 4 The hyperfine structure constants of the Li-like ions: $A = Z^3 g_I \bar{A}, B = Z^3 Q\bar{B}/[I(2I-1)]$

nlj	Z	30	41	59	92
3s	$\bar{A}$	29–03	32–03	43–03	90–03
4s	$\bar{A}$	11–03	14–03	16–03	36–03
$2p_{1/2}$	$\bar{A}$	30–03	35–03	46–03	105–02
$3p_{1/2}$	$\bar{A}$	91–04	09–03	12–03	31–03
$4p_{1/2}$	$\bar{A}$	37–04	43–04	58–04	11–03
$2p_{3/2}$	$\bar{A}$	55–04	60–04	65–04	72–04
	$\bar{B}$	10–04	11–04	12–04	17–04
$3p_{3/2}$	$\bar{A}$	14–04	16–04	18–04	22–04
	$\bar{B}$	37–05	41–05	48–05	62–05
$4p_{3/2}$	$\bar{A}$	70–05	77–05	84–05	8–04
	$\bar{B}$	12–05	14–05	18–05	26–05
$3d_{3/2}$	$\bar{A}$	90–05	94–05	98–05	12–04
	$\bar{B}$	63–06	74–06	84–06	11–05
$4d_{3/2}$	$\bar{A}$	40–05	44–05	49–05	58–05
	$\bar{B}$	22–06	32–06	41–06	56–06
$3d_{5/2}$	$\bar{A}$	39–05	42–05	46–05	52–05
	$\bar{B}$	29–06	31–06	35–06	40–06
$4d_{5/2}$	$\bar{A}$	16–05	17–05	18–05	21–05
	$\bar{B}$	11–06	12–06	14–06	17–06
$4f_{5/2}$	$\bar{A}$	07–05	08–05	11–05	14–05
	$\bar{B}$	37–07	42–07	49–07	63–07

One could note that accurate taking into consideration the correlation and QED corrections is important to reach the physically reasonable agreement between theoretical and experimental results. In Table 5 we list the experimental (A^{Exp}) and theoretical data on the magnetic dipole constant A (MHz) for different valent states of the ^{133}Cs atom ($I = 7/2$, $g_i = 0.7377208$). The theoretical data have been computed within the standard RHF (A^{RHF}) method, the RHF ($A^{RHF} + dA$) one with taking into consideration the PT second and higher corrections (look [5, 12, 18] and Refs therein) and the relativistic many-body PT (RMPT; A^{RMPT}) formalism (our data). In conclusions let us note that the key factors for the physically reasonable agreement between theory and experiment are connected with the correct taking into consideration the inter-electron correlation, nuclear, relativistic, radiative corrections. The difference between the RHF, RMPT and other data is provided by using different schemes for accounting of the inter-electron correlations. To reach the further improvement of the computed data one should take into account more correctly the spatial distribution of the magnetic moment inside a nucleus (the Bohr-Weisskopf effect), the nuclear-polarization corrections etc. (for example,

Table 5 The values (in MHZ) of the hyperfine structure constant A for valent states of the ^{133}Cs isotope: A^{Exp}—experiment; A^{RHF}—the RHF calculation data; A^{RHF} + dA^{RHF}—the RHF calculation data with taking into account the PT second and higher orders contributions [5]; A^{RMPT}—the RMPT calculation data [29] (look details in Refs. [5, 12, 18])

State	A^{RHF}	dA	$A^{RHF} + dA$	dA^{RMPT} Our data	A^{RMPT} Our data	A^{Exp}
$6s_{1/2}$	1426.81	864.19	2291.00	870.96	2294.45	2298.16
$7s_{1/2}$	392.05	151.99	544.04	152.45	545.480	545.90(9)
$6p_{1/2}$	161.09	131.58	292.67	130.08	292.102	291.90(13)
$7p_{1/2}$	57.68	35.53	94.21	35.64	94.317	94.35(4)
$6p_{3/2}$	23.944	25.841	49.785	26.322	50.205	50.275(3)
$7p_{3/2}$	8.650	7.605	16.255	7.920	16.590	16.605(6)
$5d_{3/2}$	–	–	–	7.802	16.422	–

within the Woods-Saxon model or relativistic mean field theory [3, 6, 15, 16, 20]. In last years this topic has been a subject of intensive interest [3–10].

Acknowledgements Authors are thankful to Prof. E. Brändas, J. Maruani, A. Tadjer, R. Pavlov for the invitation to present an invited report at the workshop QSCP-XX (Bulgaria, 2015). The help in editing the manuscript by Mr. Heike Rossel and Ms. Heena Naveen (Springer) is very much appreciated.

References

1. Khetselius OY (2009) Int J Quant Chem 109:3330; Phys Scripta T 135:014023 (2009); In: Nascimento M, Maruani J, Brändas E, Delgado-Barrio G (eds)Frontiers in quantum methods and applications in chemistry and physics, Series: Progress in theoretical chemistry and physics, vol 29. Springer, Berlin, pp 54–76
2. Nagasawa T, Naga A, Nakano M (2004) Phys Rev C 69:034322; Glushkov AV, Khetselius O, Svinarenko A (2013) Phys Scripta T153:014029
3. Blundell SA, Sapirstein J, Johnson WR (1992) Phys Rev D 45:1602; Dzuba VA, Flambaum VV, Sushkov OP (1997) Phys Rev A 56:R4357
4. Cheng K, Kim Y, Desclaux J (1979) Atom Data Nucl Data Tabl 24:11; Safranova UI, Safranova MS, Johnson WR (2005) Phys Rev A 71:052506; Yerokhin VA, Artemyev AN, Shabaev VM (2007) Phys Rev A 75:062501
5. Tomaselli M, Schneider SM, Kankeleit E, Kuhl T (1995) Phys Rev C 51:2989
6. Bieron J, Pyykkő P, Jonsson P (2005) Phys Rev A 71:012502
7. Mohr PJ (1974) Ann Phys (NY) 88:52; (1982) Phys Rev A 26:2338
8. Johnson W, Sapirstein J, Blundell S (1993) Phys Scripta T 46:184; Persson H, Lindgren I, Salomonson S (1996) Phys Rev Lett 76:204
9. Ivanova EP, Ivanov LN (1979) Atom Data Nucl Data Tabl 24:95; Ivanov LN, Ivanova EP, Aglitsky EV (1988) Phys Rep 166:315; Ivanova EP, Ivanov LN, Glushkov AV, Kramida AE (1985) Phys Scripta 32:512
10. Glushkov AV, Ivanov LN, Ivanova EP (1986) Autoionization phenomena in atoms. Moscow University Press, Moscow, pp 152–164; Glushkov AV, Ivanov LN (1992) Phys Lett A 170:33; Glushkov AV (2012) Quantum systems in chemistry and physics: progress in methods and applications. Nishikawa K, Maruani J, Brandas E, Delgado-Barrio G, Piecuch P (eds) Progress in theoretical chemistry and physics, vol 26. Springer, Berlin, pp 231–254

11. Khetselius OYu (2008) Hyperfine structure of atomic Spectra. Astroprint, Odessa
12. Glushkov AV, Ambrosov S, Loboda A, Chernyakova Yu, Khetselius OYu, Svinarenko A (2004) Nucl Phys A 734S:21; Tkach TB (2012) Photoelectr 21:14
13. Glushkov AV, Ambrosov SA, Loboda AV, Gurnitskaya EP, Khetselius OYu (2006) Recent advances in the theory of chemical and physical systems. In: Julien P, Maruani J, Mayou D, Wilson S, Delgado-Barrio G (eds) Progress in theoretical chemistry and physics, vol 15. Springer, Berlin, pp 285–300
14. Glushkov AV, Malinovskaya SV, Khetselius OYu, Loboda AV (2007) SLAC eConf C070910, vol 2. Menlo Park, CA, USA, pp 118
15. Serot B, Walecka J (1986) Advances in nuclear phys.: relativistic nuclear many body problem. Plenum Press, New York
16. Khetselius O (2012) Quantum Systems. In: Nishikawa K, Maruani J, Brandas E, Delgado-Barrio G, Piecuch P (eds) Chemistry and physics: progress in methods and applications. series: progress in theoretical chemistry and physics, vol 26. Springer, Berlin, pp 217–230
17. Martin G, Wiese W (1976) Phys Rev A 13:699; Zilitis V (1983) Opt Spectr 55:215; Schweizer W, Faßbinder P, Gonzalez-Ferez R (1999) Atom Data Nucl Data Tabl 72:33
18. Khetselius O, Florko T, Svinarenko A, Tkach T (2013) Phys Scripta T153:014037; Svinarenko A, Ignatenko A, Ternovsky V, Nikola L, Seredenko S, Tkach T (2014) J Phys: C Series. 548:012047
19. Malinovskaya S V, Glushkov A V, Khetselius O Yu, Lopatkin Yu M, Loboda A V, Svinarenko A A, Nikola L V, Perelygina T B (2011) Int J Quant Chem 111: 288
20. Malinovskaya S V, Glushkov A V, Khetselius O Yu, Svinarenko A A, Mischenko E V, Florko T A (2009) Int J Quant Chem 109: 3325

Effects of Isotope Characteristics on the Electron System Ground State Energy of Helium-Like Ions

Ch.J. Velchev, R.L. Pavlov, D. Tonev, Zh.K. Stoyanov, L.M. Mihailov, Y.D. Mutafchieva and D. Van Neck

Abstract The main goal of this work is to establish a link between the energy characteristics of the ground state of helium-like ions and the isotope characteristics of the nucleus of the system. High-precision calculations of the electron ground state energies of helium-like ions require to take into account the effects associated with nuclear characteristics and electron correlations. In our previous work, we calculated ground state electron energies, mass corrections and mass polarization effects of helium-like isoelectronic ions for the main nuclides with nuclear charge from $Z = 2$ to $Z = 118$. In the present work are discussed the results for 3833 existing isotopes in the same range of nuclear charge. The results presented are without the inclusion of the mass polarization effect in the minimization procedure. The complex dependence of the ground state energy on the mass corrections and mass polarization effects as a function of charge Z and neutron number N are studied. Staggering analysis for the ground state energy dependence on Z and N helped establish the electron characteristic dependence on the nuclear magic numbers.

1 Introduction

A computational procedure and the results for the ground state energy characteristics of the helium-like ions are presented in this paper. The computational procedure for these characteristics gives highly accurate results and allows for the establishment of

Ch.J. Velchev · R.L. Pavlov (✉) · D. Tonev · Zh.K. Stoyanov · Y.D. Mutafchieva
Institute for Nuclear Research and Nuclear Energy, Bulgarian Academy of Sciences,
72 Tsarigradsko Chaussee, 1784 Sofia, Bulgaria
e-mail: ropavlov@yahoo.fr

L.M. Mihailov
Institute of Solid State Physics, Bulgarian Academy of Sciences,
72 Tsarigradsko Chaussee, 1784 Sofia, Bulgaria

D. Van Neck
Center for Molecular Modelling, Ghent University, Technologiepark 903,
9052 Zwijnaarde, Belgium

A. Tadjer et al. (eds.), *Quantum Systems in Physics, Chemistry, and Biology*,
Progress in Theoretical Chemistry and Physics 30,
DOI 10.1007/978-3-319-50255-7_17

a number of correlations between the characteristics of isotopes and their interpretation. The results obtained using staggering analysis showcase interesting relations between the nuclear characteristics and the energy components of the electron system ground state. Such an approach to these problems is used for the first time. In our previous works [1–3], we calculated the ground state electron energies, mass correction and mass polarization effects and have shown specific properties of helium isoelectronic ions with nuclear charge for the main nuclides from $Z = 2$ to $Z = 118$. The calculations in this work are made for the 3833 isotopes whose characteristics are presented in [4]. The presented theoretical research allows for the development of a procedure with no free parameters, that need to be determined by experimental results. This is a serious advantage in the theoretical studies based on the energy characteristics of helium-like ions. The numerical values are obtained by solving the two-electron Schrödinger equation using a discrete variation-perturbation approach developed by the authors and based on the explicitly correlated wave functions. The procedure is presented in Sect. 2. This approach takes into account the motion of the nucleus and yields accurate values for the electron characteristics. The results are presented without the inclusion of the mass polarization in the minimization procedure. The comparison with the most recent experimental data from [4] shows that our results are in better agreement with experimental results than the considered until now most accurate theoretical results [5]. This allows for their use in theoretical studies of low [6] and high [7] temperature plasma processes; ion-ion and ion-atom interactions [8]; qualitative and quantitative structural investigations in chemistry and astronomy [9]; determining chemical composition, speed and temperature of objects using spectral measurements; nanotechnology; fusion; research on fundamental interactions by examining the processes in the atom electron shell and its structure [10]; the study of Bose-Einstein condensates [11, 12]; studies that use atomic physics methods [13]; analysis of the relationship between atomic nuclei properties and atom characteristics [14, 15]; etc.

In precise calculations of the ground state energy of helium-like ions, best results are obtained with approaches based on explicitly correlated wave functions (ECWF) [16]. The most accurate results for the ground state energy of the main isotopes from the line of stability of nuclei, with charge from $Z = 2$ to $Z = 10$, are obtained with different types of ECWFs [17–43]. New ECWFs with complicated analytical form are introduced in [44–49].

2 Method

2.1 Hylleraas Wave Function

The ECWF proposed by Hylleraas [17–21] is formulated in terms of Hylleraas coordinates:

$$s = r_1 + r_2; \quad t = r_2 - r_1; \quad u = r_{12} = |\vec{r}_2 - \vec{r}_1| \, , \tag{1}$$

where $\vec{r}_1$ and $\vec{r}_2$ are the radius-vector of two electrons in the Decarts-coordinate system with the zero point in the nuclear center and r_1 and r_2 are their magnitudes.

The generalized Hylleraas-type 2-electron wave functions we consider have the form [16]

$$|\Psi\rangle = \sum_{STU} C_{STU}|STU\rangle \,, \tag{2}$$

where

$$\langle \vec{r}_1\vec{r}_2|STU\rangle = \frac{1}{\pi\sqrt{2}} \mathrm{e}^{-s/2} s^S t^T u^U \,. \tag{3}$$

The S, T, U are positive integers, with T even as a consequence of the symmetry requirement of the spatial wave function.

The phase space in these Hylleraas coordinates is well known. For a normalizable function $F(s, t, u)$ symmetric in t one has

$$\begin{aligned} &\int d^3r_1 d^3r_2 \delta(r_1 + r_2 - s)\delta(r_2 - r_1 - t)\delta(r_{12} - u)F(s, t, u) \\ &\quad = 2\pi^2 \int_0^{+\infty} ds \int_0^s du \int_0^u dtu(s^2 - t^2)F(s, t, u) \,. \end{aligned} \tag{4}$$

The overlap of two Hylleraas wave functions is therefore given by

$$\langle STU|S'T'U'\rangle = \int_0^{+\infty} ds \int_0^s du \int_0^u dt\, u(s^2 - t^2)\mathrm{e}^{-s} s^{S+S'} t^{T+T'} u^{U+U'} \,. \tag{5}$$

With the standard integral

$$\begin{aligned} I(k, l, m) &= \int_0^{+\infty} ds \int_0^s du \int_0^u dt\, \mathrm{e}^{-s} s^k t^l u^m \\ &= \frac{(k + l + m + 2)!}{(l + 1)(l + m + 2)} \,, \end{aligned} \tag{6}$$

the overlap matrix becomes

$$\langle STU|S'T'U'\rangle = I(\mathscr{S} + 2, \mathscr{T}, \mathscr{U} + 1) - I(\mathscr{S}, \mathscr{T} + 2, \mathscr{U} + 1)) \,, \tag{7}$$

where $\mathscr{S} = S + S'$, $\mathscr{T} = T + T'$, $\mathscr{U} = U + U'$. To avoid numerical roundoff error it is advisable to calculate this difference analytically

$$\begin{aligned} &\langle STU|S'T'U'\rangle = \\ &\quad = (\mathscr{S} + \mathscr{T} + \mathscr{U} + 5)! \frac{2(2\mathscr{T} + \mathscr{U} + 6)}{(\mathscr{T} + 1)(\mathscr{T} + 3)(\mathscr{T} + \mathscr{U} + 3)(\mathscr{T} + \mathscr{U} + 5)} \,. \end{aligned} \tag{8}$$

2.2 Matrix Elements of the Hamiltonian

The non-relativistic Hamiltonian is the sum of the kinetic energy, the Coulomb attraction to the central charge, and the inter-electron repulsion. In atomic units:

$$H = T + V_c + V_e = -\frac{1}{2}(\nabla_1^2 + \nabla_2^2) - Z(\frac{1}{r_1} + \frac{1}{r_2}) + \frac{1}{|\vec{r}_1 - \vec{r}_2|} . \tag{9}$$

The potential terms are easily written in Hylleraas coordinates. Combined with the phase space element one has

$$u(s^2 - t^2)V_c = (-Z)4su , \tag{10}$$

$$u(s^2 - t^2)V_e = s^2 - t^2. \tag{11}$$

The kinetic term requires a bit more work. In general a matrix element can be written as

$$\int d^3r_1 d^3r_2 \psi_L(\vec{r}_1, \vec{r}_2)[-\frac{1}{2}(\nabla_1^2 + \nabla_2^2)]\psi_R(\vec{r}_1, \vec{r}_2)$$
$$= \frac{1}{2}\int d^3r_1 d^3r_2 [(\nabla_1\psi_L) \cdot (\nabla_1\psi_R) + (\nabla_2\psi_L) \cdot (\nabla_2\psi_R)] . \tag{12}$$

The conversion to s, t, u variables proceeds with the chain rule,

$$\frac{\partial}{\partial(x_\alpha)_i} = \left(\frac{\partial s}{\partial(x_\alpha)_i}\right)\frac{\partial}{\partial s} + \left(\frac{\partial t}{\partial(x_\alpha)_i}\right)\frac{\partial}{\partial t} + \left(\frac{\partial u}{\partial(x_\alpha)_i}\right)\frac{\partial}{\partial u}$$
$$= \frac{(x_\alpha)_i}{r_\alpha}\left(\frac{\partial}{\partial s} + (-1)^\alpha \frac{\partial}{\partial t}\right) + (-1)^\alpha \frac{(x_2)_i - (x_1)_i}{r_{12}}\frac{\partial}{\partial u} , \tag{13}$$

$\alpha = 1, 2, \;\; i = 1, 2, 3.$

Using L, R indices to indicate that the derivative operators act on the left or right wave function, one obtains

$$\nabla_{1L}\cdot\nabla_{1R} + \nabla_{2L}\cdot\nabla_{2R}$$
$$= (\frac{\partial}{\partial s_L} - \frac{\partial}{\partial t_L})(\frac{\partial}{\partial s_R} - \frac{\partial}{\partial t_R}) + (\frac{\partial}{\partial s_L} + \frac{\partial}{\partial t_L})(\frac{\partial}{\partial s_R} + \frac{\partial}{\partial t_R}) + 2\frac{\partial}{\partial u_L}\frac{\partial}{\partial u_R}$$
$$+\frac{r_1^2 - \vec{r}_1 \cdot \vec{r}_2}{r_1 r_{12}}\{(\frac{\partial}{\partial s_L} + \frac{\partial}{\partial t_L})\frac{\partial}{\partial u_R} + \frac{\partial}{\partial u_L}(\frac{\partial}{\partial s_R} + \frac{\partial}{\partial t_R})\} \tag{14}$$
$$+\frac{r_2^2 - \vec{r}_1 \cdot \vec{r}_2}{r_2 r_{12}}\{(\frac{\partial}{\partial s_L} - \frac{\partial}{\partial t_L})\frac{\partial}{\partial u_R} + \frac{\partial}{\partial u_L}(\frac{\partial}{\partial s_R} - \frac{\partial}{\partial t_R})\} .$$

After simplification, using e.g.

$$\frac{r_1^2 - \vec{r}_1.\vec{r}_2}{r_1 r_{12}} \pm \frac{r_2^2 - \vec{r}_1 \cdot \vec{r}_2}{r_2 r_{12}} = \frac{(r_1 \pm r_2)(2r_1 r_2 \pm [r_{12}^2 - r_1^2 - r_2^2])}{2r_1 r_2 r_{12}}, \tag{15}$$

the operator of kinetic energy can be rewritten as

$$T = T_{ss} + T_{tt} + T_{uu} + T_{su} + T_{tu}, \tag{16}$$

where

$$u(s^2 - t^2)T_{ss} = u(s^2 - t^2)\frac{\partial}{\partial s_L}\frac{\partial}{\partial s_R}, \tag{17}$$

$$u(s^2 - t^2)T_{tt} = u(s^2 - t^2)\frac{\partial}{\partial t_L}\frac{\partial}{\partial t_R}, \tag{18}$$

$$u(s^2 - t^2)T_{uu} = u(s^2 - t^2)\frac{\partial}{\partial u_L}\frac{\partial}{\partial u_R}, \tag{19}$$

$$u(s^2 - t^2)T_{su} = s(u^2 - t^2)\{\frac{\partial}{\partial s_L}\frac{\partial}{\partial u_R} + \frac{\partial}{\partial u_L}\frac{\partial}{\partial s_R}\}, \tag{20}$$

$$u(s^2 - t^2)T_{tu} = t(s^2 - u^2)\{\frac{\partial}{\partial t_L}\frac{\partial}{\partial u_R} + \frac{\partial}{\partial u_L}\frac{\partial}{\partial t_R}\}, \tag{21}$$

i.e. $T_{\mu\nu} = f(h, \mu, \nu)D_{\mu\nu}$, where $f(h, \mu, \nu)$ is square function of (h, μ, ν), $D_{\mu\nu}$ are derivative operators of first order and $h, \mu, \nu = s, t$ *or* u.

It is now possible to calculate the matrix elements of the Hamiltonian. Using the notations of Eqs. (7), (8) and $\mathscr{V} = \mathscr{T} + \mathscr{U}$ $\mathscr{W} = \mathscr{S} + \mathscr{V}$, one finds the matrix elements of kinetic energy:

$$\langle STU|S'T'U'\rangle = (\mathscr{W} + 5)!\frac{2(\mathscr{T} + \mathscr{V} + 6)}{(\mathscr{T} + 1)(\mathscr{T} + 3)(\mathscr{V} + 3)(\mathscr{V} + 5)}, \tag{22}$$

$$\langle STU|T_{ss}|S'T'U'\rangle = \{SS'(\mathscr{W} + 3)! - \frac{S + S'}{2}(\mathscr{W} + 4)! + \frac{1}{4}(\mathscr{W} + 5)!\} \times\frac{2(\mathscr{T} + \mathscr{V} + 6)}{(\mathscr{T} + 1)(\mathscr{T} + 3)(\mathscr{V} + 3)(\mathscr{V} + 5)}, \tag{23}$$

$$\langle STU|T_{tt}|S'T'U'\rangle = TT'(\mathscr{W} + 3)!\frac{2(\mathscr{T} + \mathscr{V} + 2)}{(\mathscr{T} - 1)(\mathscr{T} + 1)(\mathscr{V} + 1)(\mathscr{V} + 3)}, \tag{24}$$

$$\langle STU|T_{uu}|S'T'U'\rangle = UU'(\mathscr{W} + 3)!\frac{2(\mathscr{T} + \mathscr{V} + 4)}{(\mathscr{T} + 1)(\mathscr{T} + 3)(\mathscr{V} + 1)(\mathscr{V} + 3)}, \tag{25}$$

$$\langle STU|T_{su}|S'T'U'\rangle = \{(SU' + US')(\mathscr{W} + 3)! - \frac{U + U'}{2}(\mathscr{W} + 4)!\} \times\frac{2}{(\mathscr{T} + 1)(\mathscr{T} + 3)(\mathscr{V} + 3)}, \tag{26}$$

$$\langle STU|T_{tu}|S'T'U'\rangle = (TU' + UT')(\mathscr{W} + 3)!\frac{2}{(\mathscr{T} + 1)(\mathscr{V} + 1)(\mathscr{V} + 3)}, \tag{27}$$

$$\langle STU|V_c|S'T'U'\rangle = (-Z)(\mathscr{W}+4)!\frac{4}{(\mathscr{T}+1)(\mathscr{V}+3)}\,, \tag{28}$$

$$\langle STU|V_e|S'T'U'\rangle = (\mathscr{W}+4)!\frac{2(\mathscr{T}+\mathscr{V}+5)}{(\mathscr{T}+1)(\mathscr{T}+3)(\mathscr{V}+2)(\mathscr{V}+4)}\,. \tag{29}$$

2.3 Scaling

If the Hylleraas wave function in (3) is subjected to a coordinate scaling transformation

$$\langle \vec{r}_1\vec{r}_2|\Psi_\alpha\rangle = \sum_{STU} C_{STU}\ \mathrm{e}^{-\alpha s/2}(\alpha s)^S(\alpha t)^T(\alpha u)^U\,, \tag{30}$$

then the overlap, kinetic, and potential matrix elements have definite scaling,

$$\langle \Psi_\alpha|\Psi_\alpha\rangle = \langle \Psi|\Psi\rangle/\alpha^6\,, \tag{31}$$

$$\langle \Psi_\alpha|T|\Psi_\alpha\rangle = \langle \Psi|T|\Psi\rangle/\alpha^4\,, \tag{32}$$

$$\langle \Psi_\alpha|V|\Psi_\alpha\rangle = \langle \Psi|V|\Psi\rangle/\alpha^5\,. \tag{33}$$

The expectation value of the energy then becomes

$$\frac{\langle \Psi_\alpha|H|\Psi_\alpha\rangle}{\langle \Psi_\alpha|\Psi_\alpha\rangle} = \frac{\alpha^2\langle \Psi|T|\Psi\rangle + \alpha\langle \Psi|V|\Psi\rangle}{\langle \Psi|\Psi\rangle}\,. \tag{34}$$

2.4 Variational Procedure

The energy will be minimized by varying the Hylleraas wave function to both the expansion coefficients C_{STU} and the scaling factor α.

One first determines, for fixed α, the best linear combination. Variation with respect to the expansion coefficients in (34) leads to the generalized eigenvalue problem,

$$(\alpha^2[T] + \alpha[V])C_\alpha = E_\alpha[M]C_\alpha\,. \tag{35}$$

where the square symmetric matrices $[T]$, $[V]$, $[M]$ are the kinetic, potential, and overlap matrices as derived above in the basis of the $|STU\rangle$ states, and C_α is the column matrix containing the expansion coefficients. The lowest eigenvalue E_α^0 defines a function of one variable, the minimum of which can be obtained using standard techniques. This minimum is finally the best variational approximation for the ground state energy.

In practice one calculates the matrices $[T]$, $[V]$ and $[M]$ only once. We use Lövdin's orthogonalization [50]. Solving the generalized eigenvalue problem in Eq. (35) proceeds by first diagonalizing the overlap matrix $[M]$, with resulting eigenvalues d_i and eigenvectors Z_i. In a next step one constructs the matrices $[T']_{ij} = (Z_i^T[T]Z_j)/\sqrt{d_i d_j}$ and $[V']_{ij} = (Z_i^T[V]Z_j)/\sqrt{d_i d_j}$. This has the advantage that it is possible to introduce a threshold for the eigenvalues of $[M]$. A large basis of polynomials may be almost linearly dependent in the limited volume around the atom. This results in very small eigenvalues d_i of $[M]$, and numerically unstable results. Such instabilities can be simply avoided by omitting the corresponding eigenvectors Z_i in the construction of $[T']$ and $[V']$. Finally one has to solve, for various values of α, the lowest eigenvalue of the eigenvalue problem

$$(\alpha^2[T'] + \alpha[V'])C'_\alpha = E_\alpha C'_\alpha \ . \tag{36}$$

This determines the function E_α^0 of which the minimum has to be determined.

3 Mass Corrections

In a system described by a two-electron Schrödinger equation, taking into account nuclear motion entails two mass corrections [51, 52]:

(i) As in the one-electron case, the mass correction

$$\varepsilon_1 = -\frac{\varepsilon}{1+\varepsilon}E_0 \approx -\varepsilon E_0 \ , \tag{37}$$

where $\varepsilon = m_e/M$ (m_e is the electron mass and M is the nucleus mass). The accounting of mass correction leads to increasing of the atomic energy approximately by $\varepsilon|E_0| = (m_e/M)|E_0|$, independently of the atomic state.

(ii) An additional perturbation correction, 'mass polarization'

$$\varepsilon_2 = \varepsilon \int \nabla_1 \Psi^*(\vec{r}_1, \vec{r}_2)\nabla_2\Psi(\vec{r}_1, \vec{r}_2)d\vec{r}_1 d\vec{r}_2 \ , \tag{38}$$

is different for various atomic states, as it depends on the mutual disposition and space correlation of the electrons.

Within ECWF approach, the mass corrections ε_1 and ε_2 are positive. For example, the order of ε_1 for Helium is 1.22×10^{-4} [17, 18], and the order of ε_2 is 2.18×10^{-5} [22–24]. There exist experimental data for some values of the nuclear charge Z. Regarding the ground state energy of He and isoelectronic He ions, variational solving of two-particle Schrödinger equation using trial ECWF, gives energy values, which are lower than the experimentally observed ones. In order to be able to compare these values to the experimental data, obligatory condition is to add mass corrections ε_1 and ε_2 to ground state energy. Exact coincidence with the experiment is expected after addition of relativistic and *QED* corrections as well.

For the ground state, the electrons are located at relatively small distance from each other, and this effect may be significant. The term ε_2 can be derived by perturbation from the unperturbed ψ_0. Following the procedure described in Sect. 2, we obtain:

$$(\nabla_1\psi_L)\cdot(\nabla_2\psi_R) = \frac{1}{2}\{\frac{(r_1^2+r_2^2-r_{12}^2)}{r_1r_2}\frac{\partial\psi_L}{\partial r_1}\frac{\partial\psi_R}{\partial r_2} + \frac{(r_2^2-r_1^2-r_{12}^2)}{r_1r_{12}}\frac{\partial\psi_L}{\partial r_1}\frac{\partial\psi_R}{\partial r_{12}} + \\ + \frac{(r_1^2-r_2^2-r_{12}^2)}{r_2r_{12}}\frac{\partial\psi_L}{\partial r_{12}}\frac{\partial\psi_R}{\partial r_2}\} - \frac{\partial\psi_L}{\partial r_{12}}\frac{\partial\psi_R}{\partial r_{12}}, \tag{39}$$

$$\mathscr{E} = \int(\nabla_1\psi_L)\cdot(\nabla_2\psi_R)d\mathbf{r}_1d\mathbf{r}_2 = 2\pi^2\int_0^\infty ds\int_0^s du\int_0^u dt\psi_L \\ \times(\mathscr{E}_{ss}+\mathscr{E}_{tt}+\mathscr{E}_{uu}+\mathscr{E}_{st}+\mathscr{E}_{su}+\mathscr{E}_{su}^A+\mathscr{E}_{tu}+\mathscr{E}_{tu}^A)\psi_R u(s_2-t_2), \tag{40}$$

where

$$u(s^2-t^2)\mathscr{E}_{ss} = u(s^2+t^2-2u^2)\frac{\partial}{\partial s_L}\frac{\partial}{\partial s_R}, \tag{41}$$

$$u(s^2-t^2)\mathscr{E}_{tt} = -u(s^2+t^2-2u^2)\frac{\partial}{\partial t_L}\frac{\partial}{\partial t_R}, \tag{42}$$

$$u(s^2-t^2)\mathscr{E}_{uu} = -u(s^2-t^2)\frac{\partial}{\partial u_L}\frac{\partial}{\partial u_R}, \tag{43}$$

$$u(s^2-t^2)\mathscr{E}_{st} = u(s^2+t^2-2u^2)\{\frac{\partial}{\partial s_L}\frac{\partial}{\partial t_R} - \frac{\partial}{\partial t_L}\frac{\partial}{\partial s_R}\}, \tag{44}$$

$$u(s^2-t^2)\mathscr{E}_{su} = t(s^2-u^2)\{\frac{\partial}{\partial s_L}\frac{\partial}{\partial u_R} - \frac{\partial}{\partial u_L}\frac{\partial}{\partial s_R}\}, \tag{45}$$

$$u(s^2-t^2)\mathscr{E}_{su}^A = s(t^2-u^2)\{\frac{\partial}{\partial s_L}\frac{\partial}{\partial u_R} + \frac{\partial}{\partial u_L}\frac{\partial}{\partial s_R}\}, \tag{46}$$

$$u(s^2-t^2)\mathscr{E}_{tu} = -s(t^2-u^2)\{\frac{\partial}{\partial t_L}\frac{\partial}{\partial u_R} - \frac{\partial}{\partial u_L}\frac{\partial}{\partial t_R}\}, \tag{47}$$

$$u(s^2-t^2)\mathscr{E}_{tu}^A = -t(s^2-u^2)\{\frac{\partial}{\partial t_L}\frac{\partial}{\partial u_R} + \frac{\partial}{\partial u_L}\frac{\partial}{\partial t_R}\}. \tag{48}$$

3.1 Matrix Elements

For the matrix elements of mass polarization ε_2 by analogy with the kinetic energy, we obtain:

$$\langle STU|\mathcal{E}_{ss}|S'T'U'\rangle = \frac{(\mathcal{W}+3)!}{2}\frac{[(S-S')^2-(\mathcal{V}+4)^2-(\mathcal{W}+4)]\mathcal{U}}{(\mathcal{T}+1)(\mathcal{T}+3)(\mathcal{V}+3)(\mathcal{V}+5)}, \quad (49)$$

$$\langle STU|\mathcal{E}_{tt}|S'T'U'\rangle = 2(\mathcal{W}+3)!\frac{TT'\mathcal{U}}{(\mathcal{T}-1)(\mathcal{T}+1)(\mathcal{V}+1)(\mathcal{V}+3)}, \quad (50)$$

$$\langle STU|\mathcal{E}_{uu}|S'T'U'\rangle = -(\mathcal{W}+3)!\frac{2UU'(\mathcal{T}+\mathcal{V}+4)}{(\mathcal{T}+1)(\mathcal{T}+3)(\mathcal{V}+1)(\mathcal{V}+3)}, \quad (51)$$

$$\langle STU|\mathcal{E}_{st}|S'T'U'\rangle = 0, \quad (52)$$

$$\langle STU|\mathcal{E}_{su}|S'T'U'\rangle = 0, \quad (53)$$

$$\langle STU|\mathcal{E}^{A}_{su}|S'T'U'\rangle = (\mathcal{W}+3)!\frac{(S-S')(U-U')+\mathcal{U}(\mathcal{V}+4)}{(\mathcal{T}+1)(\mathcal{T}+3)(\mathcal{V}+3)}, \quad (54)$$

$$\langle STU|\mathcal{E}_{tu}|S'T'U'\rangle = 0, \quad (55)$$

$$\langle STU|\mathcal{E}^{A}_{tu}|S'T'U'\rangle = -2(\mathcal{W}+3)!\frac{TU'+T'U}{(\mathcal{T}+1)(\mathcal{V}+1)(\mathcal{V}+3)}, \quad (56)$$

where by analogy with Eqs. (16–21)

$$\mathcal{E}_{\mu\nu} = f_{\mathcal{E}}(h,\mu\nu)D^{\mathcal{E}}_{\mu\nu}.$$

Here $f_{\mathcal{E}}(h,\mu\nu)$ are again square functions and $D^{\mathcal{E}}_{\mu\nu}$ are derivative operators; the index A notes anti-commutator.

In order to obtain mass polarization ε_2 from (38), after scaling

$$\langle\Psi_\alpha|\mathcal{E}|\Psi_\alpha\rangle = \langle\Psi|\mathcal{E}|\Psi\rangle/\alpha^4, \quad (57)$$

where $\langle\Psi|\mathcal{E}|\Psi\rangle$ is sum of the matrix elements (49–56), we substitute in (57) the values of the expansion coefficients C_{STU} and scale parameter α determined by variational equation (36).

The nuclear masses used in our computations were derived from recent tables of mass excess Δ [53], using the known relation:

$$M_{NUC} = \Delta + A - Zm_e.$$

As $\varepsilon > 0$, and $E_0 < 0$ in (37) while the integral in (38) is positive, both ε_1 and ε_2 are positive and decrease the absolute value of E_0. Thus for a given element (with atomic number Z and uncorrected energy E_0), the absolute value of the corrected energy E'_0 increases with increasing isotope mass M. It will be seen however that ε_2 is about two orders of magnitude smaller than ε_1.

4 Results

The developed method was originally applied [3] to calculate the ground state energy of helium-like electron-nuclear systems with isotopes from the stability line (Table 1) as their nucleus.

These results are in excellent agreement with the theoretical and experimental results [4] (the relative error for Helium is $2 \times 10^{-4}\%$) and allow for connecting the ground state energy characteristics and the properties of the isotope, which for these particular systems is the nucleus of the ion. This leads to the idea to compute the energy characteristics of the helium-like electron-nuclear systems, using as a nucleus the 3833 isotopes included in [53], in order to investigate the dependence of these properties from the isotope properties.

Table 1 Stable isotopes

$^{4}_{2}He$	$^{7}_{3}Li$	$^{9}_{4}Be$	$^{11}_{5}B$	$^{12}_{6}C$	$^{14}_{7}N$	$^{16}_{8}O$	$^{19}_{9}F$	$^{20}_{10}Ne$	$^{23}_{11}Na$	$^{24}_{12}Mg$
$^{27}_{13}Al$	$^{28}_{14}Si$	$^{31}_{15}P$	$^{32}_{16}S$	$^{35}_{17}Cl$	$^{40}_{18}Ar$	$^{39}_{19}K$	$^{40}_{20}Ca$	$^{45}_{21}Sc$	$^{48}_{22}Ti$	$^{51}_{23}V$
$^{52}_{24}Cr$	$^{55}_{25}Mn$	$^{56}_{26}Fe$	$^{59}_{27}Co$	$^{58}_{28}Ni$	$^{63}_{29}Cu$	$^{64}_{30}Zn$	$^{69}_{31}Ga$	$^{74}_{32}Ge$	$^{75}_{33}As$	$^{80}_{34}Se$
$^{79}_{35}Br$	$^{84}_{36}Kr$	$^{85}_{37}Rb$	$^{88}_{38}Sr$	$^{89}_{39}Y$	$^{90}_{40}Zr$	$^{93}_{41}Nb$	$^{98}_{42}Mo$	$^{97}_{43}Tc$	$^{102}_{44}Ru$	$^{103}_{45}Rh$
$^{106}_{46}Pd$	$^{107}_{47}Ag$	$^{114}_{48}Cd$	$^{115}_{49}In$	$^{120}_{50}Sn$	$^{121}_{51}Sb$	$^{130}_{52}Te$	$^{127}_{53}I$	$^{132}_{54}Xe$	$^{133}_{55}Cs$	$^{138}_{56}Ba$
$^{139}_{57}La$	$^{140}_{58}Ce$	$^{141}_{59}Pr$	$^{142}_{60}Nd$	$^{146}_{61}Pm$	$^{152}_{62}Sm$	$^{153}_{63}Eu$	$^{158}_{64}Gd$	$^{159}_{65}Tb$	$^{164}_{66}Dy$	$^{165}_{67}Ho$
$^{166}_{68}Er$	$^{169}_{69}Tm$	$^{174}_{70}Yb$	$^{175}_{71}Lu$	$^{180}_{72}Hf$	$^{181}_{73}Ta$	$^{184}_{74}W$	$^{187}_{75}Re$	$^{192}_{76}Os$	$^{193}_{77}Ir$	$^{195}_{78}Pt$
$^{197}_{79}Au$	$^{202}_{80}Hg$	$^{205}_{81}Tl$	$^{208}_{82}Pb$	$^{209}_{83}Bi$	$^{209}_{84}Po$	$^{201}_{85}At$	$^{222}_{86}Rn$	$^{223}_{87}Fr$	$^{226}_{88}Ra$	$^{227}_{89}Ac$
$^{232}_{90}Th$	$^{231}_{91}Pa$	$^{238}_{92}U$	$^{237}_{93}Np$	$^{244}_{94}Pu$	$^{243}_{95}Am$	$^{247}_{96}Cm$	$^{247}_{97}Bk$	$^{251}_{98}Cf$	$^{247}_{99}Es$	$^{257}_{100}Fm$
$^{258}_{101}Md$	$^{252}_{102}No$	$^{265}_{103}Lr$	$^{265}_{104}Rf$	$^{269}_{105}Db$	$^{272}_{106}Sg$	$^{274}_{107}Bh$	$^{276}_{108}Hs$	$^{279}_{109}Mt$	$^{278}_{110}Ds$	$^{283}_{111}Rg$
$^{280}_{112}Cn$	$^{287}_{113}Uut$	$^{289}_{114}Fl$	$^{291}_{115}Uup$	$^{290}_{116}Lv$	$^{292}_{117}Uus$	$^{294}_{118}Uuo$				

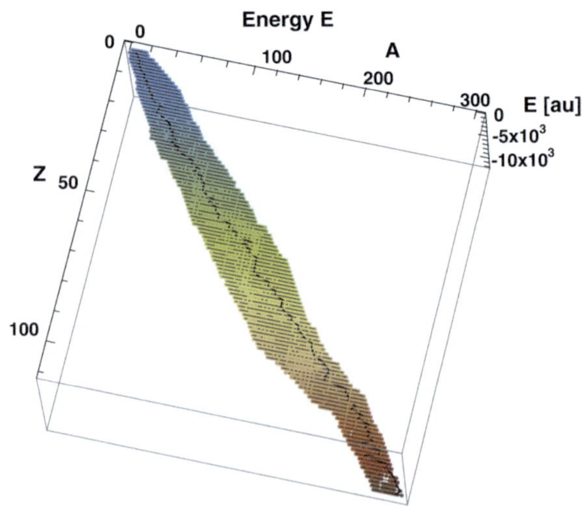

Fig. 1 Helium-like electron-nuclear systems ground state energy depending on the nuclear charge Z and its mass number A. Isotopes from the line of stability (*thick line*)

The ground state energy values of helium-like electron-nuclear systems are presented in Fig. 1. The ground state energy is presented as the sum of the energy calculated in a coordinate system fixed to the nucleus E_0 (energy obtained by taking into account only the charge of the nucleus in classical quantum mechanics) and energies corresponding to corrections of different types. As far as our views are within the classical quantum mechanics and we do not take into account the shape of the nucleus, the corrections added to E_0 are the mass correction ε_1 and mass correlation ε_2 [51].

The numerical results for E_0 are shown on Fig. 2. By definition E_0 should not depend on the number of neutrons in the nucleus but only on Z. This means that, when a three-dimensional graph of the dependence of E_0 from Z and N of the helium-like electron-nuclear systems (Fig. 3a) is plotted and is rotated so as to observe it from the positive direction of the Z-axis, the two-dimensional surface representing the dependence of E_0 from Z and N should look like a one-dimensional curve, as shown in Fig. 3b.

The results for the mass correction ε_1 and mass correlation ε_2 are shown on Figs. 4 and 5. The dependence of the ground state energy E from the nuclear charge Z and the mass number A in the helium-like electron-nuclear systems is due to the dependence of ε_1 and ε_2 from N, because E_0 does not depend on the number of neutrons, only on the nuclear charge that coincides with the atomic number.

As far as the mass correction ε_1 is determined by the expression

$$\varepsilon_1 = \frac{1}{M} E_0 \,, \tag{58}$$

it is clear that it is proportional to the reciprocal of A. For the light elements the dependence $\varepsilon_1(N)$ has a curve similar to the dependence for carbon presented on Fig. 6. For the elements in the middle of the periodic table the $\varepsilon_1(N)$ dependence

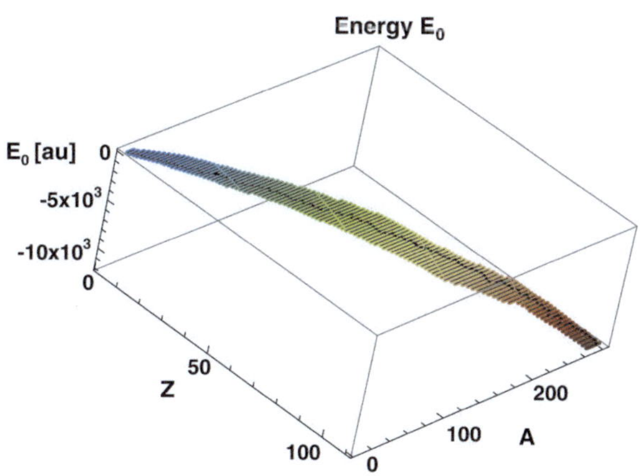

Fig. 2 Helium-like electron-nuclear systems ground state energy depending on the nuclear charge Z and mass number A. Isotopes from the line of stability (*thick line*)

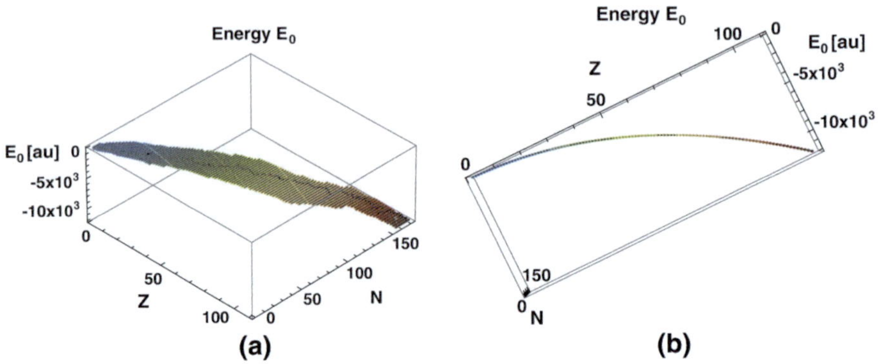

Fig. 3 Dependence of E_0 from Z and N. **a** General view, **b** view from the positive direction of the axis Z

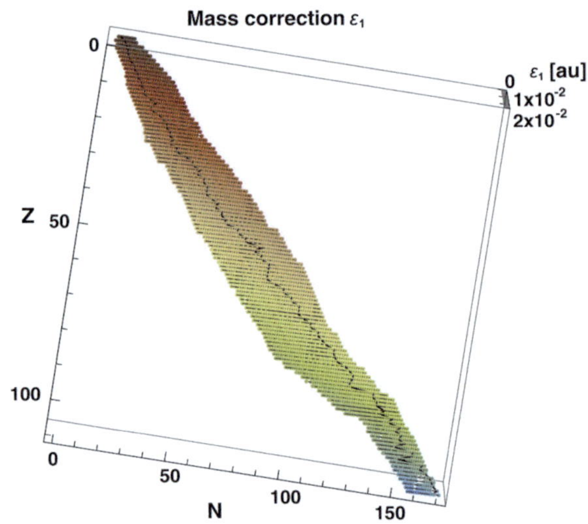

Fig. 4 Mass correction ε_1 dependence from Z and A

is similar to the one for antimony presented on Fig. 7. For the heavier elements the $\varepsilon_1(N)$ dependence is similar to the one for nobelium presented on Fig. 8.

For the mass correlation ε_2 dependence from N is determined by the dependence of $\frac{1}{M}$ from N times the integral I

$$I = \int \nabla_1 \psi \nabla_2 \psi \, d\vec{r}_1 \, d\vec{r}_2 = 2\pi^2 \int_0^\infty ds \int_0^s du \int_0^u dt \left\{ u\left(s^2 + t^2 - 2u^2\right) \left[\left(\frac{\partial \psi(s,t,u)}{\partial s} \right)^2 - \left(\frac{\partial \psi(s,t,u)}{\partial t} \right)^2 \right] - u\left(s^2 - t^2\right) \left(\frac{\partial \psi(s,t,u)}{\partial u} \right)^2 + 2\frac{\partial \psi}{\partial u} \left[s\left(t^2 - u^2\right) \frac{\partial \psi(s,t,u)}{\partial s} + t\left(u^2 - s^2\right) \frac{\partial \psi(s,t,u)}{\partial t} \right] \right\}. \tag{59}$$

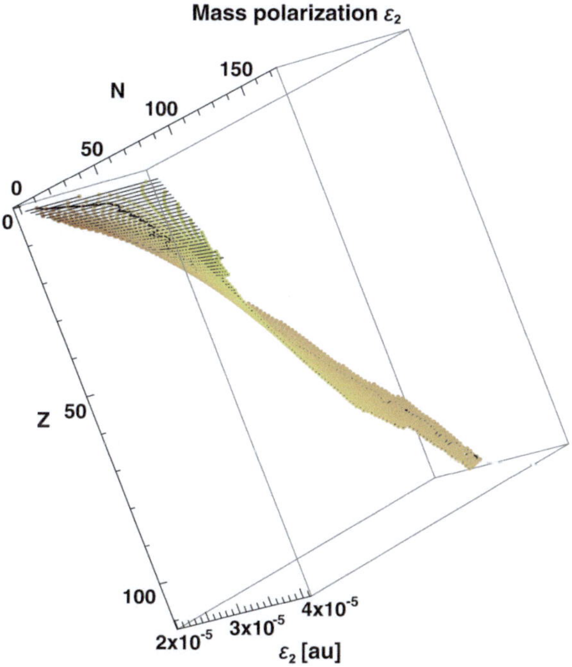

Fig. 5 Mass correlation ε_2 dependence from Z and A

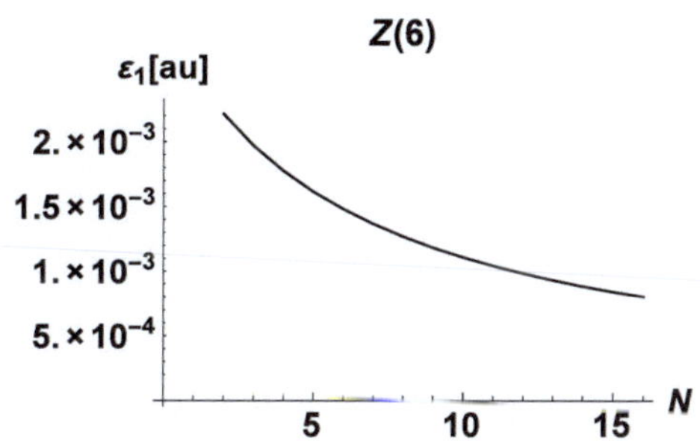

Fig. 6 The mass correction $\varepsilon_1(N)$ for carbon

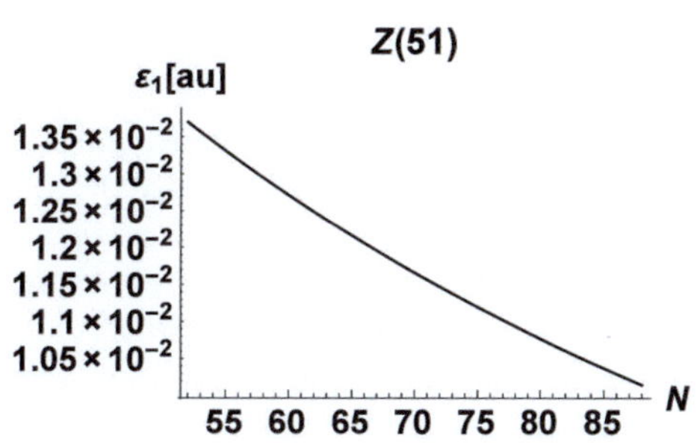

Fig. 7 The mass correction $\varepsilon_1(N)$ for antimony

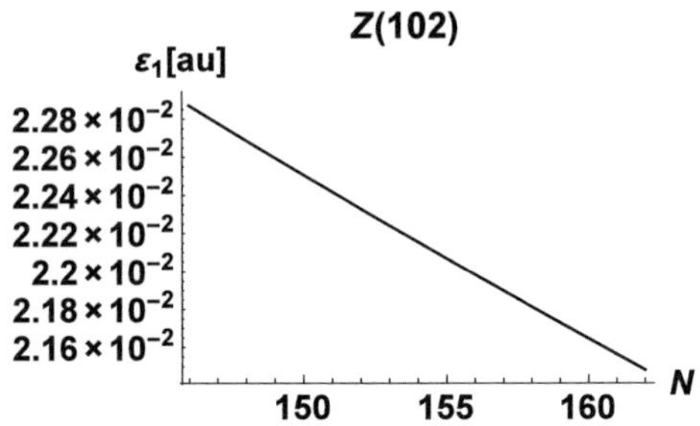

Fig. 8 The mass correction $\varepsilon_1(N)$ for nobelium

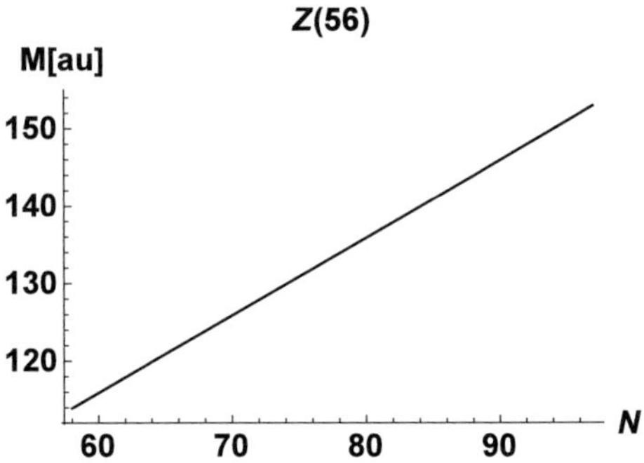

Fig. 9 Dependence $M(N)$ for barium

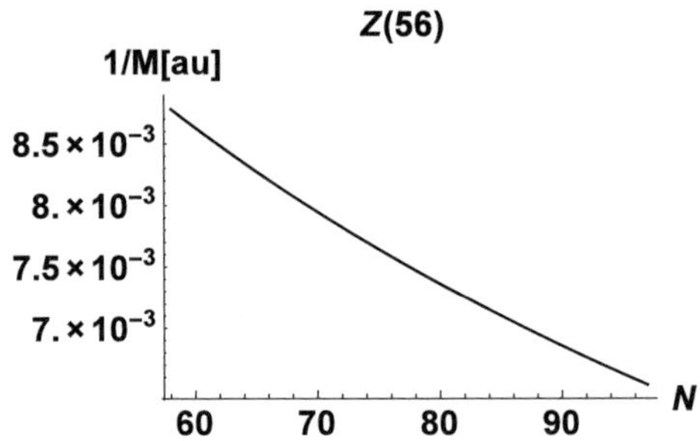

Fig. 10 Dependence $\frac{1}{M}(N)$ for barium

Both dependences have similar graphics. In Fig. 9 is presented the dependence of M from N for barium, while in Fig. 10 is presented the dependence of $\frac{1}{M}$ from N for barium. Figure 11 presents the dependence of integral (59) from N for barium. In Fig. 12 is presented the dependence of the mass correlation ε_2 from N for barium.

From the presented graphs can be seen that the analysis of the dependence of the ground state energy from the nuclear charge Z and the number of neutrons N for the helium-like electron-nuclear systems is not easy by conventional means. Impressive results are obtained by the implementation of staggering analysis that is possible because of the high precision of the calculation results.

The results for isotopes of the stability line are presented in [5]. The analysis used staggering of the fifth order, calculated by the formula

Fig. 11 Dependence $I(N)$ for barium

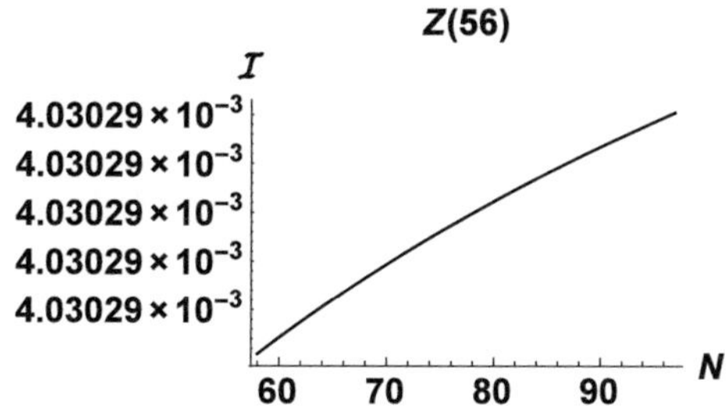

Fig. 12 Dependence $\varepsilon_2(N)$ for barium

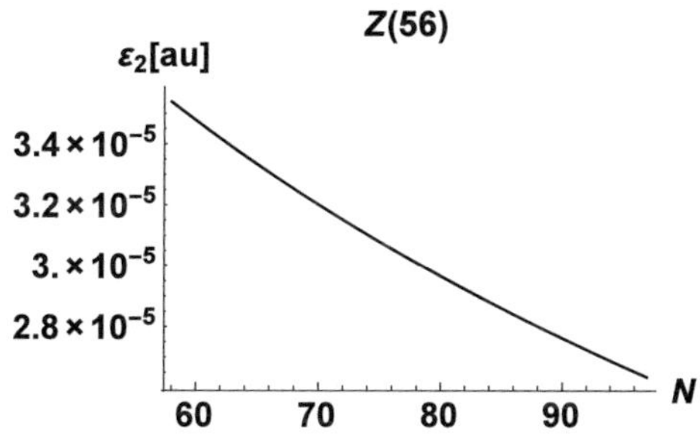

$$\begin{aligned}\mathrm{Stag}(W) = 10E(W) - 5E(W+1) - 10E(W-1) - E(W-3) + \\ + E(W+2) + 5E(W-2)\,,\end{aligned} \tag{60}$$

where W is the isotope. The overall analysis is a serious and extensive problem. We will note only some of the main features of the behavior of the staggering in terms of A for specific isotopes. This approach naturally splits the isotopes into even and odd. Figure 13 shows the staggering of $_{20}Ca$ and $_{95}Am$ in terms of A. For the light elements two areas can be seen: in the first one (the left side of Fig. 13c) the lines are far apart; in the second one (the right side of Fig. 13c) the lines are at minimum distance from one another. For the heavy elements the lines are further away at minimum distance but are closer to one another for more isotopes. The slope of the lines is considerably steeper for the light elements than for the heavy ones.

5 Conclusion

Up to now was not known the dependence of the characteristics of the ground state electron energy in helium-like ions on the isotopic composition of the nucleus. Interesting links obtained with our staggering analysis of the numerical data shows that further research in this area could lead to a new understanding of the relationship between the electron and nuclear systems. The dependence of the characteristics of the ground state electron energy on the isotopic composition of the nucleus is indicative of the derivation of the formula of Bethe-Salpeter [51], which retains all its

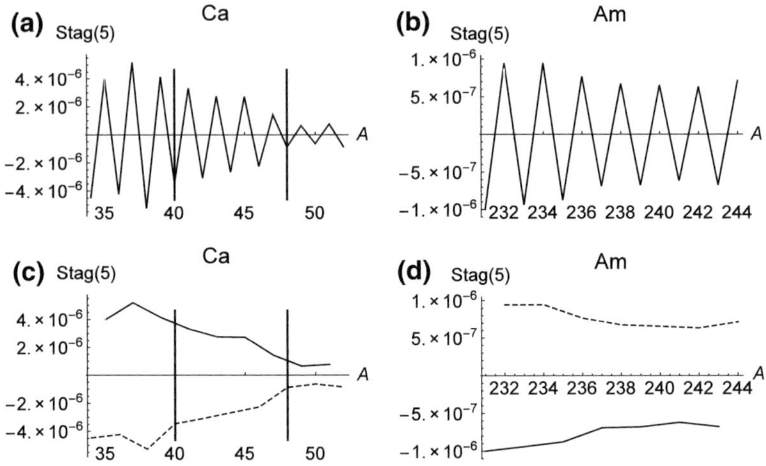

Fig. 13 *Vertical lines* are placed at locations corresponding to the magic numbers in terms of N. **a** Staggering values for $_{20}Ca$, **c** staggering values for $_{95}Am$. **b** and **d** The natural splitting of even (*dashed line*) and odd (*solid line*) isotopes

relevance today. The amazing thing about this is that the use of only four values obtained from the analysis of experimental data yields a formula for the ground state energy of helium-like electron-nuclear systems that shows a striking coincidence with results obtained from the best non-relativistic theoretical studies [5]. Our results prove the correctness of the theoretical principles that underlie the procedure we developed and the relevance of the separation of the ground state energy into its relevant components. The continuation of these studies would allow to develop new methods for the analysis of isotopes and to improve the study of electron-nuclear systems.

Acknowledgements The authors would like to thank Prof. Anna Georgieva, Prof. Svetla Drenska and Prof. Nikolay Minkov for the fruitful discussions. This work was supported by the Bulgarian National Science Fund under contract No. DFNI-T02/1. This work was supported by the Bulgarian National Science Fund under contract No. DFNI-E 01/2 and by a NUPNET-NEDENSAA project funded by the Bulgarian Ministry of Education and Science.

References

1. Pavlov RL, Maruani J, Mihailov LM, Velchev ChJ, Dimitrova-Ivanovich M, Stoianov JK, Van Neck D (2008) In: Dimitrova S (ed) Nuclear theory, vol 27. Sofia, BM Trade Ltd, p 249
2. Pavlov R, Mihailov L, Velchev Ch, Dimitrova-Ivanovich M, Stoyanov Zh, Chamel N, Maruani J (2010) J Phys Conf Ser 253:012075
3. Velchev Ch, Pavlov R, Mihailov L, Chamel N, Stoyanov Zh, Mutafchieva Y, Ivanovich M (2014) C R Acad Bulg Sci 67(11):1513–1520

4. Kramida A, Ralchenko Yu, Reader J, NIST ASD Team (2013) NIST Atomic Spectra Database (ver. 5.1). National Institute of Standards and Technology, Gaithersburg, MD. http://physics.nist.gov/asd. Accessed 26 Aug 2014
5. Velchev ChJ (2015) Study of quantum characteristics of heliumoid electron–nuclear systems. Dissertation, Institute for Nuclear Research and Nuclear Energy, BAS
6. Presnyakov LP, Shevelko VP, Yanev RK (1986) Elementarnyye protsessy s uchastiyem mnogozaryadnykh ionov (Elementary processes involving multiply charged ions). Energoatomizdat, Moskva
7. Vainshtein L.A, Shevelko VP (1986) Struktura i kharakteristiki ionov v goryachey plazme (Structure and characteristics of hot plasma ions) Nauka, Moskva
8. Beranyi D, Hock G (eds) (1988) High-energy ion-atom collisions. Springer, Berlin
9. Burke P (ed) (1983) Atoms in astrophysics. Plenum Press, New York
10. Dubetsky B (2014) arXiv:1407.7287. Accessed 29 July 2014
11. Lee DM (1996) The extraordinary phases of liquid ^3He. Nobel Media AB. http://www.nobelprize.org/nobel_prizes/physics/laureates/1996/lee-lecture.html. Accessed 19 Jan 2016
12. Cornell EA, Wieman CE (2001) Bose-Einstein condensation in a dilute gas; the first 70 years and some recent experiments. Nobel Media AB. http://www.nobelprize.org/nobel_prizes/physics/laureates/2001/cornell-lecture.html. Accessed 19 Jan 2016
13. Va'vra J (2014) Phys Lett B 736:169–173
14. Muhin KN (1993) Eksperimentalnaya yadernaya fizika (Experimental Nuclear Physics), vol 1. Energoatomizdat, Moskva
15. Muhin KN (1993) Eksperimentalnaya yadernaya fizika (Experimental Nuclear Physics), vol 2. Energoatomizdat, Moskva3
16. Rychlewski J (ed) (2003) Explicitly correlated wavefunctions in chemistry and physics. In: Progress in theoretical chemistry and physics, vol 13. Kluwer, Dordrecht
17. Hylleraas EA (1928) Z Phys 48:469
18. Hylleraas EA (1929) Z Phys 54:347
19. Hylleraas EA (1930) Z Phys 60:624
20. Hylleraas EA (1930) Z Phys 63:291
21. Hylleraas EA (1930) Z Phys 65:209
22. Pekeris CL (1958) Phys Rev 112:1649
23. Pekeris CL (1959) Phys Rev 115:1216
24. Pekeris CL (1962) Phys Rev 126:1470
25. Schwartz HM (1956) Phys Rev 103:110
26. Schwartz HM (1960) Phys Rev 120:483
27. Thakkar AJ, Koga T (1994) Phys Rev A 50:854–856
28. Hart JF, Herzberc G (1957) Phys Rev 106:79–82
29. Kinoshita T (1957) Phys Rev 105:1490–1502
30. Kinoshita T (1959) Phys Rev 115:366–374
31. Coolidge AS, James HM (1937) Phys Rev 51:855–859
32. Jolly P (1979) Int J Quant Chem 16:1149
33. Koga T (1990) J Chem Phys 93:3720
34. Koga T (1992) J Chem Phys 96:1276
35. Koga T (1995) Z Phys D34:71
36. Koga T (1996) J Chem Phys 104:6308
37. Bartlett JH (1937) Phys Rev 51:661
38. Freund DE, Huxtable BD, Morgan JD III (1984) Phys Rev A 29:980
39. Baker JD, Freund DE, Hill RN, Morgan JD III (1990) Phys Rev A 41:1247
40. Drake GWF (1988) Nucl Instrum Methods Phys Res Sect B 31:7
41. Willets L, Cherry IJ (1956) Phys Rev 103:112
42. Wang PSC (1967) J Chem Phys 47:229
43. Frankowski K, Pekeris CL (1966) Phys Rev 146:46
44. James HM, Coolidge AS (1936) Phys Rev 49:688
45. Burke EA (1963) Phys Rev 130:1871

46. Luchow A, Kleindienst H (1994) Int J Quant Chem 51:211
47. King FW (1997) J Mol Struct Theochem 400:7
48. Yan Z-C, Tambasco M, Drake GWF (1998) Phys Rev A 57:1652
49. Thakkar AJ, Koga T, Tanabe T, Teruya H (2002) Chem Phys Lett 366:95
50. Löwdin PO (1950) J Chem Phys 18(3):365
51. Bethe HA, Salpeter EE (1957) Quantum mechanics of one- and two-electron atoms. Academic Press, New York
52. Gombas P (1950) Theorie und Losungsmethoden des Mehrteilchenproblems der Wellenmechanik. Springer, Basel
53. Tuli JK (2005) Nuclear wallet cards. National Nuclear Data Center. http://www.nndc.bnl.gov/wallet/. Accessed 19 April 2006

Part IV
Biochemistry and Biophysics

Computational Study of Michellamines: Naphthylisoquinoline Alkaloids with Anti-HIV Activity

Liliana Mammino and Mireille K. Bilonda

Abstract Michellamine A, michellamine B and michellamine C are naphthylisoquinoline alkaloids with anti-HIV activity isolated from the leaves of *Ancistrocladus korupensis*, a plant indigenous in Cameroun. Michellamine B and michellamine C are atropisomeric of michellamine A, and michellamine B is the most active. The molecules consist of two units, each containing an isoquinoline moiety and a naphthalene moiety. This work presents the results of a conformational study of michellamine A performed *in vacuo* and in three solvents (chloroform, acetonitrile and water), at the (HF/6-31G(d,p) and DFT/B3LYP/6-31+G(d,p) levels. The potential energy profiles for the rotations of single bonds between moieties were determined to identify relevant conformers corresponding to different orientations of the moieties. The harmonic vibrational frequencies of the conformers were calculated *in vacuo*. The conformational preferences are influenced by intramolecular hydrogen bonds (having a stabilizing effect) and by the mutual orientations of the various moieties. The lowest energy conformers contain two cooperative O–H···O intramolecular hydrogen bonds. The isoquinoline moiety prefers to be perpendicular to the naphthalene moiety in the same unit. All the properties obtainable from a conformational study are considered: relative energies in vacuo and in solution, dipole moments, HOMO-LUMO energy gaps, free energy of solvation in the solvents considered, and vibrational frequencies and predicted vibrational spectra in vacuo.

Keywords Alkaloids · Anti-HIV compounds · Intramolecular hydrogen bonding · Michellamines · Mutual orientations of aromatic moieties · Naphthylisoquinoline alkaloids · Solute-solvent interactions

Electronic supplementary material The online version of this chapter (doi:10.1007/978-3-319-50255-7_18) contains supplementary material, which is available to authorized users.

L. Mammino (✉) · M.K. Bilonda
Department of Chemistry, University of Venda, Thohoyandou, South Africa
e-mail: sasdestria@yahoo.com

A. Tadjer et al. (eds.), *Quantum Systems in Physics, Chemistry, and Biology*,
Progress in Theoretical Chemistry and Physics 30,
DOI 10.1007/978-3-319-50255-7_18

1 Introduction

The Human Immunodeficiency Virus (HIV) is a retrovirus that infects cells of the immune system, destroying or impairing their functions [1]. The HIV infection causes the Acquired Immune Deficiency Syndrome (AIDS). AIDS has become a serious health problem worldwide since the discovery of the first case in 1981 and is one of the most fast-spreading diseases. According to the WHO-HIV/AIDS 2014 reports [2], 36.9 million people lived with HIV worldwide by the end of 2013, 2.0 million were newly infected during 2013 and 1.2 million died of AIDS (the lowest death-rate since the peak in 2005, when 3.1 million died). The highest proportion of people living with HIV was in sub-Saharan Africa: ≈24.7 million, corresponding to ≈70% of the worldwide total of infected people.

The main problem is the absence of a cure for the disease. Currently-used medicines can slow the growth of the virus or prevent its replication; but they are expensive and thus not accessible to all patients. Furthermore, drug resistant strains, due to mutations of the virus, have already been observed among patients [3]. Finding an effective and durable chemotherapy for the disease is an urgent matter. The task may involve the use of innovative combinations of drugs with diverse mechanisms of action and different ranges of anti-HIV activities [3], and also the search for new antiviral drugs with more potent activity. Natural products are a rich potential source of new drugs with new modes of action.

The study reported in this paper considered michellamine A, a naphthylisoquinoline alkaloid isolated from *Ancistrocladus korupensis*, a plant of the *Ancistrodaceae* family [3]. Naphthylisoquinoline alkaloids are exclusively found in tropical lianas of the *Dioncophyllaceae* and *Ancistrodaceae* families [4–7] and exhibit a variety of biological activities. Michellamines are naphthylisoquinoline alkaloids with a dimeric structure in which each unit consists of a naphthalene and an isoquinoline moiety. The naphthalene moieties of the two units are bonded to each other, so that the four moieties are connected in the isoquinoline-naphthalene-naphthalene-isoquinoline sequence (Fig. 1). Michellamine A and its atropisomers (michellamine B, michellamine C and others) shows good activity against HIV, with michellamine B having the greatest activity (atropisomers are stereoisomers arising because of hindered rotation about a single bond, where energy differences due to steric strain or other contributors create a barrier to rotation that is high enough to allow for isolation of individual atropisomers).

Chemically, michellamines are unique naphthylisoquinoline alkaloids not only because of their unprecedented ‘dimeric’ structure, but also because of their exceptionally high polarity resulting from their three phenolic OH and one secondary amino group per constituent unit [3].

Computational studies of naphthylisoquinoline alkaloids are still scarce, and no previous studies have been found on michellamine A or its atropisomers. The current study investigates the conformational preferences of michellamine A

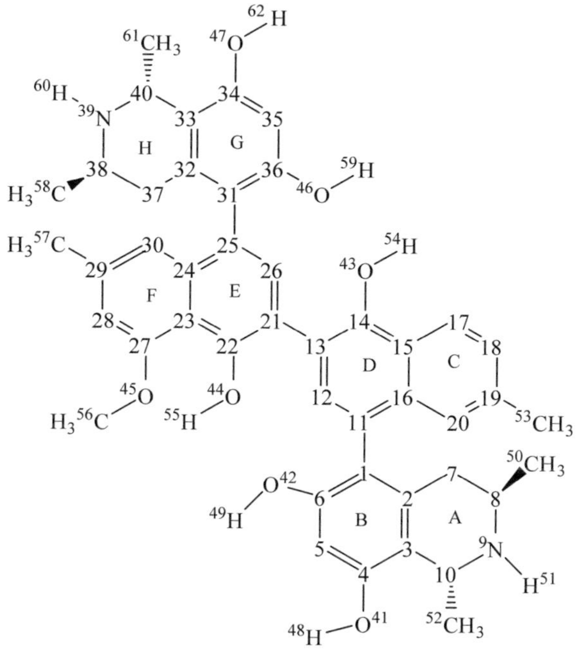

Fig. 1 Structure of the michellamine A molecule and atom numbering utilized in this work. The C atoms in the rings are represented only by the numbers denoting their positions. Only the H atoms attached to O or N are numbered separately, while the H atoms attached to a C atom are given the same number as the C atom. The rings are denoted by the letters A, B, C, D, E, F, G and H

(Fig. 1, denoted by the acronym MCA in the rest of the text). MCA was studied *in vacuo* and in three solvents with different polarities and different H-bonding abilities (chloroform, acetonitrile and water). Calculations were performed at two levels of theory: Hartree-Fock (HF/6-31G(d,p)) and Density Functional Theory with the B3LYP functional (DFT/B3LYP/6-31+G(d,p)). Calculations in solution used the PCM model.

The results show that the conformational preferences are dominantly influenced by the presence of intramolecular hydrogen bonds (IHB), followed by the mutual orientations of the four moieties. Detailed information on the properties of the calculated conformers (dipole moments, free energy of solvation in the solvents considered, HOMO-LUMO energy gaps, vibrational frequencies and predicted IR spectra, potential energy profiles for the rotation of relevant bonds, etc.) is included in the Supplementary material.

2 Computational Details

Calculations for the conformational search were performed *in vacuo* with fully relaxed geometry using two levels of theory: Hartree-Fock (HF) with the 6-31G (d,p) basis set, and Density Functional Theory (DFT) with the B3LYP functional

[8, 9] and the 6-31+G(d,p) basis set. These are the same methods utilized in a study of naphthylisoquinoline alkaloids with antimalarial activity [10]. HF is a common and moderately cheap quantum mechanical method that can yield realistic information regarding conformational analysis. Previous studies on other molecules [11, 12] showed that HF can successfully handle intramolecular H-bonding and yields HOMO-LUMO energy gaps approaching those of experiments. DFT is an alternative method to wavefunction approaches, which is often used in conformational search because it takes into account part of the correlation effects at a comparatively low cost. Among the numerous functionals available for the DFT framework, B3LYP is the most widely employed. It can provide better quality results in combination with basis sets containing diffuse functions, above all for molecular systems containing IHBs [11, 12]. Harmonic vibrational frequencies were calculated *in vacuo* at both levels to verify that the stationary points from optimization results correspond to true minima and to obtain the zero-point energy (ZPE) corrections. The frequency values were scaled by 0.9024 [13] and 0.9857 [14] respectively for HF/6-31G(d,p) and DFT/B3LYP/6-31 + G(d,p) calculations.

Calculations in solution utilized the Polarizable Continuum Model (PCM, [15–20]). In this model, the solvent is considered infinite and is modelled by a continuous isotropic dielectric into which the solute is inserted. Thus, the solute molecule is embedded in a cavity surrounded by the continuum solvent. The geometry of the cavity follows the geometry of the solute molecule, considering its solvent accessible surface. The default settings of Gaussian03 [21] were utilised, namely: Integral Equation Formalism model (IEF, [17–20]) and Gepol model for building the cavity around the solute molecule [22–24], with simple United Atom Topological Model (UAO) for the atomic radii and 0.200 $Å^2$ for the average area of the tesserae into which the cavity surface is subdivided. The SCFVAC option was selected to obtain more thermodynamic data. Calculations in solution were performed as single point (SP) calculations because the size of the molecule makes re-optimisation in solution highly expensive. Although SP calculations cannot provide information on the geometry changes caused by the solvent, they can provide reasonable information on the energetics, such as the conformers' relative energies in solution and the energy aspects of the solution process (the free energy of solvation, ΔG_{solv}, and its components).

Calculations were performed using GAUSSIAN 03, Revision D 01 [21].

All the energy values reported are in kcal/mol and all the distances are in Å. Acronyms are utilized for the calculation methods and for the media, for conciseness sake on reporting values: HF for HF/6-31G(d,p), DFT for DFT/B3LYP/6-31+G(d,p), *vac* for vacuum, *chlrf* for chloroform, *actn* for acetonitrile and *aq* for water.

3 Results

3.1 Naming of Conformers

When discussing the conformers' properties, it is important to be able to identify their geometrical characteristics clearly and easily. A system of symbols is introduced to this purpose.

Since it is convenient to be able to mention each ring individually, the rings are denoted by uppercase letters (A, B, C, D, E, F, G, H) as illustrated in Fig. 1. In order to be able to mention each of the two units of the dimeric molecular structure individually, the whole unit comprising the A, B, C, D rings is called R and the whole unit comprising the E, F, G, H rings is called S.

The most important factors stabilizing the conformers are the O–H···O IHBs and other IHB-type interactions such as O–H···π and C–H···O. Different conformers often have different combinations of these intramolecular interactions. Each IHB or IHB-type interaction is denoted by a specific lowercase letter. The letters (a–h) and their meanings are listed in Table 1 and illustrated in Fig. 2, which shows all the calculated conformers of MCA and the corresponding combinations of letters describing their characteristics in the acronyms denoting them.

The energy of the conformers is also significantly influenced by the mutual orientations of the four moieties. The mutual orientation of the R and S units is determined by the presence or absence of an IHB between them. When the IHB is present, the C14–C13–C21–C22 torsion angle is close to ±50°, depending on whether the E,F plane is 'to the left' or 'to the right' with respect to the D,C plane. When the IHB is absent, the C14–C13–C21–C22 torsion angle is close to ±90°. These four possibilities for the orientations of the two units are denoted by the

Table 1 Letters utilized in the acronyms denoting the conformers of michellamine A in this work, and their meanings

Letter	Meaning	Letter	Meaning
a	Presence of the O43–H54···O44 IHB	h	Presence of C–H61···O47
b	Presence of the O44–H55···O45 IHB	t	C14–C13–C21–C22 torsion angle close to +50°
c	Presence of the O42–H50···π interaction	v	C14–C13–C21–C22 torsion angle close to −50°
d	Presence of the O43–H54···π interaction	p	C14–C13–C21–C22 torsion angle close to +90°
e	Presence of the O44–H55···π interaction	q	C14–C13–C21–C22 torsion angle close to −90°
f	Presence of the O46–H59···π interaction	x	C24–C25–C31–C36 torsion angle close to 90°
g	Presence of C–H52···O41	y	C24–C25–C31–C36 torsion angle close to −90°

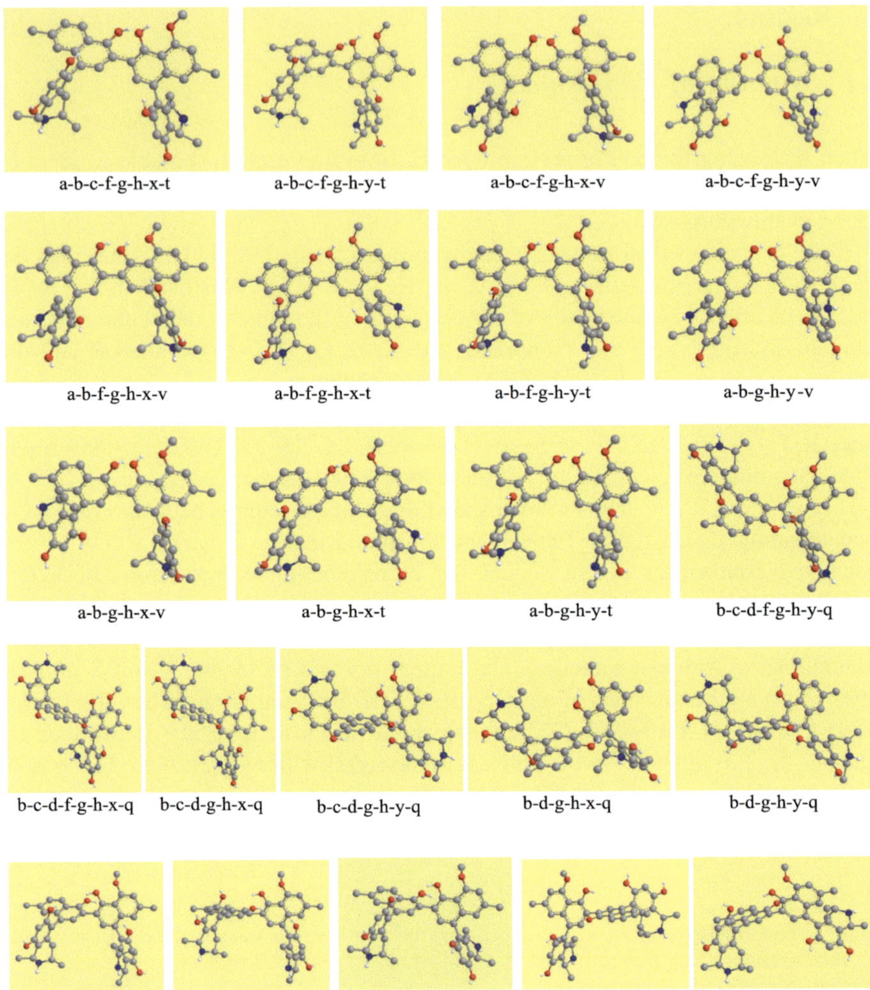

Fig. 2 Optimized geometries of the conformers of michellamine A and sets of letters identifying their characteristics in the acronyms denoting them. Geometries from HF/6-31G(d,p) results. The images illustrate the main combinations of interaction types and moieties' orientations which may be present in the conformers of michellamine A. The initial part of the acronyms is not reported under the images, because of space reasons; it is the same for all the conformers (MCA)

letters t, v, p and q (Table 1). Different combinations of moieties' orientations correspond to different IHB-type interactions. For instance, the two cooperative (a–b) IHBs are compatible with the *t* and *v* orientations between the R and S units, but not with the *p* and *q* orientations. The potential energy profile for the rotation of the C13–C21 bond between the two units shows that the minima correspond to the conformers with two consecutive IHBs.

The isoquinoline and naphthalene moieties within each unit are approximately perpendicular to each other, with the torsion angle close to ±90°. This is a known tendency for aromatic systems bonded to each other [25] and without other types of interactions (such as IHBs) which might determine different orientations [26]. Given the high number of possible combinations of different orientations and the comparatively high computational cost per conformer in relation to the size of the molecule, only the mutual orientation of the E,F and G,H moieties in the S unit was varied, whereas the mutual orientation of the A,B and C,D moieties in the R unit was kept the same for all the conformers (C6–C1–C11–C16 torsion angle close to −90°). This is justified by the expected symmetric influence of analogous orientations of the two moieties in the R and S units, and by the potential energy profile for the rotation of the C1–C11 bond. The profile shows that the orientation with C6–C1–C11–C16 close to −90° is the lowest energy one and the orientation with C6–C1–C11–C16 close to +90° corresponds to another minimum, with nearly the same energy as the −90° orientation. The two orientations are denoted by the letters x and y (Table 1 and Fig. 2). The orientations with C6–C1–C11–C16 close to +90° or to −90° do not imply different kinds of IHB-types interactions in the conformers, because of the absence of mutual interactions between the isoquinoline moieties of R and S. Thus, the consideration of different orientations of the E,F and G,H moieties in the S unit provides sufficient information on the influence of the orientations of the naphthalene and isoquinoline moiety within the same unit.

Geometries involving stacking interactions between different aromatic moieties did not prove viable on optimisation.

Examples of how the acronyms describe the characteristics of individual conformers may be functional. MCA-a-b-c-f-g-h-x-t is a conformer characterized by the presence of the O43}–H54···O44 and O44–H55···O45 IHBs (which are consecutive when present simultaneously), the O42–H50···π and O46–H59···π interactions and the C–H52···O41 and C–H61···O47 interactions, and with the C24-C25–C31–C36 torsion angle close to 90° and the C14–C13–C21–C22 torsion angle close to 50°; MCA-b-c-d-g-h-y-q is a conformer characterized by the presence of the O44–H55···O45 IHB, the O42–H50···π and O43–H55···π interactions and the C–H52···O41 and C–H61···O47 interactions, and with the C24–C25–C31–C36 torsion angle close to −90° and the C14–C13–C21–C22 torsion angle close to −90°.

3.2 *Results in Vacuo*

Table 2 reports the relative energies of the calculated conformers of MCA in *vacuo* and in the three solvents considered. The conformational preferences are influenced by IHB patterns (the number and type of IHBs present in a conformer) and by the mutual orientation of the two units and of the two moieties (naphthalene and isoquinoline) within each unit. Three types of IHB interactions may be present: O–H···O IHBs (the consecutive O43–H54···O44 and O44–H55···O45 IHBs,

Table 2 Relative energies of the calculated conformers of the michellamine A molecule HF/6-31G(d,p) results *in vacuo* and in solution and DFT/B3LYP/6-31+G(d,p) results *in vacuo*, respectively denoted as HF and DFT in the columns' headings. The results *in vacuo* are from full optimization calculations, the results in solution are from single point PCM calculations on the in-vacuo-optimized geometries. The conformers are listed in order of increasing relative energies in the HF results in vacuo

Conformer	Relative energy (kcal/mol)							
	DFT				HF			
	Vac	Chlrf	Actn	Aq	Vac	Chlrf	Actn	Aq
MCA-a-b-c-f-g-h-x-t	0.036	0.007	0.000	0.146	0.000	0.000	0.818	0.568
MCA-a-b-c-f-g-h-x-v	0.000	0.000	–[a]	0.231	0.029	0.146	0.894	0.590
MCA-a-b-c-f-g-h-y-t	0.272	0.242	0.179	0.000	0.048	0.046	0.000	0.539
MCA-a-b-c-f-g-h-y-v	0.654	2.860	1.988	0.396	0.315	0.358	0.878	0.000
MCA-b-c-d-f-g-h-y-q	2.834	2.814	2.780	2.013	1.705	1.547	2.287	1.110
MCA-b-c-d-f-g-h-x-q	3.131	3.299	3.150	–[a]	2.235	2.128	2.775	0.735
MCA-a-b-f-g-h-x-v	4.082	0.006	–[a]	0.144	4.189	2.570	0.894	–[a]
MCA-b-c-d-g-h-x-q	5.251	–[a]	3.870	2.325	4.455	3.350	3.451	2.219
MCA-a-b-f-g-h-x-t	4.426	2.860	1.989	0.409	4.578	2.879	2.619	0.766
MCA-a-b-f-g-h-y-t	4.707	14.491	13.780	14.027	4.661	2.839	2.531	0.722
MCA-b-c-d-g-h-y-q	6.196	4.936	–[a]	–[a]	5.207	3.454	3.615	1.241
MCA-c-d-e-f-g-h-y-p	10.568	11.375	11.410	–[a]	7.669	8.206	8.829	5.323
MCA-a-b-g-h-y-v	8.142	16.584	15.184	13.598	8.210	5.021	3.917	0.154
MCA-a-b-g-h-x-v	8.118	5.161	3.344	0.216	8.367	5.113	4.279	0.797
MCA-a-b-g-h-x-t	8.193	5.257	3.374	0.295	8.411	5.165	4.033	0.716
MCA-a-b-g-h-y-t	8.485	5.319	3.425	0.060	8.575	5.096	3.823	0.490
MCA-b-d-g-h-x-q	9.979	7.541	5.967	2.659	9.256	6.460	5.512	0.954
MCA-b-d-g-h-y-q	10.924	8.122	6.219	2.176	10.114	6.913	5.651	1.554
MCA-d-e-f-g-h-y-p	12.947	12.569	–[a]	9.174	10.409	10.135	9.900	6.928
MCA-c-d-e-g-h-x-p	13.750	13.077	12.413	8.745	11.014	10.058	9.999	6.188
MCA-d-e-g-h-y-p	15.758	14.062	12.812	9.126	13.049	10.943	10.381	–[a]
MCA-d-e-g-h-x-p	16.486	–[a]	13.291	9.133	14.011	11.757	11.028	6.759

[a]The calculation for this conformer did not converge in the given solvent

leading to the O43–H54···O44–H55···O45 sequence, or O44–H55···O45 alone); O–H···π interactions between the O–H in an isoquinoline moiety and the closest π system in the naphthalene moiety of the same unit (O42–H49···π, O46–H59···π) or between an O–H in the naphthalene moiety of one unit and the closest π system in the naphthalene moiety of the other unit (O43–H54···π, O44–H55···π); and C–H···O interactions within an isoquinoline moiety (C–H52···O41 and C–H61···O47).

The conformers having all the IHB-type interactions simultaneously are the lowest energy conformers. The two consecutive O43–H54···O44 and O44–H55···O45 IHBs are cooperative, as each of them strengthens the other. They have the strongest stabilizing effect, followed by a single O–H···O IHB. The H-bond lengths indicate that the IHB within the E, F naphthalene moiety (O44–H55···O45) is the strongest IHB.

The removal of the O44–H55···O45 IHB brings about the O44–H55···π interaction and the removal of the O43–H54···O44 IHB brings about the O43–H54···π interaction. Each of these removals causes an increase in the energy of the conformer. Comparison of the energies of conformers MCA-a-b-c-f-g-h-x-t and MCA-c-d-e-f-g-h-y-p shows an energy increase of ≈8 kcal/mol. In some case, the O–H···π interactions can have greater stabilizing effect than an IHB; for instance, conformer MCA-c-d-e-f-g-h-y-p, which has all the O–H···π interactions (O42–H49···π, O43–H54···π, O44–H55···π and O46–H59···π), has lower energy than some conformers having O–H···O IHBs, such as MCA-a-b-g-h-y-v. Changes in the mutual orientations of the moieties may cause energy differences of ≈1 kcal/mol.

The bond length (Å) of the O43–H54···O44 IHB is 1.753–1.761/DFT and 1.880–1.891/HF and that of O44–H55···O45 is 1.688–1.730/DFT and 1.749–1.776/HF. The O···O distance (Å) is 2.660–2.669/DFT and 2.735–2.743/HF for O43–H54···O44 and 2.560–2.589/DFT and 2.735–2.743/HF for O44–H55···O45. The O$\hat{\text{H}}$O bond angle is 148.0°/HF and 153.0°/DFT for O43–H54···O44 and 143.0°/HF and 146.0°/DFT for O44–H55···O45. For the O–H···π interactions, a bond length is not defined, as the acceptor is a whole π system and not an individual atom. It may be convenient to consider the distance between the H atom of the donor and the C atom in the aromatic system closest to it. The H···C distance thus defined is slightly shorter when the interaction involves an O–H in the naphthalene moiety of one unit and the closest π system in the naphthalene moiety of the other unit (O43–H54···π and O44–H55···π) than when it involves an O–H in the isoquinoline moiety and the closest π system in the naphthalene moiety of the same unit (O42–H49···π and O46–H59···π). The H···C distance (Å) is 2.269–2.381/DFT and 2.302–2.398/HF for O43–H54···π and O44–H55···π, and 2.317–2.382/DFT and 2.345–2.401/HF for O42–H49···π and O46–H59···π.

For the C–H···O interactions within the isoquinoline moiety (C–H52···O41 and C–H61···O47), the H···O distance is ~2.40 Å in both HF and DFT results. This value is considerably longer than the H···O distance for O–H···O IHBs, consistently with the fact that the C–H···O interaction is considerably weaker. The C···O distance for C}-\text{H52} \cdots \text{O41 and C–H61···O47 is 3.0 Å in both HF and DFT results. The $\hat{CHO}$ bond angle is ~114.0°/HF and ~116.0°/DFT for both C–H52···O41 and C–H61···O47.

The calculated harmonic vibrational frequencies of the O–H bonds highlight the frequency lowering (red shift) in correspondence to IHBs. Interesting enough, the red shift of the frequency of O43–H54 when it acts as donor in the O43–H54···O44 IHB is smaller than would be expected for an OH engaged in an IHB in which the acceptor O belongs to another OH, and is not dissimilar from the red shift of O44–H55 when engaged in the O44–H55···O45 IHB although, in this case, the acceptor is an ether O and the IHB might be expected to be weaker than with a phenol O. On the other hand, the red shift of O44–H55 when the O43–H54···O44 IHB is not present is smaller than when it is present. This can be viewed as a confirmation that the two IHBs (O43–H54···O44 and O44–H55···O45) are cooperative.

The ZPE corrections are very close for all the conformers. Their ranges (kcal/mol) are 554.6–556.1/HF and 516.0–517.5/DFT, with the greater values corresponding to lower energy conformers. The relative energies corrected for ZPE have the same trends as the uncorrected ones.

The ranges (Debye) of the values of the dipole moments in vacuo are 2.37–8.28/HF and 2.05–8.00/DFT. Conformers with O–H···O IHBs have higher dipole moment than conformers with O–H···π interactions. Conformers with only O–H···π interactions (no O–H···O IHBs) have the smallest dipole moments. For instance, MCA-c-d-e-f-g-h-y-p has all the O–H···π interactions and its dipole moment (HF) is 2.37 D, while MCA-a-b-c-f-g-h-y-v, with O–H···O IHBs, has 7.88 D dipole moment. The orientation of the OH groups is a major factor influencing the dipole moments. The same phenomenon is observed for the study of naphthylisoquinoline alkaloids with antimalarial activity [10]. Different orientations of the R and S units may cause 1–2 D difference in the dipole moments, and different orientations of the isoquinoline and naphthalene moieties within each unit may cause 1–4 D difference in the dipole moments.

The energy difference between the frontier orbitals (HOMO, Highest Occupied Molecular Orbital, and LUMO, Lowest Unoccupied Molecular Orbital) is an important molecular descriptor, as it relates to the molecule's reactivity and other properties. The HOMO-LUMO energy gap is smallest for conformers with consecutive O–H···O IHBs, slightly greater for conformers with one O–H···O IHB and highest for conformers with only the other types of interactions. Thus, the lower energy conformers also have lower HOMO-LUMO energy gaps. The estimation of the HOMO-LUMO energy gap shows marked difference between HF and DFT values. This is a known phenomenon, as DFT substantially underestimates the values of the gaps [27, 28]. However, the two methods show similar trends. The shapes of the HOMO and LUMO orbitals indicate greater concentration of electrons in the naphthalene moieties than in the isoquinoline moieties. This phenomenon was also observed for naphthylisoquinoline alkaloids with antimalarial activity [10].

3.3 Results in Solution

Of the three solvents considered, acetonitrile has higher dipole moment than water. However, its effect on a number of molecular properties (relative energies, dipole moments etc.) is intermediate between that of chloroform and that of water, as if its polarity were intermediate. Therefore, wordings of the type "as the solvent polarity increases" refer to the chloroform-acetonitrile-water sequence in terms of the effects of these solvents on the properties considered, and not of the actual values of the dipole moments of acetonitrile and water.

The relative energies of the conformers (Table 2) vary in different ways according to the types of conformers. The four lowest energy conformers are the same in all media, and are conformers with the two cooperative IHBs and the same kinds of other IHB-type interactions, and differing only by the mutual orientations of the moieties. Their relative energies remain below 1 kcal/mol in all the media, although the relative energy sequence (including the identification of the lowest energy conformer) may differ from one medium to another. For higher energy conformers, the relative energy decreases as the solvent polarity increases—which is the most common behaviour. If 3.5 kcal/mol is taken as a cautious highest threshold value for conformers which might be responsible for the biological activity, the results in water solution suggest that most conformers (all those with relative energy ≤ 3.5 kcal/mol in water solution) might be considered as potential responsible for the anti-HIV activity of MCA.

The ranges (kcal/mol) of the solvent effect (free energy of solvation, ΔG_{solv}) are 1.28–6.80/HF and 1.91–5.89/DFT in chloroform, 8.63–15.96/HF and 11.37–16.70/DFT in acetonitrile and −22.08–(−11.43)/HF and −22.84–(−13.52)/DFT in water. Table 3 reports the DFT results. ΔG_{solv} is markedly negative in water solution, while it is positive in chloroform and acetonitrile solutions. Similarly, the electrostatic component of ΔG_{solv} (G_{el}) has negative values in all the three solvents, but markedly more negative in water. All this suggests that the molecule might have greater solubility in water than in the other two solvents. The high polarity of this molecule [3] would support this suggestion. A better understanding of the phenomenon may derive from the investigation of adducts with explicit water molecules, considering also the interaction of the water molecules with the π systems of michellamine A; this will be the object of a separate work.

The presence and number of O–H···O IHBs and the O–H···π interaction appears to have an effect on ΔG_{solv}. Conformers with two O–H···O IHBs and two O–H···O interactions (a-b-c-f), with one O–H···π IHB and three O–H···π interactions (b-c-d-f), with one O–H···O IHB and two O–H···π interactions (b-c-d) and with three O–H···π interactions (c-d-e or d-e-f) have smaller ΔG_{solv} in all the solvents than conformers without any O–H···π interactions (a–b), with one O–H···O IHB and one O–H···π interaction (b–d) or with two O–H···π interactions (d–e). The conformer with all the O–H···π interactions (MCA-c-d-e-f-g-h-y-p, having four O–H···π interactions) has the smallest ΔG_{solv} in chloroform and acetonitrile and among

Table 3 Solvent effect (free energy of solvation, ΔG_{solv}) for the calculated conformers of michellamine A in the three solvents considered results from DFT/B3LYP/6-31+G(d,p) single point PCM calculations on the in-vacuo-optimized geometries

Conformers	Chlrf	Actn	Aq	Conformers	Chlrf	Actn	Aq
MCA-a-b-c-f-g-h-x-t	5.89	14.90	−13.52	MCA-c-d-e-f-g-h-y-p	7.45	16.70	–[a]
MCA-a-b-c-f-g-h-x-v	5.87	–[a]	−13.52	MCA-a-b-g-h-y-v	1.91	9.56	−21.03
MCA-a-b-c-f-g-h-y-t	5.91	14.88	−13.82	MCA-a-b-g-h-x-v	2.70	9.99	−21.67
MCA-a-b-c-f-g-h-y-v	4.82	13.98	−13.59	MCA-a-b-g-h-x-t	2.81	10.05	−21.50
MCA-b-c-d-f-g-h-y-q	5.37	14.75	−14.52	MCA-a-b-g-h-y-t	2.58	9.80	−22.00
MCA-b-c-d-f-g-h-x-q	5.76	14.87	–[a]	MCA-b-d-g-h-x-q	2.38	10.12	−21.83
MCA-a-b-f-g-h-x-v	4.31	12.61	−17.32	MCA-b-d-g-h-y-q	2.19	9.72	−22.84
MCA-b-c-d-g-h-x-q	–[a]	12.70	−17.55	MCA-d-e-f-g-h-y-p	4.89	–[a]	−17.85
MCA-a-b-f-g-h-x-t	4.15	12.32	−17.82	MCA-c-d-e-g-h-x-p	5.45	14.01	−18.05
MCA-a-b-f-g-h-y-t	4.42	12.68	−15.59	MCA-d-e-g-h-y-p	3.33	11.38	−20.86
MCA-b-c-d-g-h-y-q	3.83	–[a]	–[a]	MCA-d-e-g-h-x-p	3.44	11.37	−21.28

[a]The calculation for this conformer did not converge in the given solvent

the smallest in water. The presence or absence of C–H···O interactions does not appear to influence the values of ΔG_{solv} and G_{el}.

The dipole moment of the conformers increases with the solvent polarity.

The presence of O–H···O IHBs and of other IHB-type interactions has an influence on the changes in the HOMO-LUMO energy gap caused by the solvent. The HOMO-LUMO energy gaps of conformers with the two consecutive O–H···O IHBs (a–b), with one O–H···O IHB and one O–H···π interaction (b–d) and with three O–H···π interactions (c-d-e or d-e-f) increase as the solvent polarity increases. For conformers with one O–H···O IHB and two O–H···π interactions (b-c-d), the gap increases in chloroform and decreases in acetonitrile and water. For conformers with one O–H···O IHB and three O–H···π interactions (b-c-d-f), it increases in chloroform and acetonitrile and decreases in water. For conformers with all the O–H···π interactions (c-d-e-f), it decreases as the solvent polarity increases. The orientation of the moieties does not significantly affect the effect of the solvent on the HOMO-LUMO energy gap.

4 Discussion and Conclusions

The results of an HF/6-31G(d,p) and DFT/B3LP/6-31+G(d,p) computational study of michellamine A—a naphthylisoquinoline alkaloid with anti-HIV activity- in vacuo and in three solvents highlight the main properties of this molecule. Conformational preferences and other properties are influenced by the presence of O–H···O IHBs and other IHB-type interactions, such as O–H···π and C–H···O, and by the mutual orientations of the moieties. The O–H···O IHBs are the dominant factor determining conformational preferences and energetics. The number of calculated conformers, and the fact that other input geometries optimized to one or another of the conformers presented in this work, suggests that the conformational study is sufficiently complete and has identified all the conformers with not-excessively-high energy.

The low relative energy of many conformers in water solution suggests that a comparatively high number of conformers may be involved in the molecule's biological activity. The values of the free energy of solvation suggest greater solubility in water than in the other two solvents considered in this study, which would be consistent with the high polarity of the molecule. A separate study might consider adducts of michellamine A with explicit water molecules, to better understand its behaviour in water solution.

References

1. Global Report (2013) UNAIDS, report on the global AIDS epidemic
2. www.who.int/mediacentre/factsheets/fs360/en
3. Boyd MR (1994) J Med Chem 37:1740–1745

4. Bringmann G, François G, Aké Assi L, Schlauer J (1998) Chimia 52:18–28
5. Bringmann G, Zagst R, Schäffer M, Hallock YF, Cardellina JH II, John H, Boyd MR (1993) Angew Chem 105:1242–1243
6. Hallock YF, Manfredi KP, Dai JR, Cardellina JH II, Gulakowski RJ, McMahon JB, Schäffer M, Stahl M, Gulden KP, Bringmann G, François G, Boyd MR (1997) J Nat Prod 60:677–683
7. Bringmann G, Zhang G, Ölschläger T, Stich A, Wud J, Chatterjee M, Brun R (2013) Phytochem 91:220–228
8. Lee C, Yang W, Parr RG (1988) Phys Rev B 37:785–789
9. Harvey JN (2004) In: Nkaltsoyanis N, McGrady JE (eds) Principles and applications of density functional theory in inorganic chemistry II, Springer, p 170
10. Mammino L, Bilonda MK (2016) Theor Chem Acc 135:101. doi:10.1007/s00214-016-1843-7
11. Mammino L, Kabanda MM (2012) Int J Quantum Chem 112:2650–2658
12. Mammino L, Kabanda MM (2013) Molec Simul 39:1–13
13. Irikura K, Johnson RD III, Kacker RN (2005) J Phys Chem A 109:8430–8437
14. Merrick JP, Moran D, Radom L (2007) J Phys Chem A 111:11683–11700
15. Barone V, Cossi M (1997) J Chem Phys 107:3210–3221
16. Tomasi J, Mennucci B, Cammi R (2005) Chem Rev 105:2999–3093
17. Barone V, Cossi M, Tomasi J (1998) J Comput Chem 19:404–417
18. Cossi M, Scalmani G, Rega N, Barone V (2002) J Chem Phys 117:43–54
19. Cancès E, Mennucci B, Tomasi J (1997) J Chem Phys 107:3032–3041
20. Tomasi J, Mennucci B, Cancès E (1999) THEOCHEM 464:211–226
21. Frisch MJ, Trucks GW, Schlegel HB, Scuseria GE, Robb MA, Cheeseman JR, Montgomery JA, Vreven T, Kudin KN, Burant JC, Millam JM, Iyengar SS, Tomasi J, Barone V, Mennucci B, Cossi M, Scalmani G, Rega N, Petersson GA, Nakatsuji H, Hada M, Ehara M, Toyota K, Fukuda R, Hasegawa J, Ishida M, Nakajima T, Honda Y, Kitao O, Nakai H, Klene M, Li X, Knox JE, Hratchian HP, Cross JB, Adamo C, Jaramillo J, Gomperts R, Stratmann RE, Yazyev O, Austin AJ, Cammi R, Pomelli C, Ochterski JW, Ayala PY, Morokuma K, Voth GA, Salvador P, Dannenberg JJ, Zakrzewski VG, Dapprich S, Daniels AD, Strain MC, Farkas O, Malick DK, Rabuck AD, Raghavachari K, Foresman JB, Ortiz JV, Cui Q, Baboul AG, Clifford S, Cioslowski J, Stefanov BB, Liu G, Liashenko A, Piskorz P, Komaromi I, Martin RL, Fox DJ, Keith T, Al-Laham MA, Peng CY, Nanayakkara A, Challacombe M, Gill PMW, Johnson B, Chen W, Wong MW, Gonzalez C, Pople JA (2003) Gaussian 03. Gaussian Inc, Pittsburgh.
22. Pascual-Ahuir JL, Silla E (1990) J Comput Chem 11:1047
23. Silla E, Villar F, Nilsson O, Pascual-Ahuir JL, Tapia O (1990) J Mol Graph 8:168–172
24. Silla E, Tunon I, Pascual-Ahuir JL (1991) J Comput Chem 12:1077–1088
25. Hunter CA, Lawson KR, Perkinsa J, Urchb CJ (2001) J Chem Soc Perkin Trans 2:651–669
26. Kabanda MM, Mammino L (2012) Int J Quant Chem 112:519–531
27. https://www.wiki.ed.ac.uk/display/EaStCHEMresearchwiki/How+to+analyse+the+orbitals+from+a+Gaussian+calculation
28. Zhang G, Musgrave CB (2007) J Phys Chem A 111:1554–1561

Integrated Computational Studies on Mutational Effects of a Nylon-Degrading Enzyme

Takeshi Baba, Katsumasa Kamiya and Yasuteru Shigeta

Abstract We investigate the mutational effects on a particular mutant of the nylon oligomer hydrolase (NylB), Y170F, whose enzymatic activity is about 1/80 times smaller than that of the wild type (WT). To this aim, we have investigated them by several computational schemes and analyzed obtained data to reveal mutational effects due to Y170F. First, classical molecular dynamics (MD) simulations were performed to measure the stability of substrate-enzyme complexes. Owing to a replacement of Tyr170 by Phe170, it is found that water molecules flow to an active site, which might avoid the substrate degradation. Next, by using QM/MM Car-Parrinello molecular dynamics (QM/MM CPMD) complemented with meta-dynamics (Meta-D), we provide a detailed insight into the underlying acylation mechanism. Our results show that while in the WT the Tyr170 residue points the NH group towards the proton-acceptor site of an artificial amide bond, hence ready to react, in the Y170F this does not occur. The reason is ascribed to the absence of Tyr170 in the mutant, replaced by phenylalanine, unable to form H-bonds with the amide bond, thus resulting in an increase of the activation barrier by 11 kcal/mol. Nonetheless, despite the lack of H-bond between the Y170F and the mutant, also in this case the highest free energy barrier for the induced-fit is similar to that of WT revealed by Parallel Cascade MD (PaCS-MD) with free energy analyses. This seems to suggest that, in the induced-fit process, kinetics is

---

Electronic supplementary material The online version of this chapter (doi:10.1007/978-3-319-50255-7_19) contains supplementary material, which is available to authorized users.

T. Baba · Y. Shigeta (✉)
Department of Materials Engineering Sciences, Graduate School of Sciences, Osaka University, Osaka, Japan
e-mail: shigeta@ccs.tsukuba.ac.jp

K. Kamiya
Center for Basic Education and Integrated Learning, Kanagawa Institute of Technology, Atsugi, Japan

Y. Shigeta
Department of Physics, Graduate School of Pure and Applied Sciences, University of Tsukuba, Tsukuba, Japan

A. Tadjer et al. (eds.), *Quantum Systems in Physics, Chemistry, and Biology*,
Progress in Theoretical Chemistry and Physics 30,
DOI 10.1007/978-3-319-50255-7_19

little affected by the mutation. We also evaluated interaction energy between the substrate and amino acid residues in NylB and its changes upon the induced-fit processes by using fragment molecular orbital (FMO) method.

1 Introduction

1.1 Experimental Results for Nylon-Oligomer Degrading Enzyme

Currently, developments in separation and purification technologies for the environment have been promoted by governments, research institutes, and companies all over the world, and some of the technologies are now put into practice. These are mainly categorized as chemical, physical, and biological treatments. Among them, the bioremediation has been attracted much attention from both basic sciences and industrial uses due to high ability of microorganism for adaptive capacity toward the environment and material decomposition. Especially, the researches targeted for searching microorganism that degrade artificial compounds and their degrading enzyme have been extensively conducted.

Since polymers are one of the hard materials of the decomposition process, researches on both synthesis and degradation of the biodegradable polymers have also been actively conducted. Nylon-6 is synthesized via the ring-opening polymerization of ε-caprolactams [1]. Nylon-6 is very tough with high heat and chemical resistance and has been widely used as a raw material for the textile and plastic fiber. Annually several million tons of Nylon-6 has been produced. However, nylon factories also produce nylon-oligomers, which have low polymerization degrees of 6-aminohexanoate (Ahx) units, as industrial wastes.

Nylon-oligomer hydrolases found in the *Arthrobacter sp.* KI72, which was discovered in soil nearby nylon factories, can degrade the nylon-oligomers. Three different Nylon-oligomer hydrolases are encoded in a plasmid pOAD2, and various genetic and biochemical studies have been made [2–7]. Among them, Ahx-liner-dimer (Ald) hydrolase (NylB) has received particular attention, since NylB degrades the amide bond in a linear Ahx oligomer such as the Ald and recycles Ahx as the raw material. According to the genetic analyses, NylB has a closely resemble enzyme, NylB*, with 88% sequence identity. Despite the similarity, NylB* has reaction activity to the substrate (Ald) 1/200 times lower than NylB [3]. To enhance their reactivity and to overcome the difficulty in their crystallization, hybridizations of NylB and NylB* and random and site-directed mutagenesis works [8–13] have been made to find Hyb-24Y (one mutation of D370Y from Hyb-24), Hyb-24DY (double mutation of G181D and D370Y from Hyb-24), Hyb-24DN, and Hyb-24DNY, for which the relevant kinetic parameters, i.e. the Michaelis constant (K_M) and the rate constant (k_{cat}), could be determined (See

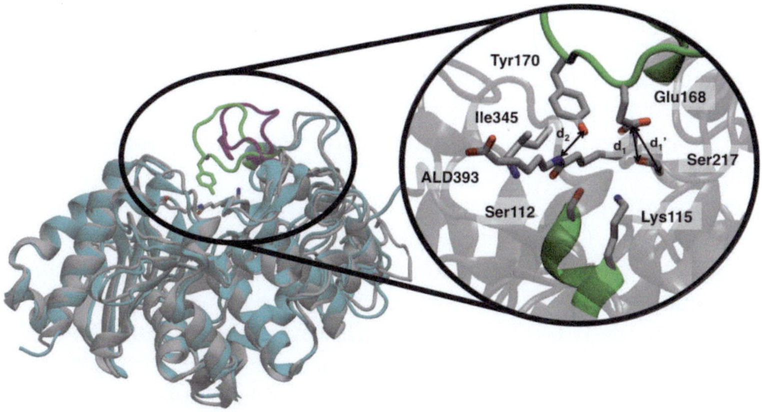

Fig. 1 X-ray structure of the NylB type nylon-oligomer hydrolase from Open form (in *cyan* and *purple*) to closed form (in *black* and *green*). Details of residues and reaction coordinates, corresponding to the region inside the *circle*. In the Y170F mutant, Tyr170 is replaced with Phe170

Table S-1) [11]. At present, Hyb-24DNY possesses the highest activity among all the known mutants (we hereafter refer it to "wild-type" (WT) for simplicity).

According to the similarities in the amino acid sequence and the structure, NylB is classified in class C of the β-lactamase family. NylB has a catalytic triad comprised of Ser112, Tyr215, and Lys115, which can promote the hydrolysis of amide and/or ester bonds [14]. X-ray crystallographic analyses reveal that the WT takes two different conformations, (i) a substrate-free form (open form) and (ii) a substrate-bound form (closed from) (See Fig. 1 for details) [9, 11]. Among the enzymes of the same family, the distinctive feature of NylB is the presence of a unique amino acid residue, Tyr170, near the catalytic triad, which forms a hydrogen bond (H-bond) with the substrate (Ald). In the present review, we focus on the role of Tyr170 and mutational effects of Tyr170 on a series of enzymatic reactions in detail.

1.2 Theoretical Analyses on Reaction Mechanism of WT

By using a reactive quantum mechanics/molecular mechanics (QM/MM) [15] Car-Parrinello molecular dynamics (CPMD) [16] based on the density functional theory (DFT) [17] enhanced with a meta-dynamics (Meta-D) approach [18], we have recently worked out its catalytic mechanism and related functional/structural specificities [19]. Specifically, the cleavage of the amide bond and the formation of the acyl-enzyme occur along the reaction path from (a) to (f) in Fig. S-1. Among these reaction steps, the reaction from (b) to (c) is the rate-limiting step and is characterized by a free energy barrier of ~21 kcal/mol. This result is in good agreement with the reported experimental value of ~17 kcal/mol estimated from the Eyring-Polanyi equation assuming the experimental k_{cat} value of 3.2 s^{-1} [20]. The resulting reaction mechanism is typical of a catalytic triad (Ser112: nucleophilic attack, Lys115: general base, Ty215: general acid) in serine proteases. However, it involves the crucial participation of a unique amino acid residue, Tyr170.

According to our preceding results, Tyr170 is expected to have two crucial functional roles, i.e. (1) a stabilization of the substrate-binding state and (2) enhancement in the proton-donation from a general acid to the substrate amide N in the tetrahedral intermediate. To know roles of a specific amino acid residue, it is worth comparing the WT with the mutant. However, the introduction of a mutation may affect various intermediate reaction step, since the enzymatic reactions undergo via complicated multistep reactions. Indeed, for validation of their mechanism, we need to consider effects of an introduction of mutation such as Y170F on both the enzymatic reaction and the substrate binding processes. In the following, we performed several theoretical analyses on the Y170F mutant in comparison with the WT to genuinely know both (1) and (2).

2 Results and Discussion

2.1 Preparation of Target Systems

Before starting both classical and first-principles molecular dynamics (MD) simulations of the NylB and mutants, model structures were constructed from the X-ray coordinates of the WT (PDB ID: 2ZMA) of NylB registered in the Protein Data Bank (PDB). Since the Ser112 had been replaced by Ala112 in the X-ray coordinates of the WT due to prevent the enzymatic reaction to a substrate, we replaced the Ala112 with Ser112 and also added missing amino acid residues by hand to yield a model structure for the WT. The WT model was first solvated in water (described by a TIP3P force field) [21] solvation box, and sodium ions were also added o neutralize the system (totally 41403 atoms in the simulation box). To obtain Y170F and E168Q mutants, we also altered the original amino acid residues, Tyr170 and Glu168, into the corresponding ones, Phe and Gln, respectively. Similarly, a Hyb-24 model was constructed by replacing Asp181, Asn266, and Tyr370 into Gly, His, and Asp from the WT model at the same time. These mutant models were also solvated and neutralized with the same manner.

All MD simulations were performed using the PMEMD module of the AMBER program package [22, 23]. For the force field parameters for proteins (the WT and mutants), the standard Amber ff99 model was adopted. On the other hand, those for a substrate (Ald), the force field parameters were generated by AMBER GAFF tool in the Antechamber program [22]. The protonation states were first determined with PROPKA, an empirical method developed by the group of Jensen [24–27], and then adjusted by direct inspection of their local hydrogen bonding environments to promote the acylation, where Lys115 is deprotonated as a general base. For the classical MD simulations, a time step was set to $\Delta t = 2$ fs with the SHAKE algorithm [28] and the particle mesh Ewald method with a cutoff of 12.0 Å was used for long-range Coulombic interaction. After a 10 ns *NPT* simulation for the WT model ($T = 300$ K and $P = 1$ atm), the box size of the 3D periodic cell was determined as $V = 77.57 \times 76.62 \times 69.30$ Å^3, which was confirmed that the substrate-enzyme complex is well separated from the periodic images, so that an influence due to NylB in neighboring cells are considered to be small. In the production runs, 25 ns *NVT* simulations were performed at $T = 300$ K. For all mutants, corresponding amino acid residues were replaced as mentioned above and the classical MD simulations were performed under the same conditions.

2.2 Classical Molecular Dynamics Simulation for Stability of Ligand Binding

To obtain structural insight into the differences among the WT, E168Q, and Y170F mutants, we analyzed statics of numbers of water molecules nearby the substrate and H-bonds between the substrate and a loop segment [29]. The representative snapshots are shown in Figs. 2. In the WT, Y170 of the loop-segment always forms H-bonds to the Ald and E168 (Fig. 2a). This H-bond network leads to the stabilization of the substrate-enzyme complex, where no water molecule easily accesses the catalytic site. On the other hand, in the E168Q and Y170F mutants (Fig. 2b, c), the Y170 residue in E168Q or the F170 residue in Y170F become far away from the catalytic site to create a cavity nearby the substrate. One or two water molecules sometimes come in the cavity. When the cavity exists, water molecules stays in the cavity and frequently access the substrate.

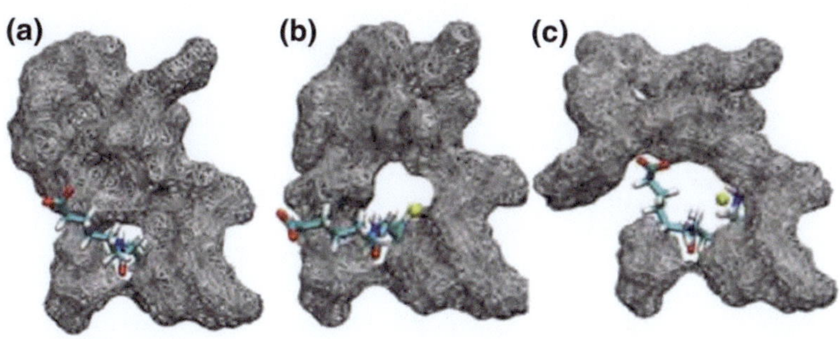

Fig. 2 Representative structure (*upper panels*) and Molecular surface (*lower panels*) of the catalytic site for **a** the wild-type, **b** E168Q, and **c** Y170F. The surface was drawn with a probe radius of 1.4 Å. In the mutants, some water molecules (*yellow ball*) exist nearby the substrate

To quantify the stability of the enzyme-substrate complex, three H-bonds between the amino acid residue in the enzyme (S112, I345, and Y170 or F170) and substrate were counted. For WT, three H-bonds form for more than 90% of the total simulation time, indicating that the enzyme-substrate complex is quite stiff. On the other hand, for the E168Q mutant, those form for about 40%. To make matters worse, three H-bonds in the Y170F mutant only forms for 10%. These results clearly show that the remarkable difference in the loop-segment dynamics between the WT and the mutants results in the formation of the cavity nearby the substrate and influx of water molecules into the cavity.

The present MD calculations indicate that the mutations (E168Q and Y170F) lead to three mutational effects on the NylB: (1) a loosening of the substrate binding, (2) a large fluctuation of the loop-segment, and (3) a loss of water exclusivity around the catalytic site. Thus, our calculations show that the movement of the loop-segment is important for regulating the reaction mechanism of NylB, in consistent with previous biochemical experiments. This result also suggests that the water molecules near the catalytic site may serve as an inhibitor for degrading the substrate. This problem will be touched in the following section.

2.3 QM/MM CPMD Simulations for Acylation Processes

The acylation reaction, which involves cleavage of an amide C-N bond and formation of chemical bonds, was simulating within a hybrid QM/MM CPMD enhanced with Meta-D approach. In our particular systems, the QM region is composed of 103 (102) atoms for the WT (the Y170F mutant) embedded in the MM region of the rest part (Fig. 3 left panel). The total net charge for the QM

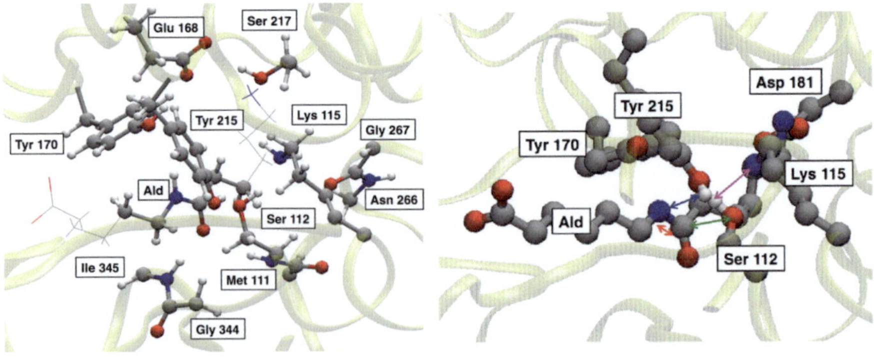

Fig. 3 Structure and main amino acid residues of NylB. (*Left*) The QM region used in our hybrid QM/MM simulations. QM atoms are shown as *balls* and *sticks*, whose colors code are *gray* for C, *red* for O, *blue* for N and *white* for H, and the *yellow ribbon* shows the secondary structure. (*Right*) The reaction coordinate (*red arrow*) and other fundamental distances (*green, blue and magenta*) used to drive and monitor the reaction

subsystem equal to -*e* and the hydrogen link atom scheme was adopted to model the interface between the QM and the MM regions.

For this analysis, the CPMD program package [30] was used for all these simulations. The electronic structure is treated at the density functional theory (DFT) level with an HCTH exchange-correlation functional, [31] where the core-valence interaction is described by Martins-Troullier pseudopotentials [32] and valence electrons are represented in a plane wave basis set with energy cut off of 70 Ry. All QM/MM CPMD simulations were done in *NVT* ensemble at $T = 300$ K and the same volume used in the classical MD simulations. The fictitious electron mass, μ, was set to 400 a.u. and the time step adopted was $\Delta t = 4$ a.u. (0.096 fs). Classical MM interactions were also described by the Amber ff99 force field. For the Meta-D simulations [18], we selected the distance between O_γ atom (OG) of Ser112 and carbonyl carbon (C1) of Ald393 (see Fig. 3 right panel) as a single CV, s_α, which properly represents the nucleophilic attack of Ser112 to the amide carbon in the substrate. The mass M_α and force constant k_α for the collective variables are typically set to 20 and 0.24 a.u. The history dependent potential was constructed by accumulation of small Gaussian penalty functions with the height and the width of 0.5 kcal/mol and 0.3 a.u., respectively, and a new Gaussian function was added every 140 steps (13.44 fs), which will be referred to as one meta-step hereafter. For the present purpose, this can be assumed to be a completion of the acylation process and the difference between the free energy barriers for acylation in the WT and Y170F mutant could be estimated.

In order to clarify the differences between the WT and the Y170F mutant during the acylation, the structural changes of the Y170F mutant during the acylation are sketched in Fig. 4a–c. Especially when the substrate accepts the H atom from Tyr215 (see Fig. 4c for Y170F mutant), it is stressed here that the WT takes almost the same configuration as depicted in Fig. 4d.

Before the acylation, the amide bond and the neighboring atoms in the Ald are initially located on the amide plane. When the acylation occurs, the H atom belonging to –OH group of Ser112 is donated to Lys115 and the O atom of it attacks the amide bond forming an sp^3 bonding configuration with the amide carbonyl C atom in the Ald, resulting in the breaking of the amide plane (see Fig. 4a). Herein, the amide N atom also tends to form an sp^3 geometry and can accept a H atom from another site. In the case of WT, –OH group of Tyr170 forms a H-bond with the amide–NH group of the Ald, and then this H-bond restricts the lone pair of N atom in the Ald to orient toward Tyr215 (see Fig. 4d). This configuration enables the N atom in the Ald to easily accept the H atom from Tyr215. In the Y170F mutant, the absence of Tyr170 and its H-bond does not allow the amide NH group to keep a configuration suitable to accept the H atom from Tyr215 (see Fig. 4b, e), resulting in a suppression of the amide bond cleavage activity.

After a few additional meta-steps, a water molecule comes from the surrounding solvent and forms an H-bond with the amide NH group of the Ald as shown in Fig. 4c, where the water molecule takes the place of Tyr170 in the WT, and the conformation of the rest part resembles closely to the WT (Fig. 4d) as mentioned before. This means that the H-bond of the water to the amide NH group must help

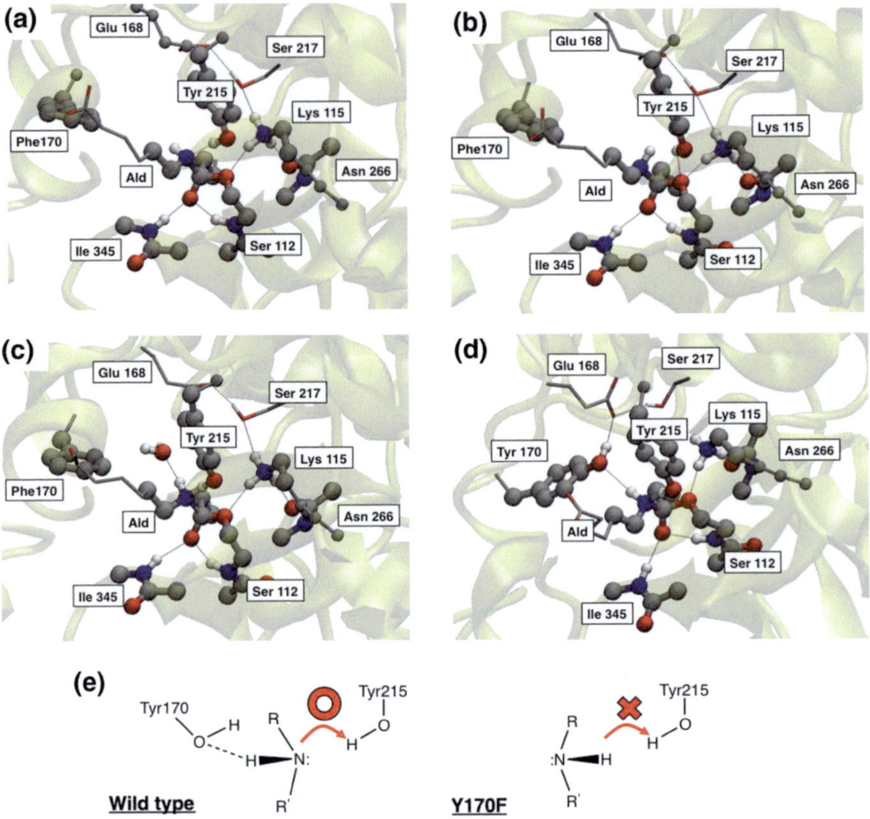

Fig. 4 Local structures and related differences between Y170F and WT relevant to the acylation process. Panels (**a**), (**b**), and (**c**) correspond to the structures taken from specific meta steps (see Fig. S-2 for detail) Panel (**d**) shows the WT for comparison. A schematic picture of the difference between Y170F and WT is shown in (**e**), where *red arrow* indicates the donation of electron to form a chemical bond

the Ald to accept the H atom from Tyr215 by controlling the direction of NH group as in the WT. These results indicate that the formation of an H-bond to the amide NH group in the Ald is indispensable to trigger the H transfer from Tyr215 and that the unique amino acid residue, Tyr170, assist the NylB in breaking the amide bond efficiently. Thus, it is expected that Tyr170 is the crucial residue that complements the catalytic triad to realize the acylation in this enzyme. Indeed, lack of Tyr170 results clearly in the suppression of the enzymatic activity by about 1/80, meaning higher free energy barrier is needed to undergo the acylation.

Generally speaking, amines have two different sp^3 configurations as shown in Fig. 4e. The conformational change between them is well known as an umbrella inversion, whose free energy barrier is about 6 kcal/mol for an ammonia molecule in the gas phase [33]. The free energy difference between the WT and the Y170F mutant regarding the activation barrier (~11 kcal/mol, according to our estimates) might be compatible with the results obtained via the umbrella inversion in a protein environment. In the previous study, it is also pointed out that the orientation of the lone pair of the N atom of the scissile amide bond is a crucial feature of enzyme catalyzed amide bond hydrolysis [34]. It is noteworthy that our present results are entirely consistent with the preceding study.

2.4 Classical Molecular Dynamics Simulation for Induced Fit Processes

In general, the most of the enzymes have strong substrate specificity, which is often likened to the relation between "lock and key" [35]. However, when a substrate attaches to an enzyme, some of the substrate-enzyme complex changes the structure to tightly interact with each other. This structural change is known as an "induced-fit" process [36]. As mentioned before, accurate X-ray crystallographic analyses and bioengineering experiments have shown that the NylB has a unique amino acid residue, Tyr170, that forms H-bond with the substrate after the substrate binding to the NylB [2–11, 14]. According to the X-ray crystallographic data, there exist two different conformations of the NylB and the NylB-Ald complex. One is a substrate-free form (open form), and the other is a substrate-bound form (closed form) [9, 11]. By comparing two structures, the closed form involves a substantial movement of the loop-segment N166–V177 towards the substrate, and the other parts overlap mutually with each other. This difference translates into the induced-fit (see Fig. 1). However, no information has been obtained concerning their formation process and how the substrate-enzyme complex is stabilized upon the induced-fit. Moreover, it is quite interesting to know the mutational effects on the induced-fit process by replacing Tyr170 by Phe170.

The PaCS-MD simulations, [37] which generate reactive trajectories from a reactant (open form, PDBid: 2ZM0) to a product (closed form, PDBid: 2ZMA), have been performed to investigate differences in the induced-fit processes of the WT and the Y170F mutant. As the reaction coordinates, the root mean square displacements (RMSD) of the loop-segment (Asn166-Val177) except for all hydrogen atoms and C_α atoms RMSD of the 111th to 116th residues are adopted in PaCS-MD simulations. All PaCS-MD simulations were carried out using the PMEMD module of the AMBER14 package again with AMBER ff99 force field [22]. After a series of PaCS-MD simulations, the free energy change due to the induced-fit was evaluated on the basis of a multiple independent umbrella sampling (MIUS) [38] with a weighted histogram analysis method (WHAM) [39]. In MIUS, 2000 reference structures taken from reactive trajectories detected in the PaCS-MD simulations were used, where biased umbrella potential with a coupling parameter $k = 0.5$ kcal/mol Å^2 was imposed on the loop-segment, apart from H atoms, the nitrogen atom of the amide bond of Ald, and the C_α atoms of the 111th to 116th

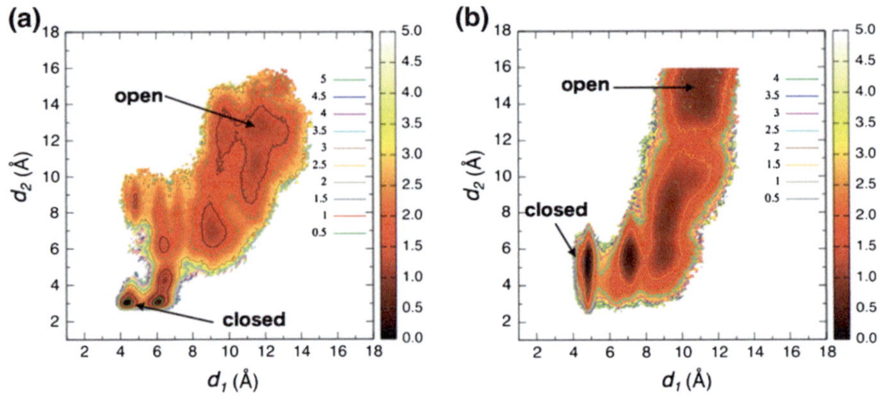

Fig. 5 Free energy landscapes for induced-fit process in the Y170F mutant (**a**) and the WT (**b**)

residues. Using WHAM, the free energy landscape (FEL) were projected onto two distinct distances: the distance between C_δ atom of Glu168 and C_α atom of Ser217, d_1, and the one between H_ζ atom of Phe170 in the Y170F mutant (O_ζ atom of Tyr170 in the WT) and N atom in amide bond of Ald393, d_2 [40].

Two-dimensional FEL along two different reaction coordinates are depicted for the WT (Fig. 5a) and the Y170F mutant (Fig. 5b) [40]. These figures clearly show that there exist two remarkable feature in these FELs. One is a difference in the number of (meta-) stable structures, i.e. seven minima in the WT, while four in the Y170F mutant. The other is the size of each basin, where these local minima are located. The global FEL minimum along d_2 direction (left-bottom area) in the Y170F mutant is wider than that in the WT. This indicates that Phe170 in the Y170F mutant more largely fluctuates than Tyr170 in the WT, resulting in the inflow of a water molecule as noted in Sects. 2.1 and 2.3. Although Glu168 is a residue common to both the WT and the Y170F mutant, Glu168 forms an H-bond with the substrate and Ser217 in the WT, but the fluctuation of Phe170 again prevents Glu168 from forming any H-bond with Ser217. Despite this remarkable structural difference, the free energy gain of the Y170F mutant due to the induced-fit is nearly identical to that of the WT. These results suggest that the mutational effect on the shape of FEL in the induced-fit process is remarkable, while that on the stabilization energy is rather small.

2.5 FMO Analyses of Enzyme-Substrate Bound States

An inter-fragment interaction energy (IFIE) analyses based on the fragment molecular orbital (FMO) method were utilized to investigate the importance of Tyr170 in the substrate specificity [41–43]. All FMO calculations were carried out as implemented in the ABINIT-MP program suite [43] at the FMO2 MP2/6-31G(d)

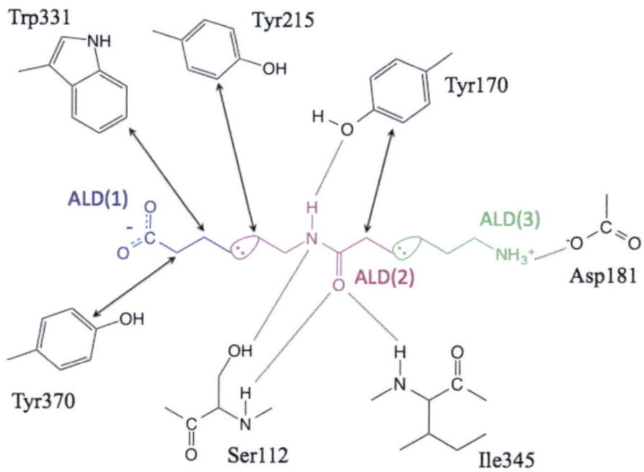

Fig. 6 **a** Schematic view of the interactions between the most relevant amino acid residues and the substrate. For clarity, the ALD is divided into three fragments, the C-terminal, the peptide site, and the N-terminal, labeled as ALD(1) (*blue*), ALD(2) (*pink*), and ALD(3) (*green*), respectively

level. To reduce the computational costs, we restrict ourselves to treat only one structure for each mutant obtained after 10 ns classical MD simulations. Moreover, we neglect surrounding water molecules except for those within a region of 15 Å around the substrate [44]. To clarify which moiety is interact with specific amino acid residues, we also divide the substrate (Ald) into three fragments (C-terminus: Ald(1), Amide-bond: Ald(2), and N-terminus: Ald(3)) as illustrated in Fig. 6. According to Table 1, the IFIE value between Asp181 and Ald(3) stronger than that between Tyr170 and Ald(2). On the other hand, the interaction between NylB and Ald(3) lacks in Hyb-24. Although the IFIE value between Phe170 in the Y170F mutant and Ald(2) becomes weaker than that between Tyr170 in the WT and Ald (2), the decrease of the IFIE is less pronounced than that to Ald(3) of the Hyb-24. These results quantitatively suggest that Tyr 170 is less important than Asp181 for the particular substrate binding. Indeed, available kinetics data on a double mutation of G181D and D370Y on Hyb-24, ($k_{cat}/K_M = 0.35$ s^{-1}mM^{-1}, $K_M = 7.1$ 1.28 mM^{-1}) and of Hyb-24Y, in which a single mutation of D370Y from Hyb-24 is present, ($k_{cat}/K_M = 0.016$ s^{-1}nM^{-1}, $K_M = 39.1 \pm 6.73$ mM^{-1}), indicate rather clearly that Asp181 contributes to enhancing the substrate specificity.

3 Conclusion

In the present review, we have demonstrated how Tyr 170 (and partly Asp181) in NylB work in the series of enzymatic reactions, (1) conformational change due to mutations, (2) the enzymatic reactions, (3) the induced-fit process, and (4) the stability of the enzyme-substrate binding by using a combination of different theoretical methods such as (a) classical MD, (b) QM/MM CPMD accelerated with Meta-D method, (c) PaCS-MD, and (d) FMO.

Our results show at which steps of this complex process Tyr170 and Asp181 are involved and which are their respective effects on the whole process. Specifically,

Table 1 The computed IFIE values between the 170th and the 181st amino acid residues and each fragment in Ald

System	Residue	IFIE[a]			
		Ald(1)	Ald(2)	Ald(3)	Total
WT[b]	Tyr170	–2.07	–6.01	0.73	–7.35
	Asp181	18.31	–2.63	–115.27	–99.59
Y170F	Phe170	–4.69	0.33	–1.44	–5.80
	Asp181	20.19	–3.05	–118.21	–101.07
Hyb-24	Tyr170	–4.87	–4.79	–4.34	–14.00
	Gly181	–0.52	0.13	1.22	0.83

[a]Unit of *kcal/mol*
[b]From Ref. [73]

Tyr170 mainly contributes to the acylation step and plays a subsidiary role in the stabilization of the enzyme-substrate complex to form H-bond with NH group in the Ald. By the free energy changes during the acylation and the induced-fit steps upon the mutation, the former process in the Y170F requires the overcoming of an activation barrier, which is 11 kcal/mol higher than the WT case, resulting in the suppression of the acylation in the Y170F mutant. On the other hand, for the induced-fit process of the Y170F mutant, although the shape of the free energy landscape changes drastically compared with the WT, the free energy barrier is nearly unchanged, suggesting that the kinetics of the induced-fit is barely affected by this mutation. This is also supported by the inter-fragment interaction energy (IFIE) analyses. Although the IFIE value between Phe170 in the Y170F mutant and the substrate is a bit lower than that between Tyr170 in the WT and the substrate, a decrement is smaller than the mutational effects due to G181D in Hyb-24. On the other hand, the IFIE analyses also suggest that Asp181 mainly contributes to the substrate specificity in the enzyme-substrate, while the influence on the hydrolytic activity is rather negligible.

Acknowledgements We would like to express acknowledgement to Profs. S. Negoro, Y. Higuchi, and H. Shibata for their valuable suggestions from the experimental side and Profs. Y. Mochizuki, T. Matsui, M. Boero, M. Nakano, and Mr H. Ando for their tight collaboration for the integrated theoretical analyses. These works were supported from the Japan Society for the Promotion of Science (JSPS) Research Fellowship for Young Scientists, Grant-in-Aid for Scientific Research (Nos. 26102525, 26107004, and 22360350). Computer resources were provided from Research Center for Computational Science, Okazaki, Japan and Center of Computational Materials Science, Institute for Solid State Physics, The University of Tokyo, Japan.

References

1. Horning EC, Stromberg VL, Lloyd HA (1952) J Am Chem Soc 74:5153
2. Kinoshita S, Kageyama S, Iba K, Yamada Y, Okada H (1975) Agric Biol Chem 39:1219
3. Okada H, Negoro S, Kimura H, Nakamura S (1983) Nature 306:203
4. Kato K, Ohtsuki K, Koda Y, Maekawa T, Yomo T, Negoro S, Urabe I (1995) Microbiology 141:2585

5. Negoro S (2000) Appl Microbiol Biotechnol 54:461
6. Nagai K, Iida K, Shimizu R, Kinugasa M, Izumi D, Kato I, Takeo M, Mochiji K, Negoro S (2014) Appl Microbiol Biotechnol, 98:8751
7. Negoro S, Shibata N, Tanaka Y, Yasuhira K, Shibata H, Hashimoto H, Lee Y, Oshima S, Santa R, Oshima S, Mochiji K, Goto Y, Ikegami T, Nagai K, Kato D, Takeo M, Higuchi Y (2012) J Biol Chem 287:5079
8. Kato K, Fujiyama K, Hatanaka HS, Priyambada ID, Negoro S, Urabe I, Okada H (1991) Eur J Biochem 200:165
9. Negoro S, Ohki T, Shibata N, Sasa K, Hayashi H, Nakano H, Yasuhira K, Kato D, Takeo M, Higuchi Y (2007) J Mol Biol 370:142
10. Ohki T, Shibata N, Higuchi Y, Kawashima Y, Takeo M, Kato D, Negoro S (2009) Protein Sci 18:1662
11. Kawashima Y, Ohki T, Shibata N, Higuchi Y, Wakitani Y, Matsuura Y, Nakata Y, Takeo M, Kato D, Negoro S (2009) FEBS J 276:2547
12. Crowley PH (1975) J Theor Biol 50:461
13. Johnson KA, Goody RS (2011) Biochemistry 50:8264
14. Negoro S, Ohki T, Shibata N, Mizuno N, Wakitani Y, Tsurukame J, Matsumoto K, Kawamoto I, Takeo M, Higuchi Y (2005) J Biol Chem 280:39644
15. Warshel A, Levitt M (1976) J Mol Biol 103:227
16. Car R, Parrinello M (1985) Phys Rev Lett 55:2471
17. Kohn W, Sham LJ (1965) Phys Rev 140:A1133
18. Laio A, Parrinello M (2002) Proc Natl Acad Sci U S A 99:12562
19. Kamiya K, Baba T, Boero M, Matsui T, Negoro S, Shigeta Y (2014) J Phys Chem Lett 5:1210
20. $\Delta G = -RT\ln k_{cat} + RT\ln(k_B\ T/h)$; A reaction rate constant: k_{cat}; Planck's constant: h; Temperature:T; Boltzmann constant: k_B
21. Jorgensen WL, Chandrasekhar J, Madura JD (1983) J Comput Phys 79:926
22. Case DA, Babin V, Berryman JT, Betz RM, Cai Q, Cerutti DS, Cheatham TE III, Darden TA, Duke RE, Gohlke H, Goetz AW, Gusarov S, Homeyer N, Janowski P, Kaus J, Kolossváry I, Kovalenko A, Lee TS, LeGrand S, Luchko T, Luo R, Madej B, Merz KM, Paesani F, Roe DR, Roitberg A, Sagui C, Salomon-Ferrer R, Seabra G, Simmerling CL, Smith W, Swails J, Walker RC, Wang J, Wolf RM, Wu X, Kollman PA (2014) AMBER 14. University of California, San Francisco
23. Salomon-Ferrer R, Goetz AW, Poole D, Le Grand S, Walker RC (2013) J Chem Theory Comput 9:3878
24. Li H, Robertson AD, Jensen JH (2005) Proteins 61:704
25. Bas DC, Rogers DM, Jensen JH (2008) Proteins 73:765
26. Olsson MHM, Søndergard CR, Rostkowski M, Jensen JH (2011) J Chem Theory Comput 7:525
27. Søndergaard CR, Olsson MH, Rostkowski M, Jensen JH (2011) J Chem Theory Comput 7:2284
28. Ryckaert J-P, Ciccotti G, Berendsen HJC (1977) J Comput Phys 23:327
29. Baba T, Kamiya K, Matsui T, Shibata N, Higuchi Y, Kobayashi T, Negoro S, Shigeta Y (2011) Chem Phys Lett 507:157
30. CPMD, http://www.cpmd.org/, Copyright IBM Corp 1990–2012, Copyright MPI für Festkörperforschung 1997-2001
31. Hamprecht FA, Cohen AJ, Tozer DJ, Handy NC (1998) J Chem Phys 109:6264
32. Troullier N, Martins JL (1993) Phys Rev B 1991:43
33. Werpetinsk KS, Cook M (1995) Phys Rev 52:R3397
34. Syrén P-O (2013) FEBS J 280:3069
35. Fischer E (1894) Ber Dt Chem Ges 27:2985
36. Koshland DE (1958) Proc Natl Acad Sci, 44:98
37. Harada R, Kitao A (2013) J Chem Phys 139:035103
38. Kästner J (2011) Comp Mol Sci 1:932
39. Souaille M, Roux B (2001) Comput Phys Commun 135:40

40. Baba T, Harada R, Nakano M, Shigeta Y (2014) J Comput Chem 35(16):1240–1247
41. Kitaura K, Sawai T, Asada T, Nakano T, Uebayasi M (1999) Chem Phys Lett 312:319–324
42. Tanaka S, Mochizuki Y, Komeiji Y, Okiyama Y, Fukuzawa K (2014) Phys Chem Chem Phys 16:10310
43. Mochizuki Y, Yamashita K, Fukuzawa K, Takamatsu K, Watanabe H, Taguchi N, Okiyama Y, Tsuboi M, Nakano T, Tanaka S (2010) Chem Phys Lett 493:346
44. Ando H, Shigeta Y, Baba T, Watanabe C, Okiyama Y, Mochizuki Y, Nakano M (2015) Mol Phys 113:319

Damage Induced by Proton Collisions on Building Blocks of Life

Marie-Christine Bacchus-Montabonel

Abstract Analysis of the mechanism involved at the molecular level in the action of radiations on biomolecular systems is of main interest in cancer therapy research. Ion-induced ionization appears at the early stage of damage before fragmentation of the biomolecule. We have investigated such a reaction step focusing on the ionization of the biomolecule by charge transfer from the incident ion. Taking account of the great impact in proton-therapy treatments, we have considered proton-induced collisions, looking at a series of different biomolecular targets, mainly the DNA and RNA building blocks with on a one hand the nucleobases uracil and thymine, and on the other hand the 2-deoxy-D-ribose sugar skeleton in both furanose and pyranose forms. The study has been extended to possible precursors of the nucleobases, 2-aminooxazole and hydantoin, recently observed in space. The question of the reactivity and stability of prebiotic compounds in astrophysical environments is indeed crucial if one considers the possibility of an exogen origin of life. In proton-rich regions of space, collisions with protons could be fundamental. The calculations have been performed by means of ab initio quantum chemistry molecular methods followed by a semi-classical collision treatment in a wide collision energy range. The present analysis provides information on the respective behavior of the different biomolecular targets which may identify qualitative trends in proton-induced damage.

Keywords Charge transfer • Ion-biomolecule collision • Ab initio calculations • Pyrimidine nucleobases • 2-deoxy-D-ribose • Prebiotic compounds

M.-C. Bacchus-Montabonel (✉)
Institut Lumière Matière, UMR5306 Université Lyon 1-CNRS,
Université de Lyon, 69622 Villeurbanne Cedex, France
e-mail: bacchus@univ-lyon1.fr

A. Tadjer et al. (eds.), *Quantum Systems in Physics, Chemistry, and Biology*,
Progress in Theoretical Chemistry and Physics 30,
DOI 10.1007/978-3-319-50255-7_20

1 Introduction

Ionizing radiations induce severe lesions to DNA [1] but, besides the direct primary radiation, important damage has been shown to be caused by the secondary particles as low-energy electrons, radicals or ions, generated along the ionizing radiation track [2]. DNA damage driven by secondary electrons or radicals has been widely investigated [3, 4], but more and more interest is devoted to collisions with ions, in particular proton and carbon ions with regard to proton- and hadron-therapy cancer treatments. Such studies may be supported by time-of-flight gas phase experiments, generally performed at keV energies [5–7] but also at lower eV energies [8, 9] combined with ab initio theoretical approaches [10, 11].

Looking to proton-induced damage, interest may be focused first on DNA and RNA building blocks, nucleobases or sugar part as well as on the phosphate group. Effectively, behaviour of nucleobase targets has been widely analysed in collision experiments in the gas-phase [6, 7] as well as in theoretical approaches [10–13], but mainly with carbon impact ions. The 2-deoxy-D-ribose sugar moiety has been also investigated experimentally [5, 8] and theoretically [14–16] with respect to the conformation of the sugar, either a five-membered furanose form in DNA, or a six-membered pyranose form preponderant in the gas phase. An exhaustive study with a comparative analysis of the behaviour of the different biomolecular targets in collision with protons could point out qualitative propensities in the proton-induced damage mechanism.

The study may be extended to possible precursors of these DNA and RNA building blocks. Numerous complex organic molecules (COM) have indeed been detected in various astrophysical environments, interstellar medium, meteorites or comets and great interest is focused on potential precursors of life [17, 18]. The question is not only to understand how these compounds could be formed in the interstellar medium or on icy grains, but also how they can survive under spatial conditions [19]. In proton-rich regions, as ionized clouds (H II region), behaviour of COM in collision with protons could be fundamental. Depending on the astrophysical environment such collisions may occur at very different temperatures [20], from 10 K ($\sim$ meV) in the interstellar medium, to 10^4 K ($\sim$ 10 eV) or more in evolved stars. They could induce the destruction of involved COM's and have to be considered in the formation and destruction reaction sequences. In order to analyse proton-induced damage in connexion with previous studies on DNA and RNA building blocks, we have considered the same process for two possible prebiotic compounds, 2-aminooxazole and hydantoin. 2-Aminooxazole has indeed been proposed recently as an intermediate in the formation of RNA, by means of an efficient and selective sequence [21], alternative to the direct unsuccessful reaction [22]. Increased interest has thus been devoted to this compound, with theoretical investigation of its formation under prebiotic conditions [23–25] and its possible observation by microwave spectroscopy [26]. 2-Aminooxazole has not been observed yet, however a precursor of uracil, hydantoin (2,4-imidazolidinedione), important in formation of poly- and oligopeptides, could be detected in recent experiments on extra-terrestrial ice [27]. We have thus investigated both biological targets in order to analyse their compared survival in proton-induced collisions.

Such collisions of an ion on a biomolecule may induce different processes: first of all, excitation and ionization of the biomolecule have to be considered, followed by the fragmentation of the ionized target. Ionization may be direct, or induced by charge transfer from the impact ion to the biomolecule. In the hypothesis of an almost instantaneous ionization process, gas-phase experiments provide fragmentation patterns [5–9] which may be analysed with regard to the theoretical evolution of the ionized target [14, 28]. But this first ionization step cannot be neglected, mainly charge transfer ionization from the incident ion to the biomolecular target. Such reaction may be studied theoretically in the framework of the molecular representation of the collisions, and we proposed an efficient quantum molecular treatment [10, 12]. The potentials and non-adiabatic coupling matrix elements between the different molecular states involved in the process are determined by ab initio quantum chemistry methods and the collision is performed by semiclassical approaches. Such treatment has been performed at the same level of theory for each target in order to develop a compared analysis and extract possible global tendencies.

2 Molecular Treatment

The charge transfer process proceeds by the evolution of the quasi-molecular system formed by the incident proton and the biomolecular target. For such molecular system, a model may be defined in the one-dimensional reaction coordinate approximation [29, 30]. The collision is thus described as a quasi-diatomic molecule moving along the reaction coordinate R associated to the distance between the proton ion and the centre-of-mass of the target. As pointed out previously, the charge transfer process is very fast and may be treated in the sudden-approximation hypothesis. Electronic transitions being thus assumed to be much faster than vibration and rotation motions, the target could be kept frozen during the collision time in the geometry of the ground state. Even if this first approach does not consider the internal motions of the molecular target, it was found to be qualitatively reliable for the charge transfer process we are considering [31, 32].

The different biomolecular targets are displayed in Fig. 1. The geometries have been optimized using CASSCF (Complete Active Space Self Consistent Field) and DFT (Density Functional Theory) methods. The uracil and thymine nucleobases are constructed around a six-membered planar ring. Different orientations of the projectile with respect to this plane could be investigated. The 2-deoxy-D-ribose (dR) presents two forms, respectively, a furanose form constructed on a five-membered ring, and a pyranose form corresponding to a six-membered ring, both structures corresponding to a non-planar ring. However, a mean plane could be defined in order to analyse, as in previous case, the different target/projectile

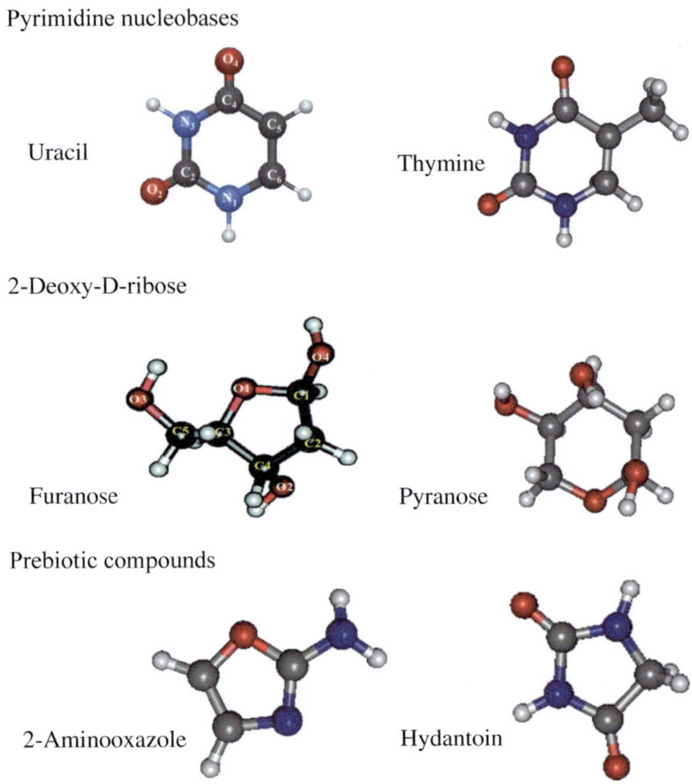

Fig. 1 Geometry of the different biomolecular targets

orientations. For both 2-aminooxazole and hydantoin, the geometry is constructed around a planar five-membered ring, and orientation with respect to this ring may be taken into account. Analysis of the anisotropy of the charge transfer process may be performed looking at a perpendicular orientation of the incident proton, or at an orientation in the ring plane.

The molecular calculations have been performed by means of the MOLPRO code [33] using the 6-311G** basis set for all atoms. Calculations were carried out in Cartesian coordinates, with no symmetries, taking account of all electrons. The potential energy curves and non-adiabatic coupling matrix elements (NACME) have been determined at the CASSCF/CASPT2 (Complete Active Space Perturbation Theory 2nd order) level of theory for R values in the 0.5–9Å distance range. In order to drive a reliable comparison, similar active spaces have been chosen for each collisional system, involving the valence orbitals of highest energy, constructed, of course, on the 1s orbital on the colliding hydrogen, and mainly on π_{C5C6} and π_{CO} orbitals for pyrimidine nucleobases, on $2p_{xyz}$ orbitals on the oxygen of the ring and of the external chain for dR, on $2p^{N}_{xyz}$, π_{CC} and π_{NC} orbitals on nitrogen and carbon of the ring for 2-aminooxazole, and on π_{CO} and $2p^{N,O}_{xyz}$ for hydantoin

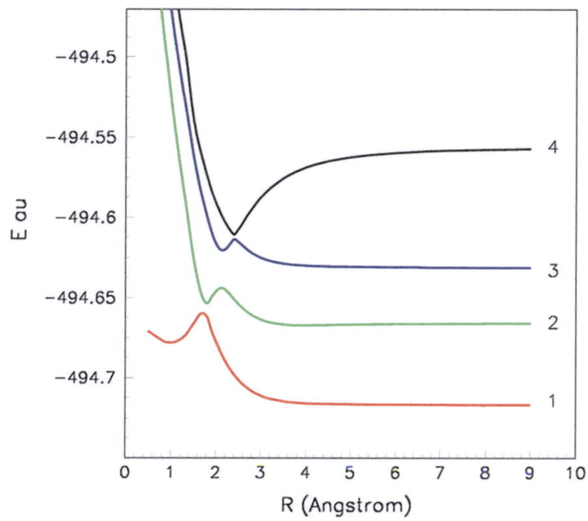

Fig. 2 Adiabatic potential energy curves of ^{1}A states of the H$^+$+ furanose collision system in the perpendicular geometry. *1* Red $\{2p_z^{O3}\ 1s^H\}$; *2* green $\{2p_z^{O1}\ 1s^H\}$; *3* blue $\{2p_{xy}^{O3}\ 1s^H\}$; *4* black $\{(2p_z^{O3})^2\}$ entrance channel H$^+$+ furanose

(see Fig. 1). The 1s orbitals on carbon, nitrogen and oxygen have been treated as frozen cores in all cases.

The charge transfer mechanism may be discussed with regard to the potential energy curves of the different molecular states involved in the collision process. In Fig. 2 is displayed the example of the proton-induced collision on dR-furanose in the perpendicular orientation. Such system shows clear evidence of a strong interaction between the entrance 4^1A{H$^+$+ furanose} channel and the highest 3^1A charge transfer channel followed by successive single excitations with charge transfer levels of lower energy. Such behaviour is quite characteristic of the charge transfer mechanism observed in the series of proton-induced collisions we have considered.

In correlation with such avoided crossings, strong non-adiabatic coupling matrix elements may be pointed out which may induce transitions between the successive molecular states. These NACME's are calculated numerically by means of the finite difference technique:

$$g_{KL}(R) = \langle \psi_K | \partial/\partial R | \psi_L \rangle = \langle \psi_K(R) | \lim_{\Delta \to 0} \frac{1}{\Delta} | \psi_L(R+\Delta) - \psi_L(R) \rangle.$$

The present expression reduces to $g_{KL}(R) = \lim_{\Delta \to 0} \frac{1}{\Delta} \langle \psi_K(R) | \psi_L(R+\Delta) \rangle$ as $|\psi_K(R)\rangle$ and $|\psi_L(R)\rangle$ eigenfunctions are orthogonal for $K \neq L$.

For numerical accuracy, a three-point calculation taking the centre-of-mass of the target as origin of electronic coordinates with a step $\Delta = 0.0012$ a.u. has been performed [34].

3 Collision Dynamics

The collision dynamics has been treated with semiclassical methods in the sudden-approximation hypothesis. The geometry of the biomolecular target has been optimized in its ground state and kept frozen in the collision treatment. This approach widely used for energies above $\sim$ 10 eV/amu [35, 36] was shown recently to drive reliable results at lower collision energies down to the eV range [37, 38]. Its extension may thus provide easily a correct order of magnitude of the cross sections in a wide collision energy range. The collision dynamics was carried out using the EIKONXS program [39] and considering all the radial coupling interactions.

As previously exhibited for collisions with carbon ions [35, 40, 41], the process is strongly anisotropic. The orientation of the impinging proton with respect to the plane (or mean plane) of the biomolecule is quite determinant. This is presented on Fig. 3a for dR and 2-aminooxazole. The charge transfer is generally favoured in the perpendicular approach where access to the molecular ring is not reduced by possible steric obstruction. In the planar orientation, the proton may interact with peripheral groups, as shown for example for the dR-furanose with a CH_2OH ring-bounded chain, or for the 2-aminooxazole NH_2 group. Such interactions appear to reduce the charge transfer cross section at lower collision energies. This is particularly significant in the case of dR-furanose, in relation with its external CH_2OH chain. At variance, for molecules with relatively regular structures with no peripheral chain, like pyranose, the charge transfer cross sections oscillate around a mean value and the anisotropic effect appears relatively weak in the whole energy domain. This orientation effect has nevertheless been taken into account in our study by calculating the cross sections in both perpendicular and planar approaches and the mean value between those values has been calculated for every molecular target. This mean value is displayed in Fig. 3b for the series of DNA building blocks and precursors 2-aminooxazole and hydantoin.

From the present results, the efficiency of electron capture in collisions with protons appears significantly enhanced for the sugar 2-deoxy-D-ribose, in particular in furanose form, than for the pyrimidine nucleobases or their precursors. The charge transfer cross sections are about three orders of magnitude higher for furanose than for hydantoin. Comparison of the different collision systems may exhibit different possible underlying factors. At first, the structure of the targets: uracil, thymine and hydantoin correspond to a similar skeleton, with the –NH–C = O peptide bond, the 2-aminooxazole is slightly different but its ring includes anyway carbon, nitrogen and oxygen atoms. On the other hand, the 2-deoxy-D-ribose ring is only composed of carbon and oxygen. The specific reactivity of nitrogen and peptide bond would induce features for nucleobases and their precursors at variance with those of sugars. Besides, both dR forms are non-planar whereas precursors 2-aminooxazole and hydantoin correspond to a planar ring like pyrimidine nucleobases, even if they are constructed on a five-membered ring. Roughly, with respect to structural considerations, the different

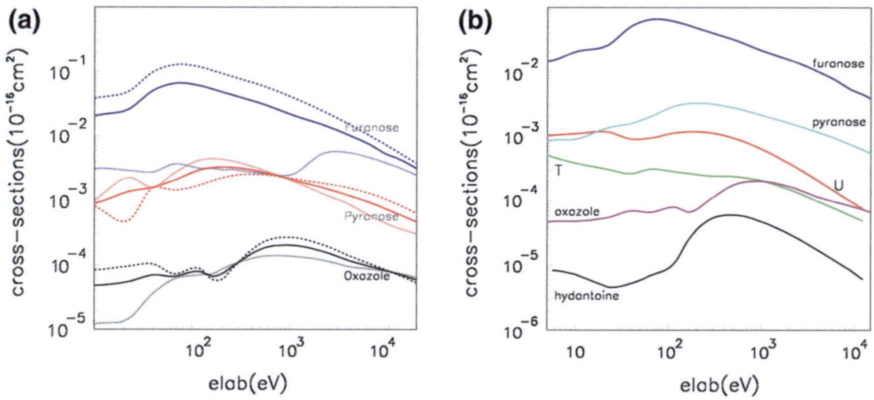

Fig. 3 Charge transfer cross sections in proton-induced charge transfer. **a** *Blue*, furanose; *red*, pyranose; *black*, 2-aminooxazole. *Dashed lines*, perpendicular approach; *dotted lines*, in-the-plane approach; *full lines*, mean value. **b** *Blue*, furanose; *light blue*, pyranose; *red*, uracil; *green*, thymine; *magenta*, 2-aminooxazole; *black*, hydantoin

targets may be separated in three groups: the nucleobases, their precursors, and the sugars, each group presenting a relative similar behaviour with collision energy. A regular lowering of the efficiency in the charge transfer process may be pointed out from furanose, pyranose, nucleobases, to the precursors 2-aminooxazole and hydantoin. Furthermore, both precursors present a significantly reduced charge transfer cross section at lower energies.

Such decrease of the charge transfer cross sections may be associated to potential consequences on the reactivity of these compounds in proton-rich environments, mainly under spatial conditions, in particular on their "resistance" to fragmentation and thus their possible survival in space. Previous studies, both experimental and theoretical, have indeed pointed out an interesting correlation between charge transfer cross sections and fragmentation yield. In collisions of carbon ions with uracil, for example, the fragmentation is almost complete at low energy for the C^{2+} impinging ion [6], with experimentally a decreasing yield for increasing energy, although theoretical charge transfer cross sections are very low at low collision energy and increase with energy [10, 12]. At variance, for the collision with C^{4+}, fragmentation is much lower, with an experimental yield almost constant with energy, in correlation with higher, almost constant theoretical charge transfer cross sections. Similar qualitative tendencies with opposite dependence for experimental fragmentation yield and charge transfer cross sections have been also pointed out for collisions with 2-deoxy-D-ribose [5]. Lower charge transfer efficiency might suggest a possible increased fragmentation of the biomolecular target leading to its destruction. Following such assumption, precursors like hydantoin or even 2-aminooxazole might be more easily destroyed than furanose in collisions with protons. This remains, of course, absolutely qualitative, but could possibly question the survival of these prebiotic species in proton-rich regions. Their implication in the reaction sequence at the origin of life would necessitate a preserved environment. Of course, all present results, experimental and theoretical, concern gas phase processes and reactions in solvated environments or at the surface of grains have to be considered. Microhydration of uracil has been shown to induce strong variations on proton-induced charge transfer [42] and such solvation effects cannot be neglected to drive conclusions on the reactions at the origin of life.

4 Concluding Remarks

We develop in this paper a comparative analysis of the charge transfer process in proton collisions on DNA sugar skeleton, pyrimidine nucleobases uracil and thymine, and their possible precursors 2-aminooxazole and hydantoin. The calculations have been carried out using ab initio quantum chemical methods followed by a semiclassical dynamics in a wide collision energy range, from eV to keV energies. Taking account of experimental and theoretical studies, a qualitative correlation with the fragmentation process may suggest an enhanced sensitivity in proton collisions for pyrimidine nucleobases and their precursors, in particular hydantoin, compared to the sugar 2-deoxy-D-ribose. As far as only gas-phase processes are concerned, such tendency could induce a potential weakness of those species in proton-rich regions of space, as ionized clouds.

Acknowledgements We acknowledge generous computational resources from the CCIN2P3 in Villeurbanne and CCRT/CINES/IDRIS under the allocation 2015-[i2014091566] made by GENCI. This work was supported by COST actions CM1204 XLIC and CM1401 Astrochemical History.

References

1. von Sonntag C (1987) The chemical basis for radiation biology. Taylor and Francis, London
2. Michael BD, O'Neill PD (2000) Science 287:1603
3. Boudaiffa B, Cloutier P, Hunting D, Huels MA, Sanche L (2000) Science 287:1658
4. Cobut V, Frongillo Y, Patau JP, Goulet T, Fraser MJ, Gay-Gerin JP (1998) Radiat Phys Chem 51:229
5. Alvarado F, Bari S, Hoekstra R, Schlathölter T (2006) Phys Chem Chem Phys 8:1922
6. de Vries J, Hoekstra R, Morgenstern R, Schlathölter T (2002) J Phys B 35:4373
7. de Vries J, Hoekstra R, Morgenstern R, Schlathölter T (2003) Eur Phys J D 24:161
8. Deng Z, Bald I, Illenberger E, Huels MA (2005) Phys Rev Lett 95:153201
9. Deng Z, Imhoff M, Huels MA (2005) J Chem Phys 123:144509
10. Bacchus-Montabonel MC, Tergiman YS (2012) Comput Theor Chem 990:177
11. Almeida D, Bacchus-Montabonel MC, Fereira da Silva F, García G, Limão-Vieira P (2014) J Phys Chem A 118:6547
12. Bacchus-Montabonel MC, Łabuda M, Tergiman YS, Sienkiewicz JE (2005) Phys Rev A 72:052706
13. Bacchus-Montabonel MC, Tergiman YS (2011) Phys Chem Chem Phys 13:9761

14. Hervé du Penhoat MA, Lopez-Tarifa P, Ghose KK., Jeanvoine Y, Gaigeot MP, Vuilleumier R, Politis MF, Bacchus-Montabonel MC (2014) J Mol Model 20:2221
15. Bacchus-Montabonel MC (2014) J Phys Chem A 118:6326
16. Bacchus-Montabonel MC (2013) J Phys Chem A 117:14169
17. Huber C, Eisenreich W, Hecht S, Wächtershäuser GA (2003) Science 301:938
18. Cooper GW, Onwo WM, Cronin JR (1992) Geochim Cosmochim Acta 56:4109
19. Ehrenfreund P, Irvine W, Becker L, Blank J, Brucato R, Colangeli L, Derenne S, Despois D, Dutrey A, Fraaije H, Lazcano A, Owen T, Robert F (2002) Rep Prog Phys 65:1427
20. Bacchus-Montabonel MC (2011) Rend Fis Acc Lincei 22:95
21. Powner MW, Gerland B, Sutherland JD (2009) Nature 459:239
22. Joyce GF (2002) Nature 418:214
23. Szabla R, Šponer JE, Šponer J, Góra RW (2013) Phys Chem Chem Phys 15:7812
24. Szabla R, Tuna D, Góra RW, Šponer J, Sobolewski AL, Domke W (2013) J Phys Chem Lett 4:2785
25. Bacchus-Montabonel MC (2015) J Phys Chem A 119:728
26. Møllendal H, Konovalov A (2010) J Phys Chem A 114:2151
27. de Marcellus P, Bertrand M, Nuevo M, Westall F, d'Hendecourt LL (2011) Astrobiol 11:847
28. López-Tarifa P, Hervé du Penhoat MA, Vuilleumier R, Gaigeot MP, Tavernelli I, Le Padellec A, Champeaux JP, Alcamí M, Moretto-Capelle P, Martín F, Politis MF (2011) Phys Rev Lett 107:023202
29. Bacchus-Montabonel MC, Talbi D, Persico M (2000) J Phys B 33:955
30. Bene E, Martinez P, Halász GJ, Vibók Á, Bacchus-Montabonel MC (2009) Phys Rev A 80:012711
31. Bacchus-Montabonel MC, Tergiman YS (2011) Chem Phys Lett 503:45
32. Bacchus-Montabonel MC, Tergiman YS (2010) Chem Phys Lett 497:18
33. Werner HJ, Knowles PJ MOLPRO Package of ab-initio programs (version 2012.1)
34. Bacchus-Montabonel MC (1987) Phys Rev A 36:1994
35. Bacchus-Montabonel MC, Tergiman YS (2006) Phys Rev A 74:054702
36. Stancil PC, Zygelman B, Kirby K (1998) In: Aumayr F and Winter HP (eds) Photonic, electronic, and atomic collisions. World Scientific, Singapore, p 537
37. Chenel A, Mangaud E, Justum Y, Talbi D, Bacchus-Montabonel MC, Desouter-Lecomte M (2010) J Phys B 43:245701
38. Linguerri R, Hochlaf M, Bacchus-Montabonel MC, Desouter-Lecomte M (2013) Phys Chem Chem Phys 15:824
39. Allan RJ, Courbin C, Salas P, Wahnon P (1990) J Phys B 23:L461. http://www.ccp6.ac.uk/downloads/eikonxs.htm
40. Bacchus-Montabonel MC (2015) Eur Phys J D 69:107
41. Bene E, Vibók Á, Halász GJ, Bacchus-Montabonel MC (2008) Chem Phys Lett 455:159
42. Bacchus-Montabonel MC, Calvo F (2015) Phys Chem Chem Phys 17:9629

How Can the Green Sulfur Bacteria in the Depths of the Black Sea Use Quantum Computing for Light Harvesting?

Deiana Drakova and Gerold Doyen

Abstract Long lasting coherence in photosynthetic pigment-protein complexes has been observed even at physiological temperatures [1–3]. Experiments have demonstrated quantum coherent behaviour in the long-time operation of the D-Wave quantum computer as well [4, 5]. Quantum coherence is the common feature between the two phenomena. However, the 'decoherence time' of a single flux qubit, the component of the D-Wave quantum computer, is reported to be on the order of nanoseconds, which is comparable to the time for a single operation and much shorter than the time required to carry out a computation on the order of seconds. An explanation for the factor of 10^8 discrepancy between the single flux qubit coherence time and the long-time quantum behaviour of an array of thousand flux qubits was suggested within a theory where the flux qubits are coupled to an environment of particles called gravonons of high density of states [6, 7]. The coherent evolution is in high dimensional spacetime and can be understood as a solution of Schrödinger's time-dependent equation. Explanations for the quantum beats observed in 2D Fourier transform electronic spectroscopy of the Fenna-Matthews-Olson (FMO) protein complex in the green sulfur bacteria are presently sought in constructing transport theories based on quantum master equations where 'good' molecular vibrations ('coloured noise') in the chlorophyll and the surrounding protein scaffold knock the exciton oscillations back into coherence [8, 9]. These 'good' vibrations are claimed to have developed in three billion years of natural selection. These theories, however, face the discomforting experimental observation that "attempts to scramble vibrational modes or to shift resonances with isotopic substitution miserably failed to affect the beating signals" [10]. As a possible way out of this dilemma we adopted the formalism of the quantum computation to the quantum beats in the FMO protein complex.

D. Drakova (✉)
Faculty of Chemistry, State University of Sofia, Sofia, Bulgaria
e-mail: nhdd@chem.uni-sofia.bg

G. Doyen
Ludwig-Maximilians-University, Munich, Germany

A. Tadjer et al. (eds.), *Quantum Systems in Physics, Chemistry, and Biology*,
Progress in Theoretical Chemistry and Physics 30,
DOI 10.1007/978-3-319-50255-7_21

Keywords Long-living coherence · Quantum biology · Green sulfur bacteria · Emergent quantum mechanics

1 Introduction

When in 2007 the New York Times published an article suggesting that plants were quantum computers, quantum information scientists exploded into laughter [11]. Their concern was the so called decoherence which is believed to be one of the fastest and most efficient processes in nature.

The experiments by Engel et al. show that long lasting coherence in photosynthetic pigment-protein complexes is observed at physiological temperatures [1–3]. Quantum coherence over hundreds of femtoseconds in the initial steps of photosynthesis, the tranfer of light quanta towards the reaction centers, in green and purple bacteria [12, 13] at room temperature [14, 15], algae at ambient temperature of 294 K [16] and even in living cells of purple bacterium [17, 18] warrant the high efficiency of the light-harvesting process. It has even been suggested that the high efficiency of photosynthesis may be associated with coherent energy and charge transfer on a long timescale [19–21].

Citing Hildner et al.: "...quantum coherence is the central aspect connecting fields as diverse as quantum computing and light-harvesting" [22].

In 2011 the first commercially available quantum computer was presented by the company D-Wave consisting of an array of more than thousand flux qubits (June, 2015), which solves optimization problems with many free parameters [4, 5]. It works as a quantum computer of the quantum annealing type for minutes and even for half an hour. However, a single flux qubit in a Ramsey interference experiment shows coherent oscillations between the two supercurrents for time of the order of 20 ns, which is 8 orders of magnitude shorter than the coherence time of the D-Wave computer [23].

This was a noteworthy result as a flux qubit is a macroscopic device with dimensions of the order of 10^4 Å. But not less noteworthy was the experimental observation of a coherent quantum behaviour of the Fenna-Matthews-Olson pigment-protein complex not only at low temperature but even at 277 K and the coherent behaviour of the light harvesting complex in the purple bacterium living cell in a dense, wet noisy environment [17, 18]. The 'time of coherent development' of the FMO-complex is picoseconds, i.e. 4 orders of magnitude shorter than that of a single flux qubit. With the help of 2D Fourier transfrom electronic spectroscopy [24–27] the amplitude of a photon echo signal as a function of the waiting time between the first two laser pulses, which excite the FMO electronically, and the third laser pulse, which stimulates the photon emission from the exciton, shows beats. The beats of the photon echo signal are due to the coherent superposition of two lowest energy excitons shared by all chlorophyll molecules and indicate coherent time development of the FMO-complex during the waiting time. In contrast to the experimental observation, an exponential decay of the photon signal with time would be expected, if interaction with the noisy environment (phonons) of the FMO-complex were dominating, as it is generally believed that electronic coherence decays on 10–100 fs timescales [28, 29].

Why is the beating between exciton states observed for picoseconds instead of a fast exponentially decaying signal without oscillations?

Semi-classical diffusive models based on stochastic dynamics have been extensively applied to study energy and charge transport in biological systems [30–32]. Within a quantum master equation approach the time development of the reduced density matrix is evaluated [33], the effect of the phonon environment is simulated by an additive Redfield term, which is dropped for the study of the time development. With a very special selection of the phonon environment (coloured noise) the beats of the photon amplitude are reproduced, but the finite final value of the photon amplitude is not explained. Furthermore the 2D Fourier transform experiment shows that no special phonons are involved in the damping of the signal, changing the FMO phonon spectrum does not influence the results [10]. Theoretical attempts based on the electron incoherent hopping process [30, 34] between pigment molecules in the photosynthetic unit have been suggested [35].

We focus on the experimental observation in the 2D Fourier transform electronic spectroscopy which demonstrates coherence by the Fenna-Matthews-Olson protein-chlorophyll complex of green sulfur bacteria over times longer than 1000 fs and even at high temperature (277 K). The amplitude of the measured photon echo signal oscillates with time, which is an indication of coherent behaviour, suggested as the basis for the extreme efficiency of the photon transfer towards the reaction center.

- Why is entanglement with the degrees of freedom of the environment not effective and not fast enough to destroy coherence?
- Despite that the photon amplitude beats are exponentially damped with time, the final value is finite, larger than zero even at high temperature. What is the reason for this?
- Missing isotope effect, random exchanges of ^{1}H atoms or of ^{12}C with their isotopes do not change the picture [36]. This procedure certainly changes the phonon spectrum, but the results on the coherent quantum beats are not influenced. Coherence is experimentally shown in bacteria [1, 2, 12, 15, 17, 18], algae [16] and green plants [37]. The structures of their protein backbones differ and hence the protein scaffolds cannot serve as the ubiquitous protection of electronic coherence, as it is suggested in Refs. [1, 2, 12, 16].
- A discussion topic is the nature of the superimposed coherent states in the protein-pigment complexes. Vibrational, electronic or vibronic states may be involved, with indications from experiment favouring their vibronic character ([13], [38–44]).

The dying out of the oscillations between the two supercurrents in the single flux qubit was explained in a coherent picture solving Shrödinger's time dependent equation in high dimensional spacetime [7]. The decisive features of the theory are summarized in the next section together with the description of a simple model of the physical picture of the processes in FMO 2D Fourier transform electronic spectroscopy. The results are presented in Sect. 3, followed by a summary and conclusions.

2 The Model and the Physical Picture

The FMO protein complex consists of seven chlorophyll molecules in a protein cavity. There is an eighth bacteriochlorophyll molecule outside the cavity whose interaction with the rest seven molecules can be neglected. The first two laser pulses in the 2D spectroscopic experiment, delayed by only 20 fs from each other, excite two excitons and lead to their coherent superposition, which results in the beating signal. The time between the first two laser pulses, i.e. the coherence time τ, in different 2D spectroscopic experiments is of the order of 20 to 500 fs and the duration of each pulse is 15–40 fs. These times are very short compared to the time needed to establish the entanglement with the gravonons (representing local version of the gravitons), the environmental excitations we take into account in addition to the excitations considered in other theories like phonons and electron-hole pairs. Therefore for the purposes of our theory the first two laser pulses are regarded as one double pulse. The third laser pulse is applied after a waiting time varying from 0 fs to 1000–1500 fs and serves to stimulate the emission of the photon, which is recorded as the photon echo after the rephasing time t.

The physical picture we describe is the following. The exciton electron wave packet is diffuse, penetrating in the protein scaffold, where as a nuclephile in a nucleophilic-like reaction the exciton electron attacks the positively polarized carbon atom of the carbonyl CO fragment. The electron can be transiently accomodated in the $3s$-affinity like orbital on the carbonyl carbon atom creating a transient negative ion resonance (NIR). Thus the electron penetrates in the carbon core region where it interacts gravitationally with the atom core. At interatomic distances the strength of the gravitational interaction is enhanced by many orders of magnitude, if hidden extra dimensions exist. In the spirit of Einstein's general relativity the gravitational interaction between masses deforms spacetime and we introduce a basis state called warp resonance for the electron (Fig. 1), where it entangles with the gravitons (the messenger particles of the gravitational field). Locally a soft gravonon structure is generated, which is constructed as a correlated motion of 3–4 atoms leading to the warping. The soft gravonons in our theory are the solution of interacting quantum harmonic oscillators centered on the 3–4 atoms close to the NIR. For more details on the theory concerning the generation of the gravonons see Refs. [7, 45].

The electromagnetic field acts as an external force on the carbon atom and, together with the thermal fluctuations, shifts it slightly in a different position creating a local deformation in the protein chain. This is no wonder since the local vibrations of the atoms in the carbonyl CO fragment are different if the carbon atom has captured transiently the electron or not. The local deformation of the protein chain occurs around the NIR, it is associated with slight increase in energy, and

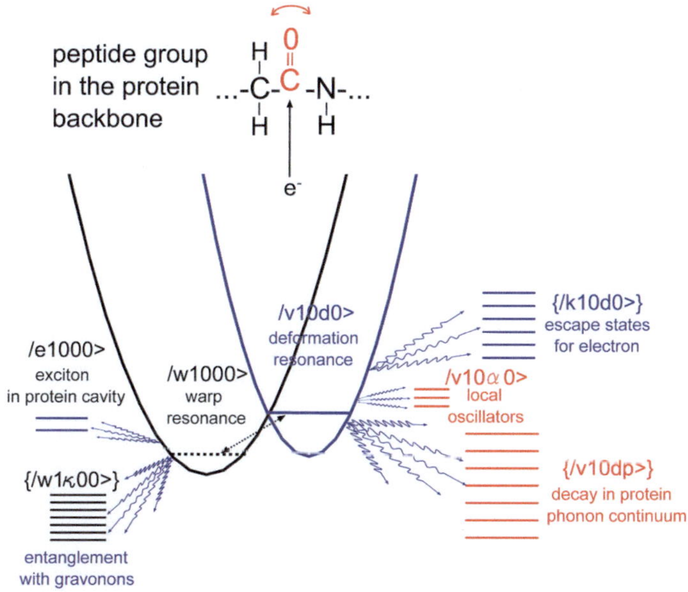

Fig. 1 Model for studying the quantum coherent behaviour of a complex of bacteriochlorophylls (FMO complex) in a protein scaffold in the green sulfur bacteria. The protein scaffold provides the carbon atom in the carbonyl *CO* fragment to capture the nuclephile, the exciton electron, in a transient negative ion resonance which relaxes and couples with phonon modes in the protein and excites local rocking and wagging motion of the carbonyl *CO* fragment. The entanglement with gravonons is effective within the NIR, but as it develops slowly compared to the coupling with the phonons, its effect during the tiny burst, induced by the double laser pulse, is neglected in the calculations. In this schematic drawing the gravonon spectrum is plotted on a scale magnified by a factor 10^8 and differs from the scale of the other spectra

is described by introducing a basis state called deformation resonance for the core motion of the carbonyl fragment. Furthermore excitation of vibrational modes like the rocking and wagging motion of the carbonyl *CO* fragment is possible which we describe by coupling the deformation resonance to two local harmonic oscillators, i.e. to local vibrational modes (Fig. 1).

In the deformation resonance coupling to the extended phonons of the protein backbone is also effective, leading to phonon excitation and energy dissipation from the local region (Fig. 1). The exciton electron has the option to decay from its localized state in the NIR in delocalized electron states, called 'escape states' because total energy has to be conserved, if phonons are excited. Electron states delocalized over the chlorophyll assembly, with the 3*s*-affinity like orbital of the carbonyl carbon atom as component, play this role. Although electron dissipation from the local region in the protein backbone is possible in the deformation resonance via phonon excitation, the interaction between the deformation resonance and the local rocking and wagging modes of the carbonyl *CO* fragment leads to reflection to and fro between the local states. Therefore, despite the dissipation into the protein backbone phonons, there remains a local non-zero electron component in the total wave packet, which explains the finite non-zero value of the photon signal in the 2D spectroscopic experiment when the deexcitation of the exciton electron is stimulated by the third laser pulse.

The gravitational interaction of the exciton electron with the carbon atom core has to be in high spacetime dimensions ($11D$) as string theory assumes. Only then gravity is strong enough at small distances and decays fast with distance (r^{-8} power law and nearly 33 orders of magnitude higher value of the gravitational constant in $11D$). The additional 7 spacial dimensions are compactified and are hidden, so that there is no discrepancy with Newton's gravitational law at distances where it was proved valid. Within the warped space the time development of the entanglement of the electron with gravonons leads to the beables.

Beables are mathematically precisely defined as a set of configurations in the expansion of the wave function containing a localized matter field and excited local gravonons. Loosely speaking beables are matter fields (e.g. atoms or electrons) in $3D$ space, entangled with gravonons, which interact with other atoms gravitationally and generate the gravonon structure in high dimensional spacetime. Therefore localized atoms, molecules or electrons in $3D$ space entangled with gravonons in high dimensions in the form of beables exist as long as the beables and the entanglement with the gravonons exist.

The word "beable" has been coined by Bell [46] as a terminology against the word observable: "The concept of 'observable' lends itself to very precise mathematics when identified with 'self-adjoined operator'. But physically, it is a rather wooly concept. It is not easy to identify precisely which physical processes are to be given the status of 'observations' and which are to be relegated to the limbo between one observation and another."

We use the expression beable in the sense of John Bell's *local* beable. According to our proposal signals in experiments in $3D$ space can be received only via beables and hence measurements are tied to beables [45]. Expressed with John Bell's words: "One of the apparent non-localities of quantum mechanics is the instantaneous, over all space, 'collapse of the wave function' on 'measurement'. But this does not bother us if we do not grant beable status to the wave function."

Already in 1927 it was revealed that an interference pattern shows up on a photographic plate only when the number of photons falling on the plate is very large [47]. The history of photon detection is nicely reviewed in Ref. [48]. Here the results of Dempter and Batho are summarized in such a way that when, during the so called 'collapse of the wave function', the photon is destroyed, there appears somewhere on the photographic plate an atom of elemental silver which will act as an embryo from which, by photographic development, a small seed of silver will grow. The silver embryo is much smaller than the electromagnetic wavelength and constitutes the beable in our picture.

Applied to the spectroscopic method used in the bacteria experiment investigated here this means that the photon of the laser pulse occupies during the interference process at least the whole volume of the cavity embraced by the protein scaffold, but if the photon is measured and hence destroyed, there appears a beable (embryo) formed in the detection devices of the experimenters.

The described mechanism of the so called collapse of the wave function has by now also been established for matter fields [49–51]. For an electron in an excitonic state of the chlorophyll it means that the wave function of this electron occupies during the interference process the whole volume of the chlorophyll, but when it hits the protein scaffold, which acts as a kind of screen, the electron becomes localized in a tiny volume of the size of the silver embryo, this being the carbon affinity level of the carbonyl group in the model developed here.

Before the laser pulse is sent in, the electron will be almost with certainty in such a beable. The laser pulse initiates then an interference process where the many particle wave function describing the photon, electron, phonon fields etc. extends over the whole volume embraced by the protein scaffold. This many particle wave function is not a beable, but, in John Bell's terminology, rather a 'limbo state'. As time passes on a matter field will somewhere entangle with gravonons and form a beable which destroys the photon. If beforehand the photon 'localizes' in a beable in the detection devices of the experimenters, it will do so proportional to the light intensity in the system.

2.1 Chooser Mechanism, Beables Created with the Chooser Mechanism

The question is why does the exciton electron become localized on a single carbonyl carbon atom of a single carbonyl *CO* fragment in the protein backbone? Quantum particles, e.g. the exciton electron, may become localized via the entanglement with gravonons in the form of beables. However, as the gravitational interaction even in high dimensional spacetime is weak, entanglement with gravonons can be effective on the energy shell alone. In the present model the exciton electron is localized because via the entanglement to gravonons a beable is generated by the local and strongly distant dependent interaction with the gravonons. In the case of on-shell coupling even a very weak coupling with the gravonons is effective. Just a few or a single carbonyl carbon atom in the protein chain provide the condition for on-shell coupling with the initial exciton electron wave packet. This is the chooser mechanism for the site of most favourable entanglement with gravonons. The chooser mechanism is responsible for the appearance of stuck particles on single sites on the detection screen in double slit experiments [52].

The exciton is a diffuse object. Its overlap with all carbonyl *CO* fragments in the protein backbone would mean that the electron might be thought to reside on all carbonyl carbon atoms in *CO* sites at the same time and would couple to all degrees of freedom all over the protein chain. Then decay in the environmental degrees of freedom is commonly predicted to be very fast. This is, however, not the case in our theory because the chooser mechanism selects a single carbonyl carbon atom on a single *CO* site where the condition for degeneracy coupling with the gravonons is fulfilled, i.e. where the chooser mechanism leading to the transient localization of the electron works. The electron can go to only a single carbonyl carbon atom in a single *CO* site, chosen by the degeneracy coupling criterion.

Beables are destroyed when the entanglement with the gravonons is changed or truncated. This occurs as a tiny burst in the time development of the beable caused for instance by the laser pulses (Fig. 2). In the tiny burst the quantum particles (atoms, molecules, electrons) are released from the entanglement with the gravonons and they burst out of the beable into 3*D* space in a 'limbo state'. All states where particles are not entangled to gravonons we call limbo states.

Within the beable the electron is localized in the warp resonance and can be measured. As the lifetime of the gravonons in the high hidden dimensions expires, the entanglement of the localized electron with the gravonons is truncated and the electron is released in 3*D* space. The electron bursts out of the entanglement with the gravonons. The tiny burst can be seen in Fig. 2. It exists for a short time before the entanglement with the gravonons is reestablished which takes 5×10^{-9}–10^{-8} s. Within the tiny burst experimentally nothing can be recorded about the electron. This situation does exist in different experiments: in double slit diffraction experiments with electrons and molecular beams we know nothing about the quantum particles during their time of flight between the source and the detection screen; in laser-induced desorption we know nothing about the adsorbate for the time of flight

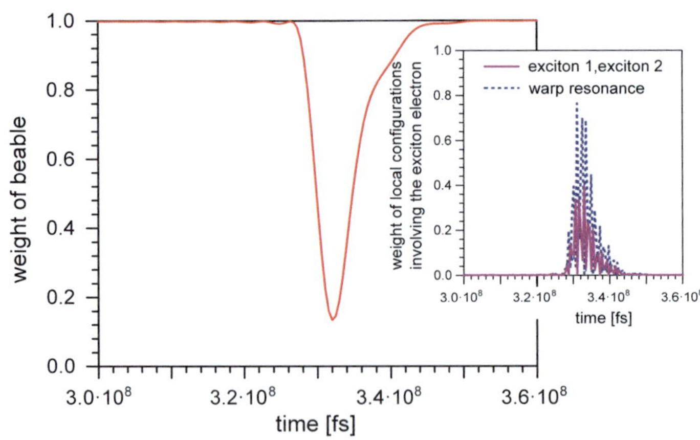

Fig. 2 Time development of the entanglement of a quantum particle, e.g. the exciton electron, with the gravonons, generating the beable. Within the beable the electron is localized in the warp resonance and can interact with the photon field, i.e. it can be measured. As soon as the lifetime of the gravonons in the high hidden dimensions expires, the electron bursts out of the entanglement with the gravonons and is released in 3*D* space. This is the tiny burst in the figure. It exists for a short time before the entanglement with the gravonons is reestablished, which takes 5×10^{-9}–10^{-8}s. The *inset* shows the bursting of the exciton electron out of the beable in three dimensional space (in the warp resonance and the excitons in the cavity) during the tiny burst

between the state with the adsorbate as a beable on the substrate surface and in the mass spectrometer, where a new beable is created.

The effect of the double laser pulse in experiment is to induce a tiny burst in the time development since the laser pulse creates an electromagnetic field, which exerts a force on the localized particle within a beable. Slight shifts of the atom positions mean changed local gravonon structure and destroyed coupling to the gravonons (very short ranged), hence the beable is destroyed. For instance, the double laser pulse in the 2D spectroscopic experiment interacting with the electric dipole due to the localized electron on the carbon atom, induces a slight shift of the carbon atom and rocking and wagging motion of the carbonyl CO^{-} fragment in the protein. This would cause such a drastic change in the gravonon structure, typical for the beable configuration existing before the laser induced tiny burst, that the entanglement with the gravonons is destroyed. This is how the gravonon structure and the beable are destroyed by the laser pulses. As the entanglement with the gravonons develops in a period of time of the order of 10^{-8} s, in the laser-pulse-induced tiny burst the electron is released from the entanglement with the gravonons, however, its coupling with other excitations, living in 4 dimensional spacetime like phonons and the vibrations of the local oscillators, gains weight. Although experimentally an electron, a photon or a molecule released from a beable cannot be directly detected during the tiny burst, we can calculate the time development further with the time dependent Schrödinger equation. This is where in our theory we start a new time development: the electron is localized in a transient NIR on a carbonyl carbon atom in a single carbonyl CO fragment, the photon is in the protein cavity, no gravonons are involved.

During the tiny burst the exciton electron is no more entangled with the gravonons, it is released in a limbo state in $3D$ space, therefore cannot be directly accessed experimentally. But everything in the 2D spectroscopic experiment occurs during the tiny burst induced by the double laser pulse. This situation lasts untill the entanglement with gravonons can develop again, giving rise to the beables. During the lifetime of the beable the exciton electron, which initially is a delocalized electron wave packet, remains localized transiently on the carbonyl carbon atom. Within this physical picture all measurable phenomena originate effectively from a very local region which then expands in $3D$ space. The chooser mechanism for the generation of the beables and the concept of the tiny burst due to the laser pulse indicate that what is measured in experiment has to be interpreted in a local picture concerning just a single carbonyl CO fragment in the protein backbone, which couples to no more than two local harmonic oscillators and the phonon excitations in the protein backbone. If more local harmonic oscillators were excited, which is possible only in the absence of the chooser mechanism, the result would be indeed fast decay of the photon beating signal. This is, however, not measured in experiment.

The method of solving Schrödinger's equation for the time development of the total wave packet is described in Ref. [45]. The Hamiltonian for the model, the formalism and the choice of the basis are presented in a more detailed article on the present topic [53].

2.2 Vibrational Energy in the Local Oscillators

An estimate of the vibrational excitation energy of the rocking and wagging motions of the carbonyl group in the protein backbone of the order of $\omega \approx 5$ meV is based on the experimental data in Ref. [2]. With this estimate at thermal energy of approximately $k_B T = 11$ meV at $T = 125$ K two local vibrations can be excited each with $\omega_1 = \omega_2 = \omega \approx 5$ meV, so that $2\omega \approx 10$ meV corresponds approximately to the thermal energy at 125 K. The rocking and wagging motions of the carbonyl group are much softer than its vibration with respect to the protein chain, hence 5 meV for the quantum of these vibrations appears physically plausible.

The energy of a local vibration can be neither 4 nor 6 meV because these values contradict the experimental curves for 77, 125 and 150 K. The estimate for the energy of the local mode $\omega \approx 5$ meV, based on the 2D Fourier transform spectroscopy experiment, corresponds to the thermal energy $k_B T$ at $T \approx 58$ K. At $T = 125$ K and $T = 150$ K the two experimental curves coincide. If by $T = 125$ K two quanta are excited, then at $T = 150$ K it is not possible to excite 3 quanta, hence 2 excited quanta would explain the coinciding curves in experiment. The two quanta 2ω should correspond to $T = 125$ K and $k_B T \approx 11$ meV, hence $\omega \approx 5$ meV. ω cannot be 4 meV ($T \approx 46$ K) because then at $T = 150$ K three vibrational quanta will be excited, whereas at $T = 125$ K only two can be excited and the two curves will not coincide. ω cannot be 6 meV ($T \approx 70$ K) either because at $T = 125$ K, $k_B T \approx 11$ meV, just a single local vibration will be excited exactly as by $T = 77$ K. The two curves at $T = 77$ K and $T = 125$ K would coincide. This is, however, not observed experimentally.

2.3 Protein Phonon Band

The protein phonon band width is of the order of 0.03 eV. This follows from the following argument. Assume a periodic constant of a simple one dimensional protein of the order of $a \approx 10$ bohr, then the maximal wave vector k_{max} is of the order of $k_{max} = \frac{\pi}{a} \approx 0.3$ bohr^{-1}. The energy corresponding to the maximal k-vector is $\omega_{p,max} = v k_{max} = 0.027$ eV (using for the sound velocity in dry air $v = 330$ m/s $= 6.6 \times 10^{12}$ bohr/s). This estimate refers to the transverse phonon modes.

3 Results for the Time Development of the Tiny Burst

We assume that the exciton electron is in a beable before the laser pulses occur (cf. Fig. 2), because the bacterium exists in reality and therefore cannot be in a limbo state. Reality consists of beables only. We now discuss what happens after the beable has been destroyed by the laser pulse.

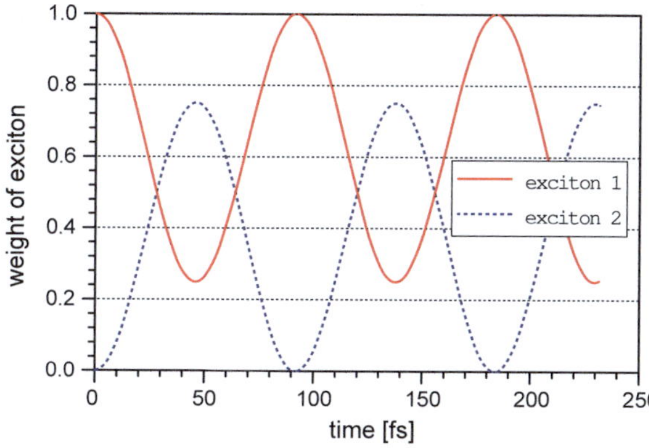

Fig. 3 Rabi oscillations between the two exciton states isolated from the environment

The first laser pulse in the 2D spectroscopic experiment excites the FMO complex electronically producing the first exciton delocalized in the protein cavity. The second laser pulse excites the second exciton and a coherent superposition between the two [54]. If no interaction with the environment exists the coherent oscillations between exciton 1 and exciton 2 will continue, the amplitude of each state changing sinusoidally with time (Fig. 3).

However, for the FMO-protein complex we have the protein cavity and the chlorophyll molecules within the cavity. Before the tiny burst the exciton electron is transiently localized in a single warp resonance on a carbon atom at a single carbonyl *CO* fragment chosen by degeneracy with the gravonons. The photon is in the cavity. The localized electron exercises a force on the carbon atom, exciting two local oscillators with the wagging and rocking movements of the carbonyl *CO* fragment in the deformation resonance. In addition to coupling to the local oscillators, coupling to the extended protein phonons within the deformation resonance allows relaxation and dissipation in delocalized phonons away from the local site. In the 2D Fourier transform electronic spectroscopy experiments these components are not observed, the measurement provides just the amplitude of the photon echo, which scales with the weight of the local configurations involving the exciton electron, as a function of time. But we can calculate the time development of the wave packet during the period of the tiny burst and it shows a strong similarity with the variations with time of the photon amplitude measured experimentally (Fig. 4).

3.1 Temperature Effects

For the solution of the time dependent Schrödinger equation at different temperatures we explicitely need the dependence of the accessible vibrational configurations on temperature, i.e. counting of the vibrational configurations in the deformation resonance, and the temperature dependence of the accessible on-shell phonon

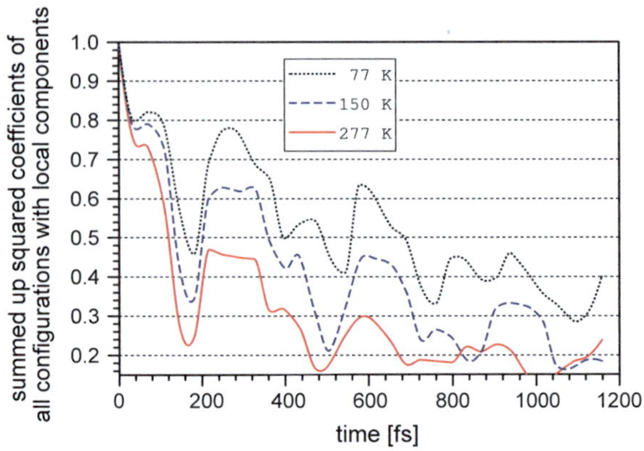

Fig. 4 Time development of the amplitude of the photon signal during the tiny burst, induced by the double laser pulse, for the model in Fig. 1 at three different temperatures of the environment. The photon signal scales with the summed weight of the field configurations with local components in the wave packet. These are configurations with the exciton electron in the warp resonance (i.e. in the beable), in the deformation resonance and the local oscillators. The initial state is the beable described in the text

configurations. The assumption is that the deformation resonance is as highly excited as the external temperature allows. This energy can then be redistributed between the local channels and the delocalized phonons.

We assume that within the deformation resonance coupling to two local quantum harmonic oscillators is effective, associated with the rocking and wagging motions of the carbonyl *CO* fragment. One local oscillator is not enough to reproduce the temperature dependence of the signal, it is not able to compete with the dissipative effect of the protein phonons. If it were just one local harmonic oscillator available for relaxation (one relaxed vibrational mode) we would get fast decay of the photon amplitude and no temperature dependence since one oscillator provides one local vibrational configuration for all experimental temperatures. If the number of coupled local oscillators and local vibrational configurations were large, i.e. the exciton electron resides transiently on many carbonyl carbon atoms in many carbonyl *CO* fragments, we would get no temperature dependence of the decay of the photon amplitude either. It would immediately drop to zero. No chooser effect and fast decay in the protein phonons all over the protein chain would be the result. Hence, the result would be fast exponential decay of the photon signal. However, in our model, starting from a delocalized exciton electron, with the chooser mechanism due to on-shell entanglement with gravonons, a localized electron in the form of a beable is generated. The local nature of the interactions in this many particle system prevents the excitation of many harmonic oscillators over the protein chain and the total decay of the photon signal.

With two local oscillators we get the exponential damping of the photon amplitude with time, the beating of the photon signal and the finite non-zero value of the photon signal at long time as shown in Fig. 4. The temperature dependence is also in satisfactory agreement with the experimental observation.

The dependence of the photon signal on the temperature is reflected by changes in the vibrational structure of the local oscillators and the accessibility of the protein phonon configurations. Figure 5 illustrates how the local oscillator vibrational structure changes with temperature if we restrict the model to two local oscillators. Einstein's model is assumed with the wagging and the rocking modes having the same energy $\omega = 5$ meV. This energy corresponds roughly to thermal energy $k_B T$ at $T = 58$ K. The energy is provided by the external heating and is available within the deformation resonance. So if the temperature in the 2D Fourier transform spectroscopy experiment equals 77 K, it means that either one or the other local oscillator can be excited, i.e. there are two degenerate vibrational configurations due to the local oscillators involved in the interaction with the deformation resonance (the first line in Fig. 5). At $T = 125$ K and $T = 150$ K the available thermal energy is sufficient to excite just 2 local vibrations which makes three degenerate local vibrational configurations (the second line in Fig. 5) and so on. Thus, with rising temperature the local vibrational structure becomes more complex. As the temperature rises 2, 3, 4 ... local vibrational configurations, which are energetically accessible and are on the energy shell with respect to the energy of the exciton wave packet, will be involved. More and more on-shell local channels get open for the wave packet when the temperature rises. This is the decisive feature of the model, which prevents the destruction of the coherent time development.

The channels for decay in delocalized protein phonons also increase, the available density of phonon configurations on the energy shell of the initial wave packet increases linearly with temperature. This is so because we assume a one-dimensional protein chain with constant density of states which provides a linear dependence on the temperature of the on-shell density of phonon configurations. If we would take the 3D Debye model for the phonons the density of on-shell phonon configurations would vary with T^3, which would result in fast dissipation in phonons, accompanied by the decay of the electron in the delocalized escape states.

Two channels open for the exciton electron compete. The first channel warrants conservation of the amplitude of the wave packet in the local region, whereas the second leads to its decay out of the local region in the delocalized protein phonons, which are not measured in the experiment and cause the exponential damping of the photon signal at short time. As time goes by the two competing effects balance each other and the photon signal gets stabilized at a finite value. The stochastic approaches try to obtain similar effects by an empirical special choice and tuning of the phonon noise such that the experimental beating signal is reproduced [33].

As the local structure in the model comprises more and more configurations with rising temperature it has a counterbalancing effect on the decay due to coupling to the delocalized phonons and warrants that the weight of local configurations with the exciton electron is retained locally. Then the third laser pulse in the experiment can interact with the localized electron in the 3s, the deformation resonance and the local oscillators and stimulate the photon emission with the amplitude retained in the local structure.

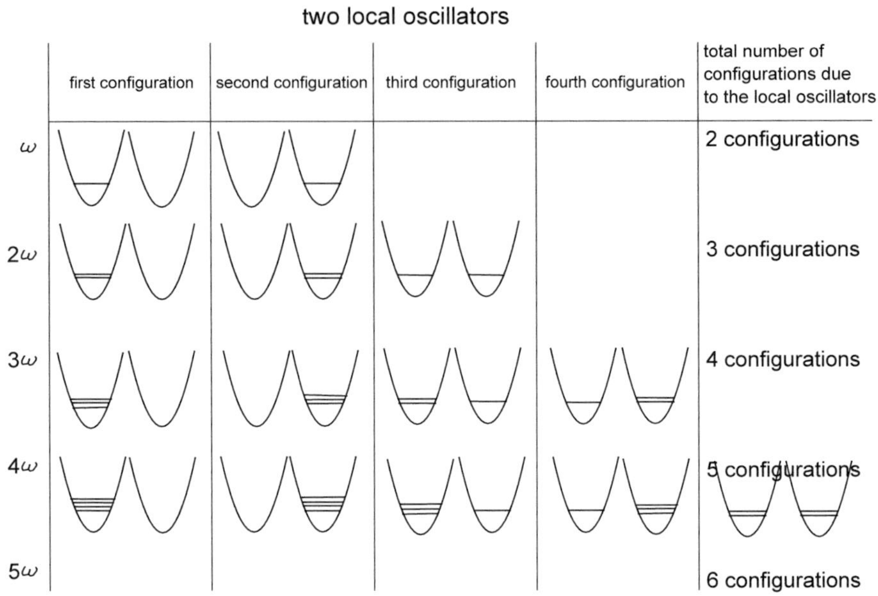

Fig. 5 Local vibrational configurations on the energy shell of the initial wave packet as a function of the excitation energy, hence as a function of the temperature, available within the deformation resonance

4 Discussion

We suggest an explanation of the questions raised in the introduction. No isotope effect on the coherences in FMO is established experimentally. In our theory it is of no significance if ^1H is replaced by ^2H or ^{12}C is exchanged with ^{13}C. The local vibrational spectrum changes, but we only need the spectrum on the energy shell of the wave packet. The local picture explains why if something changes 10 Å away from the transient negative ion resonance, for instance isotopes of hydrogen are exchanged, the change does not affect the coherent behaviour of the NIR, which is observed in experiment.

Using an argument from perturbation theory, the first order coupling between two configurations scales with the inverse of their energy difference. Hence, the on-shell contribution to the coupling dominates. Therefore the many particle configurations we take into account, are nearly on-shell with the initial wave packet since they provide the major contributions to the coupling.

Reminding the many-particle character of the field configurations in our theory we necessarily have to conclude that the nature of the states involved in the superposition, leading to the quantum beats, is vibronic, as it has been suggested by experimentalists as well.

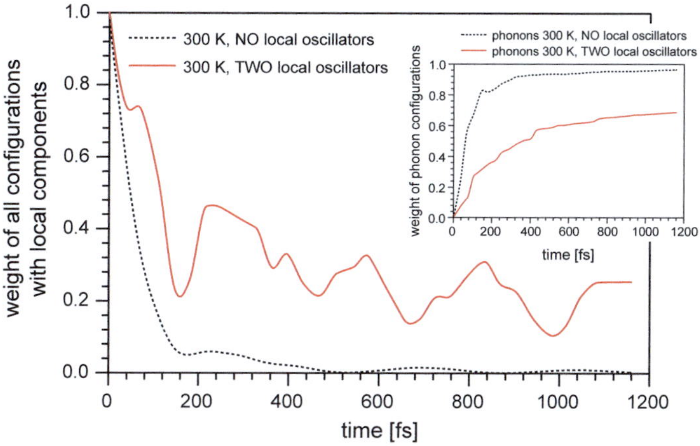

Fig. 6 Time development of the photon amplitude in a model neglecting the coupling of the exciton electron to the local oscillators with dominating coupling to the phonons (decaying *dashed curve*), corresponding to temperature 300 K, compared to the result with the local structure included (*oscillating full curve*). *Inset* development with time of the weight of field configurations involving the protein phonons with the local structure included (*full curve*) and excluded (*dashed curve*)

Let us omit the local oscillators due to the carbonyl *CO* fragment. The initial state is also changed, the electron is no more 'chosen' in the warp resonance, it is assumed to be in the somewhat diffuser deformation resonance. If in our model we would disregard the chooser mechanism and the coupling to the local *CO* oscillators and allow the exciton electron to couple to phonon continua all over the protein chain the result is similar to what has been expected before the experiment was performed (Fig. 6). The two curves in Fig. 6 are evaluated with different initial states. The curve, which reproduces the experimental result, starts with the exciton electron in a beable (which is destroyed by the photon-induced tiny burst) localized in the NIR via the chooser mechanism due to entanglement with the gravonons. For the lower curve where no chooser mechanism is operating, the initial wave packet has the electron in the deformation resonance, where it couples with the dissipative phonon environment all over the protein backbone. Ignoring the chooser mechanism leads to an exponential decay of the photon amplitude without oscillations to zero value, completely and irreversibly. Everything is lost in the protein phonons (cf. the dashed curve in the inset in Fig. 6), and this is not measured in the experiment. This resembles the prediction of conventional decoherence theory. In contrast, the chooser mechanism due to degeneracy entanglement with gravonons determines the transient localization of the electron wave packet in a beable and precludes its total decay in the environmental phonon continuum, which is the reason for retaining finite photon amplitude for long time, at least within the lifetime of the tiny burst.

All experimental observations are reproduced in a coherent picture by solving the TDSE. The chooser mechanism localizes the exciton electron within the beable, allowing it to interact with the local core movement states of the carbonyl group in addition to the dissipative interaction with the phonons in the protein backbone. In our theory these are physical phenomena within a completely coherent picture. The generation of a local beable via entanglement of the exciton electron with gravonons in the present theory is the clue to the understanding of the experiment.

5 Conclusion

The understanding of the long time coherence in the FMO protein complex is based on the localization of the exciton electron due to a localization chooser mechanism using gravitation in high dimensional spacetime. Released from the entanglement with the gravonons by the laser pulses, the electron can couple with local vibrations of the carbonyl CO group and the delocalized phonons in the protein backbone. This determines the vibronic character of the many-particle configurations leading to the beating of the photon echo signal. The two channels lead to preservation of a local non-vanishing weight of the exciton electron and to the beating photon echo signal and its exponential damping. The magic non-appearance of fast decay of the photon signal in the 2D Fourier transform electronic spectroscopic experiment is explained.

The local structure and a temperature dependent accessibility to the local structure are the clue to understand the finite final value of the photon echo signal in 2D Fourier transform electronic spectroscopy at long time. The starting point in our theory on the coherent behaviour of chlorophyll assemblies is the localized exciton electron as a beable which allows the interaction of this localized charge with the photon field. The further time development of the wave packet follows from the initial localization of the electron.

The observation that isotope substitution does not change the experimentally measured photon echo amplitude is explained in our theory. No special phonons ('colour noise', special spectral function) are needed to explain the coherent behaviour and the damped amplitude of the photon beating signal with time.

We claim that quantum mechanics needs a chooser mechanism for particle localization. Experiments can be explained within quantum mechanics only if the incoming delocalized particle wave gets localized on a local site. The solution of the time dependent Schrödinger equation in high dimensional spacetime including the entanglement to gravonons is the theory of the chooser mechanism. Local beables are the fundament of measurement. The measurement gives data on local features and this is what Schrödinger's equation with entanglement with the gravonons in high dimensional spacetime ($11D$) provides. All measured particles are local, they are local only when they entangle with gravonons as beables.

Acknowledgements We gratefully aknowledge the useful discussion and comments by G. Engel at the Gordon Center for Integrative Science, Chicago University.

References

1. Engel GS, Calhoun TR, Read EL, Ahn T, Mancăl T, Cheng Y, Blankenship RE, Fleming GR (2007) Nature 446:782
2. Panitchayangkoon G, Hayes D, Fransted KA, Caram JR, Harel E, Wen J, Blankenship RE, Engel GS (2010) Proc Natl Acad Sci USA 107:12766
3. Panitchayangkoon G, Voronine DV, Abramavicius D, Caram JR, Lewis NHC, Mukamel S, Engel GS (2011) Proc Natl Acad Sci USA 108:20908
4. Johnson MW et al (2011) Nature 473:194
5. Dickson NG et al (2013) Nat Commun 4:1903. doi:10.1038/ncomms2920
6. Drakova D, Doyen G (2015) arXiv:1220200 [quant-ph] 31 March 2015
7. Drakova D, Doyen G (2015) J Phys: Conf Series 442:012049
8. Rebentrost P, Mohseni M, Aspuru-Guzik A (2009) J Phys Chem B 113:9942
9. Mohseni M, Aspuru-Guzik A, Rebentrost P, Shabani A, Lloyd, Huelga SF, Plenio MB (2014) Quantum effects in biology. In: Mohseni YO, Engel GS, Plenio MB (eds) Cambridge University Press, Cambridge, p 159
10. Engel GS (2014) Quantum effects in biology. In: Mohseni YO, Engel GS, Plenio MB (eds) Cambridge University Press, Cambridge, p 144
11. Al-Khalili J, McFadden J (2014) Life on the edge—The coming of age of quantum biology. Bantam Press
12. Lee H, Cheng Y-C, Fleming GR (2007) Science 316:1462
13. Singh VP, Westberg M, Wang C, Dahlberg PD, Gellen T, Gardiner AT, Cogdell RJ, Engel GS (2015) J Chem Phys 142:212446
14. Harel E, Engel GS (2012) Proc Natl Acad Sci USA 109:706
15. Fidler AF, Singh VP, Long PD, Dahlberg PD, Engel GS (2013) J Phys Chem Lett 4:1404
16. Collini E, Wang CY, Wilk KE, Curmi PM, Brumer P, Scholes GD (2010) Nature 463:644
17. Dahlberg PD, Fidler AF, Caram JR, Long PD, Engel GS (2013) J Phys Chem Lett 4:3636
18. Dahlberg PD, Norris GJ, Wang C, Viswanathan S, Singh VP, Engel GS (2015) J Chem Phys 143:101101
19. Rebentrost P, Mohseni M, Kassal I, Lloyd S, Aspuro-Guzik A (2009) New J Phys 11:033003
20. Mohseni M, Rebentrost P, Lloyd S, Aspuro-Guzik A (2008) J Chem Phys 129:174106
21. Ishizaki A, Fleming GR (2009) Proc Natl Acad Sci USA 106:17255
22. Hildner R, Brinks D, van Hulst NF (2011) Nat Phys 7:172
23. Chiorescu I, Nakamura Y, Harmans CJPM, Mooij JE (2003) Science 299:1869
24. Hybl JD, Ferro AA, Jonas DM (2001) J Chem Phys 115:6606
25. Cowan ML, Ogilvie JP, Miller RJD (2004) Chem Phys Lett 386:184
26. Brixner T, Mančal T, Stiopkin IV, Fleming GR (2004) J Chem Phys 121:4221
27. Abramavicius D, Voronone DV, Mukamel S (2008) Biophys J 94:3613
28. Nagy A, Prokhorenko V, Miller RJD (2006) Curr Opin Struct Biol 16:654
29. Prezhdo OV, Rossky PJ (1998) Phys Rev Lett 81:5294
30. Förster T (1948) Naturwissenschaften 33, 166; T. Förster (1946) Ann Physik 437:55
31. Marcus RA, Surin N (1985) Biochem Biophys Acta 811:265
32. Hopfield JJ (1974) Proc Natl Acad Sci USA 9:3640
33. Shabani A, Mohseni M, Yang S, Ishizaki A, Plenio M, Rebentrost P, Aspuru-Guzik A, Cao J, Lloyd S, Silbey R (2014) Quantum effects in biology. In: Mohseni M, Omar Y, Engel G, Plenio MB (eds) Cambridge University Press, Cambridge, p 14
34. Dester D (1953) J Chem Phys 21:836
35. Şener M, Strümpfer J, Hsin J, Chandler D, Sheuring S, Hunter CN, Schulten K (2011) Chem Phys Chem 12:518
36. Hayes D, Wen J, Panitchayangkoon G, Blankenship RE, Engel GS (2011) Faraday Discuss 150:459
37. Schlau-Cohen GS, Ishizaki A, Calhoun TR, Ginsberg NS, Ballottari M, Bassi R, Fleming GR (2012) Nat Chem 4:389

38. Chenu A, Scholes GD (2015) Annu Rev Phys Chem 66:69
39. Tiwari V, Peters WK, Jonas DM (2014) Nat Chem 6:173
40. Fuller FD, Pan J, Gelzinis A, Butkus V, Senlik SS, Wilcox DE, Yocum CF, Valkunas L, Abramavicius D, Ogilvie JP (2014) Nat Chem 6:706
41. Romero E, Augulis R, Novoderezhkin VI, Ferretti M, Thieme J, Zigmantas D, van Grodelle R (2014) Nat Phys 10:677
42. Romero E, Novoderezhkin VI, van Grondelle R (2014) Quantum effects in biology. In: Mohseni M, Omar Y, Engel GS, Plenio MB (eds) Cambridge University Press, Cambridge, p 179
43. Chenu A, Christensson N, Kauffmann HF, Mančal T (2013) Sci Rep 3:2029
44. Womick JM, Moran AM (2011) J Phys Chem B 115:1347
45. Doyen G, Drakova D (2015) Found Phys 45:959
46. Bell JS (2004) Speakable and unspeakable in quantum mechanics: the theory of local beables, 2nd edn. Cambridge University Press, Cambridge, p 52
47. Dempter AJ, Batho HF (1927) Phys Rev 30:644
48. Braginski VB, Khalili FY (1992) Quantum measurement. Cambridge University Press, Cambridge, p 3
49. Hackermüller L, Uttenthaler S, Hornberger K, Reiger E, Brezger B, Zeilinger A, Arndt M (2003) Phys Rev Lett 91:090408
50. Hornberger K, Gerlich S, Haslinger Ph, Nimmrichter S, Arndt M (2012) Rev Mod Phys 84:157
51. Juffmann Th, Milic A, Müllneritsch M, Asenbaum P, Tsukernik A, Tüxen J, Mayor M, Cheshnovsky O, Arndt M (2012) Nat Nanotechnol 7:297
52. Doyen G, Drakova D (2015) J Phys: Conf Series 442:012063
53. Drakova D, Doyen G (2016) arXiv:1605.06149v1 [physics.chem-ph] 11 May 2016
54. Fleming GR, Yang M, Agarwal R, Prall BS, Kaufman LJ, Neuwahl F (2003) Bull Korean Chem Soc 24(8):1081

Part V
Fundamental Theory

The Dirac Operator in Quantum Chemistry and Physics

Jacek Karwowski

Abstract The energy spectrum of the free Dirac operator corresponding to an electron is composed of two continua: the positive one from mc^2 to $+\infty$ and the negative one from $-mc^2$ to $-\infty$ separated by the energy gap $\left(-mc^2, +mc^2\right)$. The Dirac operator in an external field may have discrete eigenvalues in the gap. These very eigenvalues and the corresponding eigenfunctions, in cases of external potentials of physical significance, correspond to physically observed states and are interesting from the point of view of both physics and chemistry. The variational description of the discrete spectrum of the Dirac operator is governed by a theorem known as the mini-max principle. There are many different modes of implementation of this theorem, ranging from simple approximations and numerical prescriptions to strict models which, in principle, can lead to a numerically exact reproduction of both the discrete part of the spectrum and the corresponding wavefunctions. Generalizations to a many-electron case create challenges related to the complicated structure of the continuous parts of the spectrum and the degeneracy of the discrete and continuous spectra leading to the so called Brown-Ravenhall disease. The evolution of our understanding of the consequences of the unusual properties of the Dirac operator over the nine decades which passed since this operator was defined is strongly coupled with the construction of specific computational models and, most important, with establishing links between the results of calculations and the physical reality. The aim of this presentation is a discussion of these topics.

Keywords Dirac operator · Dirac equation · Minimax principle · Essential spectrum · Energy gap · Complex scaling

J. Karwowski (✉)
Institute of Physics, Nicolaus Copernicus University,
Grudziądzka 5, 87-100 Toruń, Poland
e-mail: jka@fizyka.umk.pl

A. Tadjer et al. (eds.), *Quantum Systems in Physics, Chemistry, and Biology*,
Progress in Theoretical Chemistry and Physics 30,
DOI 10.1007/978-3-319-50255-7_22

1 Introduction

Paul Adrien Maurice Dirac was one of the most ingenious physicists of the 20th century. He introduced to physics (and, in a way, also to mathematics) several important, non-standard notions and useful tools. The best known include: Dirac operator, Dirac delta, Dirac bracket notation, Dirac sea. During the 1927 Solvay Conference Niels Bohr asked Dirac: *What are you working on Mr. Dirac?* The answer was: *I am trying to take the square root of something*. The success of this endeavor opened the way to the formulation of the relativistic theory of electron (1928) and to the Nobel Prize (1933). In the new theory he unified the quantum mechanics of a single electron formulated two years earlier by Erwin Schrödinger and the special theory of relativity.

In order to formulate the equation which describes the evolution of a one-electron quantum system, the Dirac equation, Dirac introduced a new mathematical construct, the square root of a second-order differential operator in the Minkowski space, known as the *Dirac operator*. Thus, the Dirac operator $\mathscr{D}$ is defined as a Hermitian, first-order differential operator which is a formal square root of a second-order differential operator, more precisely, of a linear form containing Laplacian. In other words,

$$\mathscr{D}^2 = \triangle + C^2 \tag{1}$$

where $\triangle$ is the Laplace operator and C is a real number. Dirac successfully found the square root. Here is his solution:

$$\mathscr{D} = \boldsymbol{\alpha} \cdot \mathbf{p} + \beta C, \tag{2}$$

where $\mathbf{p} = -i\nabla$, and $\boldsymbol{\alpha} = \{\alpha_1, \alpha_2, \alpha_3\}$, β are anticommuting, traceless, Hermitian matrices. Their squares are equal to the unit matrix, the smallest dimension is four and $\boldsymbol{\alpha} \cdot \mathbf{p}$ is a shorthand notation for $(\alpha_1 \mathsf{p}_1 + \alpha_2 \mathsf{p}_2 + \alpha_3 \mathsf{p}_3)$.

The motivation for introducing the Dirac operator came from physics. Quantum mechanical equation describing time evolution of a quantum system reads

$$\frac{\partial \Psi}{\partial t} = \mathsf{H}\Psi \tag{3}$$

where H is the Hamilton operator. In Lorentz-covariant equations time and coordinates (alternatively, energy and momentum) have to appear on an equal footing. Therefore a relativistic Hamiltonian has to be linear in momentum. For a free particle described by relativistic mechanics

$$\mathsf{H}^2 = c^2 \left(\mathbf{p}^2 + m^2 c^2 \right). \tag{4}$$

Therefore the *free-particle Dirac operator* has to be of the form

$$\mathsf{H} = c\,(\boldsymbol{\alpha}\cdot\mathbf{p} + \beta mc)\,. \tag{5}$$

In the case of spin 1/2 particles the dimension of the matrices has to be four.

The Schrödinger equation is scalar and it is defined in

$$\mathscr{H}^{\mathrm{S}} = L^2\left(\mathbb{R}^3\right) \tag{6}$$

Hilbert space. In order to describe spin 1/2 particles within the non-relativistic formalism, one should use the Pauli equation defined in a larger Hilbert space,

$$\mathscr{H}^{\mathrm{P}} = \mathscr{H}^{\mathrm{S}} \oplus \mathscr{H}^{\mathrm{S}} = L^2\left(\mathbb{R}^3\right) \otimes \mathbb{C}^2. \tag{7}$$

The Dirac equation, i.e., Eq. (3) with Hamiltonian given by Eq. (5), is defined in

$$\mathscr{H}^{\mathrm{D}} = \mathscr{H}^{\mathrm{P}} \oplus \mathscr{H}^{\mathrm{P}} = L^2\left(\mathbb{R}^3\right) \otimes \mathbb{C}^4 \tag{8}$$

Hilbert space. The additional, relative to the Schrödinger model, degrees of freedom are related to the electron spin. Dirac was aware of this and wrote in his 1928 paper [1]: *The α's are new dynamical variables* (...). *They may be regarded as describing some internal motions of the electron, which for most purposes may be taken to be the spin of the electron postulated in previous theories.*

The four-dimensional space in which matrices $\boldsymbol{\alpha}$ and β are defined is referred to as the *spinor space*. The theory is invariant with respect to a unitary transformation in this space. Therefore the matrices are determined up to such a transformation and for different purposes different forms of these matrices may be selected [2]. In the most common representation, known as the Dirac-Pauli one, the matrices are explicitly related to the Pauli spin matrices:

$$\boldsymbol{\alpha} = \begin{bmatrix} \mathbf{0}, & \boldsymbol{\sigma} \\ \boldsymbol{\sigma}, & \mathbf{0} \end{bmatrix}, \quad \boldsymbol{\beta} = \begin{bmatrix} \mathbf{1}, & \mathbf{0} \\ \mathbf{0}, & -\mathbf{1} \end{bmatrix}, \tag{9}$$

where $\mathbf{1}$ is 2×2 unit matrix, $\boldsymbol{\sigma}$ are three Pauli spin matrices and $\mathbf{0}$ is 2×2 matrix of zeros.

From the perspective of modern quantum chemistry, the Dirac equation is equally important as the Schrödinger equation. However, at the early stage, the relativistic effects were considered as irrelevant. This is reflected by the following quotation of Dirac [3]: *The general theory of quantum mechanics is now almost complete, the imperfections that still remain being in connection with the exact fitting in of the theory with relativity ideas. These give rise to difficulties only when high-speed particles are involved, and are therefore of no importance in the consideration of atomic and molecular structure and ordinary chemical reactions* (...). A reader interested in a broader discussion of the Dirac's views on the relativistic formulation of quantum mechanics is referred to an essay by Kutzelnigg [4].

In this paper some properties of the Dirac operator important for quantum chemical application are discussed. The literature of the subject is very rich. In particular, a recent monograph by Reiher an Wolf [5] offers probably the most complete summary of different aspects of relativistic quantum chemistry. An extensive list of references may be helpful to a reader willing to study the subject from different perspectives. Another modern presentation of the field may be found in Handbook of Relativistic Quantum Chemistry [6]. Non-standard views on the Dirac particle, in particular some possible relations between *Zitterbewegung* and the electron mass, are discussed by Mulligan [7]. Some extensions of these ideas as well as philosophical and historical aspects are presented in recent works by Maruani [8]. Hartree atomic units are used but, whenever it is convenient, the mass m and the charge q are written explicitly.

2 Dirac Operator with External Fields

Spectrum of a Hermitian operator is composed of the essential spectrum and the discrete spectrum, i.e., a set of isolated eigenvalues of finite multiplicity [9]. The energy spectrum of a free Dirac operator is given by

$$E = \pm mc^2\sqrt{1+\left(\frac{\boldsymbol{p}}{mc}\right)^2}, \quad \boldsymbol{p}^2 \in (0,\infty). \tag{10}$$

Thus,

$$E \in \left(-\infty, -mc^2\right] \cup \left[+mc^2, +\infty\right), \tag{11}$$

i.e., the essential spectrum consists of two continua: *the positive continuum* from the rest energy of the particle, mc^2, to $+\infty$, corresponds to the non relativistic energy of the free particle, and *the negative continuum* from $-mc^2$ to $-\infty$.

The negative continuum has no direct physical meaning and its interpretation was always difficult. Here is the original interpretation by Dirac [10]: *The mathematical formalism* (...) *involved a serious difficulty through its prediction of negative kinetic energy values for an electron. It was proposed to get over this difficulty making use of Pauli's Exclusion Principle* (...).. *The question then arises to the physical interpretation of the negative energy states* (...). *It was shown that one of these holes would appear to us as a particle with a positive energy and a positive charge and it was suggested that this particle should be identified with a proton. Subsequent investigation, however, have shown that this particle necessarily has the same mass as an electron.* (...) *A hole, if there were one, would be a new kind of particle unknown to experimental physics.* One can easily see that this explanation, as well as any other based on the Pauli exclusion principle, is self-contradictory. First, already in 1874 Georg Cantor [11, 12] proved that the cardinality of a set of discrete objects (integers) is strictly smaller than that of the set of real numbers (a continuum). Therefore a continuum cannot be filled in by electrons. Second, relativistic equations describing bosons also have negative continua but a reasoning based on the Pauli principle is not applicable to them. However, the Dirac sea model gave the first physical description of vacuum defined as a polarizable medium from which new particles can be created. Many concepts and terms used to describe vacuum-related phenomena in quantum field theories originate from the ideas of Dirac [13]. The problem is correctly solved in quantum electrodynamics, where the energy spectrum is bounded from below. In the Dirac theory the transformation of charge conjugation (particle—antiparticle replacement) shows that a negative energy state of the particle describes the antiparticle with positive energy.

In quantum chemical models real interactions are usually replaced by external fields which describe these interactions in a simplified, effective, way. One of the most common examples is the Born-Oppenheimer approximation, where the nuclei are assumed to form a fixed network of classical charges. In relativistic quantum mechanics a hydrogen-like atom, unsolvable analytically system of two interacting particles, an electron and a nucleus, is usually replaced by a simple problem of an electron in the external Coulomb potential. A model of an electron in an external field may be used to describe nearly all systems of chemical interest. In particular, a many-electron system may be reduced to a one-electron problem using the mean-field approach which results in Hartree-Fock or Dirac-Fock equations. The external potentials may have different transformation properties: vector, scalar, pseudoscalar, pseudovector, tensor. In the non-relativistic theories of atoms and molecules the most important are scalar V_{el} (electrostatic) and vector $\mathbf{A}$ (magnetic) potentials. In relativistic models both kinds of potentials form a single four-vector $\mathbf{A} = \{\, \mathbf{A}, i\mathsf{V}_{\text{el}}/c\}$.

The form of the Dirac operator for a particle in an external electromagnetic field may be obtained using the *minimal coupling principle* [9], i.e., by the following replacements:

$$\begin{aligned} \mathbf{p} &\mapsto \boldsymbol{\pi} \equiv \mathbf{p} + \mathbf{A}, \\ \mathsf{E} &\mapsto \mathsf{E} - \mathsf{V}_{\text{el}}, \end{aligned} \tag{12}$$

In the case of scalar potentials

$$mc^2 \mapsto mc^2 + \mathsf{V}_{\text{sc}}. \tag{13}$$

Thus, the Dirac operator with external fields reads:

$$\mathsf{H} = c\,(\boldsymbol{\alpha} \cdot \boldsymbol{\pi}) + \mathsf{V}_{\text{el}} + \left(mc^2 + \mathsf{V}_{\text{sc}}\right) \beta. \tag{14}$$

The external potentials modify spectrum of the Dirac operator. Discrete eigenvalues, corresponding to the bound states, appear in the energy gap $\left(-mc^2, +mc^2\right)$. It is worth noticing that for majority of physically significant potentials the essential spectrum of the Dirac operator remains unchanged [9]. The spectrum of a Dirac operator is shown in Fig. 1.

Since the Hamilton operator in the Dirac equation is represented by a 4×4 matrix, the wavefunction $\boldsymbol{\Psi}$ has to have 4 components. Due to its specific transformation properties under the Lorentz transformation, it is classified as a four-component *spinor* [9]. In the Dirac-Pauli representation the first two components of the wave-

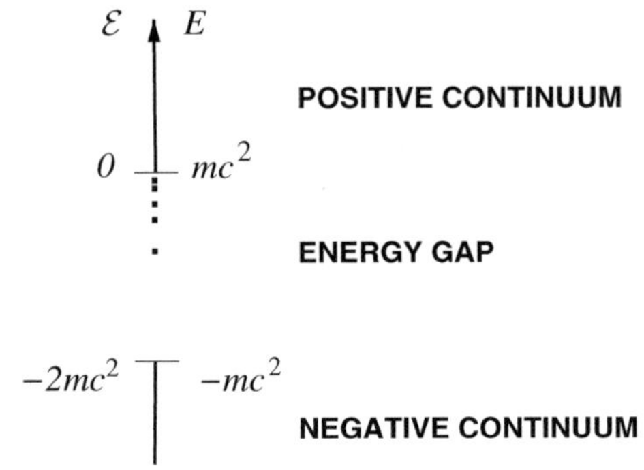

Fig. 1 Spectrum of a Dirac operator; E—the relativistic energy scale; $\mathscr{E} = E - mc^2$—the non-relativistic energy scale

function are referred to as the *large components*, $\Psi_{\rm L}$, and the last two—as the *small components*, $\Psi_{\rm S}$ and $|\Psi_{\rm L}| \sim c\,|\Psi_{\rm S}|$.

3 The Eigenvalue Problem of the Dirac Operator

Let $\mathsf{V}^{\pm} = \mathsf{V}_{\rm el} \pm \mathsf{V}_{\rm sc}$ and $\mathscr{E} = E - mc^2$. Then the eigenvalue equation of the Dirac operator (14) may be written as

$$\begin{bmatrix} (\mathsf{V}^{+} - \mathscr{E})\,\mathbf{1} & c\,(\boldsymbol{\sigma}\cdot\boldsymbol{\pi}) \\ c\,(\boldsymbol{\sigma}\cdot\boldsymbol{\pi}) & (\mathsf{V}^{-} - 2mc^2 - \mathscr{E})\,\mathbf{1} \end{bmatrix} \begin{bmatrix} \Psi_{\rm L}(\boldsymbol{r}) \\ \Psi_{\rm S}(\boldsymbol{r}) \end{bmatrix} = 0. \tag{15}$$

From here one can easily get the relation between small and large components of the eigenfunction:

$$\Psi_{\rm S}(\boldsymbol{r}) = \frac{\mathsf{R}}{2\mathscr{M}\,c}\,(\boldsymbol{\sigma}\cdot\boldsymbol{\pi})\,\Psi_{\rm L}(\boldsymbol{r}) \tag{16}$$

where

$$\mathscr{M} = m\left(1 + \frac{\mathscr{E}}{2mc^2}\right), \quad \mathsf{R} = \frac{1}{1-\mathsf{W}}, \quad \mathsf{W} = \frac{\mathsf{V}^{-}}{2\mathscr{M}\,c^2}. \tag{17}$$

By the substitution of Eq. (16) to Eq. (15) one obtains the equation defining large components of $\boldsymbol{\Psi}$

$$\left[(\boldsymbol{\sigma}\cdot\boldsymbol{\pi})\frac{\mathsf{R}}{2\mathscr{M}}(\boldsymbol{\sigma}\cdot\boldsymbol{\pi}) + \mathsf{V}^{+} - \mathscr{E}\right]\Psi_{\rm L}(\boldsymbol{r}) = 0. \tag{18}$$

Let us note that Eq. (18) is not an eigenvalue equation since both R and $\mathscr{M}$ depend on the parameter $\mathscr{E}$ (equal to the eigenvalue of Eq. (15)).

Equation (18) is a convenient starting point to developing many so called two-component methods [5] as well as to analyzes aimed at deeper understanding of the properties of the eigenvalue problem of the Dirac operator. In particular, the Dirac identity

$$(\boldsymbol{\sigma} \cdot \mathbf{B})\,(\boldsymbol{\sigma} \cdot \mathbf{C}) = (\mathbf{B} \cdot \mathbf{C}) + i\boldsymbol{\sigma} \cdot [\mathbf{B} \times \mathbf{C}]\,, \tag{19}$$

where **B** and **C** are operators, applied to Eq. (18) yields a form of the Dirac equation in which spin-free and spin-dependent terms are exactly separated [14]:

$$\left[\frac{\boldsymbol{\pi}\,\mathsf{R}\boldsymbol{\pi}}{2\mathscr{M}} + \frac{i\,\boldsymbol{\sigma}}{2\mathscr{M}}\left[\boldsymbol{\pi} \times \mathsf{R}\boldsymbol{\pi}\right] + \mathsf{V}^{+} - \mathscr{E}\right]\Psi_{\mathrm{L}}(\boldsymbol{r}) = 0. \tag{20}$$

Using identities:

$$\boldsymbol{\pi}\,\mathsf{R}\boldsymbol{\pi} = \mathsf{R}\left[\boldsymbol{\pi}^2 + \mathsf{R}(\mathbf{p}\,\mathsf{w})\boldsymbol{\pi}\right] \tag{21}$$

$$i\boldsymbol{\sigma}[\boldsymbol{\pi} \times \mathsf{R}\boldsymbol{\pi}] = \mathsf{R}\,\{(\boldsymbol{\sigma} \cdot \boldsymbol{B}) + \mathsf{R}\,\boldsymbol{\sigma}\,[(\nabla\mathsf{w}) \times \boldsymbol{\pi}]\} \tag{22}$$

we can *rewrite* Eq. (20) as

$$\left[\frac{1}{2m}\left(\boldsymbol{\pi}^2 + \boldsymbol{\sigma} \cdot\ \mathbf{B}\right) + \mathsf{H}_{\mathrm{mv}} + \mathsf{H}_{\mathrm{dso}} + \mathsf{V}^{+} - \mathscr{E}\right]\ \Psi_{\mathrm{L}}(\boldsymbol{r}) = 0, \tag{23}$$

where $\mathbf{B} = \nabla \times\ \mathbf{A}$,

$$\mathsf{H}_{\mathrm{mv}} = -\frac{1}{2mc^2}\,(\mathsf{V}^{+} - \mathscr{E})(\mathsf{V}^{-} - \mathscr{E}) \tag{24}$$

and

$$\mathsf{H}_{\mathrm{dso}} = \frac{R}{4\mathscr{M}mc^2}\,\{(\mathbf{p}\,\mathsf{V}^{-})\boldsymbol{\pi} + \boldsymbol{\sigma}\left[(\nabla\mathsf{V}^{-}) \times \boldsymbol{\pi}\right]\} \tag{25}$$

Let us note that Eq. (23) is an *exact* equation for the large components of the eigenfunction of the Dirac Hamiltonian. The operator H_{mv} is equal to the mass-velocity term of the well known Pauli expansion and $\mathsf{H}_{\mathrm{dso}}$ becomes the Darwin-spin-orbit correction if $\mathscr{E}/2mc^2$ and $\mathsf{V}^{-}/2mc^2$ are neglected relative to 1. The exact form of $\mathsf{H}_{\mathrm{dso}}$ is free of the non-integrable singularities which appear in the corresponding terms of the Pauli expansion in the case of, e.g., the Coulomb potential. Indeed, as one can easily see, if $\mathbf{A} = 0$ and $\mathsf{V}^{-} = -Z/r$ then, for $r \to 0$, $\mathsf{H}_{\mathrm{dso}}$ given by Eq. (25) is proportional r^{-2} rather than to r^{-3}, as it is in the case of the Pauli expansion.

If $\mathsf{R} = 1$, i.e., $\mathsf{V}^{-} = 0$ (or $\mathsf{w} << 1$ is neglected) then we obtain the *direct Dirac equation* [15]. The large component Eq. (18) takes the same form as the Schrödinger-Pauli equation

$$\left[\frac{(\boldsymbol{\sigma} \cdot \boldsymbol{\pi})^2}{2\mathscr{M}} + \mathsf{V}^{+} - \mathscr{E}\right]\Psi_{\mathrm{L}}(\boldsymbol{r}) = 0 \tag{26}$$

except for the energy-dependent 'mass' $\mathscr{M}$. Thus, if an exact solution of the Schrödinger equation is known, the solution of the direct Dirac equation may be obtained by a simple replacement of the parameters. Recently, a large number of papers giving this kind of solutions have been published (see, e.g., [16] and references therein). Note, that due to the energy-dependent 'Hamiltonian' in Eq. (26), functions $\Psi_{\rm L}(\boldsymbol{r})$ corresponding to different energies are not orthogonal. Orthogonal are four-component wavefunctions composed of $\Psi_{\rm L}(\boldsymbol{r})$ and $\Psi_{\rm S}(\boldsymbol{r})$. In this case the relation between large and small components simplifies to

$$\Psi_{\rm S}(\boldsymbol{r}) = \frac{(\boldsymbol{\sigma} \cdot \boldsymbol{\pi})}{2\mathcal{M}\,c}\,\Psi_{\rm L}(\boldsymbol{r}) \tag{27}$$

4 Variational Approach

Since H is unbounded from below, the eigenvalues cannot be determined by a simple minimization procedure known from the non-relativistic quantum mechanics. There are several approaches to this problem. One can transform the eigenvalue equation to a problem bounded from below by projecting onto the positive energy space. Advantages and disadvantages of this approach are broadly discussed in the literature (see e.g., [5] and references therein). In another approach, specific for operators with eigenvalues located in the energy gap of the essential spectrum, the eigenvalues are considered as the stationary points of the Rayleigh quotient

$$Q[\boldsymbol{\Phi}] = \frac{\langle \boldsymbol{\Phi}|\mathsf{H}|\boldsymbol{\Phi}\rangle}{\langle \boldsymbol{\Phi}|\boldsymbol{\Phi}\rangle}, \tag{28}$$

where $\boldsymbol{\Phi} \in \mathcal{H}^{\rm D}$ is a square-integrable four-component spinor. The stationary points of the energy hypersurface in the space of variational parameters may be found using the *minimax principle*

$$E = \min_{\{\Phi_{\rm L}\}} \left[\max_{\{\Phi_{\rm S}\}} \{Q[\boldsymbol{\Phi}]\} \right], \tag{29}$$

where $\Phi_{\rm L}$ and $\Phi_{\rm S}$ are, respectively, large and small components of Φ. The minimax principle has been formulated as a recipe for reaching the stationary points in 1986 [17, 18]. A rigorous proof has been presented 15 years later [19].

From the minimax principle one can conclude that if the small component is known exactly, then E is bounded from below in the variational space $\mathcal{H}_{\rm L}$ of the large components. Also the elimination of the small component leads to an equation in which the energy parameter is bounded from below, but then the energy parameter is not an eigenvalue, as one can see in Eq. (18). In the context of variational basis set calculations most useful is conclusion that E is bounded from below in $\mathcal{H}_{\rm L}$ if the variational space $\mathcal{H}_{\rm S}$ of the small components is *saturated*. The saturation condition may be expressed as

$$\mathcal{H}_{\rm S}^{(n)} \supset \frac{\mathsf{R}\,\alpha}{2\mathcal{M}}\,(\boldsymbol{\sigma} \cdot \boldsymbol{\pi})\,\mathcal{H}_{\rm L}^{(n)}. \tag{30}$$

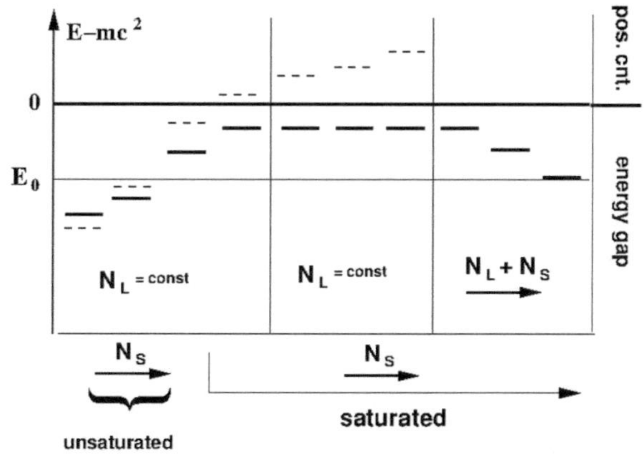

Fig. 2 Behavior of the ground-state energy (*solid*) and of a spurious root (*broken*) versus the structure of the small-component variational space; N_L, N_S—number of the basis functions in $\mathscr{H}_\mathrm{L}$ and in $\mathscr{H}_\mathrm{S}$, respectively

This implies that $\mathscr{H}_\mathrm{S}^{(n)}$ contains all functions needed for the construction of $\Phi_\mathrm{S}^{(n)}$ which, according to the Dirac equation, corresponds to $\Phi_\mathrm{L}^{(n)}$ variationally derived from $\mathscr{H}_\mathrm{L}^{(n)}$. Note, the choice of the *functional space* rather than of a specific *basis set* is essential. The results are determined by the variational space rather than by the basis set—they are invariant with respect to non-singular linear transformations of the basis. *Saturation* and *completeness* of a space have entirely different meaning. In particular, a saturated $\mathscr{H}_\mathrm{S}^{(n)}$ may contain only *one function*: $\Phi_\mathrm{S}^{(n)}$. Adding to a saturated $\mathscr{H}_\mathrm{S}^{(n)}$ more linearly independent functions does not change the computed value of E_n. This property is illustrated in Fig. 2.

The procedure described above is accurate but it is *state-specific* since it describes a single state rather than a section of the spectrum. An approximate (and frequently leading to incorrect results) kinetic balance condition

$$\mathscr{H}_\mathrm{S}^{(n)} \supset (\boldsymbol{\sigma} \cdot \boldsymbol{\pi})\, \mathscr{H}_\mathrm{L}^{(n)}, \tag{31}$$

obtained if $\mathsf{R} \to 1, \quad \mathscr{M} \to m,$ is much simpler and more frequently used. In particular, the procedures based on the kinetic balance condition are *global*, i.e., state-independent. If the kinetic balance condition is not fulfilled then spurious roots may appear, also when the basis set is saturated with respect to the state of interest. This may happen if the specific state is described correctly but the description of the remaining part of the spectrum is inadequate. This is illustrated in the following example, where also the physical nature of the spurious root is explained.

4.1 Algebraic Spectrum of a Hydrogen-Like Atom

In order to illustrate the behavior of the algebraic spectrum of the Dirac operator depending on the mutual relation between the basis sets in the large and small com-

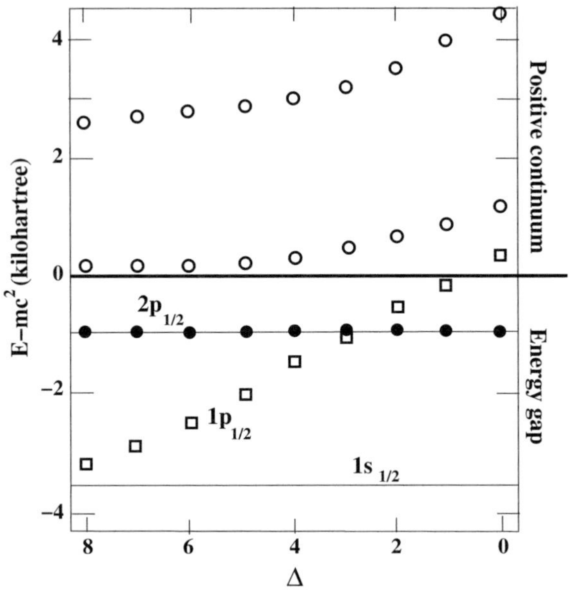

Fig. 3 Algebraic spectrum of $p_{1/2}$ states of $Z = 80$ hydrogen-like atom; *full circles*—energies of $2p_{1/2}$, squares—energies of the spurious root (denoted as $1p_{1/2}$), *open circles*—energies of the discrete states representing the positive continuum. The *horizontal lines* correspond to the exact energies of $1s_{1/2}$, $2p_{1/2}$ and to the ionization limit

ponent spaces we consider a simple example of $Z = 80$ hydrogen-like ion. The basis set in $\mathscr{H}_L$ is composed of 5 radial functions $\phi_j(r) = r^{\gamma+j}\,e^{-a\,r}$, $j = 0, \ldots, 4$, where γ and a are equal to the exact values for $2p_{1/2}$ state. We consider 9 different basis sets in $\mathscr{H}_S$, each of them composed of 6 funcions: 5 of them are the same as in $\mathscr{H}_L$ and one additional function is equal to $\phi_5^{\Delta} = r^{\gamma+5+\Delta}\,e^{-a\,r}$, where $\Delta = 0, \ldots, 8$. One should expect that the smaller is Δ, the closer to the saturation limit is the basis set in $\mathscr{H}_S$. The values of the lowest four energies are plotted versus Δ in Fig. 3. For $\Delta = 8$ a spurious root (denoted as $1p_{1/2}$) is nearly equal to the exact ground state energy. With decreasing Δ the energy of this state increases and, finally, for $\Delta = 0$, moves to the continuum. As we see, the presence of the spurious root is an artefact caused by inappropriate choice of the small component basis set.

5 Two-Electron Dirac Equation

In the case of two non-interacting electrons the two-electron Dirac Hamiltonian H_2 may be expressed in terms of the one-electron Hamiltonians H_1 as

$$\mathsf{H}_2 = \mathsf{H}_1 \otimes \mathbf{I} + \mathbf{I} \otimes \mathsf{H}_1, \tag{32}$$

where $\mathbf{I}$ is the one-electron unit operator. The spectrum of H_2 is equal to the superposition of two spectra of H_1 is shown in Fig. 1. The result of this superposition

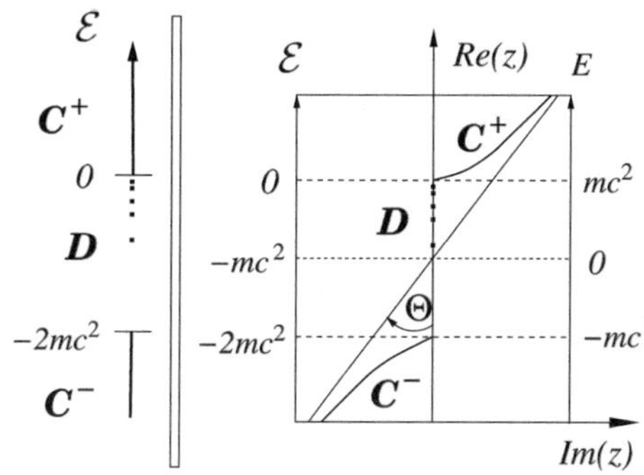

Fig. 4 Spectrum of a one-electron Dirac operator without and with complex scaling. In the complex scaled case the discrete part remained unchanged while the continua moved to the complex plane

is rather complicated and its unusual features result in serious difficulties both in computational implementations and in the interpretation of the results.

The structure of the two-electron spectrum may be seen in a transparent way using the complex scaling procedure [20]. Complex scaling, also known as *complex-coordinate rotation* (CCR), originally designed to study auto-ionizing states, appears to be a convenient tool to study spectra composed of mutually overlapping discrete and continuous sections [21, 22]. The basic for this technique theorem says that under a complex scaling of the electron coordinates

$$\boldsymbol{r} \rightarrow \boldsymbol{r}\,\mathrm{e}^{i\theta} \tag{33}$$

in the Hamiltonian, the bound-state energies do not change while the continua move to the complex plane. The effect of the CCR procedure applied to a one-electron Dirac Hamiltonian is shown in Fig. 4. In the case of two non-interacting electrons the spectrum is much more complicated. It is shown in Fig. 5. As one can see, the continuous spectrum $\mathbf{C}^{+-}$ spreads from $-\infty$ to $+\infty$ and thus, there is no gap in the essential spectrum. The discrete spectrum of the two-electron system overlaps with

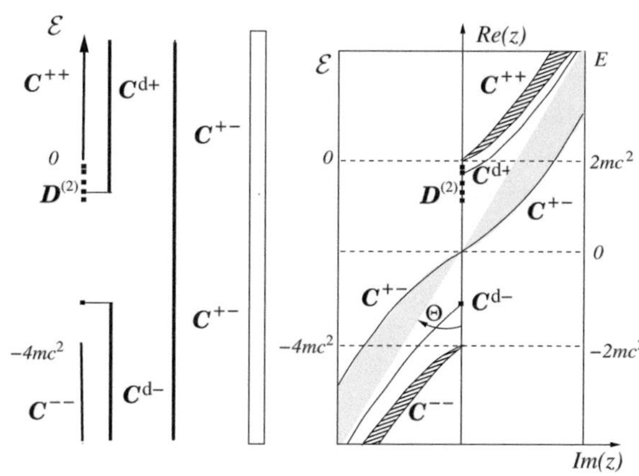

Fig. 5 Spectrum of a Dirac operator describing two non-interacting electrons without and with complex scaling. In the complex scaled case the complicated structure of the multiple continua has been revealed

$\mathbf{C}^{+-}$ continuum, referred to as the Brown-Ravenhall (BR) continuum, after Gerald E. Brown and D. Geoffrey Ravenhall who described the consequences of this property of the spectrum of a two-electron Dirac Hamiltonian [23].

If the interaction between electrons described by the non-relativistic Coulomb potential is included, we obtain the *Dirac-Coulomb* (DC) Hamiltonian. Thus, a theory based on this Hamiltonian cannot be Lorentz-covariant. An equation Lorenz-covariant up to terms proportional to α^2, the *Dirac-Coulomb-Breit* (DCB) equation, constructed Gregory Breit in 1929 [24]. Electron-electron interaction couples discrete and continuum states. Consequently, the Hamiltonian spectrum does not contain discrete energy levels, i.e., it does not have normalizable eigenfunctions. All physically bound states (including, for example, the ground state of the helium atom) are auto-ionizing. Due to the presence of the BR continuum, the energies of the auto-ionizing states are shifted relative to the corresponding bound states. The standard approach to the problem is to remove the unphysical continuum by properly constructed projection operators. In effect one obtains a well defined discrete spectrum, but the resulting model space is incomplete. For broad discussion of these issues see [5, 25] and references therein.

In principle, the minimax theorem is applicable to one-electron problems. However, usually many-electron problems are reduced to the one-electron ones by using the mean-field approach leading to the Dirac-Fock model or to density-functional-based approaches. If the Dirac-Fock equations are solved on a grid (numerically) then the exact relation between large and small component of the wavefunction is maintained and the energies are bounded from below. Therefore no positive-energy-space (PES) projection is needed. Also, if in the basis-set calculations the small component basis is saturated then we also do not need a projection. However, in other cases, in particular in high-precision multi-configuration calculations this or another kind of projection is helpful. The genuine many-electron models are based on using explicitly-correlated wavefunctions. In this case the PES projection secures correct and stable results, though by a very careful selection of the basis set one may obtain correct results also without projection [21, 22, 26, 27]. In this selection very useful can be a recently derived exact relativistic kinetic-balance condition for explicitly correlated basis functions [27].

In the context of the construction of new methods and the interpretation of the results derived from the algebraic representation of the eigenvalue problem of the DCB Hamiltonian, the techniques based on CCR are worth of some attention. In particular, a separation of the positive-energy subspace of the variational space, useful, e.g., in relativistic Hylleraas-CI, may be performed independently of the form of the variational functions [28]. It is also useful in the identification of these eigenvalues of the matrix representing the DCB Hamiltonian, which correspond to the bound states. In a discrete representation the eigenvalues corresponding to the discrete and and to the continuum state energies are mixed and may be difficult to distinguish. In the CCR spectrum, with the same basis set, one can see which eigenvalues correspond to discrete energies, which ones to the continuum states. One can also recognize to which kind of continuum a specific state belongs [21, 22, 29]. Finally, using the CCR-based language one can easily distinguish different kinds of continua and explain their role. For example, the meaning of the statement that *the non-projected DC Hamiltonian does not have bound states* is easier to explain using the CCR spectrum.

References

1. Dirac PAM (1928) The quantum theory of the electron. Part II. Proc R Soc Lond Ser A 118:351–361
2. Karwowski J, Pestka G, Stanke M, Harris F (2006) Representation of the Dirac equation and the variational principle. Int J Quantum Chem 106:3129–3139
3. Dirac PAM (1929) Quantum mechanics of many-electron systems. Proc R Soc Lond Ser A 123:714–733
4. Kutzelnigg W (2000) Perspective on Quantum mechanics of many-electron systems. Theor Chem Acc 103:182–186
5. Reiher M, Wolf A (2015) Relativistic quantum chemistry: the fundamental theory of molecular science, 2nd edn. Wiley, Weinheim
6. Liu W (ed) (2015) Handbook of relativistic quantum chemistry. Springer, Berlin, Heidelberg
7. Mulligan B (2006) Mass, energy, and the electron. Ann Phys 321:1865–1891
8. Maruani J (2015) The Dirac equation as a privileged road to the understanding of quantum matter. Quantum Matter 4:3–11
9. Thaller B (1992) The Dirac equation. Springer, Berlin
10. Dirac PAM (1931) Quantized singularities in the electromagnetic field. Proc R Soc Lond Ser A 133:60–72
11. Cantor G (1874) Über eine Eigenschaft des Inbegriffes aller reellen algebraischen Zahlen. J für die Reine und Angewandte Mathematik 77:258–262
12. Ewald WB (ed) (1996) From Immanuel Kant to David Hilbert. Oxford University Press, New York
13. Boi L (2011) The Quantum vacuum: a scientific and philosophical concept, from electrodynamics to string theory and geometry of the microscopic world. The Johns Hopkins University Press, Baltimore
14. Dyall KD (1994) An exact separation of the spin-free and spin-dependent terms of the Dirac-Coulomb-Breit Hamiltonian. J Chem Phys 100:2118–2127
15. Moore RA (1975) An alternative method of obtaining approximate solutions to the Dirac equation. Can J Phys 53:1240–1246
16. Setare MR, Haidari S (2009) Solutions of the Dirac equation for the Davidson potential. Int J Theor Phys 48:3249–3256
17. Talman JD (1986) Minimax principle for the Dirac equation. Phys Rev Lett 57:1091–1094
18. Datta SN, Deviah G (1988) The minimax technique in relativistic Hartree-Fock calculations. Pramana 30:387–405
19. Dolbeault J, Esteban MJ, Séré E (2000) On the eigenvalues of operators with gaps. Application to Dirac operators. J Funct Anal 174:208–226
20. Moiseyev N (2011) Non-Hermitian quantum mechanics. Cambridge University Press, New York
21. Pestka G, Bylicki M, Karwowski J (2006) Application of the complex coordinate rotation to the relativistic Hylleraas-CI method: a case study. J Phys B At Mol Opt Phys 39:2979–2987
22. Pestka G, Bylicki M, Karwowski J (2012) Geminals in Dirac-Coulomb eigenvalue problem. J Math Chem 50:510–533
23. Brown GE, Ravenhall DG (1951) On the interaction of two electrons. Proc Roy Soc A 208:552–559
24. Breit G (1929) The effect of retardation on the interaction of two electrons. Phys Rev 34:553–573

25. Liu W (2014) Advances in relativistic quantum mechanics. Phys Rep 537:59–89
26. Watanabe Y, Nakano H, Tatewaki H (2010) Effect of removing the no-virtual-pair approximation on the correlation energy of the He isoelectronic sequence. II. Point nuclear charge model. J Chem Phys 132:124105
27. Simmen B, Mátyus E, Reiher M (2015) Relativistic kinetic-balance condition for explicitly correlated basis functions. J Phys B At Mol Opt Phys 48:245004
28. Bylicki M, Pestka G, Karwowski J (2008) Relativistic Hylleraas configuration-interaction method projected into positive-energy space. Phys Rev A 77:044501
29. Pestka G, Bylicki M, Karwowski J (2007) Complex coordinate rotation and relativistic Hylleraas-CI: helium isoelectronic series. J Phys B At Mol Opt Phys 40:2249–2259

A Coherent Resonant Cosmology Approach and its Implications in Microphysics and Biophysics

Francis M. Sanchez

Abstract While the *reductionist* approach, best expressed in Descartes' *Method*, helped science to develop along the *objectivity* principle, the modern crisis in quantum mechanics and cosmology calls for a return to a traditional *holistic* viewpoint, where the *large* would explain the *small*. This could lead to replacing the concept of 'emergence' (where the *whole* exceeds the *parts*) by that of 'immergence' (foreseen in Mach's conjecture). This implies a temporal invariance of the cosmological parameters defined by applying the Bekenstein-Hawking holographic principle. This latter is associated with a *coherence principle* according to which any well-defined system (such as a living organism) is associated with a specific frequency, analogous to the clock of a computer. Physical laws would then be related to a computing process. This coherence principle is shown to be central in atomic physics and defines Coherent Cosmology, which can be seen as a synthesis of standard cosmology and steady-state cosmology, completed by a 'Grandcosmos' extending the observable Universe radius by a factor 10^{61} and associated with the Cosmic Microwave Background (CMB). For the observable Universe, there is a specific frequency of 10^{104} Hz, introducing a quantization of space-time 10^{61} smaller than Planck's scale. The Universe equivalent mass is expressed in terms of the main three microphysical masses: electron, proton, and hydrogen; all microphysical masses would be submultiples of it. The dimensionless 'large numbers' issued from Cosmology and Microphysics are shown to enter a Topological Axis with an emphasis for 26 dimensions, rehabilitating the Bosonic String Theory and pointing to massive gluons and superspeed signals. The Kotov non-Doppler coherent cosmic oscillation appears as an absolute clock, in holographic connection with the background. Generalized holographic conservation yields the critical condition while the trivial matter density 3/10 solves the dark energy problem. A systematic elimination of c helps to relate the physical parameters to Kotov's well-measured cosmic period: 9600.61(2) s, and c-free standard dimensional analysis confirms the invariance of the Universe horizon, matter density, and background temperature. The later appears related to the

F.M. Sanchez (✉)
Formerly at Université Paris-XI Orsay, Orsay, France
e-mail: hol137@yahoo.fr

A. Tadjer et al. (eds.), *Quantum Systems in Physics, Chemistry, and Biology*,
Progress in Theoretical Chemistry and Physics 30,
DOI 10.1007/978-3-319-50255-7_23

triple-point temperatures of H_2, O_2, and H_2O, and to mammals' temperature through Sternheimer's scale factor: $j = 8\pi^2/\ln 2$, which itself is related to the electric constant: $a \approx 137.036$. Analysis of the masses of DNA nucleotides and protein amino-acids shows a connection with Kotov's period, suggesting that DNA could be a linear hologram. The Darwinian step-by-step macroevolution theory, by unrelated random mutations and natural and sexual selection, seems then irrelevant. We have also investigated the relations between physical canonical large numbers and economic and musical numbers, hinting that the human brain may act as a multi-basis computer, favoring the universality of Intelligent Life.

Keywords Holographic principle • Steady-state cosmology • Coherent cosmology • Antimatter • Dark matter • Dark energy • Eddington theory • Combinatorial physics • Photon mass • Graviton mass • String theory • DNA • Protein • Cosmobiology

1 Introduction

There is a widespread opinion that something went wrong in contemporary standard cosmology. In a recent perspective [1] it can be read: "No one yet knows how the theoretical illnesses of cosmology will be solved, if they can be solved, or even if they need to be solved. As more conventional attempts to find solutions failed to make headway, however, it becomes tempting to try more radical ideas. As evidenced by past paradigm shifts in physics, radical ideas are often necessary for progress, and we, as a community, must be open to their exploration. Certainly, there is no point in being dogmatic about Cold Dark Matter (CDM) when there is consensus that it cannot be the full picture. Still, it should be a principled radicalism that we insist upon. Smashing the foundations of the standard cosmological model is all well and good, but the end result cannot be considered successful unless it is a truly predictive theory - one that not only fits the bulk of current and future data, but explains it as a non-trivial consequence of its deeper structure. Simply introducing additional unconstrained degrees of freedom to fit-out deviations will not do. An alternative theory should ideally strengthen the connections between cosmology and the rest of physics too, as CDM has done so ably. Theories with special constructions that disconnect the causes of cosmological phenomena from their possible consequences elsewhere look feeble. But even if evolution, rather than revolution, is needed to fix up CDM, there may still be something to recommend a more radical stance - perhaps a shake-up of our perspective, rather than our theory, is what has been needed all along".

In particular, consider the 'flatness problem', i.e., why the Horizon radius R and the Universe equivalent mass M are tied by a simple relation: $M = Rc^2/2G$. This problem is currently solved by an ad-hoc inflation step, but this introduces other theoretical difficulties [2]. Another issue is the value of the so-called dark energy density: $\Omega_\Lambda = 0.685(17)$ [3, 4]. This is compatible with the trivial value, 7/10, one obtains by combining the gravitational potential energy of a homogeneous sphere: (3/5) GM^2/R, with the above critical condition. This yields (3/10) Mc^2, letting the density 7/10 apart. This seems to indicate that cosmology may be simpler that it is ordinarily believed.

Another intriguing point concerns the Hubble constant. While a recent direct measurement involving supernovae *1a* [5] yields a value 73.8(2) km s^{-1} Mpc^{-1}, the Planck mission result [3, 4] is 67.8(9). These values are discordant but their average is very close to that tied to the so-called universe age, 13.81(5) Gyr. Such a direct correspondence is obtained in the *single time-invariant parameter* steady-state cosmology [6, 7], while the standard cosmology optimizes *six time-dependent parameters*.

The forgotten steady-state cosmology had correctly foreseen the acceleration of the galaxy recession and the critical flatness. The main argument that led to its discarding, the discovery of the Cosmic Microwave Background (CMB), was in fact not pertinent. Indeed, the steady-state cosmology is the only one that predicted correctly its temperature, from the Helium density [8]. From Oort's estimate of 10^{-30} g cm^{-3} for real matter density, and from the energy associated with each Helium atom formation, one obtains about 3 K in a single line of calculus, as shown below. By contrast, involved computations from the primordial Big Bang model, with transition from the cold to the hot Big Bang, led to temperatures between 5 and 19 K [9].

Contrary to the primordial Big Bang model, the steady-state model is highly refutable, which is a standard criterion for a theory to be scientific. Opponents to the latter proposed various arguments, which later appeared to be disputable [10]. However, it is true that the founders of the steady-state cosmology embarked in a search for a thermalizing agent such as metal or carbon whiskers [8], which were not convincing. This was the main cause for the rejection of their cosmology. But this objection also is not pertinent, because a kind of 'Grandcosmos' could play that thermalizing role, as will be explained below. So observations of CMB, which seem to confirm the standard model, could be a mere misinterpretation. It is significant that opponents concentrated their efforts on the problem of the thermal agent. The irony is that standard cosmology lead to the hypothesis of a *Multiverse* [11], which is unobservable, while a 'Grandcosmos' could be manifested by the CMB.

A delicate point in the steady-state cosmology is that, as a consequence of its basic assumption, the *Perfect Cosmological Principle,* new matter must appear to compensate for the galaxy recession. This has been seen as a violation of energy conservation. But it is not really so, since *in an invariant horizon the energy must remains invariant*. It is true that this new matter rate production is too tiny to be directly measurable. But it implies a coherence of the whole universe, implying a tachyonic physics [12] tied to quantum non-locality and, eventually, a discrete and deterministic physics [13]. A new matter apparition could be related to the strange observations reported by Arp [14].

Another dramatic observation is the non-Doppler oscillation [15], with a period $t_{cc} \approx 9600.61$ s, observed in several quasars since decades by Kotov and Lyuty. It will be shown that this is directly related to the gravitational and Fermi constants

G and G_F. So, steady-state cosmology, a highly refutable model, not only has not been refuted, but has shown to be very predictive.

There seems to be a serious objection against the steady-state model: for a given observer A, the celerity of a galaxy can exceed c when it passes across the horizon of A, while for another observer B this is not so, since the horizons A and B are different. This will be discussed in Sect. 9, where it will be assumed there is an absolute clock related to a CMB absolute space (or preferred Galilean frame).

Now, what is the meaning of the expression: 'Expanding Universe'? If one defines the Universe as the totality of everything, it is a contradiction since one cannot answer the question: 'In what is the Universe expanding?' But with the assumption of a Grand-cosmos, the situation is clearer. However, since the horizon radius is time-invariant the term 'expansion' should be replaced by 'galactic recession'. Indeed, assuming a repulsive force between two galaxies of masses m_1 and m_2 proportional to their distance l, its simplest expression is: $\sqrt{(m_1 m_2)}\ l/T^2$, where $T = R/c$ is the only free parameter in the steady-state model. This force corresponds to an exponential recession, and exceeds the gravitational force for a distance superior to $(\sqrt{(m_1 m_2)} G T^2)^{1/3}$: this is of the order of 10^6 light years, the dimension of a galaxy group. Neglecting this simple argument led to a historical misconception: Lemaître and Hubble had taken into account galaxies belonging to the Local Group, and their value of the corresponding 'Hubble constant' was underestimated by an order of magnitude. Actually, the diagram shown by Hubble was anything but a straight line and was supported by a single distant galaxy studied by Humason, his assistant at Mount Wilson observatory [8].

There remains to explain the large success of the standard, primordial Big Bang theory, called 'Λ-CDM concordance model', with a cold 'dark matter' (CDM) and a repulsive 'dark energy' tied to the so-called cosmological 'constant' Λ. In this paper we show that the two cosmologies are mutually compatible if one replaces the primordial Big-Bang phenomenon by a very fast cosmic oscillation. We first proposed this model in 2011 [16] and here we develop its implications in microphysics and in biophysics.

Section 2 is a quick reappraisal of the cosmology foundations, suggesting a model of *ultrafast computing Cosmos*, implying Diophantine physical laws.

Section 3 recalls basic c-free definitions, with proposals for the horizon radius R and cosmic oscillation period t_{cc}. This shows a dramatic symmetry between *Newton and Fermi constants*, related to the parity violation.

Section 4 is devoted to a coherence analysis of the Universe, showing that the critical condition is a mere application of the standard holography principle and leading to the *replacement of the standard inflation by a very fast oscillation*. This is tied to a quantization of length-time and under-quantization of mass, related to the vacuum energy. This latter is known to be 10^{122} larger than the visible energy, the largest discrepancy in theoretical physics.

Section 5 presents a 'Black Atom' model revealing the *tight connections between micro- and macro-physics*, related to properties of the electric coefficient $a \approx 137.036$.

Section 6 is devoted to *holographic two-step interaction*, leading to a proposal for photon and graviton masses with a gravity speed C_g exceeding c by $\approx 2.47 \times 10^{36}$.

Section 7 presents an approach of cosmology from the *point of view of a quantum system*, with a model of '*gravitational* H_2' pointing to an understanding of dark matter.

Section 8 is devoted to the *combinatorial hierarchy*, suggesting that the physical parameter values are not due to chance, contradicting the *Multiverse* hypothesis.

Section 9 shows special holographic relations, merging into a *topological axis*, this rehabilitating ultrafast bosonic string theory.

Section 10 is an introduction to *Cosmobiology*, showing that significant combinations of masses of nucleotide bases and amino-acids are related to 'magic numbers' occurring in cosmology and microphysics.

Section 11 presents the *Harmonic Principle*, confirming that fundamental laws are arithmetical in nature.

A conclusion summarizes the misconceptions that have led to the present blockage of theoretical physics and cosmology, and recalls general principles to be used in the search for a Diophantian Grand Theory.

2 A Reappraisal of Cosmology Foundations: Coherence Principle

In this paragraph, we discuss the usual reductionist approach of the Universe as an ensemble of elementary particles in statistical c-limited interactions, tied by differential equations.

According to Poincaré, the laws of physics must be invariant, a premonition of our Perfect Cosmological Principle, which adds to space homogeneity time regularity, the basis of steady-state cosmology.

More generally, the very concept of physical law implies a calculus behind it. This is in contradiction with the usual statistical interpretation of quantum physics, but will be confirmed by 'coherence analysis' (Sect. 4). Note that Poincaré was first to stress that the hypothesis of quanta is the only one that can lead to the Planck law [17]. Later on, this specialist in differential equations claimed that physics can no longer be founded solely on differential equations [18].

There are two kinds of laws: local and global. The first laws are of differential type and sensitive to boundary or initial conditions. Thus they cannot be applied consistently to cosmology, for the observable Universe is unique and, as Poincaré first noticed, free parameters would be involved [18]. The second type is that of conservation laws, with no free parameters: for instance, energy conservation in a closed system, a phenomenon that is not really understood from its classical association with a homogeneous time. But if one introduces a *coherence principle*, stating that a 'well-defined' system is vibrating with an invariant frequency f (for instance, a matter-antimatter vibration [16, 19]), then the meaning of energy conservation is that energy is associated with frequency, a more basic concept. Now, an invariant frequency is an essential requirement in holography. This technique is the most efficient way to deal with information and it can be shown to correspond to global conservation laws.

Interestingly, independently of the present 'Coherence Principle' and of the earlier arithmetic 'Holic Principle' [19], other physicists introduced a reduced 'Holographic Principle' [20], generally limited to the consideration of a single holographic unit: the Planck area. But we have shown [21, 22] that other units, especially *the linear Planck length* as well as fundamental particle and cosmic wavelengths, enter such holographic conservation relations.

But the essential point for applying holography has been overlooked: holography needs complete coherence of all waves, meaning a single frequency at work, and this is not possible if the Universe is limited by c, far too slow to insure coherence in the Universe.

In addition, the so-called 'wave-particle dualism' has never been really explained. The experimental fact is that *matter, as light, propagates by waves but is absorbed by quanta*. The simplest explanation is that rapid precursors 'analyze the situation' before the quantum effect takes place [23]. Non-locality is thus essential in wave mechanics. One cannot understand quantum physics without involving a superspeed cosmology. One of the goals of the present article is to compute this celerity.

3 The Fundamental Formula: Evidence for Ultrafast Sweeping

In some of the following definitions, c is eliminated. Here, $a \equiv \hbar c/q_e^2 \approx 137.036$ is the inverse fine-structure constant and $\lambda\!\!\!^{-}_e \equiv \hbar/m_e c \equiv ct_e$ is the reduced electron Compton wavelength. Moreover, a_G and a_w are the gravitational and electroweak analogs of a in the famous article of Carr and Rees [11], except that these authors choose the gravitational force between two protons while we consider the force between a proton and a hydrogen atom, which will be justified in our gravitational hydrogen molecule model (Sect. 7). If M and R designate the equivalent mass and radius of the Universe, and $r_H^{(0)}$ is the bare Bohr radius and m_P the Planck mass, one defines:

$$r_H^{(0)} \equiv a\lambda\!\!\!^{-}_e \tag{3.1}$$

$$a_G \equiv \hbar c/Gm_p m_H \tag{3.2}$$

$$R/2 \equiv a_G \lambda_e \qquad (3.3)$$

$$m_P^4 \equiv M m_e m_p m_H \qquad (3.4)$$

$$a_w \equiv \lambda_e^2 \hbar c / G_F \qquad (3.5)$$

$$t_{cc} \equiv \sqrt{(a_G a_w)} t_e \qquad (3.6)$$

The elimination of c is what is expected in a Coherent Universe, for this speed is clearly too small to connect such a vast space. For this reason, in order to explain the homogeneity of CMB, the standard cosmology invokes an *ad-hoc* superfast inflation. It is of course more logical to invoke quantum non-locality. Interestingly, the associated time R/c is exactly, apart a factor 2 that is explained below, the so-called Universe age. We predicted it, 18 years before its recent 0.3% determination, by using c-free dimensional analysis and the symmetrical product of electron-proton-neutron masses. The reason why this simple calculation was not made earlier is probably because the common setting $c = 1$ in formulas excluded c-free dimensional analysis. Note that the identification of the concepts of space and time was earlier criticized by Poincaré [18]. However, this 0.3% correlation also means there is something right in standard cosmology, which has to be taken into account in other models.

Since the Fermi constant G_F, the associated Fermi mass: $m_F \approx 573\ 007.33(25)\ m_e$, and the cosmic period t_{cc} are 100 times better defined than G, we have derived a more accurate value of G, G' (Table 1), whose tabulated value [3] was an average between widely spread measurements. The corresponding value for R is:

$$R \equiv 2G_F t_{cc}^2 / m_e \lambda_e^4 \equiv 2\hbar^2 / G' m_e m_p m_H \approx 13.8123 \text{ Gly} \qquad (3.7)$$

corresponding respectively to $\hbar$-free and c-free definitions.

Combined with the critical condition, this corresponds to the following symmetric relation, solving the 'Large Number Problem' and making very precise the well-known fact [11] that a_G is of order *of* W^8, W and Z being masses of the weak bosons relative to that of the electron:

$$R/2\lambda_H \equiv \sqrt{(M/m'_e)} \equiv \hbar c / G m_e m_p \approx (WZ)^4/2 \approx \sqrt{(10/\pi_{Pt})} \times 2^{137} \qquad (3.8)$$

Note that this corresponds to a special case of Eddington's formula [24]: $R/2\sigma = \sqrt{N}$, with the identifications: $\sigma \equiv \lambda_H$ and $N \equiv M/m_e$,$m'_e \equiv m_e m_p/(m_p + m_e)$ being the reduced electron mass. This could mean that the electron is a basic stuff of the Universe. One may notice that the Ptolemaic approximation: $\pi_{Pt} \approx 2 + 137/120$, appears in Eq. (3.8). This yields the more accurate value G':

Table 1 Some physical constants and particle properties

Name	Symbol	Formula	Dimension	Value	Unit
Gravitational constant (experimental)	G	$F_{gr} = Gmm'/d^2$	$M^{-1}L^3T^{-2}$	$6.6738(8) \times 10^{-11}$	$N\ m^2\ kg^{-2}$
Gravitational constant (optimized)	G'	*idem*	*idem*	6.67546×10^{-11}	*idem*
Light velocity	c	Constant in all inertial frames	LT^{-1}	299 792 458	$m\ s^{-1}$
Reduced Planck constant	$\hbar = h/2\pi$	$E = \hbar\omega$	ML^2T^{-1}	$1.05457173(5) \times 10^{-34}$	J s
Electron rest mass	m_e	Negative for positrons	M	$9.1093829(4) \times 10^{-31}$	kg
Electrical constant	$a = \alpha^{-1}$	$F_{el} = \hbar c/ad^2$	Dimensionless	137.03599907(4)	Pure number
Fermi constant	G_F	$\hbar^3/cm_F^2$	ML^5T^{-2}	$1.4358509(7) \times 10^{-62}$	$J\ m^3$
Gravitational Sanchez const.	a_G	$\hbar c/Gm_pm_H$	Dimensionless	$1.6919335 \times 10^{+38}$	Pure number
Classical electron radius	r_e	$\hbar/am_ec$	L	$2.8179403 \times 10^{-15}$	m
Compton electron radius	λ_e	$\hbar/m_ec$	L	$3.8615926 \times 10^{-13}$	m
Bare Hydrogen Bohr radius	$r_B^{(0)}$	$a\hbar/m_ec$	L	5.291772×10^{-11}	m
Hydrogen-electron mass ratio	H	m_H/m_e	Dimensionless	1837.152645	Pure number
Proton-electron mass ratio	p	m_p/m_e	Dimensionless	1836.152672	Pure number
Neutron-electron mass ratio	n	m_n/m_e	Dimensionless	1838.683659	Pure number

$$G' \approx 6.6754552 \times 10^{-11}\,\mathrm{kg}^{-1}\,\mathrm{m}^3\,\mathrm{s}^{-2} \tag{3.9}$$

This value is consistent with the relation: $p_G = m_P/2^{63.5} m_e \approx 2^{12}/\sqrt{5}$ (accurate to <140 ppm), implying a role of the Babylonian value: $\pi_{Bb} \approx 25/8$. Indeed, a systematic search on the computer reveals the following (10^{-9} accurate) relation, where *H*, *p* and *n* are the mass ratios of the hydrogen, proton and neutron to the electron (Table 1):

$$\pi_{Pt}/\pi_{Bb} \approx 6\pi^5 pH^9/p_G^4 n^7 \tag{3.10}$$

The above definitions also imply the relation:

$$\sqrt{(G'G_F)} \equiv (\lambda_e^2/t_{cc})\,\hbar/\sqrt{(m_p m_H)}, \tag{3.11}$$

showing two terms that are *both area speeds*, characteristic of the second Kepler law. This is significant of a sweeping construction-deconstruction of the Universe through a single point (that we called 'Hol' [19]), corresponding to *zero-dimension holography*. Since such a sweep is necessarily oriented this may be the source of right-left dissymmetry, called 'parity violation' in particle physics and 'chirality' in biophysics.

Note that the common belief that quantum physics is limited to the microphysics is false, since the Pauli exclusion principle also enters the calculation of a star radius via the concept of energy degeneracy. Also, if all atoms in a star are identical, a question is the limit of the star radius when the number of atoms goes to unity. The following calculation of a star radius was given by Davies [25].

A ball of gas of radius *R* remains in equilibrium if its self-gravity is balanced by the combined effect of its internal thermal pressure and its electron degeneracy pressure. This is the case if the gravitational energy per particle is comparable to the sum of the thermal and degeneracy energies. For a hydrogen gas this implies:

$$k\theta + N^{2/3}\hbar^2/m_e R^2 \sim GMm_p/R \tag{3.12}$$

with $N = M/m_p$. At low density (large *R*), the second term is small, and the temperature θ is then inversely proportional to *R*. This is the case when the star starts forming from a slowly contracting cloud of gas. Eventually, when the radius shrinks, the degeneracy term becomes important, and the temperature reaches a maximum when:

$$Gm_p^2 N/R \sim N^{2/3}\hbar^2/m_e R^2 \tag{3.13}$$

is largest. This occurs for:

$$R \sim 2\hbar^2/Gm_p^2 m_e N^{1/3} \quad (3.14)$$

which is, for N going to unity, the above redshift radius (3.3), except for a hydrogen/proton mass ratio.

It is recalled that Eddington's prediction [24] for the number of hydrogen atoms in the Universe is 136 × 2^{256}, a prediction that is consistent with the concordance value: $T \approx 13.80(5)$ Gy. Taking account of the 3/10 relative density for matter, this writes:

$$M_{mat}/m_H = (3/10)Tc^3/2Gm_H \approx 2^{256} \times 136.2(5) \quad (3.15)$$

The accuracy reaches 20 ppm if one uses G' and the neutron mass instead of Eddington's m_H So, *dark matter would in fact be similar to ordinary matter*. But as these two varieties of matter are not photon-interacting, it would mean they are vibrating in quadrature. In Sect. 7 we relate this conjecture to a similar one for antimatter.

4 Coherence Analysis: A Computing Cosmos

4.1 The General Coherence Condition

Various authors have advanced the hypothesis that the laws of physics result from a calculation process [26]. This is sustained by the properties of cellular automates [27]. Moreover, Gerard't Hooft has shown that quantum field theory can be adapted to deal with a deterministic cellular automaton [28]. This suggests that behind the so-called 'indeterminacy' of quantum physics a 'deterministic' process is at work. This induces the following 'coherence analysis', where numerical coefficients are first omitted for simplicity.

Consider a critical Universe with radius horizon R. Filling the interior sphere with observers of virtual mass m (remember the vacuum is not really empty), this forms a volume referential superseding the ordinary three-axis frame. We define a 'coherence domain' associated to the mass m as: $\lambda_m \equiv \hbar/cm$. The total mass is limited by the critical condition $M = Rc^2/2G$, so the number N_{obs} of observers is limited to the value $R\lambda_m/2l_P^2$. Note that this critical condition currently applies to a black hole, and is considered as a limitation for preventing collapse. This formula also applies to the Universe but for the latter, *galaxy recession prevents such a collapse*. Calling d the mean distance between observers, the number of observers is:

$$N_{obs} \sim (R/d)^3, \quad (4.1)$$

yielding:

$$(Rl_P)^2 \sim \lambda\!\!\!^{-}{}_m d^3. \tag{4.2}$$

This General Condition will be applied in four different ways.

4.2 *The Global Coherence Condition and the Large Number Problem*

With the global coherence condition $\lambda\!\!\!^{-}{}_m \sim R$, one gets $N_{obs} \sim (R/l_P)^2$ and:

$$d \sim \left(Rl_p^2\right)^{1/3} \sim 10^{-15}\,\mathrm{m}, \tag{4.3}$$

a result also obtained by Ng [29] considering the Universe as a 'grand parallel computer' while maintaining c-limitation. In contrast, we consider a coherent and sequential superspeed Universe. The length 10^{-15} m obtained has no significance in the standard R-variable scheme, but it is close to both the nuclear scale and classical electron radius r_e. This could be at the origin of the 'Large Number Hint', considered as a problem in the frame of the variability of R, hence the so-called 'Anthropic Principle' [11]. Note that the radius r_e^3/l_P^2 corresponds to the elimination of c between r_e and l_P. Moreover, it writes as a function of the Nambu mass: $m_N = am_e$, which plays a central role in particle physics [30]. We then introduce the following radius:

$$R' = 2\hbar^2/Gm_N^3, \tag{4.4}$$

the factor 2 coming from the fact that the associated critical mass is then very simple: $M' = m_P{}^4/m_N{}^3$. The radius R' is slightly larger than R:

$$R'/R = m_e m_p m_H/m_N^3 \approx 1.31084. \tag{4.5}$$

The simplest interpretation is to view R' as a holographic equivalent of a *Grandcosmos* beyond, as detailed below. This ratio being close to 4/3, we can assume the following half-sphere holographic quasi-conservation of the Bekenstein-Hawking entropy [20]:

$$S_{BH} = \pi(R/l_P)^2 \approx (2\pi/3)(R/r_e)^3. \tag{4.6}$$

4.3 The Single-Observer Condition: Critical Condition, General Quantization and Universe Size

With $N_{obs} \sim 1$, or the condition $d \sim R$, one obtains:

$$\lambda_m = \lambda_M = \hbar/cM = 2l_P{}^2/R \sim 10^{-95}\,\text{m}. \tag{4.7}$$

This can be seen as a 'Universe wavelength', of central importance since it enters the following *holographic form of the critical condition, $R = 2GM/c^2$*:

$$\pi(R/l_P)^2 = 2\pi R/\lambda_M. \tag{4.8}$$

It is to be noted that this goes beyond the standard limitation of length by the Planck unit l_P while the standard holographic principle [20] involves only the area l_P^2.

If one considers that any particle mass $m = M/N_m$ is a sub-multiple of the total Universe mass M then the associated wavelength λ_m is a whole multiple N_m of λ_M, which allows extending the above holographic conservation in the following manner:

$$S_{BH} = \pi(R/l_P)^2 = 2\pi R/\lambda_M = 2\pi N_m R/\lambda_m. \tag{4.9}$$

This set of circles generates the approximation of a sphere. But for this approach to be acceptable the N_m must be large numbers. So the considerable vastness of the Universe receives a justification better than the standard one, which states that the initial Big Bang conditions were adjusted to 10^{-60} or so.

Note that the characteristic mass $m_0 = \hbar/Rc \approx 2.69 \times 10^{-69}$ kg is not a quantum, but a sub-quantum: $m_0 = M/N_0$, of the total mass M, with $N_0 = (R/l_P)^2/2$. This provides an interpretation of the standard Bekenstein-Hawking entropy [20] apart for a factor $\pi/2$. This is sustained by the 2% accurate formula (on a number of the order 10^{61}):

$$\sqrt{S_{BH}} = (\pi/2)\,\hat{}\,(F/\sqrt{pn}), \tag{4.10}$$

where the exponent is the Fermi mass relative to the mean proton-neutron mass.

4.4 The Standard Coherence Condition: Grandcosmos and Vacuum Energy

In standard physics, the lower limit to a spatial dimension is the Planck length l_P. With the condition $d \sim l_P$, one obtains:

$$\lambda_m \sim R^2/l_P \sim 10^{87}\,\text{m} \sim R_{GC} \tag{4.11}$$

This defines a length of the order of a *Grandcosmos* radius defined as follows. Applying the monochromatic holographic principle to the above sphere of radius R', with l_P being the monochrome unit:

$$\pi(R'/l_P)^2 = 2\pi R_{GC}/l_P, \tag{4.12}$$

we define a radius: $R_{GC} = R'^2/2l_P \approx 6.945 \times 10^{60}\ R$.

Assuming this 'Grandcosmos' is closed by a critical condition with superspeed C, the uniformity of equivalent matter density in the Universe implies: $C/c = R_{GC}/R$. So a mass m is associated with two energies, the standard one mc^2 and the superspeed one mC^2, with a ratio $(C/c)^2 \sim 10^{122}$. This may be a solution to a main problem in contemporary physics: the vacuum energy, which shows up in the Casimir effect [31, 32] but is 10^{122} larger than the visible energy. The expression of the 'Grandcosmos' volume in terms of the Bohr radius takes the form:

$$(4\pi/3)(R_{GC}/r_H)^3 \approx a^a/\pi \ \ approx\,(1/ln2)^{\sqrt{(\mathrm{pH})}}. \tag{4.13}$$

A simple hypothesis is that the 'Grandcosmos' is the source of the cosmic microwave background. Indeed, R' appears to be linked to the Wien CMB wavelength, for one has (within 0.1%):

$$4\pi(R'/l_{Wien})^2 \approx e^a. \tag{4.14}$$

This casts a doubt on the general belief that a thermal field looses information.

4.5 The Field Coherence Condition: CMB and Biology

With the field coherence condition: $\lambda_m \sim d$, one obtains:

$$\lambda_m \sim d \sim (Rl_P)^{1/2} \sim 10^{-4}\mathrm{m}, \tag{4.15}$$

which is of the order of the CMB wavelength, but with a significant departure that will be interpreted in connection with the identification of specific cosmic parameters with biological parameters (Sect. 10). This means that:

$$N_{obs} \sim (R/l_P)^{3/2} \sim (\lambda_m/l_P)^3, \tag{4.16}$$

showing another generalization of the standard holographic principle, since now the volume of the redshift sphere (in Planck length unit) is involved.

5 The 'Black Atom' Model

The 'black atom' model [16] considers a hydrogen atom immersed in a black hole of radius R_{ba}, limiting electron circular trajectories. The intermediate space is paved with spheres of radius $r_n = n\lambda$, where $\lambda = \hbar/m_e c$ and n is an integer. The corresponding electron velocities are derived from $\hbar = m_e r_n v_n$, this implying $v_n = c/n$ (the first trajectory: $n = 1$, is excluded). Equating the corrected Bohr radius: $r_H = a\lambda\ (1 + 1/p)$—where p is the proton to electron mass ratio—with the mean radius of the spheres—limited by R_{ba}/λ_e—each with a probability proportional to n^{-2}, one obtains:

$$r_H/\lambda_e = \Sigma(1/n)/\Sigma(1/n^2). \qquad (5.1)$$

With $z \approx 0.422784335$ (the complement to 1 of Euler's constant), this defines a radius:

$$R_{ba} = \lambda_e \exp\,[(\pi^2/6 - 1)\, r_H/\lambda_e + z] \approx 1.4923 \times 10^{26}\,\text{m} \approx 15.775\ \text{Gly}. \qquad (5.2)$$

This is very close to the value of $2\hbar^2/G\ [(ad_e + 2\pi)m_e]^3$, d_e being the abnormal electron magnetic coefficient: $d_e \approx 1.001159652$. The number $a + 2\pi$ is very close to the canonical term of the Planck law: $e^{\omega} \approx 143.3249$, where $\omega = 5(1 - e^{-\omega})$ is the Wien coefficient, i.e. the ratio between the nominal wavelength $hc/k\theta$ and the Wien length. The proximity with $a + 2\pi$ suggests that a is an a angle. Indeed (within 22 ppm), $\cos a \approx 1/e$. Now, a characteristic property (within 1 ppm) is:

$$(ad_e + 2\pi)^3 \approx a^{3/2} m_n^2/m_e m_p, \qquad (5.3)$$

where the neutron and proton masses appear. So there is a relation between R_{ba}, R' and R, specifying the approximation (within 0.25%): $R_{ba} \approx (RR')^{1/2}$. This 'black atom' relation can be approximated by:

$$a/ln(2a_G) \approx (\pi^2/6 - 1)^{-1}, \qquad (5.4)$$

which specifies the earlier proposed rough relation: $a \sim ln(a_G)$ [11].

6 Holographic Two-Step Interaction

As conjectured at the end of Sect. 2, ultrafast wave precursors analyze the situation before 'deciding' where a quantum effect will arise [22]. It is to be noted that even the electromagnetic interaction is not yet really understood [33]. Consider, for simplicity, two identical systems of mass m in their ground state. The first system is characterized by a stationary wave which may be seen as the sum of diverging and converging waves: $s + s^*$, with $s = \exp\,[2\pi i f \cdot (t - r/c)]$, where f is the proper angular frequency: $mc^2/\hbar$. The second system is characterized by an analogous standing wave: $r + r^*$. Assuming that the vacuum is *not* empty, an hologram is formed: $(s + s^*)\,(r + r^*)$, which includes the *resonant* sum: $sr^* + s^*r$. So, the very presence of two systems creates an inhomogeneity in space.

Now, if the first system has an excess of energy, it means it is receiving an excess signal of a form proportional to s^*. By diffraction on the above hologram, it gives rise to: $s^*(sr^* + s^*r)$, with a resonant term r^*. Note the similarity between this holographic formalism and the unitary matrices of quantum physics: $ss^* = 1$. But the above argument shows that *convergent waves are of primordial importance, rather than the current diverging waves*. As the process is symmetrical, this leads to an oscillation. This is known as the particle exchange (implying a boson with mass m_B) associated with any interaction. But here it is assumed that the boson has a superspeed C_B. The resonance condition is that the wavelengths are identical, in analogy with Gabor's holographic microscopy condition [34]. So, for the electron:

$$\lambda_e = \hbar/m_e c = \hbar/m_B C_B. \tag{6.1}$$

In accordance with the previously discussed importance of R/r_H, one may try:

$$R/r_H = C_B/c = m_e/m_B \approx 2.47 \times 10^{36}. \tag{6.2}$$

This could define a gravitational speed, associated with a graviton mass:

$$m_{gr} = m_e r_H/R = a m_0 \approx 3.689 \times 10^{-67}\ \text{kg}, \tag{6.3}$$

where $m_0 = \hbar/Rc \approx 2.69 \times 10^{-69}$ kg is a Universe 'quantic mass' (not quantum of mass), much smaller than the Planck mass.

By extending the argument to the electroweak interaction, with characteristic mass $m_w = a_w m_e$, one may write:

$$R/r_H = C_B/c = m_w/m_B' \approx 2.47 \times 10^{36}. \tag{6.4}$$

This would define a photon mass:

$$m_{ph} = m_w r_H/R \approx 1.211 \times 10^{-55}\text{kg} \approx m_Z m_e/2^{24} m_P, \tag{6.5}$$

thus relating (within 90 ppm) to the mass of the Z neutral weak boson, which is known to be linked to the photon in electroweak theory. This value is much larger than that assessed by Maruani [35]: $m_{ph} \approx 1.355 \times 10^{-67}$ kg. But it is very close to that proposed by Marchal in terms of the 'cosmic oscillation' period $t_{cc} \approx 9600.60$ s [36]:

$$m_{ph} = \hbar/c^2 t_{cc} \sim 1.222 \times 10^{-55}\text{kg}. \tag{6.6}$$

The present-day [3, 33] *maximal* assessed value for m_{ph} is 1.8×10^{-54} kg.

7 The Universe as a Quantum System

7.1 The Basic Hydrogen Spectrum

Three years before Niels Bohr, Arthur Haas had equalized three forms of energy: the kinetic, the potential, and the quantum *nhf* using the frequency of the electron rotation: $nhv_e/2\pi r = n\hbar v_e/r$, in a 2D circular model of an electron orbiting around a proton with speed v_e on a circle of radius r [37]. In fact, from the virial theorem twice the kinetic energy ought to be considered. Neglecting first the equivalent mass correction in this two-body system, one obtains:

$$m_e v_e^2 = \hbar c/ar = n\hbar v/r_n. \tag{7.1}$$

Here, $a \approx 137.036$ is directly involved in the electric force between two elementary charges: $(q_e/r)^2 = \hbar c/ar^2$, yielding $a = \hbar c/q_e^2$ (the inverse αv_e of a is commonly called 'fine-structure constant'). Note that the so-called 'electric vacuum permittivity' and 'electric charge unit' are misleading concepts: indeed, as any electric force is a whole multiple of this unitary force, the choice of a specific unit for the electric charge is not necessary. Any electric charge is related *to a whole quantum number.*

The above relations contain the Bohr quantum relation $n\hbar = r_n m_e v_e$, yielding:

$$v_{en} = c/an, \tag{7.2}$$

$$r_n^{(0)} = n^2 a\hbar/cm_e \equiv n^2 a\lambda_e. \tag{7.3}$$

In fact, Haas was the first to apply what we call here the 'Coherence Principle' but, using the true kinetic energy, he obtained twice the correct value for r_n, and especially for the bare Bohr radius: $r_1 = r_H^{(0)} = a\lambda$. Note that with the mass correction, the real Bohr radius is: $r_H = r_H^{(0)} \times (1+ m_e/m_p) \approx r_H^{(0)} \times (H/p)$, p and H being the proton and hydrogen masses by respect to the electron mass.

7.2 The Gravitational 'Hydrogen Molecule'

Now, neglecting the electrostatic and exchange interactions, consider a proton coupled to a hydrogen molecule orbiting by gravitation on a circle of *invariant radius R*, where an electron is also circulating with velocity v_e. The gravitational potential energy can be written: $Gm_H m_p/2R$, but it can be written in the same form as above by introducing the gravitational interaction constant $a_G = \hbar c/\mathrm{G} m_H \mathrm{m}_p$. In this three-body system, the Coherence Principle gives, for $n = 1$:

$$v_e = c/2a_G, \tag{7.4}$$

$$R = 2a_G \lambda_e = 2\hbar^2/Gm_e m_H m_p \approx 13.812 \text{ Gly}, \tag{7.5}$$

which shows that the above definition is compatible with the 3-ppt accurate so-called Universe age: 13.81(5) Gy [4, 5].

By adding the standard critical condition or, equivalently, the Schwarzschild radius formula of a black hole horizon: $R = 2GM/c^2$, this can be written, using the reduced mass $m_e' = m_e m_p/(m_p+m_e)$:

$$R/2\lambda_H = \sqrt{(M/m_e')} = \hbar c/Gm_e m_p, \tag{7.6}$$

which is, as recalled above, Eddington's statistical formula [24]: $R/2\sigma = \sqrt{(M/m)}$, with the identifications: $\sigma = \lambda_H \equiv \hbar/m_H c$ and $m = m_e'$.

This responds to Carr and Rees questioning [11], who stated that current physics cannot explain the *Large Number Correlation*. Note that Eddington did not recognize this very symmetric identification because, at his epoch, the Hubble radius was under-estimated by an order of magnitude. It can be recalled that Eddington's basic argument was that in a black hole of radius *R*, the position of a particle is uncertain by a length *R*/2. If one has *N* particles, this is reduced by a statistical factor $\sqrt{N}$, yielding a reduced length $R/2\sqrt{N}$ which Eddington associated with the nuclear force range. But the above equation shows that it is rather the reduced H wavelength. The surprise comes from *N*, an equivalent number of electrons, as if everything in the Universe would be defined by electrons or if a single sweeping electron-positron pair would define it all.

7.3 Quantum Universe and Real Matter

The previous Section was limited to the case: $n = m_e R v_e/\hbar = 1$, and yielded the radius of the observable Universe. This suggests the existence of an external *Grandcosmos*, for larger values of *n*. We now assume that a single equivalent electron is associated with a celerity V_e obeying the Coherence Principle applied to the energy Mc^2:

$$m_e V_e^2 = Mc^2. \tag{7.7}$$

The question is: 'What is the corresponding quantum number $n = m_e R V_e/\hbar$'? This can be written, taking into account the above Eddington statistical relation:

$$(n\hbar/m_e R)^2 = c^2 M/m_e = (c^2 n\hbar/G m_e m_p)^2, \tag{7.8}$$

which shows a symmetry (m, $-m$) expressing the *double solution matter-antimatter*:

$$n\hbar/m_e R = \pm \hbar c^2/G m_e m_p. \tag{7.9}$$

Limiting to positive values, this leads to:

$$n = Rc^2/G m_H = 2M/m_H, \tag{7.10}$$

which is the overall number of particles (electrons + protons) in a sphere of radius R, and a natural quantum number previously used by Eddington [24]. This is a validation of the Coherence Principle, for which an equipartition of the energy $m_e V_e^2$ among the M/m_H electrons leads to an elementary kinetic term: $m_e v_e^2 = m_H c^2$, this implying:

$$v_e = c\sqrt{(m_H/m_e)}. \tag{7.11}$$

As this is not allowed by Relativity theory and as the liberation velocity at the periphery of a black hole is c, one would rather have there: $v_e \approx c$, i.e.:

$$m_H V_e^{(\mathrm{r})2} \approx Mc^2, \tag{7.12}$$

showing the way the above model can be adjusted. So, consider a reduced number of *real* hydrogen atoms with density $\Omega_{\mathrm{H}}^{(\mathrm{r})}$. The corresponding quantum number is: $n^{(r)} = 2\Omega_H^{(r)} M/m_H = m_e R V_e/\hbar$, corresponding to: $V_e = 2\Omega_H^{(r)} M\hbar/R m_e m_H$; then the kinetic term becomes:

$$m_e V_e^2 = \Omega_{\mathrm{H}}^{(\mathrm{r})\,2} Mc^2. \tag{7.13}$$

In order to satisfy Eq. (7.12), one has:

$$\Omega_{\mathrm{H}}^{(\mathrm{r})} \approx \sqrt{(m_e/m_H)} \approx 0.0233. \tag{7.14}$$

So, the apparently strange fact that the Universe is only scarcely occupied by ordinary matter would come from the large proton to electron ratio. Note that the above density is about half the standard baryonic density [3], but confirms the steady-state cosmology (SSC) model. Indeed, the SSC model does predict a thermal background, resulting from thermalization of stellar radiation. Taking for the helium mass density the standard value 0.252, this means a total helium mass of ≈ $0.252 \times 0.0233 \times M \approx 5.172 \times 10^{50}$ kg, or 7.726×10^{76} helium atoms. For each helium atom, the released energy is $(4m_H - m_{He})c^2 \approx 4.283 \times 10^{-12}$ J. Thus, the total energy is: 3.309×10^{65} J, corresponding to an energy density within the R-sphere of: 3.541×10^{-14} J m^{-3}. Equalizing this with the black body energy density: $(\pi^2/15)(kT)^4/(\hbar c)^3$, we obtain: $\theta \approx 2.616$ K, which is close enough to the CMB measured temperature: 2.726 K, to confirm the above real matter density.

Now, taking $n_m = \Omega_m M/m_e$, this defines a reduced energy respective to Mc^2:

$$(n_m \hbar/R)^2/m_e = (\Omega_m/2)^2 Mc^2 = > \Omega'_m = (\Omega_m/2)^2 \approx 0.0225, \tag{7.15}$$

which differs only by 3.7% from the above value: $\Omega_H^{(r)} \approx \sqrt{(m_e/m_H)} \approx 0.0233$, for *real* matter density.

8 The Combinatorial Hierarchy

The question arises whether there is a relation between the 3 interaction constants a, a_w, and a_G. An interesting point is the remarkable (0.56% accurate) property:

$$a_G \approx 2^{127} - 1, \tag{8.1}$$

which is a Mersenne prime number belonging to the Catalan series: indeed $127 = 2^7 - 1$, then $7 = 2^3 - 1$, and finally $3 = 2^2 - 1$, are also Mersenne prime numbers. Now their sum is $3 + 7 + 127 = 137$, which is the integer value of a, the number 137 introduced by Eddington. Note that his 'Fundamental Theory' was rejected as soon as it appeared that a is slightly distinct from 137.

The above series is known as the Combinatorial Hierarchy, which ends at the 127th power [38]. Now, 137 and a are related (within 0.12 ppm) by the formula:

$$(137^2 + \pi^2)^{1/2} \approx 137.0360157. \tag{8.2}$$

The measured value: $a \approx 137.035999074(44)$, corresponds to a value of $\pi \approx 3.140863246$. It is to be noted that π appears in the Lenz-Wyler approximation of the proton to electron mass ratio: p (1836.153) ≈ $6\pi^5$ (1836.118). This means that the increment of a relative to the 'magic number' 137 may be related to the ratio of the masses of the two most stable particles in the Universe. This resembles the gyromagnetic ratio of the electron departing from the Dirac value 2 by an expansion in terms of the product $a\pi$ [35, 39]:

$$g_0 \approx 2 + 1 / a\pi \approx 2.002322819\,(\exp 1 \approx 2.002319304). \tag{8.3}$$

Note that the Ptolemaic approximation of π: π ≈ 377/120 = 2 + 137/120 (accurate to 23 ppm) involves the 5th sum of the harmonic series:

$$\begin{aligned}
&1\\
&1+1/2 = 3/2\\
&1+1/2+1/3 = 11/6\\
&1+1/2+1/3+1/4 = 5^2/12\\
&\mathbf{1+1/2+1/3+1/4+1/5 = 137/60}\\
&1+1/2+1/3+1/4+1/5+1/6 = 7^2/20\\
&1+1/2+1/3+1/4+1/5+1/6+1/7 = 3^2 \times 11^2/420
\end{aligned}$$

The prime numbers in this series shows a recurrence of 11:

$$3; 11; 5^2; 137; 7^2; 11^2; \ldots. \tag{8.4}$$

which is the sum of the number 4 of ordinary space-time dimensions and the number 7 of hidden supergravity dimensions. Moreover, 4, 11, and 137, which are the maximal numbers of parts in an n-cutting process, yielding: n (n + 1)/2 + 1 (for n = 2, 4 and 16), are related by:

$$11^2 + 4^2 = 137. \tag{8.5}$$

In addition, 4 = 3 + 1 is the canonical relativistic partition of space-time while 11 = 10 + 1 relates 11, the number of supergravity dimensions, to 10, the superstring number.

As ancient Egyptians used only unitary fractions: $1/n$, they were probably aware of the special properties of 137. Indeed, the hypostyle room between the second and third pillars of the Amon temple in Karnak displays patterns characteristic of the Combinatorial Hierarchy and harmonic series. On each side of the main axis there are columns gathered in groups of 28, 21, 12, and 6. Now, 6 and 28 are 'perfect numbers' while 21 + 12 = 33, the prime number rank of 137. The total number of columns is then: 134 = 7 + 127 = 137 − 3, all these numbers appearing in the Catalan series (see above) and in the completed tetractys: 3 + 7 + 127 = 137.

The electric parameter a is related to the 'magic number' 137 not only through the above relation to π but also through the 'golden ratio' π:

$$137 \ln \pi \approx p \ln a / a \tag{8.6}$$

p being the proton to electron mass ratio. Equivalently, the relativistic factor in the first hydrogen orbit is (within 0.15 ppm):

$$\beta^2 = 1/(1 - 1/a^2) \approx \ln a/\ln 137. \qquad (8.7)$$

Now, a direct relation is found involving the three numbers implying the electron: a, a , and $P = m_P/m$:

$$P^{10} \approx a_w^7 (\sqrt{a})^{134}, \qquad (8.8)$$

which is accurate within 50 ppm. Splitting 10 as 3 + 7 and 134 as 7 + 127, one gets:

$$P^3 (P/a_w\sqrt{a})^7 \approx (\sqrt{a})^{134}, \qquad (8.9)$$

where the neutron to electron mass ratio n appears:

$$P/a_w\sqrt{a} \approx n^3. \qquad (8.10)$$

This relation, accurate within 90 ppm, is encountered in the systematic elimination of c in the cosmic oscillation period [16].

9 Special Holographic Conservation and Topological Axis

The following holographic expression, of type of area of a 4D-sphere: $2\pi^2 r^3$, involves very precisely the CMB wavelength: $\lambda_{CMB} = hc/k\theta_{CMB}$, yielding a temperature compatible with the measured one, $\theta_{CMB} \approx 2.7255(6)$ K:

$$2^{127} \approx 2\pi^2(\lambda_{CMB}/\lambda_e) \times (\lambda_{CMB}/\lambda_H)^2 \;\; => \;\; \theta_{CMB} \approx 2.7258204 \text{ K}. \qquad (9.1)$$

This is confirmed by the following formula involving the Fermi wavelength:

$$F^5 \equiv (\lambda_e/\lambda_F)^5 \approx 6\,(\lambda_{CMB}/\lambda_e)^3 \;\; => \;\; \theta_{CMB} \approx 2.725820(1) \text{ K}. \qquad (9.2)$$

This permits to propose the more accurate Fermi constant: $G_F \approx 1.43585090 \times 10^{-62}$ J m^3, corresponding to the Fermi-electron mass ratio: $F \approx 573007.325$. This is the value we use in what follows, with the corresponding CMB wavelength:

$$\lambda_{CMB} \approx 0.84007165 \text{ mm}. \qquad (9.3)$$

Now, the above formula $R = 2\hbar^2/Gm_e m_p m_H$ may be written in terms of a 1D-2D holographic conservation:

$$2\pi R/\lambda_e \equiv 4\pi\lambda_H\lambda_p/l_P^2, \qquad (9.4)$$

while the connection with $l_{cc} = ct_{cc}$ permits to add a 4D term implying both the Fermi and CCO wavelengths. Moreover, another 4D term clearly involves both the

CMB and neutrino (CNB) wavelengths through the characteristic ratio: $11/4 \equiv (T_{CMB}/T_{CNB})^3$, the cube of their temperature ratio:

$$2\pi R/\lambda_e \equiv 4\pi\lambda_H\lambda_p/l_P^2 \equiv 4\pi((\lambda_F l_{cc})^{1/2}/\lambda_e)^4 \approx 4\pi(\lambda_{CMB}/\lambda_e)^4 \times (11/4)^2\, p6\pi^5/H^2, \quad (9.5)$$

accurate within 0.1 ppm. This calls for a 3D holographic term, which gives the CMB nominal wavelength as a function of the hydrogen molecule:

$$2\pi R/\lambda_e \equiv 4\pi\lambda_H\lambda_p/l_P^2 \approx (4\pi/3)(\lambda_{CMB}/\lambda_{H2})^3. \quad (9.6)$$

Once more, this corresponds to a c-free calculation: starting from the constants G, $\hbar$, and the characteristic energy $k\theta_{CMB}$, one obtains a length close to the hydrogen wavelength with a geometric factor 8/3, inducing directly the above holographic relation.

Looking for a 5D term leads to the relation:

$$R/\lambda_e \approx (2\pi^2 a^3)^5 (H/6\pi^5), \quad (9.7)$$

where $2\pi^2 a^3$ is the area of a 4-sphere of radius a, and also the product of the perimeter by the area of a disk of radius a, a characteristic of 4D space. The correcting factor, involving the hydrogen to electron mass ratio H and the Lenz-Wyler approximation $6\pi^5$ for the proton-electron mass ratio, confirms the above proposed value G', a factor π being eliminated:

$$6R/\lambda_e \approx (2\pi a^3)^5 H \approx \exp(2^{26/4}). \quad (9.8)$$

This displays the bosonic dimension 26 with 1.6 ppm accuracy (see below).

According to the Holic Principle [19], the 210-D term ($2 \times 3 \times 5 \times 7 = 210$) could be relevant. Indeed one has the following relation involving the constant $k = 2R/R' = 2a^3/pH$ (with an accuracy of 15 ppm on k):

$$R/\lambda_e \approx (k)^{2\times3\times5\times7}. \quad (9.9)$$

Another geometric property is:

$$\pi(R/\lambda_e)^2 \approx \pi^{12\times13}, \quad (9.10)$$

accurate within 4.5 ppm. As $(R/\lambda)^2 \approx 2^{256}$, this implies a relation between powers of 2 and π. The number 137 also appears as relating π and p:

$$2^{1/155} \approx \pi^{1/256} \approx (2\pi)^{1/3\times137} \approx (2p)^{1/p} \quad (9.11)$$

This example shows how the consideration of cosmic properties helps to connect physical parameters.

Moreover, the c-free length defined from $\hbar$, G and the Universe mass density is, within a geometrical factor $(8\pi/3)$, very close to $\bar{\lambda}^2/l_p$, suggesting a correspondence between G and G_F:

$$\bar{\lambda}_F\, a_G\{m_e\} \equiv \bar{\lambda}_F \hbar c/Gm_e^{\,2} \approx \sqrt{(8\pi/3)}R \approx 9R'/4. \tag{9.12}$$

So, the simple technique of dimensional analysis yields the temporal invariance of the horizon radius, the background temperature and the matter mean density, leading to a justification of the Perfect Cosmological Principle.

The easiest way to display large numbers from both macro- and micro-physics on a single graph is to use a double logarithmic representation. There appears a regularity that summarizes the holographic conservation laws. By alternating micro- and macro-physical numbers, the holographic relations show the series displayed in Fig. 1. A surprise is that the numeration of the large numbers appears to be the special dimension series of string theory. One can write:

$$\bar{\lambda}_e/d \sim (R/\bar{\lambda}_e)^2 \sim (\bar{\lambda}_e/l_X)^4 \sim (\lambda_{CMB}/\bar{\lambda}_e)^8 \sim (\bar{\lambda}_e/l_W)^{16} \sim (l_{at}/\bar{\lambda}_e)^{32} \sim (\bar{\lambda}_e/l_{Gl})^{64} \sim (l_{string}/\bar{\lambda}_e)^{128} \sim 2^{256}.$$

The first two relations are well known (Weyl, Eddington, Dirac). The third one, which implies CMB, was mentioned by Davies [25]. The fourth one, that implies the intermediary boson, was noticed by Carr and Rees [11]. According to Green et al [40]:

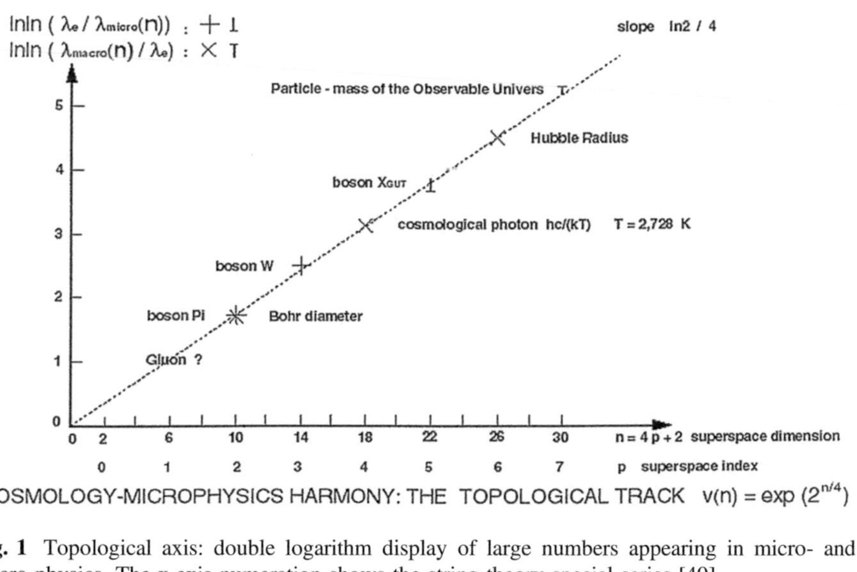

Fig. 1 Topological axis: double logarithm display of large numbers appearing in micro- and macro-physics. The x-axis numeration shows the string theory special series [40]

"In string theory, diffeomorphism anomalies arise from chiral fermions and exist only if space-time has $4p + 2$ *dimensions*". It is precisely this series that appears on the horizontal axis. Note that the gauge bosons W and X have odd p-numbers. Extrapolating to $p = 1$, this predicts a gluon mass of about 10 m_e. For $p = 7$, we get the 'holon', a new gauge boson whose mass would be that of the Universe, maybe linked to the force that repel galaxies. Note that these gauge bosons show a periodicity $\Delta n = 8$, recalling the famous Bott modulo-8 periodicity of the orthogonal group topology [41].

Note the central place of CMB on the Topological Axis. This confirms that CMB defines a special Galilean frame: its slight Doppler dissymmetry reveals an absolute speed with respect to a Grandcosmos, which can be seen as the mereging of R-radius spheres of all possible observers.

According to Polchinski [42]: *A key feature of string theory is that it is not consistent in all space-time backgrounds, but only in those satisfying certain conditions. For bosonic string theory in flat space-time, the spectrum is Lorentz-invariant only if the number of space-time dimensions D = 26. But for D > 2 the state is a tachyon.* The point $n = 26$, characteristic of bosonic string theory, relates to Hubble's radius through: $exp\ (2^{26/4}) \approx 6R/\lambda$ (to <0.7 ppt). The point $n = 10$, characteristic of superstring theory [40], which has been preferred to bosonic string theory for its suppression of tachyons, shows a remarkable micro-/macro-physical symmetry. Extending this to $n = 30$ (the 26 bosonic + 4 space-time dimensions) points to a *Grandcosmos*, correcting the general asymmetry of the scheme.

10 Cosmo-Biological Relations

In order to explain a number of striking correlations between various physical parameters, many people have invoked an *Anthropic Principle*, opening the way to the *Multiverse* conundrum. Strangely enough, tenants of the Anthropic Principle did not notice that some biological constants are related to physical ones. For instance, consider the DNA anhydrous nucleotide masses given in m_{H} units (Table 2). It can be seen that they follow the (0.3-ppt accurate) relation:

$$\mathrm{A} + \mathrm{T} \approx \mathrm{G} + \mathrm{C} - 1. \tag{10.1}$$

As each *bicodon* of the DNA chain is composed of 3 couples from the dual choice AT or GC, this means that bicodon masses are nearly invariant, differing by ±1H, 2H, 3H. This striking fact has apparently remained unnoticed by molecular biologists.

Now it can be seen that the 'bicodon' (6 nucleotides) mean mass m_{bc} is:

Table 2 The 4 DNA nucleotides by order of increasing mass (Uracil, which replaces Thymine for binding to Adenine in RNA, is not represented)

Name	Formula	Symbol	Mass/m_H
anhydrid desoxy-cytidine monophosphate (dCMP)	$C_9H_{12}N_3O_6P$	C	286.935
anhydrid desoxy-thymidine monophosphate (dTMP)	$C_{10}H_{13}N_2O_7P$	T	301.829
anhydrid desoxy-adenosine monophosphate (dAMP)	$C_{10}H_{12}N_5O_5P$	A	310.772
anhydrid desoxy-guanosine monophosphate (dGMP)	$C_{10}H_{12}N_5O_6P$	G	326.647

$$6(\mathrm{A}+\mathrm{T}+\mathrm{G}+\mathrm{C})/4 \approx 1839.27 \approx m_H/m_e(1837.15) \tag{10.2}$$

Note that the Fermi mass is 311.90 m_H, equal (within $< 2\%$) to the mean nucleotide mass, 306.55 m_H. This points to a connexion between Molecular Biology and Particle Physics, which share another common point: the distinction between right and left, referred to as optical 'chirality' or 'parity violation'.

We also found the quasi-symmetric relations (accurate within $\sim$ 0.06 ppt):

$$\mathrm{A}/\mathrm{G} \approx a/6\ surd137 - 1; \quad \mathrm{C}/\mathrm{T} \approx 137/6\sqrt{a} - 1, \tag{10.3}$$

and also (within ~8 ppm and ~20 ppm respectively):

$$\mathrm{C} \approx 137(2\pi/3),]> \tag{10.4}$$

$$\mathrm{AT}/\mathrm{GC} \approx (a/137)^{3/2}. \tag{10.5}$$

This suggests that these mass ratios have the status of mathematical constants, opening the way to further study.

On the other hand, c-free dimensional analysis based on $\hbar$, G, and t_{cc} leads to the Balmer wavelength, and based on $\hbar$, G, and $2l_{cc}$ it leads (within 0.7%) to the above quasi-invariant DNA bicodon mass:

$$\hbar^2/Gm_{bc}^3 \approx 2l_{cc}. \tag{10.6}$$

So, DNA is directly connected to the absolute cosmic clock period t_{cc}. This suggests that DNA might be a cosmic linear hologram.

Consider now the common mammal temperature: $\theta_{\mathrm{mam}} \approx 310$ K, and the triple-point temperatures of Hydrogen: $\theta_{\mathrm{H2}} \approx 13.83$ K, Oxygen: $\theta_{\mathrm{O2}} \approx 54.33$ K, and water: $\theta_{\mathrm{H2O}} \approx 273.15$ K. It can be seen they are connected to the CMB temperature by the (1%-accurate) relation:

$$\theta_{H2} \times \theta_{O2} \approx \theta_{H2O} \times \theta_{CMB}. \quad (10.7)$$

On the other hand, in the relation:

$$a/(1 + ln\, a) \approx e^{\pi}, \quad (10.8)$$

there appears the Sternheimer scaling factor [43]: $j \equiv 8\pi^2/ln\ 2 \approx a - e^{\pi} \approx e^{\pi}\ ln\ a$, which enters the canonical form: $(R/r_H)^{1/2} \approx e^{j/e}$. Then one has:

$$\theta_{mam}/\theta_{CMB} \approx j. \quad (10.9)$$

The symmetry between the Universe and Nambu radii is reinforced by considering the wavelengths associated with the mammal and triple-point water temperatures: $\lambda_{mam} \equiv hc/k\theta_{mam}, \lambda_{H2O} \equiv hc/k\theta_{H2O}$. It can be seen that the following relations hold, within 1% and 0.1% respectively:

$$(Rl_P)^{1/2} \approx \lambda_{mam}, (R' l_P)^{1/2} \approx \lambda_{H2O}. \quad (10.10)$$

Here it should be recalled that, according to Schrödinger [44], temperature plays an essential role in Life. Indeed, mammal temperature is the same for the polar bear as for the african antelope, which may appear as a waste of energy [45]. It is as if the water molecule and mammal organisms were far more important, from the conjectured 'cosmic computer viewpoint', than that of CMB. This may be seen as reverse of the 'Anthropic Principle': the Cosmos would use 'human terminals' in its computing research. This may suggest an answer to the question: '*why do we ask questions*'?

We now proceed to relations involving the amino-acids making up proteins.

Table 3 displays the masses of these amino-acids in m_H units. The mean arithmetic and geometric values of the 20 normal amino-acids are, respectively:

$$\langle 20\, AA \rangle_{arith} \approx 135.75, \quad \langle 20\, AA \rangle_{geom} \approx 132.40. \quad (10.11)$$

Adding the two abnormal amino-acids, the mean values become:

$$\langle 22\, AA \rangle_{arith} \approx 142.50, \quad \langle 22\, AA \rangle_{geom} \approx 137.71. \quad (10.12)$$

One observes that:

$$\langle 20\, AA \rangle_{arith} + 1 \approx \langle 22\, AA \rangle_{geom} - 1 \approx 136.7. \quad (10.13)$$

Now the product of the 22 AA masses is, with $P = m_P/m_e$ and $(a_w)^{1/2} = F = m_F/m_e$:

Table 3 The 20 regular amino-acids by order of increasing mass, followed by the 2 *irregulars ones*

Name	Formula	Symbol	Mass/m_H
Glycine	$C_2H_5NO_2$	Gly/G	074.48
Alanine	$C_3H_7NO_2$	Ala/A	088.40
Serine	$C_3H_7NO_3$	Ser/S	104.27
Proline	$C_5H_9NO_2$	Pro/P	114.23
Valine	$C_5H_{11}NO_2$	Val/V	116.24
Threonine	$C_4H_9NO_3$	Thr/T	118.19
Cysteine	$C_3H_7NO_2S$	Cys/C	120.22
Leucine	$C_6H_{13}NO_2$	Leu/L	130.15
Isoleucine	$C_6H_{13}NO_2$	Ile/I	130.15
Asparagine	$C_4H_8N_2O_3$	Asn/N	131.09
Aspartic acid	$C_4H_7NO_4$	Asp/D	132.08
Glutamine	$C_5H_{10}N_2O_3$	Gln/Q	145.01
Lysine	$C_6H_{14}N_2O_2$	Lys/K	145.05
Glutamic acid	$C_5H_9NO_4$	Glu/E	145.99
Methionine	$C_5H_{11}NO_2S$	Met/M	148.05
Histidine	$C_6H_9N_3O_2$	His/H	153.95
Phenylalanine	$C_9H_{11}NO_2$	Phe/F	163.91
Arginine	$C_6H_{14}N_4O_2$	Arg/R	172.85
Tyrosine	$C_9H_{11}NO_3$	Tyr/Y	179.78
Tryptophan	$C_{11}H_{12}N_2O_2$	Trp/W	202.64
Selenocysteine	$C_3H_7NO_2Se$	*Se-Cys*	*166.74*
Pyrrolysine	$C_{12}H_{21}N_3O_3$	*Py-Lys*	*253.33*

$$(137.71)^{22} \approx 3aP^2/2 \approx (aF)^6/2. \tag{10.14}$$

Coming back to the *central holographic relation* implying the absolute Kotov clock:

$$3(P/a)^2 \approx (aa_{\mathrm{w}})^3 \approx (T/t_{\mathrm{cc}})^3. \tag{10.15}$$

In addition, one can see that:

$$(3P^2)^{1/22} \approx 113.78 \approx j, \tag{10.16}$$

showing another connection between the electric constant a and the Sternheimer scale factor j through the ratio P between the Planck and electron masses. These remarks confirm that elementary masses play an unsuspected but important role in Biology.

It may be interesting to investigate whether larger biological properties are related to elementary constants occurring in micro- or macro-physics. Actually, it is found that the mass m_{00} of the largest human cell (ovocyte), with a diameter of ~120 μm [46], is about $m_P/3$ (pig ~ $m_P/4.5$, hamster ~ $m_P/10$, mouse ~ $m_P/14$), m_P

denoting the Planck mass (~ 22 μg). We found a further correlation between the number of actual protein-coding genes making up the human genome, estimated to about 20,500 from evolutionary comparisons [47], and universal constants:

$$m_{00}/2m_e \approx P/6 \approx (20890)^5 \approx (r_H/\lambda_H)^4 \approx (\lambda_{CMB}/r_H)^3 \approx [2\sqrt{(R'/R)}]^{60}, \qquad (10.17)$$

where 20,890 is close (within < 2%) to the above gene number.

11 The Harmonic Principle

In a masterly review paper, Maruani et al. [48] have shown that musical patterns occur at various levels of complexity. If macrophysics constants are related to microphysics ones and affect atoms and molecules up to biosystems, then one may expect these constants to occur also in musical patterns. Indeed, following a tradition going back to Pythagoras, we propose a *Harmonic Principle* stating that there is a connection between the canonical numbers appearing in music and physical parameters. In the Jeans classification of the 'best' musical scales [49], obtained by the so-called 'continuous fraction' analysis, there are, following the 12 degrees of the western even-tempered scale, numbers of notes of 41; 53; 306; ….

First notice that Western music involves a large number correlation: $2^{19} \approx 3^{12}$, showing a connection with the 'golden section' $\phi = (1 + \sqrt{5})/2$. One can see that, within ~ 1.5%:

$$2^{19} \approx 3^{12} \approx \phi^{137/5}. \qquad (12.1)$$

Thus, the ancestral problem of connecting the golden section with music is solved by introducing the 'magic number' 137. This is not a property of the sole Western scale, for in the ancient Han Chinese 60-interval scale one has, within ~ 0.9%: $3^{60} \approx \phi^{137} \sim 4.3 \times 10^{28}$, a large integer already quoted [19] for its special properties. Equation (12.1) also shows that the 5th harmonic sum 137/60 appears as the exponent of ϕ το yield 3, the optimal integer basis (between e and π).

Note that the optimal integer basis 3 correlates (within 12 ppm) the ratios F/a and $a/137$, where F is the Fermi to electron mass ratio (Table 1):

$$3 \approx (a/137)^{F/a}. \qquad (12.2)$$

Music experts divide the tone (about a sixth of the octave) into 9 commas, 4 forming a minor semi-tone and 5 forming a major semi-tone, yielding 9 × 6 = 54 commas in the octave. Indeed the Hindustan scale uses 53 intervals, and the perfect number 6 appears as the 137th one:

$$2^{1/53} \approx 3^{1/84} \approx 6^{1/137}. \tag{12.3}$$

It should be noted that the comma is distinguished by violinists, if not by pianists. But the presence, in the following scale, of the number 306 ≈ $p/6 \approx \pi^5$ (accurate within 20 ppm) is even more dramatic. One has,

$$3^{1836/3}(p/2a) \approx a^a \approx \exp\left[(2\pi)^3 e\right]. \tag{12.4}$$

The operational definition of the optimal basis e is that $e^{1/e}$ is maximal, and 3 is the integer closest to e. Indeed, in computer theory, it is known that using a computing basis 3 would be far more efficient than the commonly used digital basis 2, notwithstanding technical problems. In 1712, Leibnitz wrote to Christian Goldbach: '*Music is a secret exercise on numbers*'. We elaborate on this by conjecturing that the brain is a multi-basis computer using mainly the bases 2, 3, 5, and 137, which appear in the harmonic series discussed above.

Note that physical properties may also show arithmetic properties obeying an 'economy principle', i.e. they can be expressed as large numbers involving only one or two small numbers. For instance, one has (within 0.6 and 0.03%, respectively):

$$R/\lambda_e \approx 2\text{^}(2\text{^}7),\ R'/\lambda_e \approx (3\text{^}3)\text{^}(3\text{^}3). \tag{12.5}$$

Both the Cosmos R and 'Grandcosmos' R' radii exhibit 'economic numbers', expressed using the two simple integer bases 2 and 3. All this cannot be due to chance, and calls for a *Diophantine Grand Theory*.

12 Conclusions

The reductionist point of view is to consider the Universe as a mere ensemble of particles in c-limited probabilistic interaction, this resulting in a separation between scientific domains such as Cosmology and Biology. The opposite, holistic approach leads, on the contrary, to the unification of the scientific domains. Even microphysics cannot really be understood without involving Cosmology, which sheds light onto the famous problem of hidden 'local' variables. The strong arguments in favor of a 'computing Cosmos' make intelligent life likely in the whole Universe.

The misleading formalism was to choose a system of units with $c = 1$ in relativistic quantum mechanics, leading to confusion between space and time coordinates while they transform differently in the relativistic formulas. This prevented recognition of the essential role of dimensional analysis, which we related to extended holographic principles and lead us to the temporal invariance of the Universe horizon, mean density, and microwave background temperature.

While the common interpretation of quantum mechanics is of a statistical nature, our deterministic point of view is that the Cosmos 'has no choice': *the origin of the physical laws is in pure numbers*. As a result, one may discard the theory of biological evolution through natural and sexual selection of randomly occurring micromutations, for an organism is a whole entity and there is an overwhelming lack of intermediate fossils. Our conjecture is that macromutations would be monitored by a 'computing Universe' which uses $a \approx 137$ as an optimal basis. Indeed, our 'Grandcosmos' volume (in units of the Bohr radius) is precisely: $V' = a^a/\pi$.

The physical interactions have been replaced in a cosmic context, which led us to a proposal for the graviton and photon masses. Experimental refutations of de Broglie's search for a 'double solution' and of Einstein's conjecture of 'hidden *local* variables' cannot be seen as a triumph of Bohr's 'completeness' idea, which has a reductionist flavor, because these views do not include *the cosmos as the source of hidden variables* in a holistic approach.

This study was initiated by a simple idea: the conservation of geometric forms of different dimensions, in analogy with the holographic technique. This led us to very accurate relations between canonical physical ratios. The standard view is to attribute them to chance, which leads to the Multiverse conjecture and Anthropic principle. But we showed that these relations involve such special numbers as π, e, ϕ and a, as well as whole numbers occurring in string theory. This means that the traditional idea of a unique Universe should be restored, and Eddington's Fundamental Theory reassessed.

From our holographic conservation rules we have derived a 'flickering Universe' model, involving a high-frequency matter-antimatter oscillation. In fact, holographic relations reveal more than a simple geometric analogy. The related 'Coherence Principle' can be linked to the fact that holographic techniques use a coherent, monofrequency radiation. It may be inferred that waves associated with particles have some mutual coherence. In 'Coherent Cosmology' a single frequency is at work: $f = h/E \approx 10^{104}$ Hz, which can be interpreted as a matter-antimatter oscillation. 'Dark matter' could then be seen as having an oscillation in 'phase quadrature'. This can be related to de Broglie's remark on *Zitterbewegung* being a 'beat' between the electron and positron waves [50], as noted by Maruani [35]. Within this model, we have found striking relations between Newton's and Fermi's constants, and used the 10^{-6}-accurate value of the latter to propose a more accurate estimate of the former.

This leads to the idea of a 'Computing Universe' using the fundamental physical constants as optimal calculation bases. Living beings would be peripherals designed and used by this overseeing 'natural intelligence'. The non-deterministic interpretation of quantum mechanics would be replaced by hidden determinism. 'Hidden variables' would simply be, in line with Mach's intuition, the rest of the Cosmos and, of course, subject to non-locality. But strict non-locality would also be excluded, for it would involve infinite velocity. We have proposed that superspeeds are at work, such as that ($C_g > 10^{36}c$) recently proposed by Maruani [51].

Other conclusions of the present work and conjectures for future work are listed in the following. **1**. *Quantification Principle*: at the end, physical laws are arithmetical, excluding infinity and continuum. **2**. *Coherence Principle*: a unique frequency governs all phenomena in the Universe, particles and molecules, DNA chains, living cells, and whole organisms. **3**. *Vibrating Principle*: the Universe is vibrating with a periodicity $t = h/E = 2t_P^2/T$, where $E = Mc^2$ with $M = Rc^2/2G$. The period of the matter-antimatter vibration of each particle is a whole multiple of t or, equivalently, its mass is a whole sub-multiple of M. **4**. *Holographic Principle*: fundamental physical laws result from holographic conservations—in fact dimensional transfers. **5**. *Grandcosmos Principle*: an external thermostat, with radius $R_{GC} = R'^2/2l_P$, is the source of both CMB and CNB. **6**. *Computing Principle*: universal numerical constants act as computational bases for a Computing Universe. **7**. *Immergence Principle*—Inverted Anthropic Principle: Life helps in cosmic computations; biological parameters are tied to cosmic constants.

Acknowledgements The author is deeply indebted to his wife Oya for her patience and scientific assistance. Denis Gayral is thanked for his help in informatics. The author is also grateful to Valery Kotov, Christian Bizouard, Jean-Claude Pecker, Cynthia Whitney, Christian Marchal, Grigori Tomski, Christiane Bonnelle, Dominique Weigel, and Ivan Todorov for stimulating discussions. The author is especially indebted to Pr Jean Maruani for his unyielding encouragements and constructive criticisms, and also for having him invited to the QSCP workshops at Taipei and Varna. The present work was presented at QSCP-XX in Varna in September 2015.

References

1. Bull P et al (2015) Beyond CDM: Problems, solutions, and the road ahead. arXiv:1512.05356
2. Steinhardt PJ (2011) The inflation debate. Sci Am 304:36
3. Olive KA et al—Particle Data Group (2014) Review of particle physics. Chin Phys C 38:090001 (p 111)
4. Planck collaboration. Planck 2015 results. XIII. Cosmological parameters. arXiv:1502.01589v2
5. Reiss A et al (2011) Determination of the Hubble constant with the Hubble space telescope and wide field camera. Astrophys J 730:119
6. Bondi H, Gold T (1948) The steady-state theory of the expanding Universe. Mon Not R Astron Soc 108:252
7. Hoyle F (1948) A new model for the expanding Universe. Mon Not R Astron Soc 108:372
8. Hoyle F, Burbidge G, Narlikar JV (2000) Ch. 8: The cosmic microwave background: a historical account. In: A different approach to cosmology. Cambridge University Press
9. Peebles PJE (2013) Discovery of the hot big bang: what happened in 1948. arXiv:1310.2146v2
10. Kragh H (1996) Cosmology and controversy: the historical development of two theories of the universe. Princeton University Press, 500 pp
11. Carr BJ, Rees MJ (1979) The anthropic principle and the structure of the physical world. Nature 278:605
12. Feinberg G (1967) Possibility of faster-than-light particles. Phys Rev 159:1089
13. 't Hooft G. Discreteness and determinism in superstrings. arXiv:1207.3612v2
14. Arp HC (1988) Quasars, redshifts and controversies. Cambridge University Press

15. Kotov VA, Lyuty VM (1990) The 160-min periodicity in the optical and X-ray observations of extragalactic objects. Comput Rend Acad Sci Paris 310, Ser. II:743; ibid. (2010) An absolute clock in the Cosmos? Bull Crimean Astrophys Obs 103:127
16. Sanchez FM, Kotov VA, Bizouard C (2011) Towards a synthesis of two cosmologies: the steady-state flickering Universe. J Cosmol 17:7225
17. Poincaré H (1912) Sur la théorie des quanta. J de Physique 2:37
18. Poincaré H (1913) Dernières Pensées. Conference at the University of London, pp 102–103 (Flammarion)
19. Sanchez FM (1995) Holic principle. In: ANPA proceedings, vol 16, Cambridge, p 324
20. Bousso R (2002) The holographic principle. Rev Mod Phys 74:834
21. Sanchez FM, Kotov VA, Bizouard C (2009) Evidence for a steady-state, holographic, tachyonic and super-symmetric cosmology. Galilean Electrodyn 20(3):43
22. Sanchez FM (2015) The end of reductionism: coherent quantum cosmology. Galilean Electrodynamics 26(4):63
23. Sanchez FM (2006) Towards a grand unified holic theory. In: Pecker J-C, Narlikar J (eds) Current issues in cosmology. Cambridge University Press, p 257. Sanchez FM (2013) Towards coherent cosmology. Galilean Electrodyn 24(4):63
24. Durham IT (2006) Sir Arthur Eddington and the foundations of modern physics. Doctoral dissertation, p 111. arXiv:quant-ph/0603146
25. Davies PCW (1982) The accidental universe. Cambridge University Press
26. Lloyd S (2007) Programming the universe. First Vintage Books
27. Wolfram S (2002) A new kind of science. Wolfram Media Inc.
28. 't Hooft G (1992) Quantum field theoretical behavior of a deterministic cellular automaton. Nucl Phys B 386:495
29. Ng Y (2001) From computation to black holes and space-time foam. arXiv:gr-qc/0006105v5
30. Nambu H (1952) An empirical mass spectrum of elementary particles. Prog Theo Phys 7:595
31. Casimir HBG (1948) On the attraction between two perfectly conducting plates. Proc Kon Nederl Akad Wetensch B 51:793
32. Lamoreaux SK (1998) Demonstration of the Casimir force in the.6 to 6-micron range. Phys Rev Lett 81:5475
33. Okun LB (2006) Photon history, mass, charge. Acta Phys Pol B 37:565
34. Gabor D (1948) A new microscopic principle. Nature 161:777
35. Maruani J (2012) The Dirac electron … and the kinetic foundation of rest mass. Prog Theor Chem Phys (Springer) B 26:23. Maruani J (2013) The Dirac electron as a massless charge spinning at light speed …, ibid. B 27:53
36. Marchal C (2009) Physics with photons of non-zero rest mass. Int Rev Phys 3:1
37. Hermann A (1971) The genesis of quantum theory. MIT Press, Cambridge, MA, p 92
38. Bastin T, Kilmister CW (1995) Combinatorial physics. World Scientific, Singapore
39. Todorov I (2015) Pythagorean trends in quantum field theory. Opening lecture at QSCP-XX. Varna, 2015, and private communication
40. Green MB, Schwarz JH, Witten E (1987) Superstring theory. Cambridge U.P.
41. Bott R (1970) The periodicity theorem for the classical groups and some of its applications. Adv Math 4:53
42. Polchinski J (1998) String theory. Cambridge U.P., p 23
43. Sternheimer J (1994) Ondes d'échelles II. Aperçu de théories non-linéaires et d'applications biologiques, Researchgate.net/publication/279202392
44. Schrödinger E (1944) What is life. Cambridge U.P.
45. Chauvin R (1997) Le Darwinisme ou la fin d'un mythe (Ed. du Rocher)

46. Griffin J et al (2006) Comparative analysis of follicle morphology and ovocyte diameter in four mammalian species (mouse, hamster, pig, human). J Exper Clin Assist Reprod 3:2
47. Clamp M et al (2007) Distinguishing protein-coding and non-coding genes in the human genome. Proc Natl Acad Sci USA 10:19428
48. Maruani J, Lefebvre R, Rantanen M (2003) Science and Music: from the music of the depths to the Music of the spheres. Prog Theor Chem Phys (Kluwer) B 12:479, and references therein
49. Jeans J (1968) Science and Music. Dover Pub., p 188
50. de Broglie Louis (1934) l'Electron Magnétique: Théorie de Dirac, ch. 9–22. Hermann, Paris
51. Maruani J (2016) The Dirac electron: From quantum chemistry to holistic cosmology. J Chin Chem Soc (Wiley) 63:33, and references therein

The Origin and Evolution of Complex Enough Systems in Biology

Erkki Brändas

Abstract Recent criticisms of Neo-Darwinism are considered and disputed within the setting of recent advances in chemical physics. A related query, viz., the ontological thesis, that everything is physical, confronts a crucial test on the validity of reductionism as a fundamental approach to science. While traditional 'physicalism' interprets evolution as a sequence of physical accidents governed by the second law of thermodynamics, the concepts of biology concern processes that owe their goal-directedness to the influence of an evolved program. This disagreement is met by unifying basic aspects of chemistry and physics, formulating the Correlated Dissipative Ensemble, CDE, as a characterization of a 'complex enough systems', CES, in biology. The latter entreats dissipative dynamics; non-Hermitian quantum mechanics together with modern quantum statistics thereby establishing a precise spatio-temporal order of significance for living systems. The CDE grants a unitary transformation structure that comprises communication protocols of embedded Poisson statistics for molecular recognition and cellular differentiation, providing cell-hierarchies in the organism. The present conception of evolution, founded on communication with a built-in self-referential order, offers a valid argument in favour of Neo-Darwinism, providing an altogether solid response and answer to the criticisms voiced above.

Keywords Complex enough system • CES • Correlated dissipative structure • CDS • Correlated dissipative ensemble • CDE • Central nervous system • CNS • Liouville equation

---

E. Brändas (✉)
Department of Chemistry, Uppsala University, Uppsala, Sweden
e-mail: Erkki.Brandas@kemi.uu.se

A. Tadjer et al. (eds.), *Quantum Systems in Physics, Chemistry, and Biology*,
Progress in Theoretical Chemistry and Physics 30,
DOI 10.1007/978-3-319-50255-7_24

1 Introductory Remarks

1.1 Reductionism in Natural Science

A frequent understanding of natural sciences contends that biology reduces to chemistry and chemistry to physics. Even if the traditional analytic interpretation of mathematics as a language or tool to convert knowledge about nature, i.e. a formal science not being incorporated amongst the branches of physical or life sciences, some recent propositions impart that also physics can be reduced to a mathematical structure [1]. Within such views, based on strict reductionism, see more below, one argues that the present laws of physics are not only commensurate with biology, but also contain the necessary natural laws for expressing all the known biological facts.

There are numerous differences of opinion regarding the scheme outlined above, e.g. the impossibility to derive the Coulomb Hamiltonian corresponding to a molecule from first principles and the general problem of quantum chemistry to treat nuclei and electrons on a more or less equivalent basis, see Löwdin [2]. Other arguments concern the concepts of biology as the laws of physics, at present, exclude a fundamental understanding of biological processes governed by an evolved program [3, 4].

Observation and deliberate experiments instigate rational, deductive theory, formulated in the language of mathematics, with the so accumulated knowledge discussed and contained by the methods of philosophy in general and the concepts of biology in particular. Recognizing the acquired demarcation between the philosophy of science, including biology, and the enactment of reductionism as embodied in the laws of physics, the endwise connections between them should still be profound and significant. Nonetheless there seems to be a considerable gap between the reasoning and thinking between the two distinct spheres of influence as revealed by the dialogue below.

Steven Weinberg [5] in his brilliant essay *Dreams of a Final Theory. The Scientist Search for the Ultimate Laws of Nature* avers the verdict on Philosophers and of Philosophers of Science that they often carry notions of "scientific explanation" that are too strict for "real" scientists. This view is beautifully expressed in the quoted correspondence between the author of the book and his friend, the evolutionary biologist Ernst Mayr, who asserted that Weinberg's book is *a horrible example of the way physicists think*, and that it reflected a serious lack of understanding regarding the three principal classes of scientific reduction in biology, i.e. the ontological, the epistemological, and the methodological reductionism.

Weinberg answered that *the main reason I reject this categorization is that none of these categories has much to do with what I am talking about (although I suppose theory reductionism comes closest). Each of these three categories is defined by what scientists actually do or have done or could do; I am talking about nature itself.*

The present exchange between the biologist and the physicist is not brought forward only to provide an example of the distinct ways of thinking between scientists that supposedly should have a lot (chemistry) in common. The argument does confer, as already stated above, on such questions as whether life sciences in practice can be reduced to chemistry and the latter to physics. Even if merely restricted to chemistry and physics, there continue to be many disagreements, like whether quantum mechanics can fully account for all atomic and molecular structures notwithstanding the famous statement of Dirac [6], without the use of specifying experimentally derived physical and chemical facts. These aspects raise controversial questions about the doctrine of 'physicalism' imparting consequences that justify a weaker ontological hierarchy, see e.g. Weisberg, Needham, Hendry [7] for more details. However, the introduction of biology in this setting introduces an important aspect, viz., it incorporates the physicist and the biologist itself and, at the end of the day, you and me, into the picture.

Within this broader picture sits a deeper principal concern: whether, in the science and philosophy of biology, the theoretical origin and conceptual foundation of biology can be entirely reduced to the laws of physics and chemistry. As will be obvious in this paper, it is not possible to advance the case without belabouring the self-referential characteristics of living systems that do organize complex enough biological systems.

The notion of *Complex Enough System*, CES, incorporates a new entity that extends the concept of traditional chemical physical systems in that their interactions/ communications with their environments follow a teleonomic law that should be governed by an evolved program, cf. the genetic code. The mathematical formulation of this process derives from a correlated dissipative ensemble, which has a network topology of a quantum nature as well as exhibiting a Poisson-type time evolution [8]. It is important to stress that the organization incorporates self-references and consequently must be interpreted within a framework of quantum logic.[1]

By promoting a collection of established results from recent advances in chemical physics, we will, under the reading of the *Paradigm of Evolution* [9], provide its most general, yet specific features as well as embrace some important results and consequences. The argument will be arranged as follows. Included in the introductory remarks we will illustrate recent proposals and criticism regarding the physical origin of biology and various criticisms of Darwinism. In order to respond to these comments and suggestions we intersect the account by a brief exposition of current progress in theoretical chemistry and physics with a bearing on teleonomic, natural laws. In the final section we return to the criticism voiced here in the introduction as well as devising a possible solution to the controversy.

---

[1]Note that the present representation of the spatio-temporal order is neither Boolean, Bayesian, scale free, decision making or any other classical version. It is neither strictly a quantum network, cf. quantum computational schemes lacking self-referentiability. Since the actual communication network includes self-references they are denoted as Gödelian networks.

1.2 The Physicalist Origin of Biology and the Recent Criticism of Darwinism

In his last essay Ernst Mayr [4] addressed the question *What Makes Biology Unique?* His main contention and outlining of his ideas, implicated the understanding of, and agreement with, the laws of physics. While affirming that biology has the necessary characteristics of chemistry and physics, he nevertheless construed that *none of the autonomous features of biology can possibly be unified with any of the laws of physics*. In particular he confronted physical reductionism as not only necessary but also quite impossible in biology. In order to demonstrate why biology differs from physics, he recounted non-applicable ideas of the latter, like target-directed behaviour of biological functions, regulation, self-organization and adaptation, by an original clarification of the controversial concept of teleology. In addition to characterizing authentic teleomatic phenomena ruled by the natural laws of physics, he introduced the notion of so-called teleonomic processes as those governed by an evolved program. Thus, according to Mayr, there appears to be a gap in the knowledge of Nature, as the process of evolution does not seem to follow unequivocally from physics.

In addition to the principal concern whether the theoretical and conceptual foundation of biology, as voiced above, can be entirely reduced to the laws of physics and chemistry, there has been several recent inveterate criticisms of Neo-Darwinism. Although Ernst Mayr did not like the term for historic reasons, we will use it as it is practiced today, see e.g. Dawkins [10], with all its evolutionary sub-theories included, counting fundamental addenda like the genetic code and other ingredients of the Darwinian paradigm. Even if somewhat divisive, we will for simplicity refer to the term also in the context of contemporary evolution theory unless a more specific detail has to be discussed.

In particular evolutionary theory has developed with a convincing focus on genetics, where hereditary variation to a large extent depends on randomly varying mutations and in lesser part due to influences of developmental processes by environmental conditions. Within this perspective, Jablonka and Lamb [11] challenged this aspect of Darwinism. They advanced the idea of epigenetic inheritance systems (EIS), which incorporate environmental mechanisms that actively maintain differing patterns of gene expressions, structural organization etc. enabling transmission of information from mother- to daughter cells. It is important to understand that the term EIS essentially involves the inheritance of phenotypic variations which do not primarily originate in differences in DNA. Although the present insight into the evolutionary process encompasses Darwinism it also reaches out to the social and ecological domain. Although not principally in conflict with the concepts of Mayr, the authors do not propose fundamental interferences with respect to the laws of physics.

Recently, however, the materialist Neo-Darwinist comprehension of nature has been seriously criticised, scorned and even rejected, see e.g. Deacon [12], Nagel [13], Fodor and Piattelli-Palmarini [14]. Although Deacon's ambitious and challenging confrontation to validate in what way goal-directed progressions can arise from purely physical processes has met with appreciative understandings, see e.g. the positive review by Logan [15], the reactions to the more opinionated views of Fodor-Piattelli-Palmarini, accusing adaptionism of being circular, and the anti-reductionist, assertive teleological view of Nagel, have in contrast been unforgiving, see e.g. Orr [16], Ferguson [17], and Rosenberg [18].

The common denominator of the account above is the claims of incompleteness of the materialistic world-view and as a consequence criticism descends on the reductive research program including Darwin's theory of evolution. This is manifestly expressed in the opening sentence of the biological anthropologist and neuroscientist Terrence Deacon's treatise [12] viz.: *Science has advanced to the point where we can precisely arrange individual atoms ...... yet ironically we lack a scientific understanding of how sentences in a book refers to atoms, DNA, or anything at all.* To give a general name to the set of problems related to the quote above, i.e. how to integrate purpose and intention into the picture, the author introduces the concept of *ententionality*.

This limitation is zealously expressed in Thomas Nagel's recent book *MIND & COSMOS* [13] where he essentially questions the entire naturalistic world picture, including physics, chemistry, and biology extending it to incorporate the theory of evolution and even cosmology. While Nagel grants that there appears to be a scientific consensus regarding the sources of variation in the evolutionary process and that accidental genetic variation is enough once reproducing organisms have come into existence, nevertheless, there is no convincing or credible argument that explains the origin of life itself. Quoting Nagel: *given what is known about the chemical basis of biology and genetics, what is the likelihood that self-reproducing life forms should have come into existence spontaneously on the early earth solely through the operation of the laws of physics and chemistry?*

Fodor and Piattelli-Palmarini [14] offer more detailed and scrupulous criticisms claiming that the Principle of Natural Selection (PNS) is flawed and can't be fixed. Utilizing the traditional problem of free-raiders, cf. Gould and Lewontin [19], they deduce from the principle of sufficient reason that the argument concerning the mechanism of "selection for fitness" to explain the adaptation of a phenotype to its environment is circular and that phenotypes are explained not from PNS but from Natural History. Since the latter according to the authors, rather than being a theory, should be inferred as a bundle of evolutionary scenarios, therefore evolutionary biology can't be viewed as an intensional science. As a consequence, biology is a science without its own proprietary laws just like any other so-called special sciences.

1.3 Responses to the Criticism

There are numerous attempts to confront these challenges. The general scientific query whether the unity of the language of science prompts physicalism in the strict sense, i.e. if all scientific laws can be derived from the laws of physics and further that they will reduce the different branches of science to physical theory, is not at all obvious, see Brändas [20] for a recent assessment. Mayr, a devote Darwinian, claimed, as already pointed out, that physical law is not enough to define an autonomous science and philosophy of biology. On the other hand, if we want to support a physicalist origin of biology, the criticism voiced above invites detailed response.

While Deacon's exposition has not so far excited any revolutions in the minds of the citizens of the scientific community, Nagel's book in particular has been met by ferocious criticism from the leading intellectual orthodoxy. A similar controversy surrounds the deep scepticism advanced by Piattelli-Palmarini [14]. In the most recent dispute Rosenberg [18] states in defence of Darwinism that contemporary physics rules out real goals, purposes, ends and teleology in general as causal forces. By referring to adaptation as an asymmetric process (i) driven by the second law, and (ii) requiring it to be a wasteful process, using up more order in producing adaptations than the order that the adaptations constitute and maintain, Rosenberg writes:

> *It's clear that the only way to build adaptations consistent with these two requirements is to start by processes that randomly build large numbers of alternative molecular structures just through the operation of thermodynamic noise and then wait. Wait for what? For one or more molecules to turn up randomly that combines thermodynamic stability with replicability. Eventually out of shear thermodynamic noise there may come to be a molecular structure sufficient to withstand the local environment and that also encourages the emergence of copies of itself out of the atoms floating around in the thermodynamic noise. This can happen by templating, catalysing or otherwise producing copies of itself. You probably don't have to wait more than 500 million years, once the chemical constituents of the early earth were around for this to start to happen. Once it does happen, iteration of the same process will produce more and more adaptation, at greater and greater expense, just as the 2d law requires.*

In contrast to Mayr [4, 21], Rosenberg [22] views biological issues within the same epistemological framework as the philosophy of physics. Such disagreements are central to the present discussion, and it provides sustenance for analysing the quote above further. The author ends the discussion in [18] with the declaration that such processes are the only scenarios that physics will permit, and that this, in practice, is the principle of natural selection that Darwin discovered.

In Rosenberg's defence of Darwin, he makes in addition several arguments in favour of PNS including the criticism of Fodor's disjunction concept [23] as well as the important distinction between "selection for" and "selection against". At the heart of the matter lies the somewhat controversial issue of teleo-semantics, Macdonald and Papineau [24], which led Fodor and Piattelli-Palmarini to repudiate Darwinism. However, as Rosenberg points out, no causal theory whatever is so far able to account for a determinate semantic content and *ipso facto* a Darwinian theory could not do so.

It is obvious that the quote above, while defending Darwinism, or Neo-Darwinism as defined above, also confronts, not only the criticism of the materialist comprehension of Darwin's law of evolution, but also devout supporters like Mayr and Jablonka and Lamb. Thus we will have to make a twofold undertaking, i.e. defend Neo-Darwinism, while at the same time render, if possible, its physicalist origin.

1.4 Proposed Solutions to the Controversy

In view of the controversial assertions and confusing statements given above, it becomes indispensible to give an update of some relevant chemical-physics advances of significance for the present queries regarding the foundation of biological evolution. In this evaluation we start with the conceptual basis of genuine quantum chemistry, Löwdin [25, 26], and bring together appropriate findings in non-equilibrium statistical mechanics of dissipative systems, Prigogine [27], Obcemea and Brändas [28], and Yang's Off-Diagonal Long-Range Order, ODLRO [29], the structure of fermion density matrices, Sasaki [30] and Coleman [31].

To incorporate dissipative systems we mention an extension resting on a rigorous mathematical theory, i.e. the Balslev-Combes theorem [32], which is vital to understanding and appreciating non-Hermitian quantum mechanics [33–35] and its consequences for the dynamics of resonance states embedded in the continuum and their properties for higher order dynamics. These ingredients impart a prompt reordering and exposition of acquiescent microscopic self-organization [9, 20, 36] with direct bearings on Artificial Intelligence, AI, [36–40] including the formulation of Gödel's theorems in an extended logic based on the fundamental teleonomic concept of communication, extending from the microscopic- to the mesoscopic- and macroscopic levels [9].

In this undertaking we will show that Mayr's concept of teleonomy [3, 4], i.e. processes being influenced by an evolved program, is not only commensurate with, but derives from fundamental principles of physics and chemistry. It further follows that the extended framework of non-equilibrium statistical physics in combination with non-Hermitian representations give rise to a Spatio-Temporal Mnemonic structures [9], exhibiting thermally correlated structures for physical, chemical and biological evolution. As a consequence the "program state" is a sequence of transient germane state conversions of the developing, on-going dissipative (quantum and thermal) correlations, displayed by its reflexive complex enough characterization.

Since, by definition, this representation delineates CES, the latter, constituting a statistically correlated dissipative ensemble, CDE, accommodates anomalous lifetimes due to an intrinsic code forbidden protection of state decoherence [9]. The CDE operates endogenously with a characteristic built-in Poissonian type distribution, which gives way to something exogenous of ententional significance. This in turn imparts communication protocols, not only within the control of a genetic program, but also on the cellular- and higher order levels, including epigenetic inheritance programs, neurological spike trains etc., even suggesting more accurate and robust meme-like extensions, see e.g. Semon [41], Dawkins [10] and Deutsch [42].

2 The Theoretical Framework

2.1 Non-Hermitian Quantum Theory

Although the space in this document will not allow descriptions of the microscopically relevant formulations, we will try to develop a partial understanding of what needs to be explained and finally be employed in order to appreciate the consequences of the arguments and confrontations belaboured above. As is well known there appears to be a remarkable agreement between precise quantum chemical predictions and the most accurate experiments including sophisticated advanced instrumentation and precise measurements and therefore it is usually resolved that the many-body Schrödinger equation in particular and also quantum mechanics in general portray reality to an unmatched perfection. This anticipation acquiesced Per-Olov Löwdin [25] to establish a new sub-field, Quantum Chemistry, including a new journal, the International Journal of Quantum Chemistry. With the emergence of the computer revolution, the field of quantum chemistry prospered, and it is viewed today, despite the critical comments made above, as a fundamental area close to, but not synonomous with theoretical chemistry and chemical physics.

However, even with the numerous advances just mentioned, there are many inconsistencies and conundrums plaguing the fundamental formulation. Primas [43] in his thought provoking evaluation of the *myth* of universal laws brings up numerous profound problems connected with the ineptness of traditional quantum chemistry to take the necessary step from calculation to concept. Although many of the puzzles have a distinct metaphysical flavour, they aim at the deeper meaning of chemistry and a worldview, *unus mundus*, which integrates also the dimensions of semiotics. Some of the most serious ones have been discussed in [20] including in particular the well-known issues of the time irreversibility of the macro-world, the classical law of causality, not to mention questions related with general relativity, all inconsistent or incompatible with standard formulations of the microscopic domain.

Closely related is also the inquiry of the absolute nature of the second law, Sklar [44]. For example it is not possible to rigorously derive thermodynamics from microscopic dynamical laws without retaining statistical assumptions. Hence the proprietary character of the second law must be revisited and considered in the light of recent non-Hermitian quantum mechanics [33–35], possibly suggesting solutions to the "ignorance paradox", see e.g. [9, 20], and further the conclusions below.

Much could be said about the necessity to realize and accomplish a quantum mechanical extension,[2] but we refer to Refs. [9, 20, 32, 45–48] for more details. The nomenclature non-Hermitian or non-self-adjoint operators hide an important

---

[2]This extension rests on a rigorous mathematical theory, based on the Balslev-Combes theorem [32] and it is vital to understand and appreciate non-Hermitian quantum mechanics and its consequences for the dynamics of resonance states embedded in the continuum and their properties for higher order dynamics.

aspect, contrasting the guarantee to always being able to diagonalize the proper Hermitian matrix representation. A key concept here is the notion of normal operators, i.e. operators, which commute with their Hermitian conjugate. Normal operator representations lead to matrices that can be diagonalized, while general non-selfadjoint operators are not normal, necessitating the inclusion of Jordan blocks and their Segrè characteristics, i.e. the dimension of the largest block, see e.g. Löwdin [26].

2.2 *Statistical Mechanics for Dissipative Systems Far from Equilibrium*

Our particular objective befalls that of merging these aforementioned "unstable states" into a more general quantum theoretical picture, incorporating quantum statistics. While the standard spectrum of a self-adjoint Hamiltonian operator is real, a resonance eigenvalue of the extended Hamiltonian is characterized by a complex energy on the so-called second Riemann Sheet, see e.g. [20] for more details,

$$\varepsilon = E - i\epsilon; \epsilon = \gamma/2 = \hbar/2\tau \tag{1}$$

where $\hbar = h/2\pi$ with h being Planck's constant, and the complex part ϵ relating, as usual, to the full width at half maximum, γ, with τ being the life time of the "state".

Since we will be referring to quantum systems imbedded in an environment, to be specified later, we will not carry out any thermodynamic limits unless explicitly specified. It is therefore appropriate to start by specifying a general quantum theoretic system operator strictly defined as a reduced density matrix derived from an abstract N-particle fermionic wave function $\Psi(x_1, x_2, \ldots x_N)$, where in principle the variables represent space and spin and if necessary could also involve time. Strictly speaking the density matrix should contain both the light carriers, like the electrons, and the nuclei, omitting the latter temporarily, see more below. Taking the trace over all particle variables except q (usually $q = 2$) the corresponding reduced density matrix writes in the Löwdin formulation [49], i.e. the matrix normalized to the number of pairings, but also other standards exist [29, 31],

$$\Gamma^{(q)}\left(x_1, x_2, \ldots x_q | x_1', x_2', \ldots x_q'\right)$$
$$= \binom{N}{q} \int \Psi(x_1, x_2, \ldots x_q, x_{q+1}, \ldots x_N) \Psi^*(x_1', x_2', \ldots x_q', x_{q+1}, \ldots x_N) dx_{q+1}, \ldots dx_N \tag{2}$$

The energy of the state is easily expressed in terms of a suitable reduced Hamiltonian H_2 involving only standard two-particle atomic and molecular interactions of Coulomb type [31].

$$E = \mathrm{Tr}\left\{H_2\Gamma^{(2)}\right\} \tag{3}$$

In principle, as said, the system operator should comprise all the electrons and nuclei of the quantum mechanical system, but it is often convenient to reduce the atomic and molecular degrees of freedom to the relevant ones that relate to the particular physical situation under examination. In certain cases, $\Gamma^{(2)}$ shows a large eigenvalue indicating the onset of superconductivity or superfluidity [29–31], provided the quantum states are sufficiently protected from the environment so as not destroying the build-up of long-range off-diagonal correlations.

In order to deal realistically with biological systems it is necessary to incorporate the occurrence and primary importance of temperature, i.e. the influence of thermally induced correlations. In passing we note that general theories of "macroscopic quantum theory" originate from quantum correlations of electron pairs in superconductivity [50, 51], spin dynamics in condensed disordered matter [52], coherent-dissipative structures in aqueous solutions [53] and in connection with complex enough systems in biology [36]. We will define an abstract Hilbert Space through a set of suitable basis vectors—they might be the molecular degrees of freedom in a cell, or in cells like neurons that are in their rest state or excited during a perception. The relevant degrees could be light carriers like electrons or electron holes, nuclear movements like the double proton tunnelling motion in DNA or the endogenous chemicals that enable neurotransmission. In these arrangements one should be aware of the mirroring relation between light carrier correlations—carriers like those of the electrons—and the doings in the nuclear skeleton produced by the apt quantum thermal correlations that subsists in the chaotic hot and wet environment in a biological system like e.g. the brain [54, 55].

Continuing one forms a base set $| h_i\rangle$, $i = 1,2,3\ldots n$, where $n \gg N/2$ is the space dimension, based on these quasi-bosonic degrees of freedom (quasi since they might, from a quantum statistical standpoint, be paired fermions). In the following we will formally write the set as a bold face row vector $|\boldsymbol{h}\rangle$ with components $|h_i\rangle$. Note that one is dealing with a dissipative system so the dimension n will not be fixed from the beginning, but will change and fluctuate from one situation to another. For a given Hamiltonian (the full Hamiltonian H or the reduced one, H_q depending on q) the system operator and its reduced partners evolves according to the Liouville Equation

$$i\hbar\frac{\partial^{(q)}}{\partial t} = \mathcal{L}^{(q)} = \left[H^{(q)} - {}^{(q)}H\right] \tag{4}$$

In particular our N-particle system (if restricted to N electrons) involves $N(N\text{-}1)/2$ pairings with the total energy expressed as a sum of the corresponding pair energies see Eq. (3), with the fundamental electronic correlations, now hidden in a correctly N-representable and correlated $\Gamma^{(2)}$.

An interesting reduction to a statistically degenerate state, named the extreme configuration by Coleman [31], appears when the structure exhibits a large eigenvalue and the population of states become statistical [9, 20, 36]. A specific situation develops when the system supports ODLRO,[3] Off-Diagonal Long-Range Order [29], and condenses to $M = N/2$ bosons (or fermionic pairs), e.g. to a superfluid or a superconducting phase, expected to arise for most systems at sufficiently low temperatures. In this particular representation, when the wavefunction, Eq. (2), is an antisymmetrized germinal power, $\Psi(g^{N/2}) \propto g \wedge g \wedge g \ldots .g$, where $g = g_1$, the density matrix becomes essentially [20]

$$\Gamma^{(2)} = \lambda_{\mathrm{L}} |g_1\rangle\langle g_1| + \lambda_{\mathrm{S}} \sum_{k,l=1}^{n} |h_k\rangle(\delta_{kl} - \frac{1}{n})\langle h_l| \tag{5}$$

or more compactly in the representation $|\boldsymbol{g}\rangle$, see below

$$\begin{aligned} \Gamma^{(2)} &= \lambda_{\mathrm{L}} |g_1\rangle\langle g_1| + \lambda_{\mathrm{S}} \sum_{k=2}^{n} |g_k\rangle\langle g_k| \\ \lambda_{\mathrm{L}} &= \frac{N}{2} - \frac{N(N-2)}{4n}; \lambda_{\mathrm{S}} = \frac{N(N-2)}{4n(n-1)} \end{aligned} \tag{6}$$

with $\lambda_{\mathrm{L}} \to \frac{N}{2}; \lambda_{\mathrm{S}} \to 0; n \to \infty$. The basis $|\boldsymbol{g}\rangle$ is obtained from a preferred localized basis of geminals), $|\boldsymbol{h}\rangle$, i.e. of paired fermions (antisymmetric with respect to permutation of the fermionic space-spin degrees of freedom) and with $\omega = e^{i\pi/n}$, i.e.

$$|\boldsymbol{g}\rangle = |\boldsymbol{h}\rangle \boldsymbol{B}$$

$$\boldsymbol{B} = \frac{1}{\sqrt{n}} \begin{pmatrix} 1 & \omega & \omega^2 & \cdot & \omega^{n-1} \\ 1 & \omega^3 & \omega^6 & \cdot & \omega^{3(n-1)} \\ \cdot & \cdot & \cdot & \cdot & \cdot \\ \cdot & \cdot & \cdot & \cdot & \cdot \\ 1 & \omega^{2n-1} & \omega^{2(2n-1)} & \cdot & \omega^{(n-1)(2n-1)} \end{pmatrix} \tag{7}$$

Note that there is one large eigenvalue, λ_{L}, while λ_{S}, the small eigenvalue is (n-1) -degenerate. To fulfil the full trace relationship of $\Gamma^{(2)}$ one needs to account for the unpaired contributions, which however will be neglected as being unimportant in these settings. The transformation $\boldsymbol{B}$ will serve two important purposes, i.e. diagonalizing $\Gamma^{(2)}$ as well as bringing the associated thermalized conversion to classical canonical form. As will be clear in the following its factorization properties will be

[3]The celebrated concept of ODLRO was developed by Yang [29] about 15 years after the publication of the famous Bardeen - Cooper - Schrieffer theory of super-conductivity, for a comparison see [51]. The formulation does focus on the collective properties of matter at sufficiently low temperatures. For a physical system approaching zero temperature with a non-degenerate ground state the entropy goes to zero. Under specific conditions the system does develop ODLRO.

the key to formulating teleomatic processes of thermally excited complex enough biological systems.

The actual time evolution should incorporate the analytically continued representations, i.e. the resonance states of the dilated Hamiltonian [28]. As regards the latter this is a technical mathematical problem that either needs screening the Coulomb interactions, which is physically realistic, or alternatively introducing the notion of quasi-isometries converting non-scaled quasi-isometric evolution into contractive evolution of the scaled ones [56]. We will return to the actual time evolution further below, in connection with the characterization and the definition of the Correlated Dissipative Ensemble, CDE.

2.3 *The Liouville Equation and the Prigogine Energy Operator*

In order to examine the conditions that are commensurate with the thermal conditions of a living system, we return to the formulation above. As a precondition one must thermalize the extreme ensemble from isolation to a dissipative system at appropriate temperatures. The quantum-thermal correlations will be incorporated commensurate with a quantum extended canonical ensemble, adapted through a Bloch type equation using the Prigogine energy super operator [57], see also [58]. Since we will embark on a non-equilibrium description, yet close to equilibrium, the statistics of the ensemble will essentially turn out to be Poissonian.

The extension to the non-self-adjoint case requires proper dilations that maintain the analyticity of the scaling parameter in the space part of variables, i.e. $x_i \to \eta x_i$, with $\eta = |\eta| e^{i\theta}$ and θ some appropriately chosen positive angle. Hence Eq. (2) above should read (note the complex conjugate sign of the second set of coordinates)

$$\Gamma^{(q)}\left(x_1, x_2, \ldots x_q | x_1^{\prime *}, x_2^{\prime *}, \ldots x_q^{\prime *}\right) \\ = \binom{N}{q} \int \Psi(x_1, x_2, \ldots x_q, x_{q+1}, \ldots x_N) \Psi^*(x_1^{\prime *}, x_2^{\prime *}, \ldots x_q^{\prime *}, x_{q+1}^{*}, \ldots x_N^{*}) dx_{q+1}, \ldots dx_N \tag{8}$$

At the outset we have expressed our theoretical entity, i.e. a non-degenerate ground state of a molecular system, i.e. $\varrho = \Gamma^{(2)}$ of Eqs. (2) above. In passing we remark that there are two interdepending contiguous problems: (a) the thermal bath surrounding our open system and (b) the varied dynamics of, on one hand, the light fermion carriers and, on the other, the movements in the heavier nuclear skeleton. As already stated, it is unavoidable to go beyond the standard Born-Oppenheimer approximation. In principle there are essentially two ways to view the problem, e.g. as a scattering experiment, where the electrons impinge on the nuclei, see e.g. [20], or to work directly with the density matrix, where either the light fermionic portion or the nuclear degrees of freedom are traced out leaving in each case a nuclear- or an electronic dynamical problem. The crucial reading is that in both portrayals there exists a mirroring relation between the entangled subsystems entailing (I) the light fermion carriers and (II) the nuclear skeleton; see e.g. Löwdin [26] and Brändas and Hessmo [55].

Considering the relationship between (I) and (II), the formulation will be attended with a complementary understanding, reading the description of molecular systems as a dichotomy between mirroring degrees of freedom. Redefining the Liouvillian, i.e. the commutator with the Hamiltonian H, we introduce the corresponding anticommutator

$$\mathcal{L}_{\mathrm{B}}\varrho = \frac{1}{2}(H\varrho + \varrho H) \tag{9}$$

where the Prigogine energy superoperator, [57], $\mathcal{L}_{\mathrm{B}}$ is subject to the Bloch equation ($\beta = \frac{1}{kT}$)

$$-\frac{\partial \varrho}{\partial \beta} = \mathcal{L}_{\mathrm{B}}\varrho \tag{10}$$

in which k is Boltzmann's constant and T the absolute temperature. Noting the difference between Eqs. (9) and (4), it is imperative to provide a setup where the density matrix, during analytic continuation, must be represented by a complex symmetric form, i.e. $|\cdot\rangle\langle\cdot| \rightarrow |\cdot\rangle\langle\cdot^*|$, representing a complex conjugate in the bra-position. In particular for the density matrix given by Eq. (5) represented according to Eq. (8), one obtains directly the thermalized solution (note that e.g. the system I is open with respect to its coupling to system II and vice versa) at a given temperature T, choosing for simplicity the total energy threshold to be zero

$$e^{-\beta\mathcal{L}_{\mathrm{B}}}\varrho = \lambda_{\mathrm{L}} \sum_{k,l=1}^{n} |h_k\rangle e^{i\beta\frac{1}{2}(\epsilon_k+\epsilon_l)}\langle h_l| + \lambda_{\mathrm{S}} \sum_{k,l=1}^{n} |h_k\rangle e^{i\beta\frac{1}{2}(\epsilon_k+\epsilon_l)}(\delta_{kl} - \frac{1}{n})\langle h_l| \tag{11}$$

The derivation is straightforward resting on the convention that the basis functions h_l can be chosen real without restrictions. Employing the model of the nuclei as vibrating oscillators, we can e.g. use partitioning techniques to determine the complex energy of each oscillator dressed by the interactions from the other ones and the environment. Hence each oscillator yields, commensurate with the mirror theorem and the reciprocal relationship between the energy width ϵ_k and the lifetime τ_k matching the temperature of the environment

$$z_k = E_k - i\epsilon_k = -i\epsilon_k = -i\hbar/2\tau_k \tag{12}$$

where we have used the fact that the thermal excitations push the energy close to the threshold, i.e. at the zero energy level with each $E_k = 0$.

2.4 The Constructive Role of Quantum-Thermal Correlations

In this subsection we will demonstrate the constructive role of quantum—thermal correlations. As shown by Tegmark [1], quantum states in the hot and wet environment of a living system, like the human brain, are not likely to endure decoherence. Although there have been criticisms regarding the impossibility to invoke long-range coherence in biological systems, see e.g. [59], the conceivable irrelevance and disapproval of a number of coherence time relations do not solve the decoherence problem for the "survival" of quantum mechanics in complex enough biological systems. In contrast we will identify a mainly different answer to the coherence—decoherence dilemma [9, 20, 36, 52, 53, 58, 60].

One way to establish a solution to this quandary, can be done via a simple thermal scattering guide of an open system involving n bosonic or paired fermionic degrees of freedom related with a relaxation process given by the time scale τ_{rel}, assumed to be distinct from the smaller thermal timescale. The system is dissipative, as it exchanges energy and entropy with its environment. For instance the system may consist of the various building blocks in biological systems, from molecular aggregates describing the order of DNA and RNA all the way up to the whole cell having its place in the proprietary arrangement of the living being either assigned to develop material constructions or to promote communication channels for the nervous system.

The present derivation has been given at various places before, so we will condense the discussion here and referring to previous reviews for more details [20, 58]. Defining the "incoming beam" of the light carriers reaching an area σ_{tot}, activated by the correlated nuclei, and corresponding to a spherically averaged total cross section, consistent with the physical constraints of the model. The protocol describes a procedure that one typically will detect one quasi particle degree of freedom in the differential solid-angle element $d\Omega$ during the timescale $\tau_{\text{corr}} = \hbar/kT$ here, given by Heisenberg's uncertainty relation ($\tau_{\text{corr}} \approx 2.46 \cdot 10^{-14}$s at 310 K).

From conventional scattering theory one obtains the incident flux, N_{inc} of the number of particles/degrees of freedom per unit area and time as

$$N_{\text{inc}} = \frac{n}{\sigma_{\text{tot}}\tau_{\text{rel}}} \tag{13}$$

which together with number, $N_s d\Omega$, of particles scattered into $d\Omega$ per unit time being

$$\sigma_\Omega d\Omega = N_s d\Omega = \frac{d\Omega}{\tau_{\text{corr}}} = \frac{kT}{\hbar} d\Omega \tag{14}$$

Hence one gets for the total cross section

$$\sigma_{\text{tot}} = \int \sigma_\Omega d\Omega = \int \frac{N_s}{N_{\text{inc}}} d\Omega \tag{15}$$

leading to the simple relation between

$$\frac{n}{4\pi} = \frac{kT}{\hbar}\tau_{\text{rel}} = \frac{\tau_{\text{rel}}}{\tau_{\text{corr}}} \tag{16}$$

Establishing a correlated cluster of harmonic oscillators with the widths, see Eq. (1) $\gamma_l = \hbar\tau_l^{-1}$ with the (smallest) energy difference between the equidistant harmonic oscillator levels being $\hbar\tau_{\text{rel}}^{-1}$ displaying a spectrum from the zero-point energy to $\hbar\tau_{\text{corr}}^{-1}$. The quantized oscillators are in a sense reminiscent of Planck's law. Straightforward examination of the situation reveals

$$\tau_{\text{rel}} = (l-1)\tau_l = \tau_2 = \frac{n\tau_{\text{corr}}}{4\pi}; l = 2, 3 \ldots n \tag{17}$$

From Eqs. (11), (12) and (17) one gets

$$\beta\epsilon_l = \frac{2\pi(l-1)}{n} \tag{18}$$

which inserted into Eq. (11) gives, $\Gamma_{\text{T}} = e^{-\beta\mathcal{L}_{\text{B}}}\varrho$

$$\Gamma_{\text{T}} = \varrho_{\text{T}} = \lambda_{\text{L}} \sum_{k,l=1}^{n} |h_k\rangle e^{i\frac{\pi}{n}(k+l-2)} \langle h_l| + \lambda_{\text{S}} \sum_{k,l=1}^{n} |h_k\rangle e^{i\frac{\pi}{n}(k+l-2)} (\delta_{kl} - \frac{1}{n})\langle h_l| \tag{19}$$

The result (19) reveals its importance via the transformation $\boldsymbol{B}^{-1}$. Introducing the new basis $|\boldsymbol{h}\rangle\boldsymbol{B}^{-1} = |\boldsymbol{f}\rangle$ one obtains

$$\Gamma_{\text{T}} = \varrho_{\text{T}} = \lambda_{\text{L}} J^{(n-1)} + \lambda_{\text{S}} J \tag{20}$$

with J being the nilpotent operator defined by $J^{(n)} = 0; J^{(n-1)} \neq 0$ imparting the quantum transitions

$$J = \sum_{k=1}^{n-1} |f_k\rangle\langle f_{k+1}| \tag{21}$$

In passing one notes that the matrix representation of the shift operator J is an n-dimensional matrix with one's above the diagonal and the remaining elements equal to zero. In general any matrix representation of a degenerate eigenvalue partitions into blocks of various dimensions. The largest dimension occurring defines the Segrè characteristic of the degeneracy.

The result will be of crucial importance, see more below, since it implies that $\Gamma^{(N)} = |\Psi(g^{N/2})\rangle\langle\Psi(g^{N/2})|$, after reduction, analytic continuation and thermalization corresponds to $\Gamma_{\mathrm{T}}^{N} = |\Psi(f^{N/2})\rangle\langle\Psi(f^{*N/2})|$, $f = f_1$ and with $\Psi(f^{N/2})$ and $\Psi(f^{*N/2})$ orthogonal to each other.[4] We will use this result commensurate with two primary aspects, (i) *to develop relevant building blocks for complex enough biological systems* and (ii) *to use the properties of the transformation* ***B*** *in (7) as a semantic code for communication between entities on the molecular level.*

Retracing one may recognize that once "communication" is established between molecules and cells this might naturally be extended to higher order levels of semiotic exchanges. The order of organization, Eqs. (19–21), will here be referred to as a Correlated Dissipative Structure, CDS, which, as will see in the next section, will be an essential ingredient of the Correlated Dissipative Ensemble, CDE.

2.5 The Correlated Dissipative Ensemble and its Time Evolution

We have derived from Eqs. (2–5) a thermalization procedure, that through the boundary condition, Eq. (18), converted the density matrix by describing the molecular state distribution as transitions between the apt states of the system. This distribution is suitable for representing entities like cells and cellular networks. Note that the quantization condition above relates the temperature, the relevant time scales and the size of the correlated open system. The choice of zero energy level in the system (cell) is commensurate with the zero trace property of matrices like Γ_{T}. In order to maintain a "living state" characterized by (a) its dissipative coupling to the environment, (b) its metabolic processes, (c) the genetic function and (d) homeostasis for appropriate spatio-temporal regulation, one must describe the biochemical pathways by which the cell obtains energy. The *sine qua non* is both catabolism, i.e. the breakdown of molecules to generate energy, and the anabolitic synthesis of what the cell needs. As will be shown below, the CDS of the previous section imparts microscopic self-organization and serves as a proxy of the Helmholtz free energy including the functional emergence of quantum-thermal correlations.

To facilitate our aim to consider a higher order dynamics, we will utilize the CDS as base units for a specific Liouville formulation to be detailed below. Although it may seem a misnomer to call the result an ensemble, it will be shown that the higher order structure in terms of established resonance components indeed leads to a Poisson distribution of these elements defining physical communication channels between e.g. cells. In particular for the central nervous system, CNS, in the

[4]This issue will be discussed in more detail elsewhere.

presence of a special type of cells, called neurons, the dynamics is characterized by short time scale oscillations, building up pulses of light carriers (electrons), spikes that correlate the basic dissipative systems (here cells/neurons) and providing an irreducible coupling-communication between them.

There are several ways to prove the reduction processes (i) $\Gamma^{(N)} \rightarrow \Gamma^{(2)}$, (ii) $\Gamma^{(2)} \rightarrow \Gamma_{\mathrm{T}}$ and finally (iii) $\Gamma^{(N)} \rightarrow \Gamma_{\mathrm{T}}^{N}$. The step (i), linked to Yang's celebrated concept of ODLRO, [29], was independently derived by Sasaki [30] and further employed by Coleman [31] in his extreme state formulation of a wavefunction representable $\Gamma^{(2)}$. Sasaki's derivation concerns his studies of a system of fermions or bosons composed of two subsystems and the so-called Sasaki formula [30, 61], which is based on a counting argument involving the properties of the symmetric group. A simple statistical derivation was published e.g. in [9, 20, 36, 62]. Step (ii) did, in addition to the aforementioned citations, originate in the proceedings of the Resonance Workshop; held at Lertorpet in 1987, see e.g. [63] and particularly [64]. Finally the step three consists of defining an analogous $\Gamma^{(N)}$ based on f instead of g and continued analytically to $\Gamma_{\mathrm{T}}^{N} = \left|\Psi(f^{N/2})\rangle\langle\Psi(f^{*N/2})\right|$.

Although the reductions above appear complicated the result is easily combined into a higher order Liouville structure, continuing the model build-up from molecular aggregates in the cell organization to the actual cell, that we will call a Correlated Dissipative Ensemble. Defining a "cell basis" or quantum dot-like Jordan block units, cf. Eqs. (20–21), with n and N large

$$\varrho^{i} = C^{i} = \frac{1}{\sqrt{1+q^{2}}}\left\{q|f_{1}^{i}\rangle\langle f_{n}^{i}| + \frac{1}{\sqrt{(n-1)}}\sum_{k=1}^{n-1}|f_{k}^{i}\rangle\langle f_{k+1}^{i}|\right\} \tag{22}$$

$$\mathrm{Tr}\left\{\varrho\varrho^{\dagger}\right\} = 1; q = \frac{p}{1-p}; p = N/2n \tag{23}$$

each base entity, ϱ^{i}, 1 = 1,2,..m, building a basis $\boldsymbol{H}$. This entails the build-up of a higher order Liouville super operator structure based on the propagator/generator $\mathcal{P}$, see below, where in analogy with $\boldsymbol{h},\boldsymbol{g},\boldsymbol{f}$ of Sect. 2.4, the corresponding base entities (here cells) $\boldsymbol{H},\boldsymbol{G},\boldsymbol{F}$, subject to the same fundamental transformation $\boldsymbol{B}$ (where the dimension m in principle different from n but, as will be seen below, related to the one in (4), and with $\omega = e^{i\pi/m}$ as in analogy with the previous structure.

Since a degeneracy generally consists of several Jordan blocks of various orders, we will describe the propagator, as it generates the time evolution for any member of the ensemble, corresponding to the Segrè characteristic m, irreducibly coupled via J, see below, obtaining, *mutatis mutandis*, with $I = \sum_{k=1}^{m}|F_{k}\rangle\langle F_{k}|$

$$\mathcal{P} = (\omega_{0}\tau - i)I + J \tag{24}$$

$$J = \sum_{k=1}^{m-1} |F_k\rangle\langle F_{k+1}| \tag{25}$$

with ω_0 the thermal frequency, $\tau = \tau_{\text{rel}}$, the average lifetime of the cell, and τ_{corr} the fast timescale, here essentially equal to the thermal molecular motion. Note the analogy between Eq. (24), pertaining to cellular entities and the thermalization derived from Eqs. (10–11). The Liouville configuration imparts, as already said, the thermal frequency ω_0 and the correlation time $\tau_{\text{corr}} = \hbar/kT$ from environmental interactions. Hence the build-up of (24) is straightforward, save the non-conventional appearance of the operator J. In principle one can build a symmetric geminal power of "cell" functions, based on an "appropriate" bosonic degree of freedom, and then apply the Sasaki formula to obtain a result analogous to Eqs. (5–7). For simplicity we will study the consequences of the incorporation of a representative irreducible correlations instigated by J in Eq. (24).

Describing the generator $\mathcal{P}$ in dimensionless units, we obtain the cellular Q-value as

$$Q = \omega_0 \tau \tag{26}$$

cf. its use in regard to quality aspects of oscillators or resonators. Here one may determine Q as follows. From Eqs. (16–17) one finds, with $\tau = \tau_{\text{rel}}$, that

$$\frac{n}{4\pi} = \frac{kT}{\hbar}\tau_{\text{rel}} = \omega_0 \tau_{\text{rel}} = \frac{\tau_{\text{rel}}}{\tau_{\text{corr}}} = Q \tag{27}$$

and since we have defined $\tau = \tau_{\text{rel}}$ as the average timescale for the cell C_i, we can integrate over the solid angle $d\Omega$ and obtain the result

$$\int Q d\Omega = n \tag{28}$$

The fundamental result, (28) imparts important information, since the cell's Q-value signifies not only the dimension n of the *intracellular dynamics*, but it also conveyed a possible interrelation between the latter and the *intercellular correlations* implied by the m-dimensional transformation $\boldsymbol{B}$. Due to the factorizing nature of $\boldsymbol{B}$, i.e. the ensuing cyclic properties of its column vectors, the Q-value, n, of the cell basically entreats a semantic encoding/decoding inclusion, depending on the value of m, that protects simultaneously all the levels from the molecular- the super-molecular- to the cellular levels and possibly beyond.

We are now in position to define the causal propagator $\mathcal{G}(t)$ and the resolvent $\mathcal{G}_R(z)$ defined by

$$\mathcal{G}(t) = \mathrm{e}^{-i\mathcal{P}\frac{t}{\tau}}; \mathcal{G}_R(\omega) = (\omega\tau J - \mathcal{P})^{-1} \tag{29}$$

yielding directly by inserting the Liouvillian, Eq. (24),

$$\mathrm{e}^{-i\mathcal{P}\frac{t}{\tau}} = e^{-i\omega_0 t} e^{-\frac{t}{\tau}} \sum_{k=0}^{m-1} \left(\frac{-it}{\tau} \right)^k \frac{1}{k!} J^k \tag{30}$$

$$(\omega\tau I - \mathcal{P})^{-1} = \sum_{k=1}^{m} [(\omega - \omega_0)\tau + i]^{-k} J^{(k-1)} \tag{31}$$

From $F_k(t) = e^{-i\omega_0 t} e^{-\frac{t}{\tau}} F_k(0)$ one finds for the rth power of t (note that only F_1 is an eigenfunction of $\mathcal{P}$ while the remaining F_k's complete the root manifold)

$$N(t) \propto |\langle F_1 | J^r | F_{r+1} \rangle| \left(\frac{t}{r} \right)^r \frac{1}{r!} e^{-\frac{t}{\tau}} = \left(\frac{t}{r} \right)^r \frac{1}{r!} e^{-\frac{t}{\tau}} \tag{32}$$

with the definition

$$J^r = \sum_{k=1}^{m-\mathrm{r}} |F_k\rangle\langle F_{k+r}| \tag{33}$$

For the highest power $m-1$ one gets

$$dN(t) = t^{m-2} \left(m - 1 - \frac{t}{\tau} \right) N(t) dt \tag{34}$$

which imparts a modified microscopic law of evolution

$$dN(t) > 0; t < (m-1)\tau \tag{35}$$

In passing we note that Eqs. (34–35) imparts a higher-level timescale $\tau_{\mathrm{com}} = (m-1)\tau_{\mathrm{rel}}$, In other words it modifies the boundary conditions Eqs. (17–18), with $\tau_{\mathrm{corr}} \rightarrow \tau_{\mathrm{rel}}$ and $\tau_{\mathrm{rel}} \rightarrow \tau_{\mathrm{com}}$. It is remarkable that the communication bearing transformation $\boldsymbol{B}$ authorizes "messages" between the molecular-DNA-RNA-gene level and the hierarchical organization of cellular function and order. Actually the possibilities of a communication protocol is simply given by the cyclic properties of the columns of $\boldsymbol{B}$, see Eq. (7) displaying $\sqrt{12}\boldsymbol{B}$ as a diagram, where the dimensions of the cyclic vectors of $\boldsymbol{B}$ are given in the appropriate entry. We will not display the first column vector of one-dimensional units "1", and therefore we will only have 11 columns, see Ref. [20] for more details

$$\begin{array}{ccccccccccc}
 & & & & & 2 & & & & & \\
 & & & 3 & & 2 & & 3 & & & \\
 & & 4 & & & & & & 4 & & \\
 & 6 & & 3 & & 2 & & 3 & & 6 & \\
12 & & 4 & & 12 & & 12 & & 4 & & 12 \\
 & 6 & & 3 & & 2 & & 3 & & 6 & \\
 & & 4 & & & & & & 4 & & \\
 & & & 3 & & 2 & & 3 & & & \\
 & & & & & 2 & & & & &
\end{array} \tag{36}$$

observing also the manifest column symmetry of the graph. Space will not allow a more detailed discussion of the various possibilities to interpret the consequences of the present paradigm of evolution with specific reference to its teleonomic character, see e.g. Mayr [3, 4].

The present CDE representation has an interesting overlap with Trehub's retinoid model, devised for visual perceptions [65], with a straightforward parity argument based on the CDE, for the perplexing image switching of Necker's famous cube illusion [8]. The retinoid system involves neurons, especially a class called the autapse, i.e. neurons with synapses connected onto themselves. Since neurons "communicate" via spike trains, one might exploit the metaphor of a "Call Centre", seeing the communication as a number of "phone calls" between the cells during a given time, t, being multiples of their characteristic time $\tau = \tau_{\text{rel}}$. Therefore the probability that k "calls" are transferred during a given time interval, with each "telephone call" occurring with a known average (intensity) parameter $\lambda_l = (l - 1)\tau_{\text{rel}}/\tau_{\text{rel}} = (l - 1); l = 2, 3, \ldots m$, i.e. with a definite distribution for each value of l, is simply given by

$$P_{\lambda_l}(k) = \frac{(l-1)^k}{k!} e^{-(l-1)} \tag{37}$$

with the mean equal to the variance $\lambda = l - 1$. The number of calls during $\mathcal{T}_{l-1} = (l-1)\tau_{\text{rel}}$ is at maximum for $l = m$. If counting $l = 0$ as an event the "society of neurons" comprises, during $\mathcal{T}_{m-1}$, with a probability according to (37), every cell shares m communications distributed over m possible "sites" in the organism. Hence the length, m, of a message is directly matched with the variance and the mean (a well-known property of the Poisson distribution).

In summary, each cell is characterized by a Correlated Dissipative Ensemble, CDE, which, through the factorized canonical vectors of the associated transformation $\boldsymbol{B}$ incorporate nested encodings, as programmed in the process of evolution. Communication runs over several orders of time scales, from those related to the genetic code to encompass more complex semiotics, e.g. the plan for the accumulation of proteins, stored in the genetic alphabet and further conditioned via resonant mechanisms to cellular interactions depending e.g. on the cell's quality value.

3 Conclusion

3.1 Consciousness as an Irreducible Process

In this account we have so far guided our formulation to cells in general and neurons in particular, cf. the neural network of the Retinoid System. However, the present representation of a neural communication network is neither Boolean, Bayesian, decision making- or any other classical version. It is neither strictly a quantum network, cf. quantum computational schemes lacking self-referentiability. Since the nodes are self-referential and communicative they should be called a Gödelian network [8, 9, 20, 36]. Obviously a cellular environment, with precise "responsibilities", must assist the neurons, as the latter will be functioning in analogy with the example of Poisson distributed telecommunications, i.e. a "Call Centre". One such crucially supporting network is the astroglial "Master Hub" as advocated by Fernandes de Lima and Pereira [66]. Further contraptions are related to the astrocyte role of releasing neurotransmitters in the synaptic cleft, and how this is coupled to and dependent on calcium ion propagation.

In order to promote consciousness as a general and complete concept one needs to undertake the whole activity from (i) perception, (ii) awareness, (iii) cognition, and (iv) feelings and also emotions as an irreducible process of the mind, yet being characterized by fundamental time scales. It is evident that the "protracted communication hypothesis" originates already in the initial stage, i.e. about 3-7 ms, of the sensory neocortex-reorganised background due to the impact of e.g. a conditioned stimuli, CS [67]. The next stage that concerns awareness, cf. Libet's conscious awareness delay up to several hundred ms, nevertheless retrospectively creating a subjective back track. Hence the conscious mental experience cannot survive without the brain processes that give rise to it. It seems that the steps (i) perception, (ii) awareness can be separated according to specific time scales, i.e. (i) 5–10 ms, (ii) 200–500 ms, while (iii) cognition is fundamentally irreducible as regards the time scales of (i) and (ii).

Somehow the consciousness of the SELF develops its projected images over several hundreds of ms, but on "the lower level" this distinction is irreducible. In terms of the various tasks of the different key actors one might discern:

(a) *the stochastic background* for tonic firing without presynaptic input (inverse stochastic resonance), i.e. the neurons and the astrocytes are accessible for any activity produced by an external *stimulus-external* in the sense of outside the neuronal and astroglial networks.
(b) *the phasic firing* occurring after a neuron is activated due to presynaptic activity (due to e.g. the astrocytes) and its acquiring motion, giving rise to a so-called stochastic resonance, on top of any background (tonic) activity a neuron may have.
(c) *the local field potential varies as a result of synaptic activity* and reflects the sum of action potentials from neurons within a few hundred μm, while the slower ionic movements contribute over a few mm.
(d) *the synaptic activity* consisting of (mainly) spatially distributed Poisson statistics which permits semantic encoding-decoding communication (the astroglial *Master Hub*)
(e) *the spike trains* exhibit (mainly) temporally distributed Poissonian statistics, which conveys encoding decoding communication. (the neuronal *Call Center*).
(f) these spike records reflect *the astroglial input*, consisting of high-frequency fluctuations in the potential difference, and being filtered out, leaving only slower fluctuations.

(g) *the low-pass filtering* due to complex electrical properties of extracellular space being associated with subjective feelings.

The points (a–e) "take care" of the steps (i–iii), while (f–g) relate to (iv) and possibly to a longer two-second duration [66]. Nevertheless (a–g) are irreducible in the sense that without any of the six reported steps above a conscious feeling may not be "created". The point (g) should be commensurate with the explanation of the so-called SMTT (Seeing More Than There is) experiments [68]. Note that the Call-Centre analogy of the Gödelian neurological network does not document which neurons are recorded, which may be commensurate with the "dark matter problem in neuroscience" [69].

A crucial supporting neuron-astrocyte mechanism refers to the increase of cytosolic calcium ions, i.e. the regulation of the intracellular calcium flux contributing to the effect of calcium-dependent synaptic change. More than 20 years ago [53], we investigated the (still) unresolved problem of excess mobility of hydrogen (and hydroxyl) ions in aqueous solutions with K^+ and Na^+ ions (and Cl^- for the hydroxyl ion) as background. The model, based on coherent dissipative dynamics, more or less identical to the concepts above, led to predictions that actually confirmed and predicted contemporary NMR spin-echo experiments [70, 71].

However, the molecular dynamics community, heavily committed to the classical Grotthus model, did not appreciate the mechanism. Still, the temperature dependence, as revealed by high precision conductivity data, displayed the increase in the logarithm of the quotient between excess conductivities of hydrogen—and hydroxyl ions as a function of temperature in stark contrast to classical models. The mechanism suggests that hydrogen (and hydroxyl) ions might easily provide the gradient and be indirectly responsible for the mechanism of transporting Ca^{2+} ions, through aqua-porins or receptors like Ryanodine [72], since the hydrogen ion or proton can hardly move through these channels. Thus one might consider taking the pH dependence on the cytoplasmic matrix into account in the regulation of positive ion mobility in the cytosol.

It is obvious that the main part of the chemical processes that guard the evolution of a CES that is part of a life form is, even if belonging to a human brain, primarily unconscious. However, in the latter instance, the physical processes evoking awareness and thus being central to consciousness, nevertheless constitutes an irreducible phenomenon even if the authentic code processing can be partitioned into tangible temporal steps.

3.2 Regulation by an Evolved Program

Intelligence is a highly controversial term and as a result we will only use it in its most trivial sense, i.e. in the meaning to pick out or understand something, but also to discern, distinguish, differentiate and recognize. Following Mayr [3, 4], one might essentially use telos, purpose or goal, in two different contexts. On one hand referring to physical laws, like gravity or the second law of thermodynamics, as governing teleomatic processes, and on the other a programmed activity, restricted to action or behaviour owing its goal-directedness to the influence of an evolved program, as governing teleonomic actions during evolution, thereby lifting prevailing teleological thinking to a higher conceptual level. Mayr's final concern [4] relate to insufficiencies in the laws of nature, a resolution that would entrench any definition of a *bona fide* science of biology. Aided by these remarks and clarifications, including the presently highlighted progress in chemistry and physics, we will return to Rosenberg's statement, Rosenberg [18], cited in the introductory remarks.

It is obvious that the quote refers to a minimalistic amount of physical laws in order to outline a clear and unmistaken physical stance accounting for the origin of life and its subsequent evolution. Omitting primeval physical processes like nucleosynthesis, where atomic nuclei were formed in stars and then ejected to form, in the case of earth, the solar nebula from which planets form via gravitational growth, traditional narratives focus on the so-called primordial soup of the early earth. This representation did catch affirmative support due to the famous Miller-Urey experiments [73] and Miller [74] in which the prebiotic atmosphere of the earth, as believed to reflect the true environment, was carefully simulated, showed that these conditions synthesized an abundance of organic compounds including several amino acids. Later experiments, Oró [75], established the prebiotic synthesis of the nucleobase adenine.

Thus it appears that the second law of thermodynamics seems to support [18], as it is commensurate with long-established chemical experiments. The critical statement of Nagel [12], the carefully administered confrontation by Deacon [12] and the more opinionated claims by Fodor and Piatelli-Palmarini [14], do not, however, agree with Rosenberg, as neither would Mayr in his conceptual stance. Chemical reactions produce large molecules but do not provide adaptation, where the latter usually refers to the functions of a higher-level target aiming at the access of the potential benefits inherently predicting their activities. In fact the intrinsic, but indirect, causal nature of exaptations, cf. Lewontin and Gould's Spandrels of San Marco [19], could also fall into the present category. Hence, the crucial question is whether we have "at home" all the laws of nature needed to understand how evolution does indeed "produce us".

It is well known, that quantum theory and Einstein relativity are not yet considered to be satisfactory joined, notwithstanding a consistent description of some precise key properties of gravitational laws have been recently stated, see Refs [9, 20]. Furthermore the current formulation faces the incidence of well-known inconsistencies and conundrums, plaguing physicalism, transcending from the microscopic- to the macroscopic level, see e.g. Primas [43]. Neo-Darwinistic principles, i.e. natural selection and the genetic code depend on causal forces and teleonomic laws or in other words causality is needed to instigate adaptation. Hence, neither causality, despite their absence on the fundamental microscopic level, nor progression that rely on semantically encoded processes, can be ruled out by physical law. Indiscriminately banning intentionality, or rather ententionality [12], one risks "throwing out the baby and keeping the bath water".

The persistent development, as envisioned by Rosenberg [18], viz. *emergence of copies of itself out of the atoms floating around in thermodynamic noise, templating, catalysing or otherwise producing copies of itself*, mimics Darwin's belief in gradualism. Evidently, as the laws of chemistry stipulates, larger and larger assemblies of molecular and macro-molecular structures evolve for plausible biological processes—and after some 500 million years iteration of the same process will produce more and more adaptation all in concert with the second law. This conclusion is based on a trusting acceptance of the causality related term adaptation, and the subsequent statement in [18] that only processes related to the 2nd law is permitted by physics. The conviction might be inappropriate as this specific scenario does fall between the chairs of micro-macro correlates. For instance, it is natural to characterize the law of causality as a fundamental rule on the macroscopic level, while cause and effect seem to vanish in the microscopic arena. Besides, one cannot rigorously derive the second law of thermodynamics, from statistical physics, without subjectively include a loss of information, see e.g. Ref. [44]. Thus the thesis, as stated earlier that everything is physical, is not complete, Mayr [4], necessitating a consistent verification of physics beyond the domain of time-reversible and non-causal classical and quantum physics.

An extension of the picture given in [18] is consequently unavoidable in order to strictly take account of concepts like adaptability and the origin of the genetic code. Recapitulating Mayr [4], the physicalist ideas are not sufficient to probe biology with its roots in concepts rather than physical laws. In this setting belongs the criticism of gene-centered Neo-Darwinism by Jablonka and Lamb [11], directed at a more specific interpretation of evolution. For recent debates on open questions in the philosophy of biology, see Ayala and Arp [76].

To account for the disparity, discussed above, one must be able to find an extension that incorporates some form of teleonomy. As our main purpose implicates a formulation of a CES, from the microscopic- to the macroscopic level, any solution of the controversial statements and criticisms of Darwinism must be taken seriously starting with a re-evaluation of modern quantum chemistry and chemical physics as developed over the last 50–60 years. The fields have, as already stated, attracted considerable attention owing to general developments of theoretical methodologies and significant advances in computer technologies [77]. Theoretical chemical physics comprise many different branches where the significance of the presence of so-called unstable states in the continuous spectra [33–35] has led to a rigorous expansion of quantum mechanics to include irreversible quantum dynamics, including both nuclear and electronic motions beyond the Born-Oppenheimer picture [55].

The rigour and the maturity of the sub-fields have already led to the coinage of Non-Hermitian Quantum Mechanics [35] opening up new vistas for exploration, see particularly Sect. 2 above. A substantial portion of novel developments has guided extensions to fundamental areas, like non-equilibrium quantum statistics, original interpretations of the quantum-classical enigma, the inter-level formulation of subdynamics, and the subsequent derivation of CDS, the Correlated Dissipative Structure, and CDE, the Correlated Dissipative Ensemble, providing base units for microscopic self-organization. The intrinsic (Gödelian) self-referential property of the correlated ensemble [9], supports transformations that exhibit nested encodings and communication protocols through Poisson-distributed channels. Molecular aggregations like genes and cells recognize their position in the hierarchy of biological organisms via instructions through quality Q-values. Hence "communication" takes place between molecular constituents, starting from the nucleus of the undifferentiated stem cell to somatic cells, and also in connection with heritable variations in the lineage of germ cells, as well as in higher-level objects for embedded messages. In principle this includes epigenetic inheritance systems [11]. Function, adaptation and natural selection follow through encoding, storage and retrieval of teleonomic processes in the neural system, the support systems and the spinal cord or to be brief in the brain. In other words the law of self-reference leads to constraints through an evolved program containing the blue print for the appropriate structure, confined, restricted and teleonomically regulated.

It should be obvious for the reader that, with the objectives met, and with the validity of a natural teleonomic law at hand, a minor addition to the Rosenberg quote [18], offers a legitimate physicalist response to the problems of incorporating Darwinism under the laws of physics. With such a law, as mathematically formulated above, nature is provided with an intrinsic property, and hence, with this appendage Nagel's major objection concerning the vanishing likelihood, that life forms comes spontaneously into existence, should be adequately answered.

4 Afterword

We have here extended the notion of "communication" to provide syntax for semantic purposes, extracted in physically well-defined information channels, to express function, information, reception, submission, transmission, regulation and adaptation. For this purpose, we have reformulated thermodynamics to comprise merged quantum-thermal correlations, which explicitly show the important roles of temperature and biological time scales in complex enough biological systems. This development not only points the way towards microscopic self-organization, but it also imparts non-equilibrium thermodynamics with a teleonomic origin. These advances are possible in a properly defined non-Hermitian framework, allowing more general, symmetry-violating solutions to the wide-ranging Liouville master equations and related thermalization procedures.

Communication and functional understanding start already at the spatial intra-cell level, where fortified encodings or programs, linked to the undifferentiated stem cell, are down-loaded, initiated and opened with the proper "copyright license". Once molecular communication is realized and developed between cells and cellular aggregates, amongst protein segments and protein domains and in 3D arrangements of the multiple folding of multi-subunit protein complexes, Czaba et al., [78], it continues to map concepts, perceptions, and possibly thoughts on the neurological level, metaphorically represented as a "Call Center". Cellular recognition via teleonomically controlled Q-values provide boundary conditions for the material build-up of biological structures from a CES with coordinated duties and responsibilities in the organism all the way to the spinal cord and the brain.

The Poisson process is stochastic, lacking intrinsic memory, giving rise to something non-intrinsic of teleonomic implication. Consequently, synchronized neural spike trains in the nervous system, the "Call Center", obeying Poisson statistics, where classes of distributions with various λ-values, grants the necessary channel for semantic communication. Decoding–encoding is based on simple arithmetic and provides an original "language" for syntax and semantics. The process amplifies through the nervous system, with constraints due to the synthesis of proteins in synaptic transmissions, and continues through the ladders and ultimately to the perceptions of the mind.

It is a natural step to anticipate the semantic process to transcend appropriate regions of the cortex and view higher–level perceptions, "codes of codes" as protracted CDS's. These extensions suggest a more precise and complete gene-based definition of the "social gene" coined the *meme* by Dawkins [10]. Cultural evolution has recently been discussed by Deutsch [42], where he concludes that knowledge changes the structure of the Universe. A complex meme should instigate both micro- and macro-evolution, and possibly extend to and beyond the social arena. In particular, it ascertains general aspects of communication from the spatio-temporal molecular and cellular level of a CES through higher-order levels, e.g. the social, the ecological, and even the cosmological rank. Evolution thus comprises also the non-material world, including mathematics, linguistics, fine arts, music, science, etc. By combining the superoperator algebra defined in Eqs. (4) and (9), it is easy to see the connection with the recent work of Schmeikal on geometric Clifford algebra of space-time, establishing novel relations between logic and geometry [79]. Note that context-dependence of genetic information pushing semantic information in prebiotic matter has recently been reviewed by Küppers [80].

Acknowledgements I am grateful to the local organisers of QSCP XX, Prof. Alia Tadjer and Prof. Rossen Pavlov, Sophia, Bulgaria, for generous hospitality and for running an excellent symposium. I thank Arnold Trehub, Bernd Schmeikal and Alfredo Pereira Jr. for their specific insight. This work has over time been supported by the Swedish Natural Science Research Council, the Swedish Foundation for Strategic Research, The European Commission and the Nobel Foundation.

References

1. Tegmark M (2003) Parallel Universes. Sci Am
2. Löwdin P-O (1988) The mathematical definition of a molecule and molecular structure. In Maruani J (ed) Molecules in physics, chemistry, and biology, vol 2. Kluwer Academic Publishers, p 3

3. Mayr E (1974) Teleological and teleonomic: a new analysis. Boston Stud Philos Sci 14:91
4. Mayr E (2004) What makes biology unique?. Cambridge University Press, New York
5. Weinberg S (1994) Dreams of a Final theory: the scientists search for the ultimate laws. Vintage Books, Random House, Inc. New York
6. Dirac PAM (1929) Quantum mechanics of many-electron systems. Proc R Soc A123 (792):714
7. Wiesberg M, Needham P, Hendry R (2011) Philosophy of chemistry, the stanford encyclopedia of philosophy. Zalta E. N. (ed). http://plato.stanford.edu/archives/sum2012/entries/reduction-biology/
8. Brändas EJ (2015) Proposed explanation of the phi phenomenon from a basic neural viewpoint. Quantum Biosyst 6(1):160
9. Brändas EJ (2015) A zero energy universe scenario: from unstable chemical states to biological evolution and cosmological order. In: Nascimento MAC, Maruani J, Brändas EJ, Delgado-Barrio G (eds) Frontiers in quantum methods and applications in chemistry and physics, vol 29. Springer, Dordrecht, p 247
10. Dawkins R (1976) The selfish gene. Oxford University Press, New York
11. Jablonka E, Lamb M (2005) Evolution in four dimension—genetic, epigenetic, behavioral, and symbolic variation in the history of life. The MIT Press, Cambridge
12. Deacon TW (2012) Incomplete nature: how mind emerged from matter. W. W. Norton & Company, New York, London
13. Nagel T (2012) MIND & COSMOS: why the materialist neo-darwinian conception of nature is almost certainly false. Oxford University Press, Oxford, New York
14. Fodor J, Piattelli-Palmarini M (2010) What Darwin got wrong. Farrar, Strauss and Giroux, New York
15. Logan RK (2012) Review and précis of Terrence Deacon's incomplete nature: how mind emerged from matter. Information 3:290
16. Allen Orr H (2013) Awaiting a New Darwin. The New York Review of Books, vol 60, p 2
17. Ferguson A (2013) The Heretic. The Weekly Standard, vol 18, p 27
18. Rosenberg A (2013) How Jerry Fodor slid down the slippery slope to Anti-Darwinism, and how we can avoid the same fate. Eur J Philos Sci 3(1):1
19. Gould SJ, Lewontin RC (1979) The spandrels of San Marco and the Panglossian paradigm: a criticism of the adaptationist programme. Proc R Soc Lond B Biol Sci 205:581
20. Brändas EJ (2012) Examining the limits of physical theory: analytical principles and logical implications. In: Nicolaides CA Brändas EJ (eds) Unstable states in the continuous spectra, Part II: Interpretation, theory, and applications. Advances in Quantum Chemistry, vol 63. Elsevier, Amsterdam, p 33
21. Mayr E (2001) What evolution is. Basic Books New York
22. Rosenberg A (1985) The structure of biological science. Cambridge University Press, Cambridge
23. Fodor JA (1980) Special sciences, or the disunity of science as a working hypothesis. Read Philos Psychol 1:120
24. Macdonald G, Papineau D (eds) (2006) Teleosemantics: new philosophical essays. Oxford University Press Inc, New York, p 1
25. Löwdin P-O (1967) Program. Nature of quantum chemistry. Int J Quantum Chem 1:1
26. Löwdin P-O (1998) Linear algebra for quantum theory. John Wiley & Sons, New York
27. Prigogine I (1996) The end of certainty: time, chaos, and the new laws of nature. The Free Press, New York
28. Obcemea CH, Brändas EJ (1983) Analysis of Prigogine's theory of subdynamics. Ann Phys 151:383
29. Yang CN (1962) Concept of off-diagonal long-range order and the quantum phases of liquid helium and of superconductors. Rev Mod Phys 34:694
30. Sasaki F (1965) Eigenvalues of fermion density matrices. Phys Rev 138B:1338
31. Coleman AJ (1963) Structure of fermion density matrices. Rev Mod Phys 35:668

32. Balslev E, Combes JM (1971) Spectral properties of many-body Schrödinger operators with dilatation-analytic interactions. Commun Math Phys 22:280
33. Nicolaides CA, Brändas EJ (eds) (2010) Unstable states in the continuous spectra, Part I: analysis, concepts, methods, and results. Advanced Quantum Chemistry, vol 60, p 1
34. Nicolaides CA, Brändas EJ (eds) (2012) Unstable states in the continuous spectra, part ii: interpretation, theory and, applications. Advanced Quantum Chemistry, vol 63, p 1
35. Moiseyev N (2011) Non-Hermitian quantum mechanics. Cambridge University Press, New York
36. Brändas EJ (2011) Gödelian structures and self-organization in biological systems. Int J Quantum Chem 111:1321
37. Lucas JR (1961) Minds Machines and Gödel. Philosophy 36:112
38. Seel M, Ladik J (1986) The tragicomedy of modern theoretical biology. In: Weingartner P, Dorn G (eds) Foundations of Biology: A Selection of Papers Contributed to the Biology Section of the 7th International Congress of Logic, Methodology and Philosophy of Science. Verlag Hölder-Pichler-Tempsky, Vienna, p 145
39. Penrose R (1994) Shadows of the mind: a search for the missing science of consciousness. Oxford University Press, Oxford
40. Feferman S (2011) Gödel's incompleteness theorems, free will and mathematical thought. In: Swinburne R (ed) Free will and modern science. Oxford University Press for the British Academy, London, p 102
41. Semon R (1904) Die Mneme. W. Engelmann, Leipzig
42. Deutsch D (2011) The beginning of infinity. Viking, Penguin, New York
43. Primas H (1983) Chemistry, Quantum mechanics and reductionism. perspectives in theoretical chemistry. Springer, Berlin
44. Sklar L (1993) Physics and chance philosophical issues in the foundations of statistical mechanics. Cambridge University Press, Cambridge
45. Rittby M, Elander N, Brändas E (1983) Scattering in view of the Titchmarsh-Weyl Theory. Int J Quantum Chem 23:865
46. Hehenberger M, McIntosh HV, Brändas E (1974) Weyl's theory applied to the Stark effect in the hydrogen atom. Phys Rev A 10:1494
47. Brändas E, Froelich P (1977) Continuum orbitals, complex scaling and the extended virial theorem. Phys Rev A 16:2207
48. Howland JS (1983) Complex scaling of ac Stark Hamiltonians. J Math Phys 24:1240
49. Löwdin P-O (1955) Quantum theory of many-particle systems. I. Physical interpretations by means of density matrices, natural spin orbitals, and convergence problems in the method of configuration interaction. Phys Rev 97:1474
50. Dunne LJ, Murrell JN, Brändas EJ (1990) Off-diagonal long-range order from repulsive electronic correlations in the ground state of a two-dimensional localized model of a high-T_C cuprate superconductor. Phys C 169:501
51. Brändas EJ, Dunne LJ (2014) Bardeen-Cooper-Schrieffer (BCS) theory and Yang's Concept of Long-Range Order (ODLRO). Mol Phys 112:694
52. Chatzidimitriou-Dreismann CA, Brändas EJ, Karlsson E (1990) quantum correlation effects in the spin dynamics of Gd at high temperatures in the light of complex dilation theory. Phys Rev Rapid Commun B 42:2704
53. Chatzidimitriou-Dreismann CA, Brändas EJ (1991) Proton delocalization and thermally activated quantum correlations in water: complex scaling and new experimental results. Ber Bunsenges Phys Chem 95:263
54. Carlson BC, Keller JM (1961) Eigenvalues of density matrices. Phys Rev 121:659
55. Brändas EJ, Hessmo B (1998) Indirect measurements and the mirror theorem. Lect Notes Phys 504:359
56. Kumicák J, Brändas EJ (1993) Complex scaling and Lyapunov converters. Int J Quantum Chem 46:391
57. Prigogine I (1980) From being to becoming, Freeman. W. H Freeman and Company, San Fransisco

58. Brändas EJ (1995) Relaxation processes and coherent dissipative structures. In: Lippert E, Macomber JD (eds) Dynamics during spectroscopic transitions. Springer, Berlin, p 148
59. Salari V, Tuszynski J, Rahnama M, Bernroider G (2011) Plausibility of quantum coherent states in biological systems. J Phys Conf Ser 306:1
60. Brändas EJ (2011) Some comments on the problem of decoherence. Int J Quantum Chem 111:215
61. Coleman AJ, Yukalov VI (2000) Reduced Density Matrices. Coulson's Challenge. Lecture notes in chemistry, vol 72. Springer , Berlin, pp 1–282
62. Brändas EJ, Chatzidimitriou-Dreismann CA (1991) On the connection between certain properties of the second-order reduced density matrix and the occurrence of coherent-dissipative structures in disordered condensed matter. Int J Quantum Chem 40:649
63. Brändas EJ, Elander N (1989) Resonances the unifying route towards the formulation of dynamical processes—Foundations and applications in nuclear, atomic and molecular physics. Lecture notes in chemistry, vol 325. Springer, Berlin, pp 1–564
64. Reid CE, Brändas EJ (1989) On a theorem for complex symmetric matrices and its relevance in the study of decay phenomena. Lecture notes in chemistry, vol 325. Springer, Berlin, p 475
65. Trehub A (1991) The cognitive brain. MIT Press
66. Fernandes de Lima VM, Pereira Jr. A (2016) The plastic glial-synaptic dynamics within the neurophil: a self-organizing system composed of polyelectrolytes in phase transition. Neural Plasticity, 1
67. Freeman WJ (2006) A cinematographic hypothesis of cortical dynamics in perception. Int J Psychophysiol 60(2):149
68. Trehub A (2007) Space, self, and the theatre of consciousness. Conscious Cogn 16:310
69. Shoham S, O'Connor DH, Segev R (2006) How silent is the brain: is there a "dark matter" problem in neuroscience? J Comp Physiol A Neuroethol Sens Neural Behav Physiol 192 (8):777
70. Pfeifer R, Hertz HG (1990) Activation energies of the proton-exchange reactions in water measured with the ^{1}H-NMR Spin Echo Technique. Ber Bunsenges Phys Chem 94:1349
71. Weingärtner H, Chatzidimitriou-Dreismann CA (1990) Anomalous H^+ and D^+ conductance in H_2O–D_2O mixtures. Nature 346:548
72. Futaqi D, Kitano K (2015) Ryanodine-receptor-driven intracellular calcium dynamics underlying spatial association of synaptic plasticity. J Comput Neurosci 39(3):329
73. Miller SL, Urey HS (1959) Organic compound synthesis on the primitive earth. Science 130 (3370):245
74. Miller SL (1953) A production of amino acids under possible primitive earth conditions. Science 117(3046):528
75. Oró J (1961) Mechanism of synthesis of adenine from hydrogen cyanide under possible primitive earth conditions. Nature 191(4794):1193
76. Ayala JF, Arp R (eds) (2010) Contemporary debates in philosophy of biology. Wiley-Blackwell, Chichester, West Sussex
77. Brändas EJ (2014) The statement of the goals of the international society for theoretical chemical physics. Int J Quantum Chem 114:961
78. Csaba G, Birzele F, Zimmer R (2009) Systematic comparison of SCOP and CATH: a new standard for protein structure analysis. BMC Struct Biol 9:23
79. Schmeikal B (2015) Four forms make a universe. Adv Appl Clifford Algebras 25:1
80. Küppers B-O (2016) The nucleation of semantic information in prebiotic matter. In: Domingo E, Schuster P (eds) Quasispecies: from theory to experimental systems. currrent topics in microbiology and immunology, vol 392. Springer International Publishing Switzerland, p 23

Author Index

A. Tadjer et al. (eds.), *Quantum Systems in Physics, Chemistry, and Biology*,
Progress in Theoretical Chemistry and Physics 30,
DOI 10.1007/978-3-319-50255-7

Subject Index

A. Tadjer et al. (eds.), *Quantum Systems in Physics, Chemistry, and Biology*,
Progress in Theoretical Chemistry and Physics 30,
DOI 10.1007/978-3-319-50255-7